Evolutionary Developmental Biology of Invertebrates 2

Andreas Wanninger

Editor

Evolutionary Developmental Biology of Invertebrates 2

Lophotrochozoa (Spiralia)

Springer

Editor
Andreas Wanninger
Department of Integrative Zoology
University of Vienna
Faculty of Life Sciences
Wien
Austria

ISBN 978-3-7091-1870-2   ISBN 978-3-7091-1871-9   (eBook)
DOI 10.1007/978-3-7091-1871-9

Library of Congress Control Number: 2015947925

Springer Wien Heidelberg New York Dordrecht London
© Springer-Verlag Wien 2015

Cover illustration: Scanning electron micrograph of an early trochophore larva of the tusk shell, *Antalis entalis*, a scaphopod mollusk. See Chapter 7 for details

Printed on acid-free paper

Springer-Verlag GmbH Wien is part of Springer Science+Business Media (www.springer.com)

Preface

The evolution of life on Earth has fascinated mankind for many centuries. Accordingly, research into reconstructing the mechanisms that have led to the vast morphological diversity of extant and fossil organisms and their evolution from a common ancestor has a long and vivid history. Thereby, the era spanning the nineteenth and early twentieth century marked a particularly groundbreaking period for evolutionary biology, when leading naturalists and embryologists of the time such as Karl Ernst von Baer (1792–1876), Charles Darwin (1809–1882), Ernst Haeckel (1834–1919), and Berthold Hatschek (1854–1941) realized that comparing ontogenetic processes between species offers a unique window into their evolutionary history. This revelation lay the foundation for a research field today commonly known as Evolutionary Developmental Biology, or, briefly, EvoDevo.

While for many of today's EvoDevo scientists the principle motivation for studying animal development is still in reconstructing evolutionary scenarios, the analytical means of data generation have radically changed over the centuries. The past two decades in particular have seen dramatic innovations with the routine establishment of powerful research techniques using micromorphological and molecular tools, thus enabling investigation of animal development on a broad, comparative level. At the same time, methods were developed to specifically assess gene function using reverse genetics, and at least some of these techniques are likely to be established for a growing number of so-called emerging model systems in the not too distant future. With this pool of diverse methods at hand, the amount of comparative data on invertebrate development has skyrocketed in the past years, making it increasingly difficult for the individual scientist to keep track of what is known and what remains unknown for the various animal groups, thereby also impeding teaching of state-of-the-art Evolutionary Developmental Biology. Thus, it appears that the time is right to summarize our knowledge on invertebrate development, both from the classical literature and from ongoing scientific work, in a treatise devoted to EvoDevo.

Evolutionary Developmental Biology of Invertebrates aims at providing an overview as broad as possible. The authors, all renowned experts in the field, have put particular effort into presenting the current state of knowledge as comprehensively as possible, carefully weighing conciseness against level of detail. For issues not covered in depth here, the reader may consult additional textbooks, review articles, or web-based resources, particularly on well-established model systems such as *Caenorhabditis elegans* (www.wormbase.org) or *Drosophila melanogaster* (www.flybase.org).

Evolutionary Developmental Biology of Invertebrates is designed such that each chapter can stand alone, and most chapters are dedicated to one phylum or phylum-like taxonomic unit. The main exceptions are the hexapods and the crustaceans. Due to the vast amount of data available, these groups are treated in their own volume each (Volume 4 and Volume 5, respectively), which differ in their conceptual setups from the other four volumes. In addition to the taxon-based parts, chapters on embryos in the fossil record, homology in the age of genomics, and the relevance of EvoDevo for reconstructing evolutionary and phylogenetic scenarios are included in Volume 1 in order to provide the reader with broader perspectives of modern-day EvoDevo. A chapter showcasing developmental mechanisms during regeneration is part of Volume 2.

Evolutionary Developmental Biology of Invertebrates aims at scientists that are interested in a broad comparative view of what is known in the field but is also directed toward the advanced student with a particular interest in EvoDevo research. While it may not come in classical textbook style, it is my hope that this work, or parts of it, finds its way into the classrooms where Evolutionary Developmental Biology is taught today. Bullet points at the end of each chapter highlight open scientific questions and may help to inspire future research into various areas of Comparative Evolutionary Developmental Biology.

I am deeply grateful to all the contributing authors that made *Evolutionary Developmental Biology of Invertebrates* possible by sharing their knowledge on animal ontogeny and its underlying mechanisms. I warmly thank Marion Hüffel for invaluable editorial assistance from the earliest stages of this project until its publication and Brigitte Baldrian for the chapter vignette artwork. The publisher, Springer, is thanked for allowing a maximum of freedom during planning and implementation of this project and the University of Vienna for providing me with a scientific home to pursue my work on small, little-known creatures.

This volume covers the animals that have a ciliated larva in their life cycle (often united as Lophotrochozoa), as well as the Gnathifera and the Gastrotricha. The interrelationships of these taxa are poorly resolved and a broadly accepted, clade-defining autapomorphy is lacking. Spiral cleavage is sometimes assumed as the ancestral mode of cleavage of this grouping, and therefore the clade is named Spiralia by some authors, although others prefer to extend the term Lophotrochozoa to this entire assemblage. Aside from the taxon-based chapters, this volume contains a chapter that highlights similarities and differences in the processes that underlie regeneration and ontogeny, using the Platyhelminthes as a case study.

Tulbingerkogel, Austria   Andreas Wanninger
January 2015

Contents

Gnathifera

1

Andreas Hejnol

Chapter vignette artwork by Brigitte Baldrian.
© Brigitte Baldrian and Andreas Wanninger.

A. Hejnol
Sars International Centre for Marine Molecular
Biology, University of Bergen,
Thormøhlensgate 55, Bergen 5008, Norway
e-mail: andreas.hejnol@uib.no

A. Wanninger (ed.), *Evolutionary Developmental Biology of Invertebrates 2: Lophotrochozoa (Spiralia)*
DOI 10.1007/978-3-7091-1871-9_1, © Springer-Verlag Wien 2015

INTRODUCTION

The taxon Gnathifera was erected based on morphological data by Ahlrichs (1995, 1997). The taxon comprises the Gnathostomulida and Syndermata (which unites Rotifera, Acanthocephala, Seisonida) (Fig. 1.1). With the discovery of *Limnognathia maerski* (Kristensen and Funch 2000), the taxon Micrognathozoa has been included into the Gnathifera. The name Gnathifera is based on the presence of a complex jaw apparatus in the pharynx of all groups, except Acanthocephala (Sørensen 2003; Funch et al. 2005). Gnathifera are tiny, bilaterally symmetric animals that live in aquatic habitats. Only the parasitic acanthocephalans reach body lengths of up to 80 cm. The acanthocephalans have lost many morphological characters as adaptations to their parasitic lifestyle, including the jaw apparatus and the digestive tract.

Gnathifera has been placed in the Spiralia, often affiliated with the Platyzoa (Funch et al. 2005; Dunn et al. 2008; Witek et al. 2008, 2009; Hejnol et al. 2009; Wey-Fabrizius et al. 2014). However, the sister group of Gnathifera remains unclear. Since gene sequences of most of the gnathiferan species seem to evolve fast, it remains difficult to resolve with confidence the internal relationships using molecular data (Witek et al. 2008; Wey-Fabrizius et al. 2014). Figure 1.1 illustrates the likely phylogenetic relationships as a consensus phylogeny that is based on recent molecular as well as morphological data. In all gnathiferans, fertilization is internal and the development direct. The parasitic acanthocephalans have evolved additional dispersal stages that allow infection and transitions between hosts.

GNATHOSTOMULIDA

Gnathostomulids are wormlike, microscopic, marine, interstitial animals that are covered with a monociliary epidermis. There are about 100 described species that are ordered into two taxa, the elongated Filospermoidea and the more compact Bursovaginoida (Sterrer 1972; Sørensen et al. 2006). The animals have a mouth opening that contains the pharyngeal bulb with the cuticular jaw structure. Gnathostomulids have a blind gut – some species possess a "temporal anus" (Knauss 1979) – an anterior brain, a ventral ganglion that is affiliated with the mouth, and one to three pairs of basiepithelial nerves (Kristensen and Nørrevang 1977; Müller and Sterrer 2004).

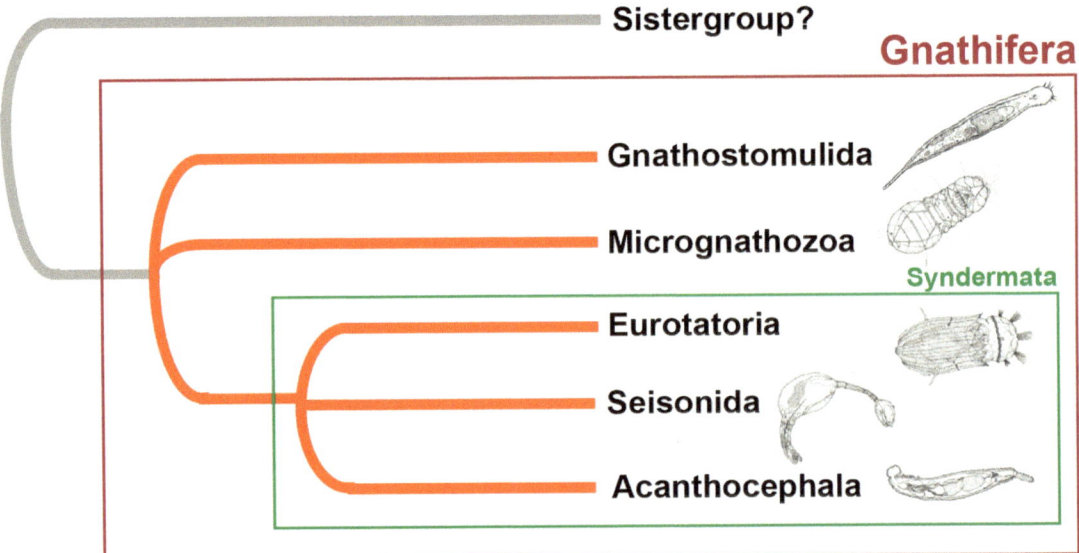

Fig. 1.1 Gnathiferan phylogenetic relationships. Phylogenetic relationships of gnathiferan taxa according to recent molecular and morphological studies (Images modified from Funch et al. 2005)

Early Development

For only one gnathostomulid species – *Gnathostomula jenneri* – the cleavage pattern has been described (Riedl 1969). Eggs (55 µm in diameter) are laid by body rupture and start cleaving 2 h later (22 °C). The first and second cleavages are "nearly equal" and meridional (Fig. 1.2A, B). Riedl (1969) describes a spiral, dexiotropic arrangement of the micromeres that results from the third round of cell divisions (Fig. 1.2C). The spindle is alternating in angle and thus follows the pattern of a typical spiralian embryo (Fig. 1.2D). Riedl (1969) also described the appearance of an "annelid cross" on day 4 of development, which is a specific arrangement of blastomeres in spiral embryos (see Chapters 6, 7, and 9) but is nowadays not considered of phylogenetic significance (see Maslakova et al. 2004). Day 4 is also the time when gastrulation is completed, where two larger cells – which Riedl (1969) interprets as the left and right mesendoblast – are visible (Fig. 1.2E). The hatchling of *G. jenneri* is 100 µm in length and lacks the jaws but has a primordium of the pharynx (Fig. 1.2F). The juvenile has a gut, but no lumen is visible.

Postembryonic Development

During postembryonic development, organ systems are successively developed until the juvenile is about 200 µm long. This is in striking contrast to other gnathiferan taxa that have eutely (Riedl 1969).

Gnathostomulid development needs urgent reinvestigation with modern methods. The described presence of a spiral cleavage program is unique to the taxon Gnathifera and makes gnathostomulids a key taxon for the understanding of the development and evolution of other gnathiferan taxa (Hejnol 2010).

MICROGNATHOZOA

The freshwater species *Limnognathia maerski* is the only micrognathozoan described so far (Kristensen and Funch 2000). With a size of only 100 µm in length, it is one of the smallest metazoans. No male individuals have been observed so far. The females have paired gonads, of which one develops a single oocyte at a time. Nothing is known about the reproduction mode, development, or interactions with its habitat, the submerged moss of freshwater springs on Disko Island, Greenland.

SYNDERMATA

The Syndermata are characterized by their so-called intrasyncytial lamina, a skeletal structure that is located inside the syncytial epidermis (Clément and Wurdak 1991). The taxon is comprised of three monophyletic groups: the Eurotatoria (comprising the Monogononta and Bdelloidea), the Seisonida, and the parasitic Acanthocephala (Fig. 1.1). The phylogenetic interrelationship of these taxa is currently under debate (see above).

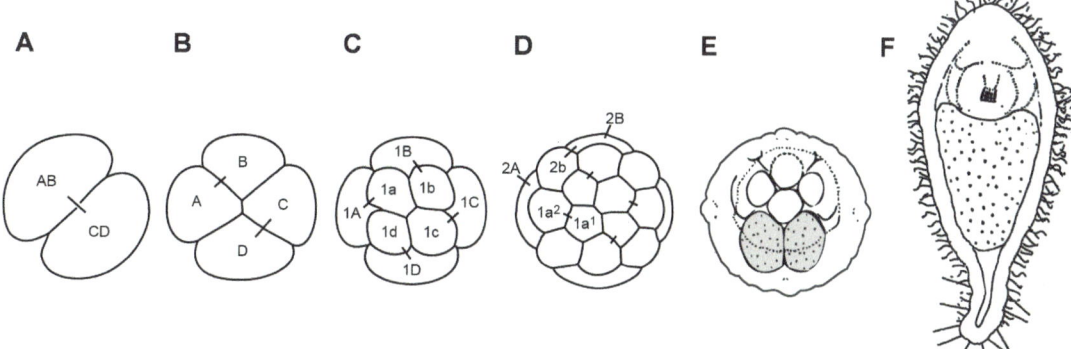

Fig. 1.2 Cleavage program of the gnathostomulid *Gnathostomula jenneri* after Riedl (1969). The naming of the blastomeres is according to Riedl (1969) who applied the spiralian nomenclature to the blastomeres. All stages are a view on the animal pole: (**A**) two-cell stage, (**B**) four-cell stage, (**C**) eight-cell stage, (**D**) 16-cell stage, and (**E**) approx. 64-cell stage. The shaded blastomeres are the mesendoblasts. (**F**) Hatchling with ciliated epidermis

Rotifera

Rotifers ("wheel bearers") are aquatic, tiny (0.05–1 mm long) metazoans that are free-living and abundant in the marine and freshwater environment. There are around 2,000 described species, of which about 200 species live in the marine environment. Rotifers use the ciliated rows of cells at the anterior end (corona) for locomotion. The animals have an anterior brain, a lateral pair of nerves, and a through gut (except *Asplanchna*) with an anterior mastax that carries the jaw apparatus. The body has a so-called hemocoel, a fluid-filled cavity that is not surrounded by epithelia and is thus not a true coelom.

Rotifers are divided into two groups, the Monogononta and the Bdelloidea, which mainly differ in their arrangements of the gonads. Bdelloid rotifers possess two bilaterally arranged gonads, while the monogononts possess only one gonad that is located medially. Both taxa furthermore differ in their mode of reproduction. While monogononts are either gonochoric (dwarf males) or parthenogenetic, bdelloid rotifers have no males and are exclusively parthenogenetic (Flot et al. 2013). Similar to all members of the Gnathifera, rotifers have direct development and no larval stage. As for other groups that have been placed together as "Aschelminthes" (or "Nemathelminthes"), rotifers are said to have a fixed number of cells when hatching (eutely)

(Clément and Wurdak 1991). Several monogonont species have been studied regarding cell numbers (see references in Gilbert 1989). There have been several studies regarding the development of rotifers, which have mainly focused on monogonont species (see Table 1.1). Based on the previous descriptions, the early development of monogonont and bdelloid rotifers does not differ much and will be described together in the following section. Several authors (Siewing 1969; Nielsen 2005) mention that rotifers have a modified spiral cleavage.

Cleavage

Rotifers deposit an oval, fertilized egg that is protected by a resistant shell. In some representatives (*Brachionus*, *Keratella*), the female carries the fertilized egg until hatching. The embryo gives off one polar body before the onset of cleavage, and the position of the polar body marks the animal pole of the embryo. The first cell division is unequal, dividing the egg into a large blastomere (called CD by some authors), whose descendants will give rise to the germovitellarium and other internal tissues. The smaller blastomere – named AB – will mainly contribute to the ectoderm (Fig. 1.3A). In all rotifer species investigated so far, the AB blastomere divides equally into blastomeres A and B. The blastomere CD divides unequally into the smaller blastomere C and the larger blastomere D (Fig. 1.3B).

Table 1.1 Studies of rotifer development

Reference	Species	Stages and methods
Boschetti et al. (2005)	*Macrotrachela quadricornifera* (Bdelloidea)	Early cleavage (confocal microscopy)
Tessin (1886)	*Eosphora digitata* (Monogononta)	Whole development, light microscopy
Zelinka (1892)	*Mniobia* (*Callidina*) *russeola* (Bdelloidea)	Whole development, light microscopy, histology
Jennings (1896)	*Asplanchna herrickii* (Monogononta)	Whole development, light microscopy, histology
Tannreuther (1919)	*Asplanchna sieboldii* (Monogononta)	Cleavage and gastrulation of the male
Tannreuther (1920)	*Asplanchna sieboldii* (Monogononta)	Whole development, light microscopy, histology
Nachtwey (1925)	*Asplanchna priodonta* (Monogononta)	Whole development, light microscopy, histology
Lechner (1966)	*Asplanchna girodi* (Monogononta)	Whole development, histology, UV ablation experiments
Pray (1965)	*Lecane cornuta* (Monogononta)	Whole development, light microscopy
Mrázek (1897)	*Asplanchna* (Monogononta)	Whole development, light microscopy
Car (1899)	*Asplanchna brightwellii* (Monogononta)	Cleavage, gastrulation, light microscopy

The size of blastomere C is similar to that of A and B. This four-cell stage looks similar to the four-cell stages of unequal cleaving spiralians, which also display a larger blastomere (D) and three smaller blastomeres (Chapter 7). The polar body is located in the center of the four blastomeres where all blastomeres are in contact with each other (Fig. 1.3B). The polar body marks the animal pole of the embryo, and interestingly, during the subsequent cell divisions, bdelloids and monogononts differ in the further spatial interrelationships of blastomeres and polar body. While in monogononts the polar body is located first on the tip of the small blastomeres that are formed until the 16-cell stage (Jennings 1896; Nachtwey 1925; Lechner 1966), the polar body in the bdelloid species is located closely to the large blastomeres that will gastrulate (Zelinka 1892). This difference seems to be fundamental at first glance, since many nomenclatures, e.g., the spiralian nomenclature developed by Conklin (1897), use the position of blastomeres in relation to the polar body for the naming of the cells. Following this rule, in monogononts the large cell that forms the germovitellarium would be a vegetal macromere, while in bdelloids this homologous blastomere would constitute an animal micromere. Consequently, the names of the individual homologous cells would change dramatically. However, since the fate of the individual blastomeres is comparable between both rotifer taxa and in both groups the polar body will eventually lie at the future anterior pole of the embryo, the polar body cannot be used for naming the cells in the rotifer embryo, and thus other criteria, such as the fate of the blastomeres, have to be applied.

In both embryos, however, the four blastomeres form tiers, which may be called quadrants, since they are descendants of the blastomeres of the four-cell stage and lie in one row with the mother cells (Fig. 1.3C, D).

Gastrulation

Gastrulation in rotifers begins with the internalization of the large blastomere D at the 16-cell stage (Fig. 1.4A). This is consistent between all rotifer species investigated so far. The ectodermal cells contribute to the internalization by epibolic overgrowth of blastomere D, which gives off two very small blastomeres that later undergo apoptosis (Fig. 1.4B). During internalization, blastomere D divides, with one daughter cell giving rise to the vitellarium and the other to the germarium of the adult (Fig. 1.4B). After immigration of the D blastomere, endodermal cells follow the D blastomere that will form the digestive tract of the adult (Fig. 1.4C). In monogonont embryos, the polar body is transferred by these cellular movements to the future anterior pole of the embryo. While the origin of the germovitellarium and the digestive tract has been described in rotifers, the origin of mesodermal tissues such as musculature and nephridia remains unknown (Fig. 1.4D). Only a gross description of the final fates of the early blastomeres is currently available for rotiferans (Fig. 1.5).

Organogenesis

Endodermal cells form the tube of the digestive tract and differentiate into the stomach, also called mastax (Fig. 1.4D). The former blastopore forms the opening that differentiates into the mouth on the anterior ventral side (Fig. 1.4D). The ectoderm on the dorsal side undergoes more cell proliferation and will give rise to the cerebral ganglion of the adult. These cells will get internalized from the dorsal ectoderm to form the brain of the adult (Fig. 1.4D). Future epidermis cells close the immigration site and differentiate into a syncytial integument. At this stage, the differentiation of the major parts of the digestive tract (mouth opening, pharynx, and mastax) is visible. In some embryos, the formation of an oviduct has been described (Lechner 1966). The last phase of organogenesis begins with the complete closure of the site of immigration of the cerebral ganglion precursors. The cilia of the ciliary bands start beating and the pseudocoel becomes visible. Before hatching, the juvenile extends along the anterior-posterior axis and hatches. The hatchling possesses a functional excretory system with protonephridia. Due to the eutelic condition, growth is achieved by extension of cell volume and not by further cell proliferation.

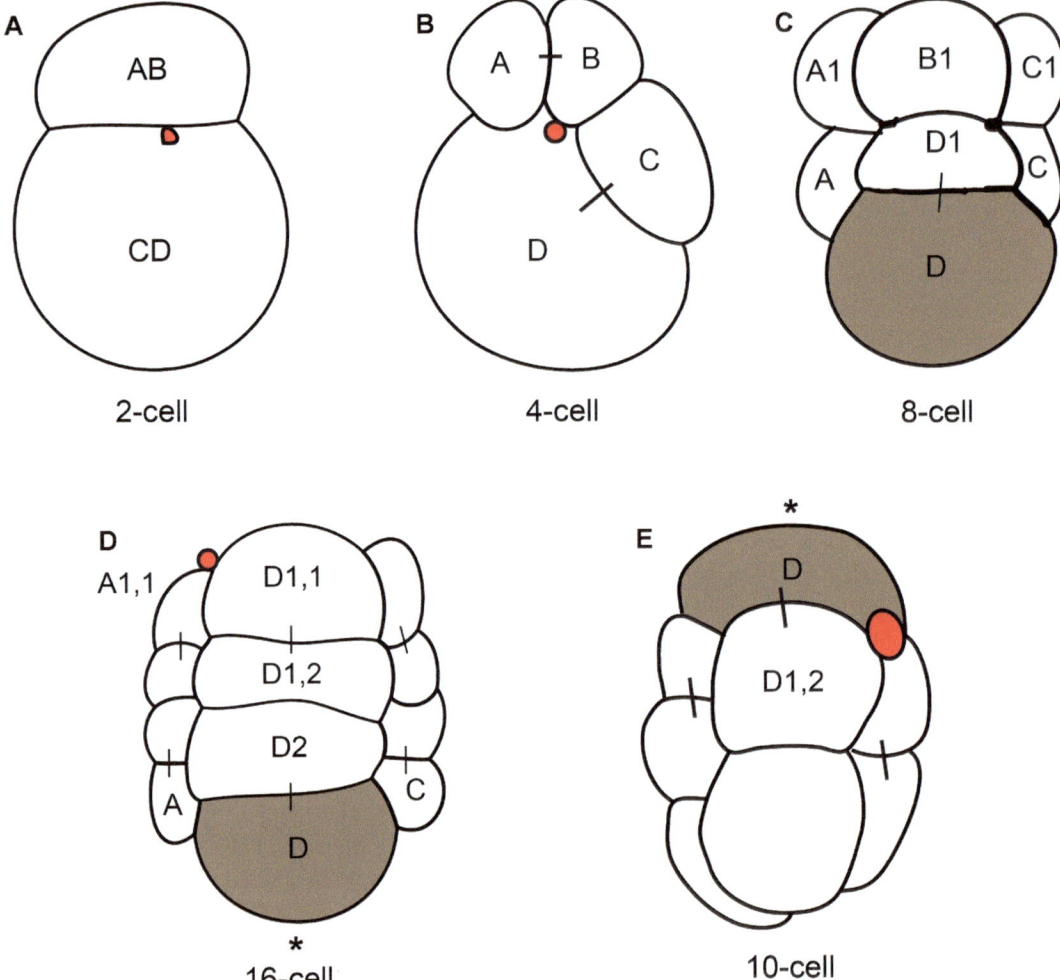

Fig. 1.3 Rotifer cleavage up to the 16-cell stage. (**A**) Two-cell stage. Polar body shaded in *red*. (**B**) Four-cell stage. (**C**) Eight-cell stage with blastomere D that will form the germovitellarium shaded in *brown*. (**D**) 16-cell stage of the monogonont *Asplanchna* after Tannreuther (1919) and Lechner (1966). The future anterior pole of the embryo is marked with an *asterisk*. (**E**) 10-cell stage of the bdelloid *Callidina* (Zelinka 1892) which shows reverse polarity. Note the polar body close to the site of gastrulation that also corresponds to the future anterior pole

Experimental Embryology

The only study using experimental approaches to investigate rotifer embryos is the work of Lechner (1966) on *Asplanchna*. He used UV irradiation to destroy early blastomeres during cleavage. The results are either a complete destruction of the embryo or – results after less or partial irradiation – partial differentiation of the embryo. After partial ablation of individual blastomeres, the remaining cells were still able to differentiate to late cellular fates, such as ciliated cells, epidermal cells, and cells of the digestive tract. Depending on the degree and timepoint of the irradiation – the work does not clearly state the duration and target – the embryo either does not gastrulate, arrests after gastrulation, or even manages to hatch out of the egg shell with internal organs nearly intact. Lechner's (1966) series of experiments indicates that cell fates are determined early and that blastomeres can differentiate into their final fates without induction of functional neighbor cells. The rotifer embryo can thus be

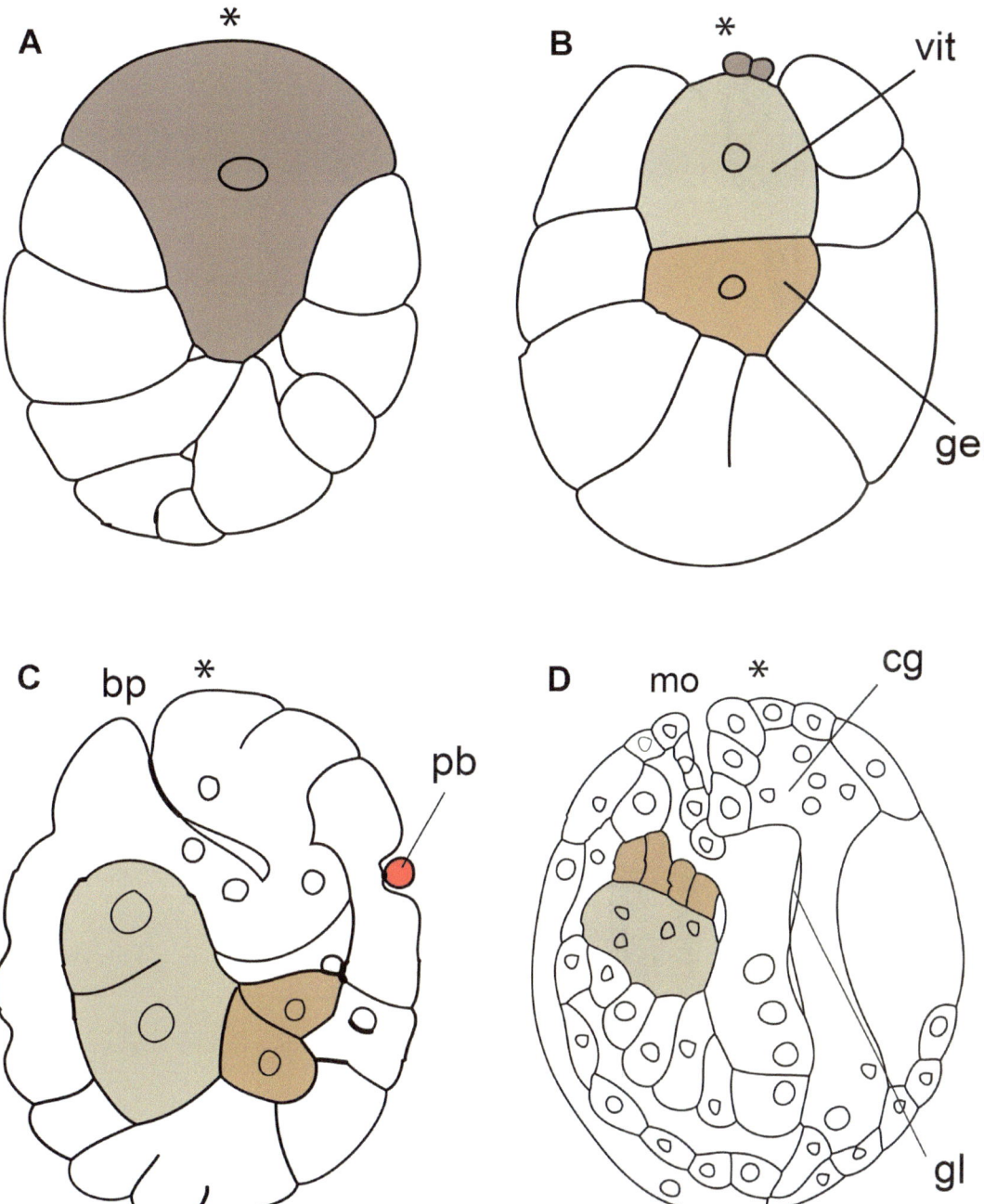

Fig. 1.4 Rotifer gastrulation. Gastrulation summarized based on studies of different rotifer species (*Asplanchna girodi*, *Asplanchna priodonta*, *Callidina russeola*). (**A**) Blastomere D (*shaded in dark brown*) before internalization at the future anterior pole of the embryo. (**B**) The large blastomere has divided into the germarium precursor (*ger, shaded in light brown*) and the blastomere that will form the vitellarium (*vit, shaded in brown gray*). Small, degenerating blastomeres that have been given off from blastomere D before internalization are *shaded in dark brown*. (**C**) Late gastrulation stage after further endodermal cells have immigrated through the blastopore (*bp*). The polar body (*pb, shaded in red*) is closer to the future anterior pole (*asterisk*). (**D**) Later stage with organ systems already present. The mouth opening (*mo*) is anterior. The cells that will form the cerebral ganglion (*cg*) have been internalized. The gut lumen (*gl*) is visible

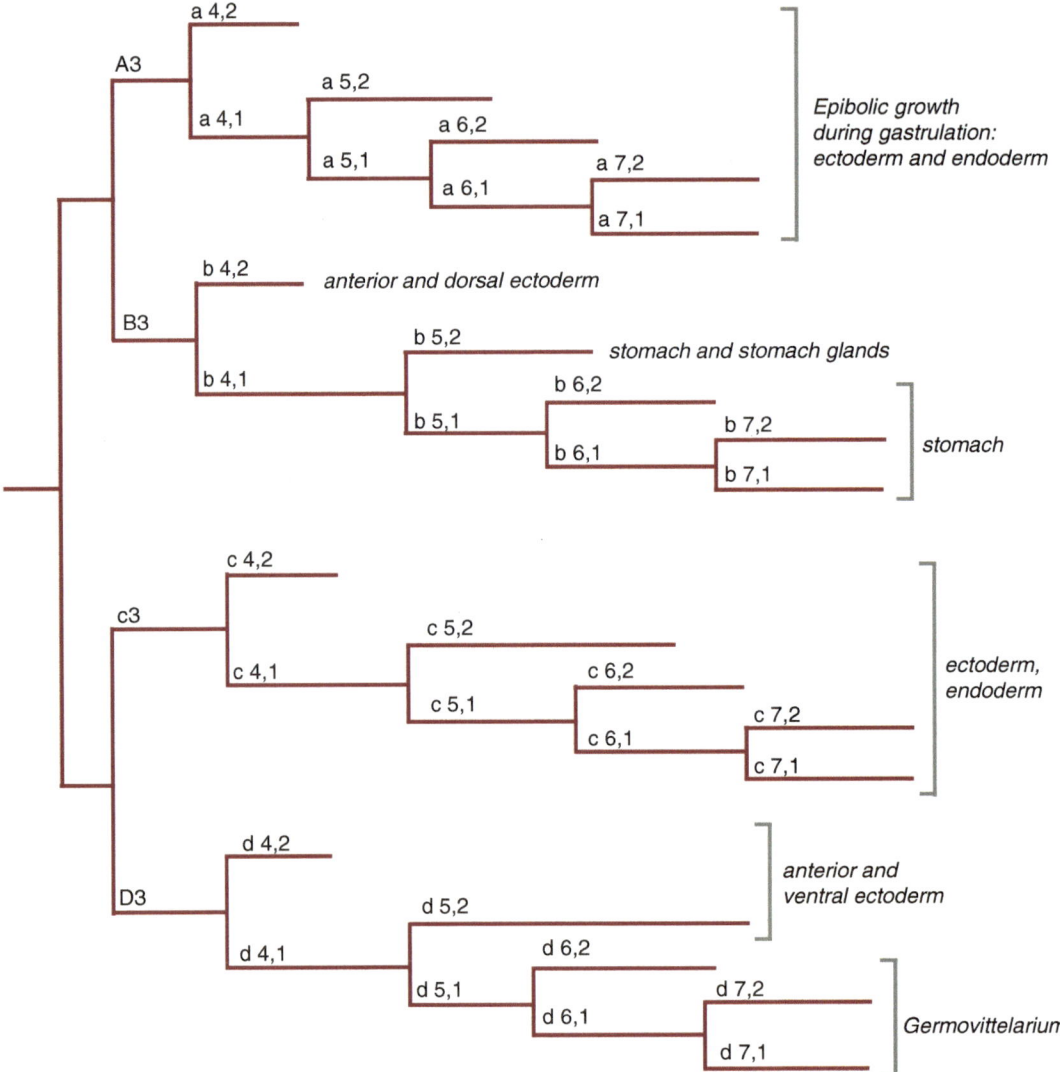

Fig. 1.5 Rotifer fate map. The fate map of the monogonont *Asplanchna girodi* reconstructed after Lechner (1966)

seen as a classic "mosaic" embryo in which cells can differentiate autonomously. However, more detailed and sophisticated experiments are necessary to characterize the underlying molecular mechanisms and early inductional processes that are likely to be active also in this embryo.

Molecular Approaches

Molecular studies on rotifer development are rare. Smith and coworkers (Smith et al. 2010) studied the gene expression of the germ line markers *vasa* and *nanos* in the monogonont

rotifer *Brachionus plicatilis*. The first ubiquitously expressed *vasa* ortholog becomes restricted to the germovitellarium line and is later expressed weakly in the vitellarium and stronger in the oocytes and germ cells (Smith et al. 2010). The same has been found for the *nanos* ortholog (Smith et al. 2010). The recent sequencing of the first rotifer genome of the bdelloid *Adineta vaga* (Flot et al. 2013) and reports of RNA interference in a monogonont rotifer (Snell et al. 2011) can provide tools to study the molecular mechanisms of rotifer development in more detail. Here, bdelloid rotifers provide a unique system to study the reduced tetraploidy (Flot et al. 2013) and the multiplication of many key regulatory genes and signaling pathways and their impact on the development of an embryo. The bdelloid species *Adineta vaga* has many more homeobox gene families with four copies than tetraploid vertebrates; this could be explained by the gene conversion that these animals undergo to eliminate mutations under the absence of sexuality (Flot et al. 2013). The Hox gene cluster is atomized and a posterior Hox gene is missing.

Seisonida

The development of this peculiar rotifer lineage has not been studied. The separate sex adults deposit fertilized eggs close to their attachment site directly on the host the leptostracan crustacean *Nebalia*. Only two species have been described, *Seison nebaliae* and *Seison annulatus*.

Acanthocephala

Acanthocephalans are a gnathiferan taxon that is exclusively parasitic, and all 1,100 species described show many adaptations to the parasitic lifestyle (Meyer 1933). These wormlike animals live as adults in the digestive system of vertebrates (final host) and take up nutrients through the integument. Acanthocephalans have a reduced digestive tract and evolved reproductive strategies that allow them to infest the intermediate host (crustaceans) and final host (Fig. 1.6). Acanthocephalans have separate sexes, males and females copulate, and eggs are fertilized internally.

Development

The development of acanthocephalans has been studied in about ten species by several authors (Schmidt 1985). The most detailed description of the cell lineage of the embryo that gives rise to an early stage called the acanthor (often called "larva"), which then infests the intermediate host, has been conducted by Hamann (1891), Kaiser (1893), and Meyer (1928, 1933).

The oval-shaped, fertilized egg has two polar bodies that mark the future anterior pole of the animal (Fig. 1.6A). The first cell division is equatorial to the longitudinal egg axis and gives rise to the animal blastomere AB and the slightly larger vegetal blastomere CD (Fig. 1.6B). The two blastomeres divide equally in the next cell division round and produce the animal blastomere B3, the median blastomeres A3 and C3, and the vegetal blastomere D3 (Fig. 1.6C). The following cell divisions are unequal. Blastomere D3 is the first to divide unequally into the micromere D4.1 and the macromere D4.2. The other macromeres follow the same division pattern in that they give off smaller micromeres in the direction of the polar bodies. In different species, the embryo becomes syncytial at different timepoints, ranging from the four-cell to the 36-cell stage. A clear fate of the blastomeres has not been determined. Interestingly, up to the 25-cell stage of *Gigantorhynchus gigas* (Meyer 1928), only macromeres divide – the micromeres are arrested. The quartet of macromeres forms thus "mother cells" of all micromeres before the micromeres start to divide again in the 25-cell stage.

Previous studies agree that during further development the micromeres condense and become internalized to form the "central nuclear mass" (Fig. 1.6G). Several authors described this as a form of gastrulation, since these condensed nuclei will be part of all the internal tissues except the epidermis. The embryo begins to form additional – up to four – egg layers of different density that are composed out of polysaccharides and other material.

Acanthocephalan Life Cycle

The acanthor is the infectious stage that possesses a boring organ (aclid organ) that is used to penetrate the gut of the intermediate host (crustacean). The acanthor is first surrounded by several membranes, and it only hatches when it has been taken up by the intermediate host (Fig. 1.6I). The acanthor – after penetrating the gut and entering the mesenteron – can be silent for hours or days. The acanthor begins to grow in size and the organs of the worm are formed by the cells in the membranes, thus forming the acanthella (Fig. 1.6J). In the acanthella, the brain begins

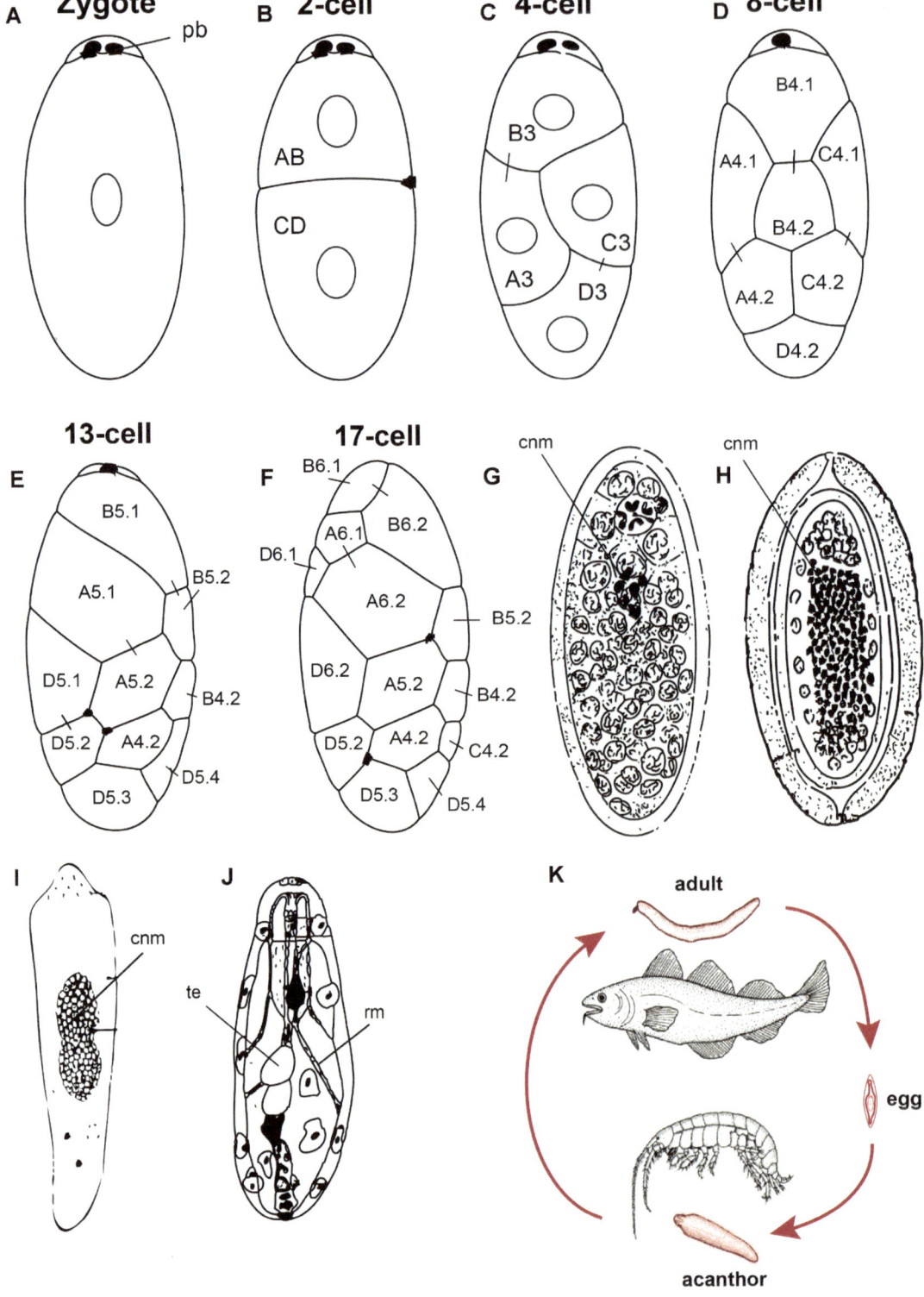

to form, and the inner cell mass starts to form musculature (retractor muscle) and ovaries or testes. Finally, hooks are formed on the proboscis. The worm – now called cystacanth – possesses all structures of the adult and is ready to infect the main host when the arthropod infested with a cystacanth is consumed by the vertebrate host.

Taken together, the data currently available on gnathiferan ontogeny show that all species investigated to date are direct developers. Some of the taxa are eutelic (Eurotatoria), while others show postembryonic development. Acanthocephalans have evolved a complicated life cycle that allows the species to transfer between hosts and reproduce in masses. The cleavage pattern differs between groups. While spiral cleavage has been described in gnathostomulids, this has so far not been confirmed for rotifers or acanthocephalans.

OPEN QUESTIONS

- Detailed cell lineage studies in all gnathiferan taxa
- Organogenesis in all major taxa
- The basic molecular mechanisms that trigger the development of gnathiferan groups
- Developmental gene expression of all major gene families in all gnathiferan subtaxa

References

Ahlrichs WH (1995) Ultrastruktur und Phylogenie von *Seison nebaliae* (Grube 1859) und *Seison annulatus* (Claus 1876). PhD thesis, Georg August University Göttingen

Ahlrichs WH (1997) Epidermal ultrastructure of *Seison nebaliae* and *Seison annulatus*, and a comparison of epidermal structures within the Gnathifera. Zoomorphology 117:41–48

Boschetti C, Ricci C, Sotgia C, Fascio U (2005) The development of a bdelloid egg: a contribution after 100 years. Hydrobiologia 546:323–331. doi:10.1007/S10750-005-4241-Z

Car L (1899) Die embryonale Entwicklung von *Asplanchna brightwellii*. Biol Zentbl 19:59–74

Clément P, Wurdak E (1991) Rotifera. In: Harrison FW, Ruppert EE (eds) Microscopic anatomy of invertebrates. Wiler Liss, New York, pp 219–297

Conklin EG (1897) The embryology of *Crepidula*. J Morphol 13:1–226

Dunn CW, Hejnol A, Matus DQ et al (2008) Broad phylogenomic sampling improves resolution of the animal tree of life. Nature 452:745–749. doi:10.1038/nature06614, nature06614 [pii]

Flot JF, Hespeels B, Li X et al (2013) Genomic evidence for ameiotic evolution in the bdelloid rotifer *Adineta vaga*. Nature 500:453–457. doi:10.1038/nature12326

Funch P, Sørensen MV, Obst M (2005) On the phylogenetic position of Rotifera – have we come any further? Hydrobiologia 546:11–28

Gilbert JJ (1989) Rotifera. In: Adiyodi KG, Adiyodi RG (eds) Reproductive biology of invertebrates. Wiley, Chichester, pp 179–199

Hamann O (1891) Monographie der Acanthocephalen (Echinorhynchen). Jena Z Naturw 25:113–231

Hejnol A (2010) A twist in time – the evolution of spiral cleavage in the light of animal phylogeny. Integr Comp Biol 50:695–706. doi:10.1093/icb/icq103

Hejnol A, Obst M, Stamatakis A et al (2009) Assessing the root of bilaterian animals with scalable phylogenomic methods. Proc R Soc Ser B 276:4261–4270. doi:10.1098/rspb.2009.0896

Jennings HS (1896) The early development of *Asplanchna herrickii* De Guerne. Bull Mus Comp Zool 30:1–117

Kaiser J (1893) Die Acanthocephalen und ihre Entwickelung. Bibl Zool 2:1–374

Knauss E (1979) Indication of an anal pore in Gnathostomulida. Zool Scr 8:181–186

Kristensen RM, Funch P (2000) Micrognathozoa: a new class with complicated jaws like those of Rotifera and Gnathostomulida. J Morphol 246:1–49

Fig. 1.6 Acanthocephalan development and life cycle. All embryonic stages are oriented with the polar bodies (*pb*) up; cleavage pattern of *Giganthorhynchus gigas* after Meyer (1928). (**A**) Fertilized zygote. (**B**) Two-cell stage. (**C**) Four-cell stage. (**D**) Eight-cell stage. (**E**) 13-cell stage. (**F**) 17-cell stage. (**G**) Syncytial embryo with multiple cells. The central nuclear mass (*cnm*) begins to form by immigration of nuclei at the future anterior pole of the embryo. (**H**) Later stage embryo. The central nuclear mass is established and an additional layer has been formed around the embryo. (**I**) Hatched acanthor from the arthropod host. Hooks are visible at the anterior pole. Central cell mass is still undifferentiated. (**J**) Developed acanthella (male). The central cell mass has formed retractor muscles (*rm*) and the gonads (*te*, testes). (**K**) Life cycle of the acanthocephalan *Echinorhynchus gadi* (for description, see text)

Kristensen RM, Nørrevang A (1977) On the fine structure of *Rastrognathia macrostoma* gen. et sp.n. placed in Rastrognathiidae fam.n. (Gnathostomulida). Zool Scr 6:27–41

Lechner M (1966) Untersuchungen zur Embryonalentwicklung des Rädertieres *Asplanchna girodi* De Guerne. Roux' Arch f Entwicklungsmech 157:117–173

Maslakova SA, Martindale MQ, Norenburg JL (2004) Fundamental properties of the spiralian developmental program are displayed by the basal nemertean *Carinoma tremaphoros* (Palaeonemertea, Nemertea). Dev Biol 267:342–360. doi:10.1016/j.ydbio.2003.10.022

Meyer A (1928) Die Furchung nebst Eibildung, Reifung und Befruchtung des *Gigantorhynchus gigas*. Zool Jb Anatomie 50:117–218

Meyer A (1933) Acanthocephala. Akademische Verlagsgemeinschaft, Leipzig

Mrázek A (1897) Zur Embryonalentwicklung der Gattung *Asplanchna*. Sitz-Ber Kgl Böhm Gesell Wiss 58:1–11

Müller MCM, Sterrer W (2004) Musculature and nervous system of *Gnathostomula peregrina* (Gnathostomulida) shown by phalloidin labeling, immunohistochemistry, and cLSM, and their phylogenetic significance. Zoomorphologie 123:169–177

Nachtwey R (1925) Untersuchungen über die Keimbahn, Organogenese und Anatomie von *Asplanchna priodonta* Gosse. Z Wiss Zool 126:239–492

Nielsen C (2005) Trochophora larvae: cell-lineages, ciliary bands and body regions. 2. Other groups and general discussion. J Exp Zool B Mol Dev Evol 304:401–447. doi:10.1002/jez.b.21050

Pray F (1965) Studies on the early development of the rotifer *Monostyla cornuta* MÜLLER. Trans Am Microsc Soc 84:1965

Riedl RJ (1969) Gnathostomulida from America. Science 163:445–452

Schmidt G (1985) Development and life cycles. In: Crompton D, Nickol B (eds) Biology of the Acanthocephala. Cambridge University Press, Cambridge, pp 273–305

Siewing R (1969) Lehrbuch der vergleichenden Entwicklungsgeschichte der Tiere. Verlag Paul Parey, Hamburg/Berlin

Smith JM, Cridge AG, Dearden PK (2010) Germ cell specification and ovary structure in the rotifer *Brachionus plicatilis*. EvoDevo 1:5. doi:10.1186/2041-9139-1-5

Snell TW, Shearer TL, Smith HA (2011) Exposure to dsRNA elicits RNA interference in *Brachionus manjavacas* (Rotifera). Mar Biotechnol (NY) 13:264–274. doi:10.1007/s10126-010-9295-x

Sørensen MV (2003) Further structures in the jaw apparatus of *Limnognathia maerski* (Micrognathozoa), with notes on the phylogeny of the Gnathifera. J Morphol 255:131–145

Sørensen MV, Sterrer W, Giribet G (2006) Gnathostomulid phylogeny inferred from a combined approach of four molecular loci and morphology. Cladistics 22:32–58

Sterrer W (1972) Systematics and evolution within the Gnathostomulida. Syst Zool 21:151–173

Tannreuther GW (1919) Studies on the rotifer *Asplanchnia ebbesbornii*, with special reference to the male. Biol Bull 37:194–207

Tannreuther GW (1920) The development of *Asplanchna ebbesbornii* (Rotifer). J Morphol 33:389–419

Tessin G (1886) Über Eibildung und Entwicklung der Rotatorien. Z Wiss Zool 44:273–302

Wey-Fabrizius AR, Herlyn H, Rieger B, Rosenkranz D, Witek A, Welch DBM, Ebersberger I, Hankeln T (2014) Transcriptome data reveal Syndermatan relationships and suggest the evolution of endoparasitism in Acanthocephala via an epizoic stage. PLoS One 9:e88618. doi:10.1371/journal.pone.0088618

Witek A, Herlyn H, Meyer A, Boell L, Bucher G, Hankeln T (2008) EST-based phylogenomics of Syndermata questions monophyly of Eurotatoria. BMC Evol Biol 8:345. doi:10.1186/1471-2148-8-345

Witek A, Herlyn H, Ebersberger I, Mark Welch DB, Hankeln T (2009) Support for the monophyletic origin of Gnathifera from phylogenomics. Mol Phylogenet Evol 53:1037–1041. doi:10.1016/j.ympev.2009.07.031

Zelinka C (1892) Studien über Räderthiere. III. Zur Entwicklungsgeschichte der Räderthiere nebst Bemerkungen über ihre Anatomie und Biologie. Z Wiss Zool 53:1–159

Gastrotricha

2

Andreas Hejnol

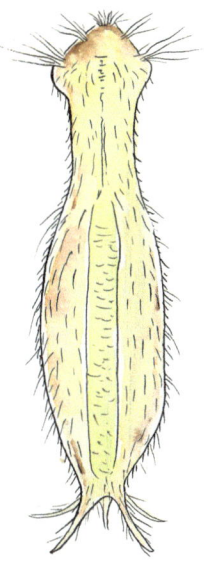

Chapter vignette artwork by Brigitte Baldrian.
© Brigitte Baldrian and Andreas Wanninger.

A. Hejnol
Sars International Centre for Marine Molecular
Biology, University of Bergen,
Thormøhlensgate 55, 5008 Bergen, Norway
e-mail: andreas.hejnol@uib.no

A. Wanninger (ed.), *Evolutionary Developmental Biology of Invertebrates 2: Lophotrochozoa (Spiralia)*
DOI 10.1007/978-3-7091-1871-9_2, © Springer-Verlag Wien 2015

INTRODUCTION

Gastrotricha is a clade of aquatic microscopic animals that are among the smallest metazoans. Gastrotrichs mainly live in the marine and freshwater interstitial environment and on the surface of aquatic plants. There are around 700 described species of gastrotrichs, which are divided into the marine taxon Macrodasyida and the marine and freshwater taxon Chaetonotida. So far, no fossils have been assigned to the taxon Gastrotricha. All gastrotrichs are direct developers, and the adults measure between 60 and 600 μm, with some exceptions growing to a size of several millimeters. The acoelomate body is wormlike, bilaterally symmetric, and sometimes covered with scales and a cuticle (Fig. 2.1; Ruppert 1991). The ventral side is covered with locomotory cilia ("gaster," belly; "thrix," hair). The animals have a brain that is mainly composed of an elaborated dorsal commissure, whereas a ventral commissure is present in some species (Schmidt-Rhaesa 2015).

The brain is connected via lateral axon tracts or neurite bundles to the posterior end of the body (Schmidt-Rhaesa 2015). The alimentary canal has an anterior mouth with a muscular pharynx, a midgut that is composed of less than 100 cells and ends with a posterior anal opening (Ruppert 1991). Most species lack an ectodermal hindgut and some even lack an anal opening (*Urodasys*). No specialized hemal or respiratory system is present in gastrotrichs, but the excretory system is composed of one to six pairs of protonephridia (Ruppert 1991). Gastrotrichs possess characteristic adhesive tubes that are composed of two to three gland cells that secrete adhesive substances ("dual-gland system"). Gastrotrichs can be dioecious or protandric hermaphrodites, and some obligate parthenogenetic species have been described (Hummon 1974). Fertilization is internal and eggs are deposited through a female opening or body rupture. The phylogenetic position of Gastrotricha is still under debate, and affiliations with the Platyzoa,

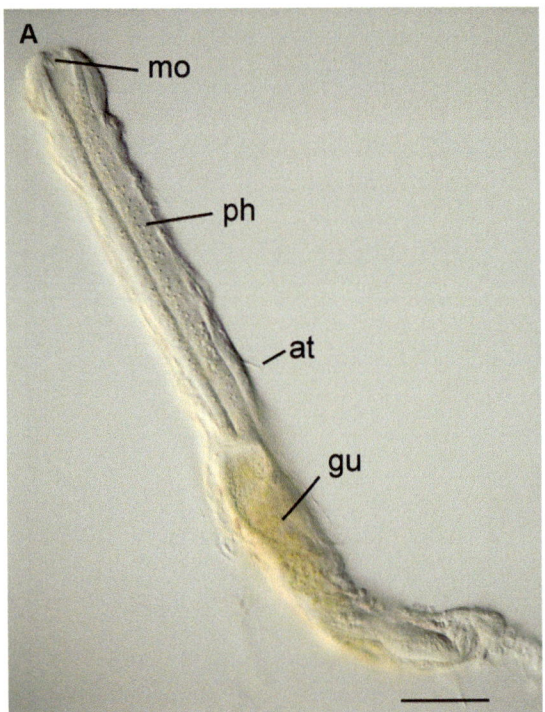
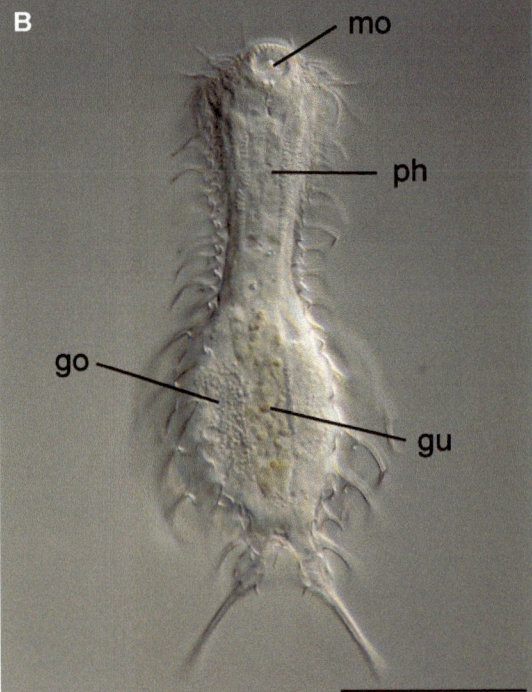

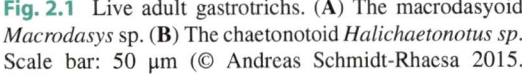

Fig. 2.1 Live adult gastrotrichs. (A) The macrodasyoid *Macrodasys* sp. (B) The chaetonotoid *Halichaetonotus sp.* Scale bar: 50 μm (© Andreas Schmidt-Rhaesa 2015.

All Rights Reserved). *mo* mouth opening, *ph* pharynx, *at* adhesive tube, *gu* gut, *go* gonad

a suggested monophylum within the Spiralia, have been hypothesized based on the results of large-scale phylogenomic analyses (Dunn et al. 2008; Hejnol et al. 2009).

EARLY AND LATE DEVELOPMENT

There are only a few studies of gastrotrich development. Besides the most comprehensive studies of the chaetonotoid species *Lepidodermella squamata* by Sacks (1955) and the macrodasyoid species *Turbanella cornuta* by Teuchert (1968), other data are very scarce (de Beauchamp 1929; Brunson 1949; Swedmark 1955). Gastrotrich development has been video recorded (Teuchert 1975; Schnabel et al. 2006), and the results confirm that the early development is highly stereotypic (Schnabel et al. 2006). So far, no developmental gene expression studies have been published.

Eggs are deposited with an egg shell that is smooth in marine species and can contain spines or other appendages in the freshwater species. The fertilized eggs are attached to sand grains or to algae with the use of secretory glands of the adult. Both gastrotrich groups differ regarding their development in terms of duration and their cleavage patterns.

Development of Chaetonotoida

The most detailed description of chaetonotoid development is that of the parthenogenetic freshwater species *Lepidodermella squamata* (Sacks 1955) that confirmed the partial descriptions of the early and late cleavage made by Ludwig (1875), de Beauchamp (1929), and Brunson (1949).

The cleavage pattern of *Lepidodermella squamata* is similar to that of *Neogossa antennigera* (de Beauchamp 1929). Eggs are deposited as fertilized zygotes and with an egg shell. At 22 °C the second meiotic division occurs after egg deposition, and a polar body is extruded and localizes to the longitudinal end of the oval egg, marking the animal pole of the embryo (Fig. 2.2A). The first cell division is equal and

along the animal-vegetal axis and gives rise to the animal blastomere AB and the vegetal blastomere CD (Fig. 2.2A). In the next division round, AB is the first that divides slightly unequally into the smaller blastomere A and the larger blastomere B, followed by the division of CD into the vegetal blastomeres C and D (Fig. 2.2B). After these divisions, the animal blastomeres are positioned about 90° twisted to the vegetal ones (Fig. 2.2B). In the next round of cell division to the eight-cell stage, all blastomeres divide perpendicular to the animal-vegetal axis. The first cell to divide is AB, followed by CD (Fig. 2.2C). The four animal blastomeres are located in the furrows between the four vegetal blastomeres. In the subsequent division rounds, the animal blastomeres cleave before the vegetal blastomeres (Fig. 2.2D). The 32-cell stage has a blastocele, and gastrulation starts with the immigration of two sister cells, A5.3 and A5.4 (Fig. 2.2E; Sacks 1955). The animal pole of the embryo will form the anterior region of the juvenile. The site where the two cells entered the blastocele will later form the stomodeum of the adult (Fig. 2.2F). During later development, a posterior opening is formed that gives rise to the proctodeum (Fig. 2.2F, G). The organ systems begin to form 6 h after egg deposition, and the embryo undergoes an extension, which bends the future juvenile inside the egg shell (Fig. 2.2H). The endodermal gut connects the anterior stomodeum and the posterior proctodeum to each other. A pharynx is clearly visible after 9 h of oviposition. Cell differentiation of epidermal scales, cilia, gut cells, and caudal forks occurs between 9 and 23 h of development. The juvenile hatches after 23–31 h.

Development of Macrodasyoida

The literature about macrodasyoid gastrotrich development is very scarce. The main work stems from Teuchert (1968, 1975) on several species, with the main description of *Turbanella cornuta*. A short note on *Macrodasys* had been published earlier (Swedmark 1955). The development in macrodasyoid gastrotrichs takes longer than in

Fig. 2.2 Development of the chaetonotoid *Lepidodermella squamata* after Sacks (1955) and the macrodasyoid *Turbanella cornuta* after Teuchert (1968). Comparative stages of the embryos of *L. squamata* (*left*) and *T. cornuta* (*right*). Animal pole to the *top*, ventral to the *left*. (**A, A'**) Two-cell stage. (**B, B'**) Four-cell stage. (**C, C'**) Eight-cell stage. (**D, D'**) 16-cell stage. (**E, E'**) Gastrulation. (**F, F'**) Stomodeum formation. (**G, G'**) Digestive tract visible. (**H, H'**) Late stage of a juvenile in the egg shell. *bp* blastopore, *st* stomodeum, *pr* proctodeum. Identically colored blastomeres indicate their homology and demonstrate the differences between the cleavage patterns of the two species

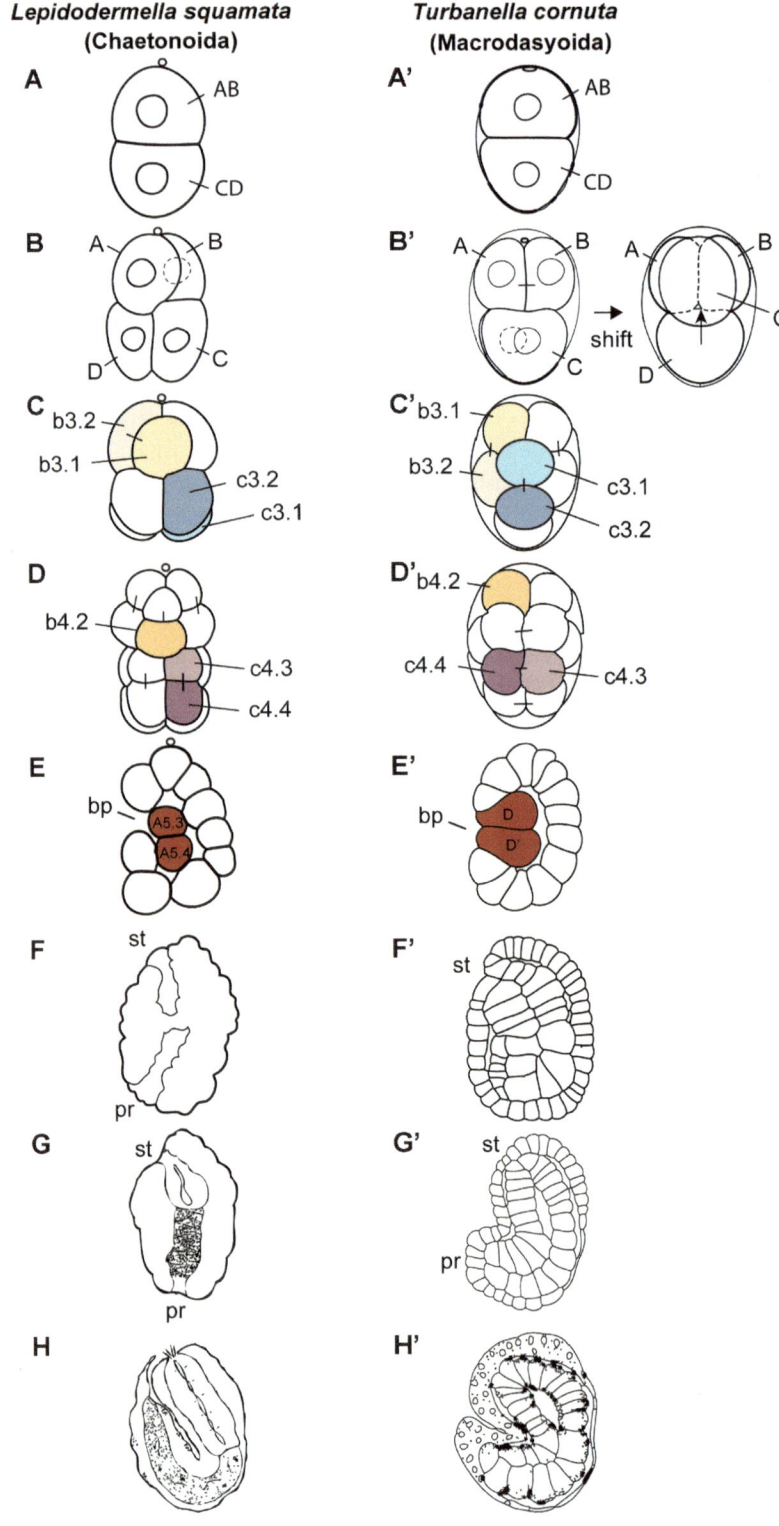

Lepidodermella squamata (Chaetonoida)

Turbanella cornuta (Macrodasyoida)

the investigated chaetonotoid species. *Turbanella cornuta* develops in 3–4 days until hatching, and *Macrodasys caudatus* up to 10 days (Teuchert 1968). The eggs are fertilized internally, and two polar bodies are extruded sequentially after egg deposition at the animal pole of the embryo (Fig. 2.2A'). The first cleavage is equatorial and equal and divides the embryo into the animal blastomere AB and the vegetal blastomere CD (Fig. 2.2A'). In the next cleavage round, blastomere AB divides before CD. Both cells divide meridionally, but the cleavage planes are perpendicular to each other. In the early four-cell stage, the animal blastomeres A and B lie between the vegetal blastomeres C and D (Fig. 2.2B'). After the cell divisions, one of the vegetal blastomeres, C, moves toward the animal pole, leaving blastomere D at the vegetal pole (Fig. 2.2B'). According to Teuchert (1968), C exclusively forms ectoderm, while D – besides some mesodermal and ectodermal portions – will give rise to all endoderm. After the shift of the C blastomere, the embryo begins the next round of divisions. The animal blastomeres divide first along the animal-vegetal axis, and the dorsally located cell C divides along the same axis (Fig. 2.2C'). The last cell to divide in this round is the ventral blastomere D. In the next round of divisions, all cells divide perpendicular to the previous cleavage plane (Fig. 2.2D'). At the 30-cell stage, 11–12 h after egg deposition, two ventral descendants of blastomere D begin to immigrate into the blastocele and will later form the endoderm (Fig. 2.2E'). According to Teuchert (1968), the mesodermal cells follow after the cells have undergone another division. The other blastomere identities and the later cell divisions in the *T. cornuta* embryo remain unclear. The blastopore elongates, closes from the posterior end, and forms the opening of the stomodeum (Fig. 2.2F'). The embryo is now 40–50 h old. The proctodeum will be formed later at a new site at the posterior tip of the extended embryo (Fig. 2.2G').

According to Teuchert (1968), mesodermal bands are formed by mesodermal precursors that gastrulate after the two endodermal precursors. The mesodermal bands will likely form the musculature of the adult. The anterior ectoderm proliferates and ectodermal cells will contribute to the brain. Cell differentiation and organ system formation begins 50–55 h after egg deposition. The juvenile hatches after 4 days at 22 °C.

Comparison Between Chaetonotoid and Macrodasyoid Early Development

The described cleavage patterns for chaetonotoid gastrotrichs are consistent, and only minor differences have been detected (Sacks 1955). The differences between chaetonotoid and macrodasyoid gastrotrichs are more fundamental (Fig. 2.2). Teuchert points out that one of the major differences between the embryos of both taxa is the origin of the cells that form the stomodeum (Teuchert 1968). In chaetonotoids, the animal ("oral") cells form the blastopore and the endoderm, while in macrodasyoids, the endoderm and blastopore are formed from the vegetal cells ("caudal," i.e., posterior). This fundamental difference led Teuchert to turn Sacks' descriptions of the *Lepidodermella squamata* embryo 180°, so that in both embryos the endoderm seemingly originates from the vegetal pole (Teuchert 1968). However, video recordings of *L. squamata* development (Schnabel et al. 1997) show that indeed Sacks' (1955) observations are correct. Other differences in the development of both groups concern the angle during early cleavage (see different arrangements in Fig. 2.2) and how the animal blastomeres get shifted against the vegetal ones in the early embryo. In macrodasyoid embryos, the spindles place the blastomeres between the vegetal ones, while in chaetonotoid embryos the cells shift after the placement on top of the sister cell.

In both groups, the entire endoderm seems to be formed by two gastrulating cells. The mesoderm is formed by precursors that gastrulate later – possibly at different positions in the embryo.

ORGANOGENESIS

Organogenesis in gastrotrichs has not been described in detail. Sacks' work about *Lepidodermella squamata* describes the formation of the digestive tract (Fig. 2.2; Sacks 1955). No data about mesoderm or nervous system development is available for a chaetonotoid species. Teuchert (1968) provides data about the development of some organ systems in the macrodasyoid *Turbanella cornuta*. The brain is formed by ectodermal thickenings at the anterior end of the animal (Fig. 2.2). Cells proliferate from the epidermis to the inside of the embryo and become arranged in a crescent-shaped dorsal commissure above the pharynx. The *T. cornuta* embryo has one pair of protonephridia anlagen, which get multiplied in a serial arrangement after hatching (Teuchert 1968). The pharynx is formed by two cell rows that will form a tube composed of 70–80 cells. The lumen-less gut is composed of approximately 20 large cells that are arranged in a row. During later development of the embryo, a small lumen becomes visible. The mesoderm is composed of two lateral bands. Since gastrotrichs are acoelomate, these cells will mainly give rise to the musculature of the animal. Teuchert (1968) also mentions that it is possible that posterior cells of the mesodermal bands might form the gonads. The cells of the lateral mesodermal bands proliferate and fill the space on the dorsal and ventral sides of the gut. Teuchert (1968) assumes that these cells will form the longitudinal dorsal and ventral musculature. Another mesodermal derivative that is specific to macrodasyoid gastrotrichs is the so-called "Y-organ" of unknown function (Ruppert 1991; Teuchert 1968).

The differences between both groups and the overall paucity of data on gastrotrich development make it difficult to reconstruct the ancestral mode of development for gastrotrichs and at present render detailed comparisons with other larger taxonomic units difficult.

OPEN QUESTIONS

- Reinvestigation of macrodasyoid development regarding fate map and cell lineage
- Virtually all aspects of organogenesis, including neuromuscular development
- Molecular characterization including gene expression patterns of basic developmental processes, cell type differentiation by investigation of transcription factors, signaling cascades, and structural genes in both gastrotrich groups

References

Brunson R (1949) The life history and ecology of two North American gastrotrichs. Trans Am Microsc Soc 68:1–20

de Beauchamp P (1929) Le développement des Gastrotriches. Bull Soc Zool Fr 54:549–558

Dunn CW, Hejnol A, Matus DQ, Pang K, Browne WE, Smith SA, Seaver E, Rouse GW, Obst M, Edgecombe GD, Sorensen MV, Haddock SH, Schmidt-Rhaesa A, Okusu A, Kristensen RM, Wheeler WC, Martindale MQ, Giribet G (2008) Broad phylogenomic sampling improves resolution of the animal tree of life. Nature 452:745–749

Hejnol A, Obst M, Stamatakis A, Ott M, Rouse GW, Edgecombe GD, Martinez P, Baguñá J, Bailly X, Jondelius U, Wiens M, Müller WEG, Seaver E, Wheeler WC, Martindale MQ, Giribet G, Dunn CW (2009) Assessing the root of bilaterian animals with scalable phylogenomic methods. Proc Roy Soc Ser B 276:4261–4270

Hummon W (1974) Gastrotricha. In: Giese A, Pearse J (eds) Reproduction of marine invertebrates. Academic, New York

Ludwig H (1875) Ueber die Ordnung Gastrotricha Metschn. Z Wiss Zool 16:193–225

Ruppert EE (1991) Gastrotricha. In: Harrison FW, Ruppert EE (eds) Microscopic anatomy of invertebrates: aschelminthes. Wiley-Liss, New York

Sacks M (1955) Observations on the embryology of an aquatic gastrotrich, *Lepidodermella squamata* (Dujardin, 1841). J Morphol 96:473–495

Schmidt-Rhaesa A (2015) Gastrotricha. In: Schmidt-Rhasa A, Harzsch S, Purschke G (eds) Structure and evolution of invertebrate nervous systems. Oxford University Press, Oxford (in press)

Schnabel R, Hutter H, Moerman D, Schnabel H (1997) Assessing normal embryogenesis in *Caenorhabditis elegans* using a 4D microscope: variability of development and regional specification. Dev Biol 184:234–265

Schnabel R, Bischoff M, Hintze A, Schulz AK, Hejnol A, Meinhardt H, Hutter H (2006) Global cell sorting in the *C. elegans* embryo defines a new mechanism for pattern formation. Dev Biol 294:418–431

Swedmark B (1955) Développment d'un Gastrotriche Macrodasyoide, *Macrodasys affinis* Remane. CR Acad Sci Paris 240:1812–1814

Teuchert G (1968) Zur Fortpflanzung und Entwicklung der Macrodasyoidea (Gastrotricha). Z Morphol Tiere 63:343–418

Teuchert G (1975) Organisation und Fortpflanzung von *Turbanella cornuta* (Gastrotricha). Film C1176. Institut für den wissenschaftlichen Film, Göttingen

Platyhelminthes

3

Teresa Adell, José M. Martín-Durán, Emili Saló, and Francesc Cebrià

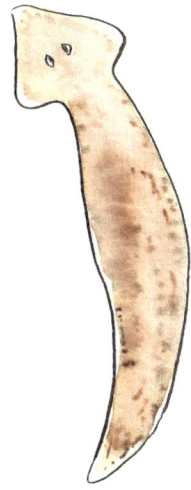

Chapter vignette artwork by Brigitte Baldrian.
© Brigitte Baldrian and Andreas Wanninger.

T. Adell • E. Saló (✉) • F. Cebrià
Department of Genetics, Faculty of Biology,
Institute of Biomedicine, University of Barcelona,
Av. Diagonal 643, edifici Prevosti, planta 1,
Catalunya, Barcelona 08028, Spain
e-mail: esalo@ub.edu

J.M. Martín-Durán
Sars International Centre for Marine
Molecular Biology, University of Bergen,
Thormøhlensgate, 55, Bergen 5008, Norway

INTRODUCTION

The phylum Platyhelminthes comprises dorsoventrally flattened worms commonly known as flatworms (from the Greek *platys*, meaning flat, and *helminthos*, meaning worm) (for a general overview of this phylum, see Hyman 1951; Rieger et al. 1991). Platyhelminthes are one of the largest animal phyla after arthropods, mollusks, and chordates and includes more than 20,000 species, more than half of which are parasitic flatworms. Free-living flatworms (classically referred to as 'Turbellaria') live in a large variety of habitats, from freshwater springs, rivers, lakes, and ponds to the ocean and moist terrestrial habitats. Their size ranges from microscopic worms to the 30 m long tapeworms found in the sperm whale. Free-living flatworms are most often white, brown, grey, or black; polyclads (marine flatworms) and terrestrial species usually display bright colours and patterns. Molecular phylogenetic studies place the Platyhelminthes within the Spiralia clade. The most recent internal phylogenies support the subdivision of the Platyhelminthes into two main groups: the earliest branching lineages grouped into the paraphyletic 'Archoophora' and the more divergent monophyletic Neoophora (Laumer and Giribet 2014; Riutort et al. 2012). The 'Archoophora' includes those groups with endolecithal eggs. They are exclusively free-living organisms and are classified into three orders: Catenulida, Polycladida, and Macrostomida (Fig. 3.1). The Neoophora includes all groups with ectolecithal eggs. It comprises several free-living orders, together with the parasitic groups (the classes Trematoda, Cestoda, and Monogenea) united under the monophyletic Neodermata.

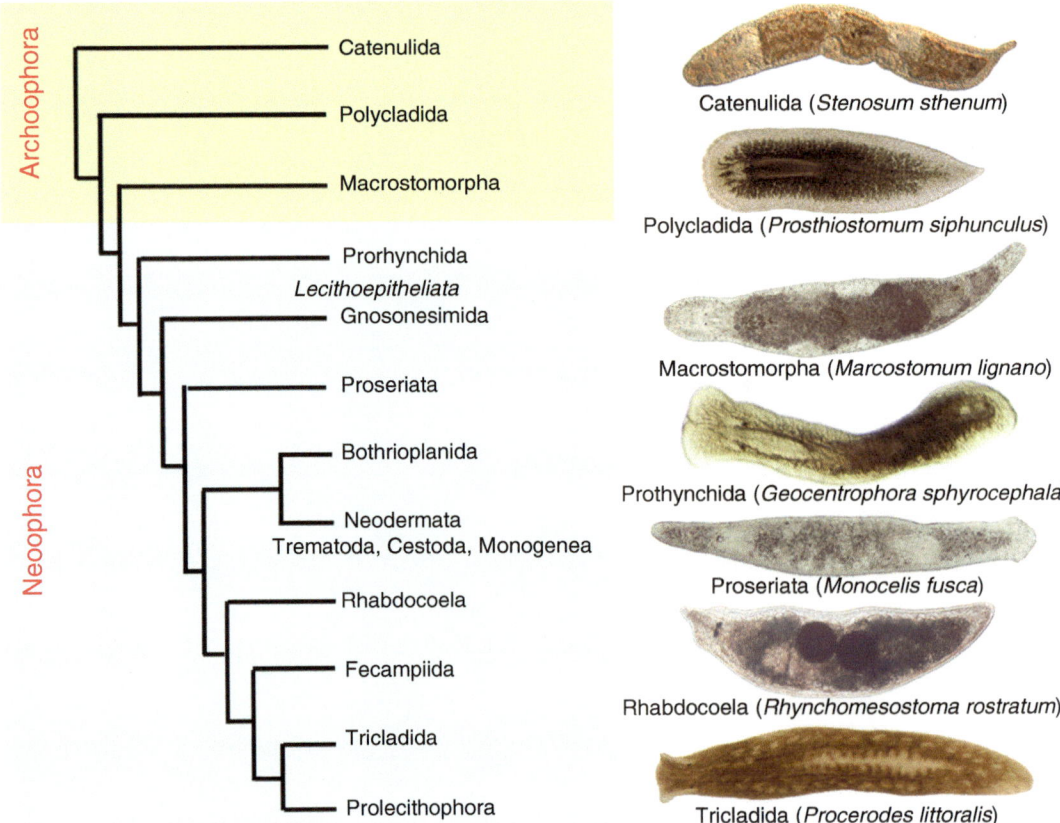

Fig. 3.1 Phylogenetic consensus tree of Platyhelminthes. The phylogenetic classification is based on Laumer and Giribet (2014), in which the former Lecithoepitheliata appears as a paraphyletic group, now divided into two new orders (Prorhynchida and Gnosonesimida). For simplicity, the old Lecithoepitheliata name is used here. Images of representative species are shown (Adapted from Martín-Durán and Egger (2012))

Platyhelminths lack a coelom; circulatory, skeletal, and respiratory systems; and a definitive anus. They have a central nervous system (CNS), a blind gut, and flame-bulb protonephridia. The space between the organs is filled with mesenchyme (or parenchyma). Their epidermis is a monolayer of multiciliated cells that provide locomotion. A characteristic trait of this phylum is the presence of a large number and variety of adhesive secretions from specialized cells and organs. Platyhelminths have an extreme morphological plasticity, and some are capable of regenerating a whole organism from a small piece or can change their dimensions continuously according to the availability of food. This property is related to the existence of a unique population of adult totipotent stem cells, termed 'neoblasts', which provide the cellular basis for this morphological plasticity.

Platyhelminths are with few exceptions hermaphroditic (with cross-fertilization), with rather complex reproductive systems. However, some free-living species have the ability to reproduce asexually by fission and subsequent regeneration; moreover, for some parasitic flatworms, asexual multiplication is an obligatory part of their life cycle. In contrast to other animals, most platyhelminths have eggs that are devoid of yolk (ectolecithal eggs), which is supplied by specialized yolk cells. This trait is one of the main causes for the greatly modified embryonic development observed in many flatworms.

This modified embryonic development and their high regenerative capacity (see Chapter 4) makes them of widespread biological interest. This chapter reviews the embryonic development of the different subtaxa of Platyhelminthes from the perspective of evolutionary developmental biology. Chapter 4 uses Platyhelminthes as a model phylum to study the regenerative potential across the animal kingdom, with a particular focus on the CNS and the photoreceptors, highlighting the pivotal role played by the neoblasts. The role of evolutionarily conserved signalling pathways (Wnt, BMP, Hippo) in controlling adult plasticity, maintenance of axial polarity, and growth is also discussed in Chapter 4.

COMPARATIVE PLATYHELMINTH DEVELOPMENT

The embryonic development of flatworms has long attracted the attention of embryologists and phylogeneticists. The structure of the oocyte was used as one of the main features to classify Platyhelminthes: entolecithal species (with eggs that contain all the yolk needed for development) were classified as Archoophora, since this condition is considered basal, whereas ectolecithal species (with eggs that must take up yolk from outside) were classified as Neoophora, as they required the invention of specialized yolk cell-producing organs called vitellaria (a detailed and recent revision can be found in Martín-Durán and Egger 2012).

Since the advent of modern molecular phylogenetic analyses, the inclusion of Platyhelminthes in the Spiralia has been well supported and widely accepted (Dunn et al. 2008). The spiral cleavage pattern of Platyhelminthes is characterized by an oblique orientation of the spindles with respect to the embryonic animal-vegetal axis. Due to the stereotypic nature of this cleavage pattern, it is possible to identify and name each cell according to an established nomenclature. Depending on the species, spiral cleavage may be either equal or unequal. In the latter case, one of the blastomeres at the two-cell stage (blastomere CD) and four-cell stage (blastomere D) is typically larger than the others. A feature of the spiral development is the early specification of blastomeres (determinative development). Comparative embryology shows that polyclads follow a relatively stereotypical spiral cleavage pattern, while in some neoophorans, spiral cleavage is replaced by an irregular and disperse cleavage, referred to as *Blastomerenanarchie* (Thomas 1986; Baguñà and Boyer 1990). Currently, it is assumed that the acquisition of ectolecithal eggs involved the modification of the cleavage pattern. The observation that quartet spiral cleavage patterns can still be recognized in some neoophorans (lecithoepitheliates and proseriates), despite the presence of extra-embryonic yolk cells within the egg, agrees with the view that spiral cleavage most likely constitutes the plesiomorphic cleavage pattern in Platyhelminthes (Thomas 1986; Baguñà and Boyer 1990). Interestingly, the striking diversity of platyhelminth embryology

Table 3.1 Comparison of the main embryonic traits of different subtaxa of Platyhelminthes

	Egg type	Cleavage pattern	Larva/direct development
Catenulida	Entolecithal	Spiral (early)	Luther's larva and direct development
Polycladida	Entolecithal	Spiral	Müller's, Goette's, and Kato's larva and direct development
Macrostomorpha	Entolecithal	Spiral (early)	Direct development
Lecithoepitheliata	Ectolecithal	Spiral	Direct development
Proseriata	Ectolecithal	Spiral (early)	Direct development
Bothrioplanida	Ectolecithal	Disperse	Direct development
Rhabdocoela	Ectolecithal	Irregular	Direct development
Fecampiida	Ectolecithal	Irregular	Direct development
Prolecithophora	Ectolecithal	Disperse	Direct development
Tricladida	Ectolecithal	Disperse	Direct development (the embryo has been considered a 'cryptic larva')
Neodermata	Ectolecithal	Disperse	Larva and direct development

contrasts with the similarity of adult body plans observed, at least among the free-living groups (Fig. 3.1).

Although most of the available data corresponds to classical descriptive embryological approaches, molecular techniques are beginning to be implemented in representative species from several taxa of Platyhelminthes, such as the polyclads and the triclads (Rawlinson 2010; Lapraz et al. 2013). The following sections discuss the main findings on embryonic development in each taxon of flatworms, focusing on early development. Table 3.1 summarizes the most important known embryonic traits of each group. Emphasis is placed on embryonic features with evolutionary significance and topics that deserve further attention.

Archoophorans

The 'Archoophora' (Catenulida, Macrostomorpha, and Polycladida) is a paraphyletic group including all flatworms with endolecithal eggs. Archoophorans exhibit quartet spiral cleavage, at least during the early zygotic divisions.

Catenulida

Knowledge of embryonic development in catenulids is scarce. The only points worthy of note here are that their early development looks similar to polyclads and, although most of them hatch as a juvenile, a so-called Luther's larva has been described for *Rhynchoscolex simplex* (Reisinger 1924). The significance of the existence of this larval stage within Platyhelminthes will be discussed later.

Polycladida

The Polycladida consists of large, almost exclusively marine animals. It is divided into two suborders: the Cotylea, with a prominent sucker posterior to the female genital opening, and the Acotylea, which lack a sucker. Polyclad flatworms provide an interesting system to explore the evolution of development within Platyhelminthes and Spiralia, because, unlike most other flatworms, they undergo spiral cleavage similar to that seen in some other spiralian taxa and some lineages also display indirect development through the formation of Müller's and Goette's larvae. To date, the most comprehensive description of their early embryonic development comes from the species *Hoploplana inquilina* (Surface 1907; Boyer et al. 1998). Recently, a new species, *Maritigrella crozieri*, has been established as a suitable species that can be maintained in the laboratory (Rawlinson 2010; Lapraz et al. 2013).

Polyclad cleavage is spiral, usually with no size difference between the cells of the four-cell stage, although the D cell can be slightly larger in some species. However, the cleavage pattern differs from that of canonical spiralians in that the micromeres of the fourth quartet (4a–4d) are atypically large and the macromeres (4A–4D) are smaller. During gastrulation, the animal micromeres 4a–4c and 4A–4D become internalized completely through epiboly over the vegetal macromeres (Fig. 3.2). Micromere 4d produces

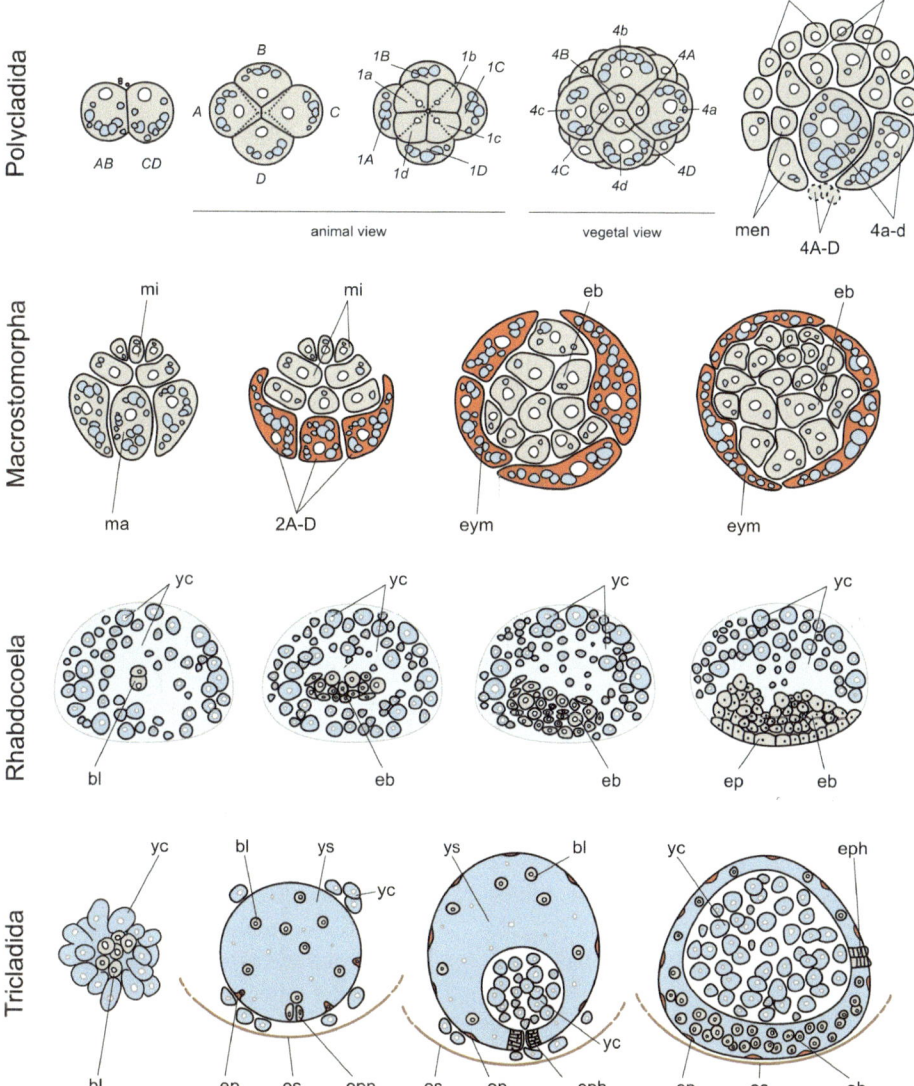

Fig. 3.2 Embryonic development of Polycladida, Macrostomorpha, Rhabdocoela, and Tricladida. Polyclads exhibit a conserved quartet spiral mode of development, although the macromeres (*4A–D*) are smaller than the micromeres (*4a–d*). Gastrulation occurs through epiboly of the animal micromeres over the vegetal macromeres. Another peculiarity of polyclad development is that macromeres *4A–D* and micromeres *4a–c* degenerate and the whole endoderm and a large part of the mesoderm originate from the *4d* micromere. Macrostomids show the typical quartet spiral cleavage pattern up to the eight-cell stage. Afterwards, the vegetal macromeres *2A–D* flatten and cover the embryo. This transient membrane is later replaced by the definitive epidermis. The rest of the blastomeres remain in the inner region and form the embryonic blastema, in which organogenesis takes pace. In rhabdocoeles, the first cell division gives rise to an animal micromere and a vegetal macromere. Proliferation of these two initial cells forms an embryonic blastema, which will be located on the future ventral side of the embryo. The epidermis differentiates from this embryonic blastema, which will ingest external yolk cells. In triclads, no signals of spiral cleavage are observed, but blastomeres cleave in a disperse manner and remain isolated within the yolk mass. Afterwards, some blastomeres differentiate into the primary epidermis and embryonic pharynx, which will be used to ingest the yolk cells. After yolk ingestion, the remaining blastomeres proliferate and differentiate into the definitive organs. In all schemes, an idealized cross section of the animal-vegetal axis is presented (animal to the *top*, vegetal to the *bottom*). Yolk granules are in *light blue*, hull cells in *orange*, and embryonic cells in *grey*. *bl* blastomere, *eb* embryonic blastema, *ec* ectoderm, *ep* epidermis, *eph* embryonic pharynx, *epp* embryonic pharynx primordium, *es* egg shell, *eym* embryonic yolk mantle, *ma* macromere, *mec* mesoectoderm, *men* mesoendoderm, *mi* micromere, *yc* yolk cell, *ys* yolk syncytium (Adapted from Martín-Durán and Egger (2012))

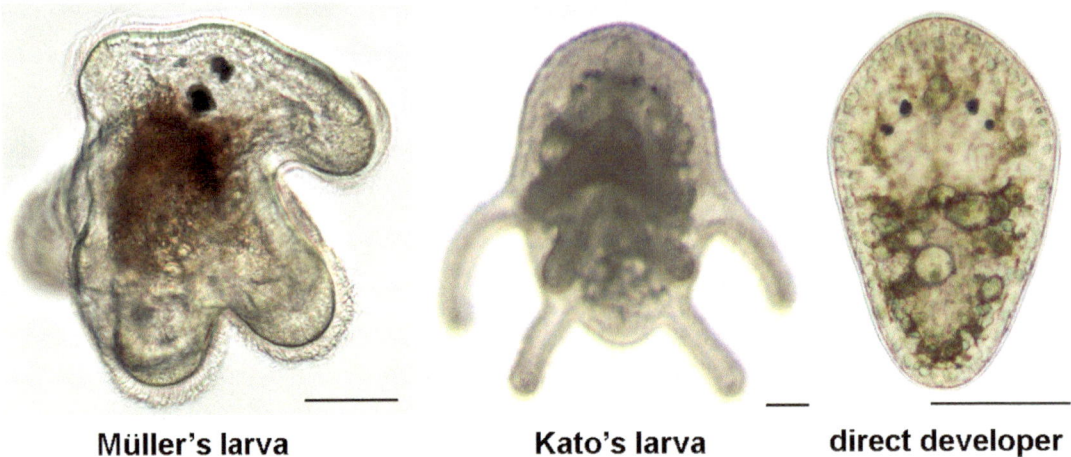

Müller's larva   **Kato's larva**   **direct developer**

Fig. 3.3 Larval types and juveniles of Polycladida. Müller's larva of *Prosthiostomum siphunculus*, Kato's larva of *Planocera reticulata*, and directly developing juvenile of *Pseudostylochus obscurus* (Adapted from Martín-Durán and Egger (2012))

both endoderm and mesoderm. Micromeres 4a–c are the main yolk-containing cells that are incorporated into the gut. The macromeres degenerate (Surface 1907; Thomas 1986; Boyer et al. 1996). This differs from a suggested spiral archetype (e.g., mollusks and annelids; see Chapters 7 and 9), where 4d makes only mesoderm and 4a–c and the macromeres form the endoderm. The apical micromeres give rise to the brain and the epidermis. An inner cell layer is formed by the progeny of 4d (called mesentoblast) and the animal micromere 2b (mesectoderm) (van den Biggelaar et al. 1997; Boyer et al. 1998). The musculature, gland cell, protonephridial system, and pharynx musculature are derived from this inner layer. The dorsoventral (DV) and anterior-posterior (AP) body axes are established by the mesentoblasts and their progeny, which form a ventral plate of cells. A series of blastomere ablation studies showed that polyclad development is determinative (Boyer et al. 1998).

Later development shows much variation from almost direct to indirect with free-swimming larvae, which are classified according to the number of lobes and eyes that are present. Many polyclads have a spherical Müller's larva with eight lobes and three eyes. Some species have Goette's larva or Kato's larva which only differ from each other in the number of lobes and eyes (Gammoudi et al. 2011, 2012). Some species directly develop into small juvenile polyclads (Fig. 3.3).

Macrostomorpha

The Macrostomorpha comprises small flatworms mainly from the taxon Macrostomida. To date, studies on their embryonic development are restricted to the genus *Macrostomum*. *Macrostomum lignano* has recently been introduced as a new model organism for studying their development and evolution, since it can be easily reared in the laboratory, produces a multitude of eggs year round, and has a very short generation time (2–3 weeks at 20 °C). Its embryogenesis has been described in detail (Morris et al. 2004). Early cleavage, up to the eight-cell stage, occurs in a typical spiral cleavage pattern. In later stages, development starts to deviate from this pattern, mainly because of the formation of an external yolk mantle from the four vegetal yolky macromeres, the so-called hull cells, which cover the embryo (Willems et al. 2009). This process was called 'inverse epiboly' by Thomas (1986). Within the yolk mantle, the remaining blastomeres form a proliferating mass, the embryonic primordium, from which the definitive structures, such as the epidermis that will replace the hull cells, and the gut primordium will arise (Fig. 3.2). After 4–7 days (at 20 °C), the juvenile hatches.

Hull cells are not only found in macrostomids but in all flatworm taxa, with the exception of polyclads. The presence of hull cells is the most obvious deviation from the typical spiral developmental pattern in Platyhelminthes. By definition, hull cells are large, yolk-rich embryonic cells that surround the blastomeres after the early divisions. Although hull cells look similar and occupy the same position within flatworm embryos, they do not share any ontogenetic origin, indicating that they are not homologous structures. Hull cells may originate from the vegetal macromeres (as in *Macrostomum*), from animal micromeres (as in proseriates or lecithoepitheliates), or from vitellaria yolk cells (as in triclads, prolecithophorans, and rhabdocoeles).

Neoophorans

The common feature of Neoophora is the laying of ectolecithal eggs. Neoophora have oogonia that are divided into the germarium, which produces the oocytes, and the vitellarium, which produces the yolk cells. Thus, the egg contains both oocytes and extra-embryonic yolk cells, and all Neoophora embryos have developed mechanisms to engulf the external yolk cells, usually by forming a temporary epidermis known as hull membrane. Ectolecithality determines the highly divergent mode of development within this group and with respect to the rest of the spiralians. Whereas some of them retain a kind of quartet spiral cleavage (Lecithoepitheliata and Proseriata), others (Bothrioplanida, Rhabdocoela, Fecampiida, Prolecithophora, and Tricladida) have a disperse mode of cleavage that, in the case of the Tricladida, has been classically called *Blastomerenanarchie* (Seilern-Aspang 1958). Juveniles directly develop from an embryonic blastema. Obviously, such divergent cleavage patterns result in divergent gastrulation and lineage fate processes.

Lecithoepitheliata and Proseriata
Lecithoepitheliata and Proseriata retain a kind of stereotypical quartet spiral cleavage pattern, just up to the eight-cell stage in Proseriata. As in other spiralians, and in contrast to the situation observed in polyclads, the macromeres are large and give rise to the endoderm (Reisinger 1972). In Lecithoepitheliata, gastrulation involves epiboly of the micromeres over the vegetal macromeres at the 25–30-cell stage. The micromeres then flatten and differentiate into the hull membrane, which engulfs a portion of the yolk. In Proseriata, no pattern is discernable in late cleavage stages. In both groups, the inner mass of blastomeres differentiates into an embryonic blastema that occupies the future ventral side of the embryo, where the definitive organs and the epidermis that will replace the hull membrane will differentiate (Reisinger 1972).

Bothrioplanida, Rhabdocoela, Fecampiida, and Prolecithophora
The embryonic development of Bothrioplanida, Rhabdocoela, Fecampiida, and Prolecithophora species has not been studied in detail. However, since they share essential features, they are considered here together. These four groups show no signs of a spiral quartet. In Rhabdocoela and Prolecithophora, the first cell division gives rise to an animal micromere and a vegetal macromere, which have been proposed to be homologous to blastomeres AB and CD of canonical spiralians (Hartenstein and Ehlers 2000; Younossi-Hartenstein et al. 2000). Late cleavage does not follow a regular pattern and results in an embryonic blastema from which a transitory epidermis differentiates to engulf the external yolk cells (Fig. 3.2). Afterwards, the embryonic blastema differentiates to form the definitive epidermis and organ primordia, as observed in all other neoophoran flatworms. All directly hatch as juvenile worms.

Tricladida
Tricladida includes marine, freshwater, and terrestrial species. Due to the amazing regenerative abilities of the adults in some species, Tricladida is by far the best-studied platyhelminth taxon, not only at the morphological but also at the molecular level. The majority of species analysed with respect to their potential to regenerate are freshwater animals (Brøndsted 1969), but embryological studies have been reported from representatives

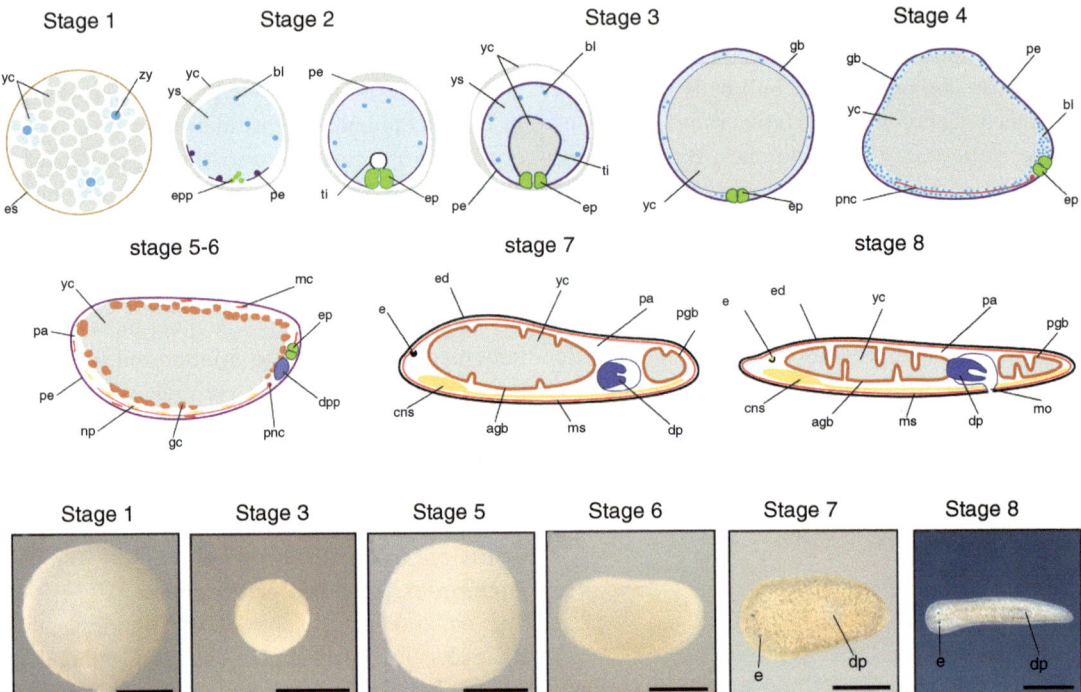

Fig. 3.4 Overview of embryogenesis in the planarian *Schmidtea polychroa*. Stage 1, blastomeres divide within the mass of yolk cells. Stage 2, each zygote gives rise to a yolk-feeding embryo with its transient embryonic structures (i.e., primary epidermis, embryonic pharynx). Stage 3, due to the ingestion of yolk cells, the embryo grows in size, and the initial yolk-derived syncytium becomes restricted to the periphery of the embryo, forming the so-called germband, composed mainly of proliferating blastomeres. Stage 4, blastomeres in the germband start differentiating, and the definitive nerve cords become visible. Stages 5–6, specification of definitive axial identities and differentiation of the adult cell types and organs (definitive pharynx and brain primordium) start. Stage 7, the definitive organs arise and develop to acquire the mor- phology observed in stage 8 embryos and hatchlings. *agb* anterior gut branch, *bl* blastomere, *bp* brain primordium, *cns* central nervous system, *dp* definitive pharynx, *dpp* definitive pharynx primordium, *e* eye, *ed* epidermis, *ep* embryonic pharynx, *epp* embryonic pharynx primordium, *es* egg shell, *gb* germband, *gc* gastrodermal cell, *mc* muscle cell, *mo* mouth, *ms* muscle, *np* neural precursor, *pa* parenchyma, *pe* primary epidermis, *pgb* posterior gut branch, *pnc* pioneer nerve cord, *ti* temporary intestine, *vnc* ventral nerve cord, *yc* yolk cells, *ycI* type I yolk cells, *ycII* type II yolk cells, *ys* yolk syncytium, *zy* zygote. Images of planarian embryos corresponding to different stages are shown. Anterior is to the *left*. Scale bars: 500 μm except for stage 8 where it is 1 mm (Adapted from Martín-Durán et al. (2012a))

of all three habitat types. Whereas the freshwater planarian *Schmidtea mediterranea* has emerged as the model for the study of regeneration mechanisms and stem-cell biology, its sister species *Schmidtea polychroa* has become the model of choice for embryological studies (reviewed in Martín-Durán et al. 2012a).

Triclads lay cocoons from which one to ten juveniles may hatch (Tekaya et al. 1999). The cocoons contain several zygotes surrounded by a large number of extra-embryonic yolk cells (Cardona et al. 2005), which fuse and form a syncytium after deposition (Figs. 3.2 and 3.4). The early blastomeres become embedded inside the syncytium, where cleavage takes place. As in the Prolecithophora and Bothrioplanida, *Blastomerenanarchie*-type cleavage is observed with no canonical pattern and none of the blastomeres remaining next to the other. After several divisions, transient structures organize, such as the embryonic epidermis (the hull membrane) and the embryonic pharynx, through which the yolk syncytium is engulfed (Figs. 3.2 and 3.4). The rest of the blastomeres remain in the periphery of the embryo in an undifferentiated state. It is known that they express stem-cell-associated gene markers such as *vasa* and *tudor* (Solana and Romero 2009). Once the embryo has internalized

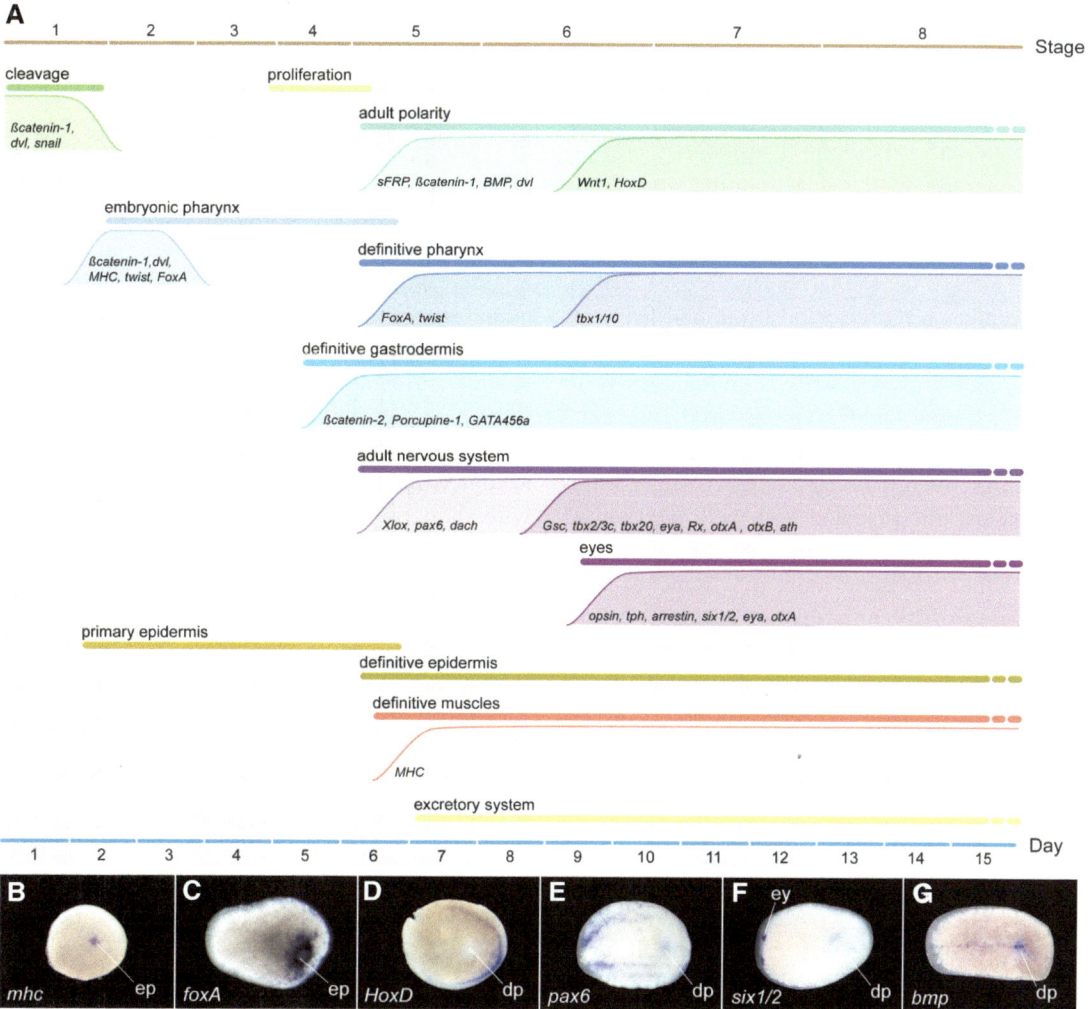

Fig. 3.5 Gene expression during planarian embryonic development. (**A**) Overview of major developmental events (*coloured bars*) and associated gene expression (*coloured curves*) with respect to stages and days of development at 20 °C (Modified from Martín-Durán et al. (2012a)). *Broken lines* to the *right* indicate continued expression of the respective genes in adult animals. (**B–G**) Expression pattern of specific genes analysed by whole mount in situ hybridization (stage 2 in **B**, stage 5 in **C**, stage 6 in **D–F**, and stage 7 in **G**). See main text for references

enough yolk, the blastomeres migrate and differentiate into the definitive structures that will replace the embryonic ones, that is, the definitive epidermis, pharynx, nervous, digestive, and muscle systems (Fig. 3.4). The process of organogenesis was formerly described as involving the formation of three main ventral primordia (an anterior brain primordium, a central pharynx primordium, and a posterior primordium), as in other neoophoran flatworms. However, recent studies on *S. polychroa* demonstrate that the appearance of the definitive structures does not occur in any specific AP position (Cardona et al. 2005; Martín-Durán and Romero 2011). Molecular studies show that the stages at which the primordia of the definitive organs appear, the so-called five to six stages, are the ones in which the molecular markers of axial polarity (Wnt and BMP pathway elements) appear (Marín-Durán et al. 2010), supporting consideration of this stage as the point at which the body plan of the adult is established (for a review of gene expression during *Schmidtea polychroa* embryogenesis, see Fig. 3.5). Interestingly, it is at these stages when triclads start to display an ability to regenerate, suggesting that neoblasts appear at the same time as other definitive cell types. The most obvious event of adult organogenesis is the formation of the definitive pharynx primordium ventrally to the degenerating embryonic pharynx (Martín-Durán and Romero 2011). The pharynx primordium, which strongly expresses the gut marker *foxA* (Martín-Durán et al. 2010), first occupies a posterior position, and it moves to more central regions as the embryo elongates anterior-posteriorly. Simultaneously to the formation of the pharynx, gastrodermal cells (*GATA456a-positive* cells) appear scattered through the margins of the embryo and eventually enclose the ingested yolk cells and define the definitive gastrodermis, which is already divided into the three main gut branches. The appearance of the musculature seems to follow a similar dynamic, with muscular precursors (*myosin heavy chain-positive*, mhc, cells and cells immunolabeled by the specific antibody TMUS-13) being first specified diffusely throughout the embryo and then progressively forming an orthogonal grid as the embryo elongates and flattens (Cardona et al. 2005; Martín-Durán et al. 2010). Neurogenesis starts with the formation of a definitive neural primordium, made of two bilaterally symmetrical condensations of neural progenitors. As embryogenesis continues, this primordium develops into the anterior bilobed brain and the two main ventral nerve cords (Cardona et al. 2005). In contrast to studies on the embryonic development of the digestive and muscular systems, and despite the great amount of data regarding adult nervous system regeneration (see Chapter 4), gene expression studies on planarian embryonic neurogenesis are still lacking. As a first step, the embryonic development of the photoreceptors was characterized at the genetic level (Martín-Durán et al. 2012b). Simultaneously to the early formation of the brain, an anterior pair of eye precursor cells (*opsin*- and tryptophan hydroxylase (tph)-positive cells, detected by the antibody against arrestin) appears beneath the dorsal epidermis, which develops into the eyes of the juvenile (Martín-Durán et al. 2012b). As in adult eye regeneration,

this process seems to be controlled by the genes *sine oculis* and *eye absent* but appears to be independent of *Pax6*, a proposed master regulator of animal eye development (see Chapter 4).

Although the formation of the hull membrane is a common trait in Platyhelminthes embryos, the presence of a transient embryonic pharynx, with the associated neural structures, is only observed in triclads. An interesting question is thus the extent to which this complex transient yolk-feeding embryo can be considered an encapsulated larva and whether its formation can be considered a process of gastrulation. Although it had traditionally been accepted that triclads, and other Platyhelminthes with disperse cleavage, lack true gastrulation movements, the anatomy of their embryos indicates that well-defined tissues belonging to the ectoderm (primary epidermis), mesoderm (pharyngeal muscle), and endoderm (temporary intestine) are present at this time. Moreover, recent molecular studies have demonstrated the expression of evolutionarily conserved gastrulation-related genes during the formation of the embryonic pharynx, including *snail, twist, foxA,* and *β-catenin* (Martín-Durán et al. 2010). These findings suggest that ancient mechanisms of early cell fate specification are still present in triclad embryos and, therefore, that planarian development can be separated into two morphogenetic stages: a first gastrulation process and a subsequent 'metamorphosis'. During gastrulation, the three germ layers segregate and establish the primary organization of the feeding embryo or 'cryptic' larva (Cardona et al. 2006). Metamorphosis then establishes the definitive adult body plan through the involvement of totipotent blastomeres. During this second stage, the morphogenetic and patterning mechanisms are similar to those used during regeneration and homeostasis in the adult, e.g., in the use of the Wnt and BMP signalling pathways to establish the AP and the DV axis (Martín-Durán et al. 2010) (see Chapter 4). The cryptic larva may be considered a vestigial larva, or, since several embryos share the same maternal resources in each cocoon, it could be the evolutionary

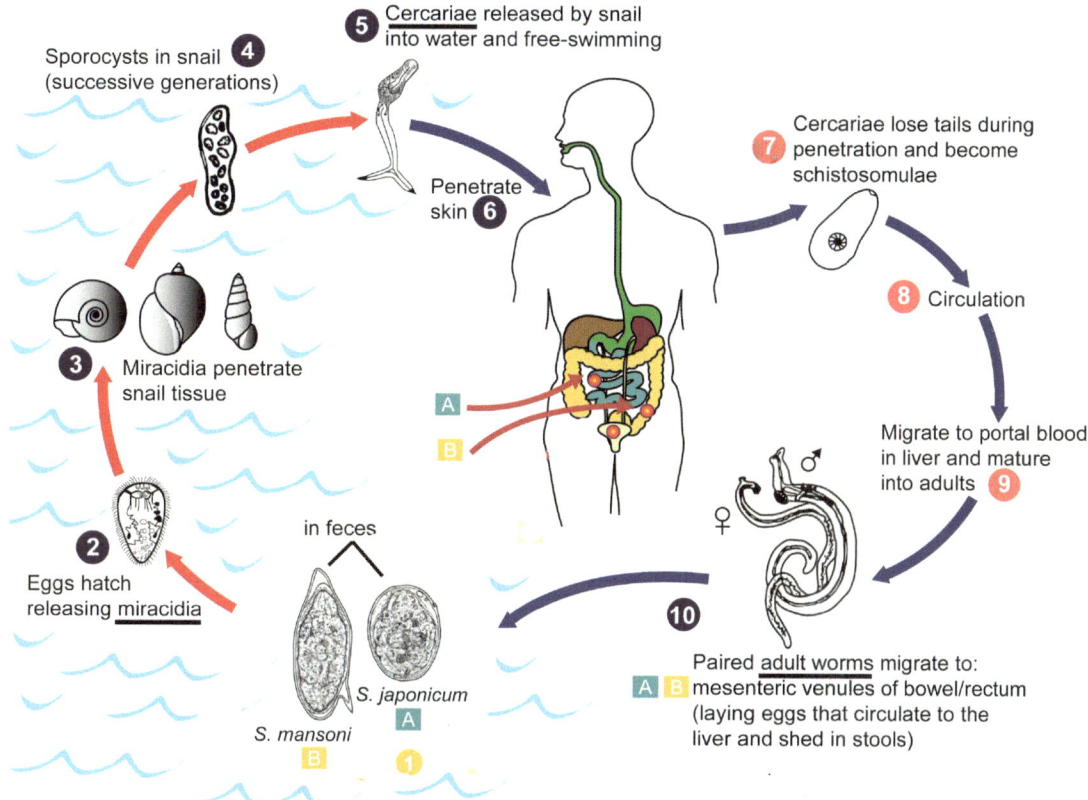

Fig. 3.6 Life cycle of *Schistosoma*. Eggs are expelled with faeces (*1*). Under optimal conditions, miracidia (*2*) hatch from the eggs and swim and penetrate the snail intermediate host (*3*). The stages in the snail include two generations of sporocysts (*4*) and the production of cercariae (*5*). Upon release from the snail, the infectious cercariae swim, penetrate the skin of the human host (*6*), and become schistosomulae (*7*). The schistosomulae migrate through several tissues and stages to their residence in the veins (*8*, *9*). Adult worms in humans reside in the mesenteric venules in various locations, which seem to be specific for each species (*10*). The females deposit eggs in the small venules of the portal and perivesical systems. The eggs are moved progressively towards the lumen of the intestine (*S. mansoni* and *S. japonicum*) and are eliminated with faeces (*1*) (Adapted from Wikimedia Commons)

result of their competition for the limited maternal yolk (Cardona et al. 2006). Further discussion about the gastrulation and the direct versus indirect mode of development in Platyhelminthes is presented below.

Neodermata

Neodermata comprises the parasitic flatworms, classified in three taxa: Monogenea, Cestoda, and Trematoda. The body wall of Neodermata, an unciliated syncytium called the neodermis, represents the defining morphological criterion of the clade. Ontogenetically, the neodermis is formed secondarily underneath an earlier developing primary ciliated epidermis, which is replaced during development. Other specializations of the Neodermata, a result of its adaptation to the endoparasitic lifestyle, are the reduction or absence of a gut, a simplified nervous system, and the existence of a complex life cycle that involves several larval stages inhabiting different animal hosts (Fig. 3.6). Their life cycle has evolved to give rise to a large number of descendants. Embryological studies of these taxa are scarce, since most of the effort is focused on the understanding of the life cycle of the species which represent a public health problem, either directly producing human diseases, such as schistosomiasis (Fig. 3.6), or affecting human resources, e.g., as pests in fish farms.

Cestoda

Cestoda are commonly known as tapeworms, and about 4,000 species have been described. Almost all vertebrate species may function as potential hosts. Tapeworms usually require at least two hosts, and humans represent the primary host of several species, with infestation occurring through uptake of uncooked meat such as pork (*Taenia solium*), beef (*Taenia saginata*), or fish (*Diphyllobothrium* spp.). Adult tapeworms are parasitic in the digestive tract of vertebrates, and often one of the intermediate hosts is an invertebrate. Tapeworms usually have long flat bodies composed of many reproductive units or proglottids. This represents a specific trait of Cestoda, since each proglottid contains the male and female reproductive structures and can reproduce independently; therefore, it could be considered that a tapeworm is actually a colony of proglottids. Tapeworms completely lack a digestive system, and, as in Trematoda, there are no external motile cilia. In contrast to monogeneans and trematodes, however, their entire surface is covered with small projections that resemble the microvilli of the vertebrate small intestine. As in the intestine, these microvilli serve to expand the absorptive surface. This represents an adaptation for a tapeworm, since it must absorb all of its nutrients across the tegument. Tapeworms have well-developed muscles, and their excretory and nervous systems are similar to those of other flatworms. They have no special sense organs but do have modified cilia that function as sensory endings in the tegument. One of their most specialized structures is the scolex, the organ of attachment, which usually has suckers and spiny tentacles.

Tapeworms have a complex life cycle that begins with an infectious larva, called a coracidium in aquatic tapeworms and oncosphere in terrestrial tapeworms. This simplified passive larva enters the primary host by being ingested. The larva is extremely small, containing only a few embryonic cells (50–100) surrounded by several protecting envelopes. The inner envelope, called

the embryophore, may retain its ciliation. Upon hatching in the gut of the primary host, the larva sheds its protective envelopes and penetrates through the intestinal wall into the body cavity of the host. Once inside the body cavity, this primary larva metamorphoses into a secondary larva, called the cysticercoid.

A well-known example is the beef tapeworm *Taenia saginata*, the adult form of which lives in the human small intestine and can attain a length of 20 m, folded back and forth in the host intestine. Fertilized eggs shed from the human host are ingested by cows, the intermediate hosts, where six-hooked larvae, the oncospheres, hatch and migrate to skeletal muscles where they encyst to become 'bladder worms' or cysticercoids. Each of these juveniles remains quiescent until the uncooked muscle is eaten by humans, where it attaches to the intestine and matures. More dangerous to humans is the species *Taenia solium*, or pork tapeworm, because humans can act also as secondary hosts. Cysticercoids can move from the muscle to the brain and cause cysticercosis, which can be lethal.

Embryogenesis of *Taenia* and *Hymenolepis* species has been described at the morphological level (Hartenstein and Jones 2003). The general mode of flatworm development can still be discerned in both of them. The oocytes are surrounded by a hard outer capsule formed within the ovary. They divide totally and unequally. Similar to what has been described for the embryos of many neoophoran flatworms, no spiral cleavage pattern is apparent. During the early cleavage divisions, large macromeres separate from the micromeres and give rise to two syncytial sheaths surrounding the embryo. The inner envelope is considered as the primary epidermis or embryophore, which is only ciliated in coracidium larvae of aquatic tapeworms. Like in free-living neoophoran platyhelminths, non-morphogenetic movements that resemble a classical gastrulation are observed. During organogenesis the hooklets, specific structures that cestode larvae use to penetrate the epidermis of the host, are differentiated, together with the muscle and the nervous system. Oncosphere larvae have a complex system of muscle fibres connected to the hooklets, and this enables them to generate coordinated movement to enter the body cavity of the host. A pair of neurons located in the anterior tip of the larva sends axons posteriorly and branch profusely along the deep muscle fibres to innervate the musculature (Hartenstein and Jones 2003).

Despite its extreme reduction in overall cell number and different cell types, the oncosphere larvae of cestodes exhibit similarities to the larvae or juveniles of other flatworms, notably an anterior 'brain' (consisting of only two neurons in cestode larvae), as well as other resemblances during ontogenesis. A common feature of many cestode and trematode species is the simultaneous apoptotic death of several micromeres to form the embryonic envelopes and the multiplication and differentiation of other blastomeres. This genetically programmed cell death leads to a simplification of the infective larval stage and likely represents an ontogenetic adaptation to the parasitic life strategy (Młocicki et al. 2010).

Mesocestoides corti represents a different model for studying cestode biology and development. It is particularly interesting because of its intermediate larval stage (tetrathyridium), composed of a scolex and an unsegmented body that is able to proliferate asexually by longitudinal fission in the intermediate hosts. This enables to study asexual reproduction, proglottid formation, and strobilar development, leading to the formation of serially arranged genital organs, one in each segment (Koziol et al. 2010).

Trematoda

Trematodes are all parasitic flukes, and as adults they are almost all found as endoparasites in vertebrates. However, they share several characteristics with free-living Platyhelminthes, such as a well-developed alimentary canal and similar reproductive, excretory, muscle, and nervous systems.

Trematodes have a complex life cycle. The intermediate host is a mollusk, and the definitive host is a vertebrate. In some species, a second and sometimes even a third intermediate host is involved. The genus *Schistosoma* has been the subject of several studies, since it is the causative agent of schistosomiasis, which ranks as one of the major infectious diseases in the world and is widely prevalent in developing countries. As in all trematodes, the life cycle of *Schistosoma* is very complex (Fig. 3.6). It has separate sexes that live for several years in blood vessels. Inside the host vasculature, each female can release about 300 eggs per day. When eggs are excreted into fresh water by the primary host, a ciliated larva called miracidium hatches in about 5 days. In snails of the genus *Biomphalaria*, miracidia undergo cycles of asexual reproduction, leading to the emergence of a second larval type, the cercaria. Cercariae swim actively and penetrate the definitive host, again through the skin to enter blood vessels, where they develop into adults (Jurberg et al. 2009). Through this complex cycle, a single zygote of parasitic flatworms can give rise to an enormous number of progeny. The disease is caused not by the transmitted eggs but rather by chronic inflammation that occurs when eggs released by female worms get trapped in small blood vessels of the liver or other organs (Fig. 3.6).

Schistosomes are neoophorans. Therefore, the yolk cells, produced by the vitellarium, are found outside the oocyte. In general, their embryology resembles that of free-living flatworms. Schistosome eggs are not retained in the female uterus and remain uncleaved when laid. Cleavage starts in the blood of the primary host and produces two blastomeres that differ slightly in size. During the first cleavages, the vitelline cells fuse into a single yolk syncytium. Subsequent cleavages are asynchronous and form a discoidal embryonic primordium with an animal-vegetal axis. This grows to form the stereoblastula, which comprises a central proliferating embryonic primordium surrounded by an outer envelope, produced by the migration of three to four macromeres. At this point, cellular differentiation begins, giving rise to an inner envelope and the appearance of early organ primordia. Afterwards, organogenesis takes place, with the differentiation of the neural mass, the epidermis, the musculature, and the miracidial glands. Hatching releases a fully formed larva that displays muscle contraction, beating cilia, and flame cells.

Although almost nothing is known about the molecular control of trematode embryogenesis, a recent report demonstrated the importance of TGFb signalling during egg development. It was shown that the protein encoded by SmInAct, the *Schistosoma* homolog of activin, is produced only when females are paired with males in an immunologically competent setting (Freitas et al. 2007). In this study, RNA interference showed that SmInAct plays a crucial role in egg development.

Recently, the early development of an Aspidogastrea species (*A. limacoides*) has been studied at the ultrastructural level (Świderski et al. 2011). The subclass Aspidogastrea is a small group of parasitic flatworms generally considered to belong within the Trematoda, although several differential traits call this into question. Aspidogastreans occur worldwide in marine and freshwater environments, where they use a mollusk as an obligate host. Although in some cases a vertebrate is a facultative or obligate final host, their entire life cycle may occur within the mollusk. In contrast to what is found in trematodes and cestodes, there are no asexual generations in known life cycles of aspidogastreans. Also in contrast to other parasites, it shows a rather direct development, since it produces a 'larva' called a cotylocidium that exhibits the fundamental organization of the adult. The eggs of *A. limacoides* are ectolecithal and contain a fully developed cotylocidium when laid, revealing some degree of ovoviviparity. The first cleavage is equal, but subsequent cell divisions are unequal and asynchronous. These cleavages end in the formation of the early embryo, which is composed of several blastomeres of different sizes. The largest blastomeres, the three macromeres, undergo a very early fusion of their cytoplasm, forming a syncytial layer of the embryonic envelope directly beneath the egg shell. The presence of this early and unique envelope is the most important difference between Aspidogastrea and other Neodermata. The lack of an inner embryonic envelope in aspidogastreans may be explained by the close contact between the vitelline syncytium and the differentiating blastomeres, which provides a direct passage for nutrients. Moreover, the existence of an inner ciliated epithelium such as in some cestodes is not required in Aspidogastrea since the different organization of the life cycle makes a locomotor function unnecessary. Almost simultaneously with blastomere multiplication, the progressive degeneration of some micromeres begins, which is a common trait in neodermatans.

EVOLUTIONARY IMPLICATIONS

In the following sections, the most reliable embryonic characters are summarized and discussed, putting forward evolutionary hypotheses and future directions that will be useful for developmental and evolutionary studies.

Spiral Cleavage in Ectolecithal Embryos

Current phylogeny supports the hypothesis that quartet spiral cleavage is the ancestral developmental mode in Platyhelminthes. The basal position of Polycladida in the most recent phylogenetic analysis further supports this view (Laumer and Giribet 2014). The developmental deviations found in polyclads, such as the degeneration of fourth quartet macromeres and of micromeres 4a–c, are probably apomorphies of this group, since they are not present in those taxa of neoophoran Platyhelminthes that retain a quartet spiral cleavage. Further study of the embryonic development in the Catenulida will be essential to understand the evolutionary history of cleavage patterns in the Platyhelminthes.

The presence of external yolk cells in neoophorans does not necessarily imply the loss of spiral cleavage, as exemplified in lecithoepitheliates and proseriates. The complete absence of spiral cleavage in the rest of the Neoophora (Neodermata, Bothrioplanida, Rhabdocoela, Fecampida, Prolecithophora, and Tricladida) might thus be explained by a single evolutionary event at the base of this group. However, the significant differences between the cleavage in rhabdocoeles and the remaining taxa suggest that they could have evolved independent adaptations to their ectolecithal condition. Whether the disperse cleavage in the remaining taxa shares a common origin remains to be clarified. A recent study demonstrated the absence of centrosomes in the triclad *Schmidtea mediterranea* and in the neodermatan *Schistosoma mansoni* (Azimzadeh et al. 2012), suggesting that this loss occurred concomitantly with the loss of spiral cleavage and the emergence of disperse cleavage in the ancestor of triclads and schistosomes. This hypothesis will need to be further tested by studying the presence of centrosomes in other groups of flatworms.

It remains unclear how the changes in the ancestral quartet spiral cleavage affect cell fates during early embryogenesis. Polyclads follow a determinative mode of cleavage, whereby the loss of blastomeres during early development cannot be compensated by the remaining ones (Boyer et al. 1998). Due to experimental difficulties, ablation experiments are still lacking in the neoophorans, and thus it remains unclear whether disperse cleavage is determinative or regulative.

Gastrulation in Neoophoran Platyhelminthes

The ancestral mode of gastrulation in Platyhelminthes seems to be the epibolic movement of animal micromeres over the vegetal-most blastomeres, as observed in polyclads, lecithoepitheliates, and other phyla with quartet spiral cleavage. However, this mode of gastrulation is not conserved in neoophorans, since it is influenced by the amount of yolk and the disperse cleavage. This raises the question of how they gastrulate or indeed whether they gastrulate at all. In the groups with disorganized cleavage, such as the triclads, in which the blastomeres are not attached to each other (*Blastomerenanarchie*), whether or not a true gastrulation occurs is still an open question. An attempt to answer this question requires that we first define gastrulation. If we consider that gastrulation involves a series of coordinated cell and tissue movements that lead to the formation of the distinct cell layers of an embryo (ectoderm, mesoderm, and endoderm) and to a basic body plan, we must accept that the

formation of the planarian feeding embryo meets these criteria. Despite the 'anarchic' situation that might be apparent at first sight, defined groups of blastomeres are formed during planarian development. Blastomeres that form the primary epidermis, the embryonic pharynx, or remain undifferentiated in the syncytium each have a specific molecular profile. Thus, the expression of *βcatenin-1*, a master regulatory gene for the formation of the endomesoderm across metazoans, is restricted to the cells that form the embryonic pharynx, and these cells express genes related to specification of the mesoderm (*twist* and *myosin heavy chain*) and endoderm (*foxA*) (Martín-Durán et al. 2010, 2012a). This segregation of different groups of blastomeres is also directly related to the establishment of the primary embryonic organization, which represents the scaffold along which the definitive embryo will develop.

Related to gastrulation, one of the longstanding questions concerning the evolution of spiralian development relates to the origin of the mesoderm. In many spiralians, the 4d cell (the mesentoblast) produces the endomesoderm and some of the endoderm. The second and third quartet of micromeres, which generate largely ectodermal derivatives, also provide a second source of mesoderm referred to as ectomesoderm (Boyer et al. 1998). In polyclads, three peculiarities that deviate from other spiralians are found: the complete endoderm is formed by 4d, the $4d^2$ cell and not 4d forms the mesentoblast, and the macromeres are smaller than the micromeres and do not contribute to any embryonic tissue formation. On the other hand, although only preliminary data are available, current observations indicate that the mesoderm is formed differently in *Macrostomum*, since the progenitor of the mesoderm precursor (4d cell) forms a hull cell and is then lost. Since the only data about the origin of the endomesoderm in the groups that have lost spiral cleavage is the formation of the transient embryonic pharynx of triclads from the early segregation of a subset of blastomeres, further studies are necessary to determine whether the molecular signals that specify endomesodermal tissues are conserved in Platyhelminthes despite the evolution of such divergent developmental modes. In particular, attention should be paid to the role of pathways such as canonical Wnt signalling in archophoran and neoophoran species.

Correlation Between Zygotic, Embryonic, and Adult Polarities

Spiralian development relies on maternal determinants and cell-cell interactions to specify the different cell lineages and axial organizers during cleavage. In some taxa, the axial organizer, the D quadrant and its descendants, is already defined as early as in the four-cell embryo, due to the transmission of maternal determinants to one of the four cells through asymmetric cell divisions. These embryos are unequal-cleaving spiralians. However, other spiralian lineages are equal-cleaving, and therefore, early cleavage does not lead to specification of the D quadrant. Nevertheless, it is the process of cell-cell communication that sets up the embryonic organizer, during which one of the vegetal blastomeres contacts the animal micromeres and becomes specified as the D quadrant (van den Biggelaar and Guerrier 1979). Regardless of the type of cleavage, in spiralians the descendants of the D quadrant will occupy a dorsoposterior position in the embryo and will guide the cell lineages of the A, B, and C quadrants to follow their specific positions. In this way, the animal-vegetal axis of the oocyte represents the future anterior-posterior axis of the embryo. The extent to which these events are conserved in Platyhelminthes, however, remains unclear. In *Macrostomum*, this relationship seems to be conserved (Seilern-Aspang 1957). In polyclads, although the D quadrant is specified as in other equal-cleaving spiralians, a 90° shift of the anterior-posterior axis in relation to the animal-vegetal axis has been described (Younossi-Hartenstein and Hartenstein 2000). In the neoophoran groups that retain spiral cleavage (lecithoepitheliates and proseriates), the D quadrant is also specified, but the animal-vegetal axis corresponds to the definitive dorsoventral axis, where the animal pole becomes the ventral side.

As expected, the neoophoran groups that exhibit disperse cleavage do not show any apparent axial correlation. However, conclusions are difficult to draw, since the groups with disperse cleavage, such as triclads, have yet to be studied in detail. The existence of stereotypic cell-cell interactions between blastomeres and any inductive signal between them is unlikely to occur in triclads. However, some studies have shown a primary polarity in the oocyte (Benazzi 1950; Anderson and Johann 1958), and the presence of maternal determinants, like *βcatenin-1*, has also been demonstrated (Martín-Durán et al. 2010).

To conclude, no specific correlation between the animal-vegetal axis and the anterior-posterior axis is present in Platyhelminthes, except for macrostomids and polyclads. Clearly, the appearance of ectolecithal development led to dramatic changes in the specification of the embryonic polarity. However, a combination of morphological and molecular studies in representatives of each taxon will be required to resolve the mechanism of axial establishment and its evolutionary significance in Platyhelminthes.

Direct Versus Indirect Development: Was a Larva Part of the Ancestral Platyhelminth Life Cycle?

Indirect development has been described in three groups of free-living platyhelminths: catenulids with Luther's larva; polyclads with Müller's, Goette's, and Kato's larva; and fecampiids. However, Luther's larva of catenulids has only been observed in a single species, and it is virtually identical to the adult, suggesting that catenulids would be better considered as direct developers (Martin-Duran and Egger 2012). Similarly, the existence of a true larval stage in fecampiids is under debate. Polyclad larvae are the only nonparasitic platyhelminths that show unique features not found in the adult worm, such as the lobes and an apical organ, which are resorbed during metamorphosis. In Müller's and Goette's larvae, a dorsoventral flattening and an increase in the number of eyes also take place during metamorphosis (Ruppert 1978).

Nevertheless, most parts of the larval body are retained during postembryonic development, similar to, e.g., polychaete annelids or most mollusks (see Chapters 7 and 9).

All other free-living Platyhelminthes undergo direct development. However, the presence of an intracapsular larva has been proposed in triclads. This is based on the presence of specific organs in the early yolk-feeding embryo (embryonic pharynx and primary epidermis) that are later replaced by the definitive organs in the hatchling planarians in a process that could be considered a metamorphosis. The difference in the morphology of the triclad embryo and juvenile is supported by the existence of a different molecular profile (Martín-Durán et al. 2010). Nevertheless, the absence of indirect development in the lineages that branched off between polyclads and triclads indicates that the polyclad larvae and the putative intracapsular larva of triclads are likely independent adaptations of both lineages.

All neodermatan flatworms have one or more larval stages, but these are probably adaptations to their parasitic life style and not homologous to the polyclad larvae. Therefore, only polyclads and neodermatans show a prototypical larval stage, and the existence of cryptic larvae has been proposed for triclads. The distribution of indirect development in the current internal phylogeny of Platyhelminthes (Laumer and Giribet 2014) together with the presence of direct development in the Gastrotricha, a proposed sister taxon to flatworms (Dunn et al. 2008; Struck et al. 2014), suggests that direct development may be the ancestral life style in the Platyhelminthes.

Organogenesis in Platyhelminthes

Little attention is paid in this chapter to late embryogenesis of Platyhelminthes, when organogenesis takes place. The main reason is that, besides for Tricladida (see above), not much is known about this process either at the descriptive or the molecular level in the majority of the flatworm lineages. Nevertheless, although early development among different Platyhelminthes is highly variable, there seem to be significant commonalities in the formation of the adult body plan in all neoophorans and macrostomorphs (Martín-Durán and Egger 2012). In these groups, the definitive cell types and tissues develop from embryonic anlagen rather late in embryogenesis, after the extra-embryonic yolk cells have been incorporated into the embryo (in the Neoophora) or the hull membrane has covered the embryo (in the Macrostomorpha). As an example of neoophoran organogenesis, organ development seems to occur quite diffusely in the triclad *Schmidtea polychroa*, with the different cell types forming simultaneously throughout the entire germband. Only the brain and the pharynx primordia appear as a specific cluster of cells in the anterior and posterior region, respectively. In the Macrostomorpha, definitive organs also develop from an embryonic primordium (Morris et al. 2004). In this group, however, a close association between myoblasts and neuroblasts during differentiation has been described, perhaps reflecting the intricate functional relationship of muscle and nerve cells in the body wall musculature (Reiter et al. 1996).

Polyclads are an exception to the other flatworms since they exhibit a typical determinative spiralian development, in which cell types are specified early in ontogeny and proper organogenesis starts right after gastrulation. This situation is thus more similar to other studied spiralians and can be considered to be the ancestral state to all Platyhelminthes. Studies are, however, scarce and have mostly focused on the formation of the larval neuromuscular system (Younossi-Hartenstein and Hartenstein 2000; Bolaños and Litvaitis 2009; Rawlinson 2010, 2014; Semmler and Wanninger 2010). In indirect developing polyclads, an initial apical helicoid muscle develops into an orthogonal grid of longitudinal and circular muscles, which corresponds to the adult body wall musculature. Longitudinal muscles seem to derive from the vegetal 4d blastomere (endomesoderm), while circular fibres develop from the 2b micromere (ectomesoderm) (Boyer et al. 1996). The development of the dorsoventral parenchymatic muscles occurs after hatching in the planktonic phase and seems to be associated with the metamorphosis into the juvenile, as this process likely controls the dorsoventral flattening of the larva. Direct developing polyclads seem to lack the initial apical helicoid muscle, but the embryo hatches with a muscular pattern similar to the one observed in the adult. Neuronal development has been studied in indirect developing species, in particular regarding the development of the apical organ. Neurogenesis starts with the specification of a few apical serotonin- and FMRFamide-like-positive cells that establish the brain and apical plate primordia and the formation of posterior neuronal cell bodies. From the brain, dorsolateral projections extend posteriorly to meet the posterior neurons, and ventro-lateral projections innervate the mouth and larval lobes. How this larval neural pattern metamorphoses into the definitive adult nervous system remains unclear.

Although much effort has been made by many embryologists for over a century, it is clear that most of the work still remains to be done. It will be essential to investigate the embryogenesis of representatives of each platyhelminth group at the molecular level to understand the evolution of the body plan of this diverse phylum and to compare it with other metazoans. The position of Platyhelminthes within the spiralians and their diversified mechanism of development, for which molecular information is mainly missing, make the study of flatworm embryology one of the most exciting fields of research in EvoDevo. Moreover, the intriguing issue of the ontogenetic origin of the neoblasts still remains to be solved.

OPEN QUESTIONS

- Are the divergences from the stereotypical spiral cleavage pattern observed in the Polycladida also present in the Catenulida?
- Does the *Blastomerenanarchie* mode of development share molecular features with canonical spiralian development?
- To what extent are major developmental pathways conserved in the canonical spiralian polyclads and the most divergent neoophoran flatworms?

- What is the developmental origin of neoblasts (adult stem cells)?
- How do the different developmental strategies observed in Platyhelminthes relate to the different regenerative capacities of each lineage?

Acknowledgements We thank Bernhard Egger for providing platyhelminth images showed in Fig. 3.1, the schemes of platyhelminth embryogenesis in Fig. 3.2, and the images of larvae in Fig. 3.3. We thank Iain Patten for advice on the English. This work was supported by grant BFU2012-31701 (Ministerio de Economía y Competitividad, Spain) to F.C; grant BFU2008-01544 (Ministerio de Economía y Competitividad, Spain) to ES and TA; grant 2009SGR1018 (Agència de Gestió d'Ajuts Universitaris i de Recerca) to ES, FC, and TA; and grant AIB2010DE-00402 (Ministerio de Economia y Competitividad Accion Integrada). J.M.M-D. is supported by Marie Curie intra-European fellowship 329024.

References

Anderson JM, Johann JC (1958) Some aspects of reproductive biology in the freshwater triclad turbellarian, *Cura foremanii*. Biol Bull 115:375–383

Azimzadeh J, Wong ML, Downhour DM, Sánchez Alvarado A, Marshall WF (2012) Centrosome loss in the evolution of planarians. Science 335:461–463

Baguñà J, Boyer BC (1990) Experimental embryology in aquatic plants and animals. Plenum Press, New York, pp 95–128, Chap Descriptive and experimental embryology of the Turbellaria: present knowledge, open questions and future trends

Benazzi M (1950) Ginogenesi in tricladi di acqua dolce. Chromosoma 3:474–482

Bolaños DM, Litvaitis MK (2009) Embryonic muscle development in direct and indirect developing marine flatworms (Platyhelminthes: Polycladida). Evol Dev 11:290–301

Boyer BC, Henry JQ, Martindale MQ (1996) Dual origins of mesoderm in a basal spiralian: cell lineage analyses in the polyclad turbellarian *Hoploplana inquilina*. Dev Biol 179:329–338

Boyer BC, Henry JQ, Martindale MQ (1998) The cell lineage of a polyclad turbellarian embryo reveals close similarity to coelomate spiralians. Dev Biol 204:111–123

Brøndsted HV (1969) Planarian regeneration. Pergamon Press, Oxford

Cardona A, Hartenstein V, Romero R (2005) The embryonic development of the triclad *Schmidtea polychroa*. Dev Genes Evol 215:109–131

Cardona A, Hartenstein V, Romero R (2006) Early embryogenesis of planaria: a cryptic larva feeding on maternal resources. Dev Genes Evol 216:667–681

Dunn CW, Hejnol A, Matus DQ, Pang K, Browne WE, Smith SA, Seaver E, Rouse GW, Obst M, Edgecombe

GD et al (2008) Broad phylogenomic sampling improves resolution of the animal tree of life. Nature 452:745–750

Freitas TC, Jung E, Pearce EJ (2007) TGF-beta signaling controls embryo development in the parasitic flatworm *Schistosoma mansoni*. PLoS Pathog 3:e52

Gammoudi M, Noreña C, Tekaya S, Prantl V, Egger B (2011) Insemination and embryonic development of some Mediterranean polyclad flatworms. Invertebr Reprod Dev. doi:10.1080/07924259.2011.611825

Gammoudi M, Egger B, Tekaya S, Noreña C (2012) The genus *Leptoplana* (Leptoplanidae, Polycladida) in the Mediterranean basin. Redescription of the species *Leptoplana mediterranea* (Bock,1913) comb. nov. Zootaxa 3178:45–56

Hartenstein V, Ehlers U (2000) The embryonic development of the rhabdocoel flatworm *Mesostoma lingua* (Abildgaard, 1789). Dev Genes 210:399–415

Hartenstein V, Jones M (2003) The embryonic development of the bodywall and nervous system of the cestode flatworm *Hymenolepis diminuta*. Cell Tissue Res 311:427–435

Hyman LH (1951) The invertebrates. II. Platyhelminthes and rhynchocoela. The acoelomate bilateria. McGraw-Hill, New York

Jurberg AD, Gonçalves T, Costa TA, de Mattos AC, Pascarelli BM, de Manso PP, Ribeiro-Alves M, Pelajo-Machado M, Peralta JM, Coelho PM, Lenzi HL (2009) The embryonic development of *Schistosoma mansoni* eggs: proposal for a new staging system. Dev Genes Evol 219:219–234

Koziol U, Domínguez MF, Marín M, Kun A, Castillo E (2010) Stem cell proliferation during in vitro development of the model cestode Mesocestoides corti from larva to adult worm. Front Zool 7:22. doi:10.1186/1742-9994-7-22

Lapraz F, Rawlinson KA, Girstmair J, Tomiczek B, Berger J, Jékely G, Telford MJ, Egger B (2013) Put a tiger in your tank: the polyclad flatworm *Maritigrella crozieri* as a proposed model for evo-devo. Evodevo 4:29. doi:10.1186/2041-9139-4-29

Laumer CE, Giribet G (2014) Inclusive taxon sampling suggest a single, stepwise origin of ectolecithality in Platyhelminthes. Biol J Linn Soc 111:570–588

Martín-Durán JM, Egger B (2012) Developmental diversity in free-living flatworms. EvoDevo 3:7. doi:10.1186/2041-9139-3-7

Martín-Durán JM, Romero R (2011) Evolutionary implications of morphogenesis and molecular patterning of the blind gut in the planarian *Schmidtea polychroa*. Dev Biol 352:164–176

Martín-Durán JM, Amaya E, Romero R (2010) Germ layer specification and axial patterning in the embryonic development of the freshwater planarian *Schmidtea polychroa*. Dev Biol 340:145–158

Martín-Durán JM, Monjo F, Romero R (2012a) Planarian embryology in the era of comparative developmental biology. Int J Dev Biol 56:39–48

Martín-Durán JM, Monjo F, Romero R (2012b) Morphological and molecular development of the eyes during embryogenesis of the freshwater planarian *Schmidtea polychroa*. Dev Genes Evol 222:45–54. doi:10.1007/s00427-012-0389-5

Młocicki D, Swiderski Z, Conn DB (2010) Ultrastructure of the early embryonic stages of *Corallobothrium fimbriatum* (Cestoda: Proteocephalidea). J Parasitol 96:839–846

Morris J, Nallur R, Ladurner P, Egger B, Rieger R, Hartenstein V (2004) The embryonic development of the flatworm *Macrostomum* sp. Dev Genes Evol 214:220–239

Rawlinson KA (2010) Embryonic and post-embryonic development of the polyclad flatworm *Maritigrella crozieri*; implications for the evolution of spiralian life history traits. Front Zool 7:12

Rawlinson KA (2014) The diversity, development and evolution of polyclad flatworm larvae. EvoDevo 5:9

Reisinger E (1924) Die Gattung Rhynchoscolex. Z Morphol Ökol Tiere 1:1–37

Reisinger E (1972) Die Evolution des Orthogons der Spiralier und das Archicoelomatenproblem. Z Zool Syst Evolutionforsch 10:1–43

Reiter D, Ladurner P, Mair G, Salvenmoser W, Rieger R, Boyer B (1996) Differentiation of the body wall musculature in *Macrostomum hystricinum marinum* and *Hoploplana inquilina* (Plathelminthes), as models for muscle development in lower Spiralia. Rouxs Arch Dev Biol 205:410–423

Rieger R, Tyler S, Smith JPS III, Rieger G (1991) Platyhelminthes: Turbellaria. In: Harrison FW, Bogitsh BJ (eds) Microscopic anatomy of invertebrates, vol 3. Wiley-Liss, New York, pp 7–140

Riutort M, Álvarez-Presas M, Lázaro E, Solà E, Paps J (2012) Evolutionary history of the Tricladida and the Platyhelminthes: an up-to-date phylogenetic and systematic account. Int J Dev Biol 56:5–17

Ruppert EE (1978) A review of metamorphosis of turbellarian larvae. Settlement and metamorphosis of marine invertebrates. Elsevier, New York, pp 65–81

Seilern-Aspang F (1957) Die Entwicklung von *Macrostomum appendiculatum* (Fabricius). Zool Jahrb Anat 76:311–330

Seilern-Aspang F (1958) Entwicklungsgeschichtliche Studien an paludicolen Tricladen. Arch EntwMech Org 150:425–480

Semmler H, Wanninger A (2010) Myogenesis in two polyclad platyhelminths with indirect development, *Pseudoceros canadensis and Stylostomum sanjuania*. Evol Dev 12:210–221

Solana J, Romero R (2009) SpolvlgA is a DDX3/PL10-related DEAD-box RNA helicase expressed in blastomeres and embryonic cells in planarian embryonic development. Int J Biol Sci 5:64–73

Struck TH, Wey-Fabrizius AR, Golombek A, Hering L, Weigert A, Bleidorn C, Klebow S, Iakovenko N, Hausdorf B, Petersen M, Kück P, Herlyn H, Hankeln T (2014) Platyzoan paraphyly based on phylogenomic data supports a noncoelomate ancestry of Spiralia. Mol Biol Evol. doi:10.1093/molbev/msu143

Surface FA (1907) The early development of a polyclad, *Planocera inquilina*. Proc Acad Nat Sci Phila 59:514–559

Swiderski Z, Poddubnaya LG, Gibson DI, Levron C, Młocicki D (2011) Egg formation and the early embryonic development of *Aspidogaster limacoides* Diesing, 1835 (Aspidogastrea: Aspidogastridae), with comments on their phylogenetic significance. Parasitol Int 60:371–380

Tekaya S, Sluys R, Zghal F (1999) Cocoon production, deposition, hatching and embryonic development in the marine planarian *Sabussowia dioica* (Platyhelminthes, Tricladida, Maricola). Invertebr Reprod Dev 35:215–223

Thomas MB (1986) Embryology of the Turbellaria and its phylogenetic significance. Hydrobiologia 132:105–115

Van den Biggelaar JA, Guerrier P (1979) Dorsoventral polarity and mesentoblast determination as concomi-tant results of cellular interactions in the mollusk *Patella vulgata*. Dev Biol 68:462–471

Van den Biggelaar JAM, Dictus WJAG, van Loon AE (1997) Cleavage patterns, cell-lineages and cell speci-fication are clues to phyletic lineages in Spiralia. Semin Cell Dev Biol 8:367–378

Willems M, Egger B, Wolff C, Mouton S, Houthoofd W, Fonderie P, Couvreur M, Artois TJ, Borgonie G (2009) Embryonic origins of hull cells in the flatworm *Macrostomum lignano* through cell lineage analysis: developmental and phylogenetic implications. Dev Genes Evol 219:409–417

Younossi-Hartenstein A, Hartenstein V (2000) The embryonic development of the polyclad flatworm *Imogine mcgrathi*. Dev Genes Evol 210:383–398

Younossi-Hartenstein A, Ehlers U, Hartenstein V (2000) Embryonic development of the nervous system of the rhabdocoel flatworm *Mesostoma lingua* (Abilgaard, 1789). J Comp Neurol 16:461–474

Regeneration and Growth as Modes of Adult Development: The Platyhelminthes as a Case Study

4

Francesc Cebrià, Emili Saló, and Teresa Adell

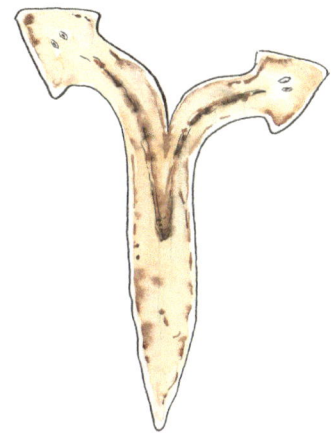

F. Cebrià • E. Saló (✉) • T. Adell
Department of Genetics, Faculty of Biology,
Institute of Biomedicine, University of Barcelona,
Av. Diagonal 643, edifici Prevosti, planta 1,
Catalunya, Barcelona 08028, Spain
e-mail: esalo@ub.edu

A. Wanninger (ed.), *Evolutionary Developmental Biology of Invertebrates 2: Lophotrochozoa (Spiralia)*
DOI 10.1007/978-3-7091-1871-9_4, © Springer-Verlag Wien 2015

ADULT DEVELOPMENTAL PROCESSES: REGENERATION IN PLATYHELMINTHES

Some species of Platyhelminthes have become model systems in which to study whole-body regeneration in adults. Before describing how this capacity is distributed and varies within the phylum, however, it is important to introduce the adult pluripotent stem cells that confer this remarkable ability in flatworms, the so-called neoblasts.

Neoblasts: The Pluripotent Stem Cells of Adult Platyhelminthes

A common feature that characterises the phylum Platyhelminthes is the presence of neoblasts, the only cell type that proliferates in adult somatic tissues. The term was coined by Harriet Randolph in 1892 to refer to undifferentiated embryonic-like cells observed during the formation of new mesoderm after fission in earthworms (Annelida). Later, in 1897, she extended the term neoblasts to similar staining observed in smaller cells present in planarians (Tricladida) (for a historical review, see Baguñà 2012; Rink 2013; Adell et al. 2014). The term 'neoblast' refers to adult stem cells that, in triclads, are scattered throughout the flatworm body and are only absent in the animal head tip and the pharynx. Neoblasts are the only cells that exhibit mitotic activity and the capacity to differentiate into all somatic and germinal cell types. Neoblasts can be easily identified by their small size (7–12 μm in diameter), spindle shape, and the presence of a single filopodium and a high nuclear to cytoplasmic ratio (Fig. 4.1). At the ultrastructural level, the neoblasts contain chromatoid bodies close to the nucleus (Auladell et al. 1993); these round structures lack membranes and are composed of ribonucleoproteins, and they resemble the germ granules found in the germ cells of other animals. Germ granules are thought to function as centres for post-transcriptional regulation of mRNA (Extravour and Akam 2003), and several studies have shown that genes involved in post-transcriptional regulation and chromatin

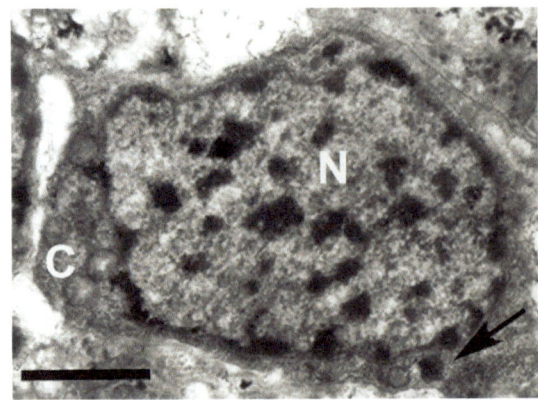

Fig. 4.1 Electron micrograph of a planarian neoblast. A neoblast is a small cell (close to 10 μm) with a large nucleus (*N*) and a small cytoplasm (*C*). *Arrow* indicates the chromatoid bodies usually observed in this cell type. Scale bar: 0.5 μm

modification are also present in neoblasts (Rossi et al. 2007; Fernández-Taboada et al. 2010; Rouhana et al. 2010; Labbe et al. 2012; Onal et al. 2012; Wagner et al. 2012). Although the ontogenetic origin of neoblasts is not clear, it has recently been suggested that they correspond to primordial stem cells (PriSCs) (Solana 2013). These PriSCs are thought to constitute stem cells intercalated between the zygote and the germ line. Depending on the reproductive mode and regenerative capabilities of the different animals, the PriSCs may give rise to mainly primordial germ cells or other types of somatic cells. The pluripotency of the neoblasts was first hypothesised many decades ago (reviewed in Brøndsted 1969; Baguñà 2012; see also, Baguñà et al. 1989a; Ladurner et al. 2000; Newmark and Sánchez-Alvarado 2000). However, it has only recently been shown that an individual neoblast is capable of differentiating into all cell types in the planarian *Schmidtea mediterranea* (Wagner et al. 2011).

Although all Platyhelminthes possess neoblasts, not all of them are capable of regenerating. In Platyhelminthes that do not regenerate, neoblasts are probably required for the homeostatic cell turnover observed in adult animals. As in the case of their free-living relatives, pluripotent stem cells are fundamental in the development of parasitic Platyhelminthes, where they are called germinative, germinal, or regenerative cells. Despite the different nomenclature, the basic processes of cell proliferation and differentiation are shared between free-living and parasitic Platyhelminthes. Neoblasts are considered fundamental for the complex plastic life cycles of parasitic species, which are subject to regulation by host signals. In cestodes, systems have been developed for the short-term culture of neoblast populations from asexually multiplying larval stages (Brehm 2010). Despite some similarities, the germinative cells do not contain chromatoid bodies and they do not express *vasa* and *piwi* gene orthologs, suggesting fundamental differences between these and other neoblast cell populations (Collins et al. 2013; Koziol et al. 2014). In *Taenia solium*, there is a germinative region posterior to the apex of the scolex, the neck, in which stem cells proliferate continuously, differentiate into all cell types, and migrate to the tegument (Merchant et al. 1997). In trematodes, two populations of neoblast-like cells have been identified in the species *Schistosoma mansoni*. Somatic stem cells have been identified in adult schistosomes (Collins et al. 2013) and in the sporocyst, the larval stage that lives in the intermediate mollusk host (Wang et al. 2013). These data suggest that schistosome adult stem cells persist through its life cycle. Transcriptomic analysis supports the existence of a molecular signature in free-living Platyhelminthes and Trematoda, suggesting an ancient role for these genes in regulating the stem cell population of Platyhelminthes (Collins et al. 2013; Wang et al. 2013).

Regenerative Capabilities

Several species of Platyhelminthes are known to have amazing regenerative capabilities, understood as the capacity to regrow parts of their bodies lost after a traumatic amputation. Among the most popular current animal models for regeneration, triclads such as *Schmidtea mediterranea* and *Dugesia japonica* (Newmark and Sánchez Alvarado 2002; Agata 2003; Saló 2006; Gentile et al. 2011), as well as the macrostomid *Macrostomum lignano* (Egger et al. 2006; Morris et al. 2006), occupy a prominent position. In fact, several species of triclads and polyclads have been linked to the history of regeneration research since its very beginning, as Pallas already described these striking abilities in 1774 (Pallas 1774; for a historical revision, see Brøndsted 1969). The fact that some species are capable of regenerating a complete animal from a tiny piece of their bodies (something quite unique among bilaterians) led Dalyell to claim that planarians were 'almost immortal under the edge of the knife' (Dalyell 1814).

Regeneration has quite a broad distribution in the animal kingdom (Birnbaum and Sánchez Alvarado 2008; Bely and Nyberg 2010; Poss 2010), since many phyla contain species with some regenerative capability. However, it is also true that closely related species can display a very different regenerative potential (e.g., the ability to regenerate entire limbs of some amphibians versus the complete absence of regeneration in most other vertebrates). Also, regeneration can occur at different biological levels of organisation: cell, tissue, organ, structure, or the whole body (Bely and Nyberg 2010). Thus, on the one hand, regeneration is broadly distributed, but on the other, the distribution and potential of this biological property is uneven. This has raised the question of whether regeneration is an ancestral trait that was lost multiple times through evolution (Bely 2010; Bely and Nyberg 2010) or, based on some taxon-specific components, that it appeared independently in different regeneration-competent lineages (Garza-García et al. 2010).

Based on the observation that different models use the same conserved signalling pathways or genetic programmes to regenerate similar structures, placing regeneration as an ancestral trait at the base of metazoans seems plausible. In this sense, the loss of the capacity to reactivate those conserved pathways after tissue loss could explain the loss of regeneration in many species. Recently, studies on annelids have provided some examples of evolutionary loss of regeneration (Bely 2010). However, if regeneration is not ancestral but evolved independently in different lineages, one might expect to identify taxon-specific factors, genes, or programmes in those regeneration-competent species. Not many examples of such specific regeneration-promoting genes have been described (Poss 2010). One of them is *Prod1*, which appears to be restricted to salamanders and is essential for limb regeneration in these animals (Garza-García et al. 2010). As more transcriptomic and genomic data become available for a larger number of species with and without regenerative abilities, comparative studies should provide insights into the evolutionary origin of regeneration.

As in other phyla, not all species of Platyhelminthes display the same regenerative abilities (Fig. 4.2). Free-living flatworms fall in one of three different categories. Some groups, such as many rhabdocoels and some lecithoepitheliates, cannot regenerate at all. Other groups can regenerate some organs or individual parts but not others. For example, some species of triclads and macrostomids can regenerate the posterior region but not the anterior part with the brain and eyes. Finally, some species of catenulids, polyclads, macrostomids, and triclads can regenerate any organ and structure (reviewed in Egger et al.

2007). Among model flatworm species for regeneration studies, *Schmidtea mediterranea* and *Dugesia japonica* can regenerate a whole animal from almost any tiny piece of their bodies, whereas *Macrostomum lignano* can regenerate as long as the brain and pharynx are present. Even among triclads, which include many species that can regenerate a whole animal from a tiny piece of their bodies, some species show more restricted regenerative abilities than others. In fact, up to five to eight types of triclads have been proposed according to their ability to regenerate (Sivickis 1930; Teshirogi et al. 1977). This classification is mainly based on the capacity to regenerate a head (with the cerebral ganglia), which varies depending on the level of amputation along the anterior-posterior (AP) axis. Thus, some species will regenerate a head from any body piece, others will regrow the anterior region only if amputated in front of the pharynx, and others will never regenerate the head if amputated behind the cephalic ganglia. *Dendrocoelum lacteum*, *Procotyla fluviatilis,* and *Phagocata kawakatsui* are good examples of

Fig. 4.2 Phylogenetic consensus tree of Platyhelminthes. The phylogenetic classification is based in Laumer and Giribet (2014), in which the former Lecithoepitheliata appears as a paraphyletic group, now divided into two new orders (Prorhynchida and Gnosonesimida). For simplicity, the old Lecithoepitheliata name is used here. The regenerative ability of each order is indicated next to it (Adapted from Egger et al. (2007))

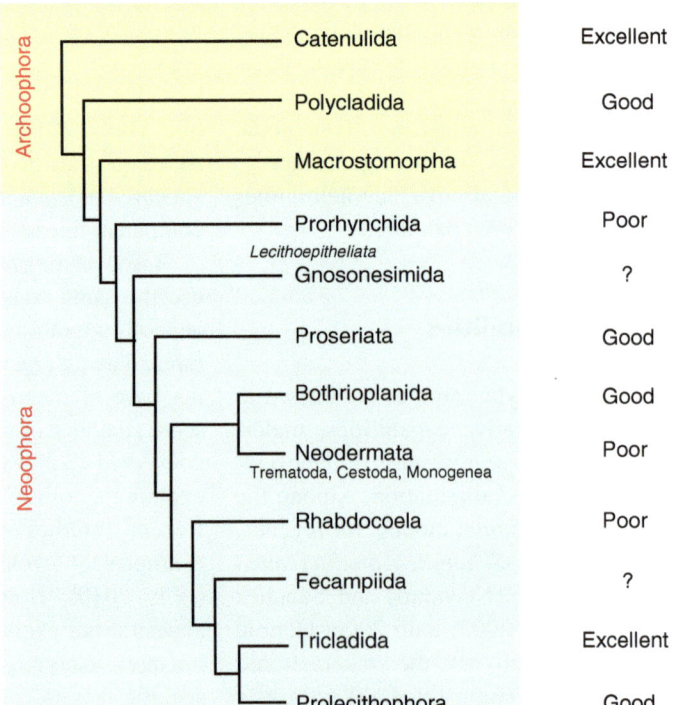

	Regenerative ability
Catenulida	Excellent
Polycladida	Good
Macrostomorpha	Excellent
Prorhynchida	Poor
Lecithoepitheliata Gnosonesimida	?
Proseriata	Good
Bothrioplanida	Good
Neodermata (Trematoda, Cestoda, Monogenea)	Poor
Rhabdocoela	Poor
Fecampiida	?
Tricladida	Excellent
Prolecithophora	Good

triclads in which postpharyngeal amputation generates tail pieces that are not capable of regenerating a new head region. In contrast, the anterior half can fully regenerate a new tail. Remarkably, it has recently been shown that the silencing of a single gene, a *β-catenin* homolog, is able to rescue anterior regeneration in these tail pieces, indicating that they retain regenerative potential but that, under natural conditions, this programme is not activated (Liu et al. 2013; Sickes and Newmark 2013; Umesono et al. 2013).

Although parasitic species of Platyhelminthes outnumber free-living flatworms, much less is known about their regenerative capabilities. Senft and Weller (1956) cultured immature specimens of *Schistosoma mansoni* and observed that in some cases in which the animals had been damaged in the isolation procedure, they could regenerate the missing parts in vitro. Also, during the life cycle of schistosomes, infectious cercariae transform into schistosomula (the larval stage of *Schistosoma* in their vertebrate hosts) after they lose their tails when they enter their definitive host. In vitro, tails of detached cercariae are able to grow some tissues resembling a head in shape although they die within a few days (Coultas and Zhang 2012). It has been also described that the cestode *Diphyllobothrium erinacei* can regenerate. Isolated larvae of this species were subject to different types of injuries and amputations and were able to regenerate after they had been transferred to their intermediate and definitive hosts (Vorontsova and Liosner 1960). More recently, it has been shown that primary cells from *Echinococcus multilocularis* are able to form cell aggregates from which young metacestode (larval stage of tapeworms) vesicles develop in vitro (Spiliotis et al. 2008). Nevertheless, further experiments are needed to determine the regenerative capabilities of larval and adult parasitic Platyhelminthes.

The regeneration potential within this phylum is often associated with the reproductive mode. Platyhelminthes are, with few exceptions, hermaphroditic (Hyman 1951). However, asexual reproduction by fission occurs in some catenulids and macrostomids, and it is quite common among triclads (Hyman 1951; Reuter and Kreshchenko 2004). As expected, regenerative abilities are less pronounced in taxa containing species with sexual reproduction than in taxa that reproduce asexually, as regeneration is absolutely required for survival of the mother animal during asexual reproduction (Egger et al. 2007). Some species, such as *Microstomum lineare*, combine sexual and asexual modes of reproduction. In these flatworms, a good regenerative capacity is seen during the asexual stage of their life cycle, but this capacity is reduced with the appearance of sexuality (Reuter and Kreshchenko 2004).

Asexual reproduction by fission may occur by paratomy or architomy. During paratomy, the differentiation of the organs, such as the new cephalic ganglia or eyes, precedes the fissioning itself. Often, the 'mother' worm is transformed into a chain of zooids that will later split into several independent 'daughter' flatworms. In contrast, during architomy, fissioning precedes the differentiation of the new organs of the daughter flatworm. In fact, no external signs can be distinguished before the fission process starts. Despite this lack of external signs and organ differentiation before fissioning, some molecular changes have been described in animals committed to asexual reproduction (Bueno et al. 2002). Architomy is common in triclads, whereas paratomy is often found in several species of catenulids and macrostomids (Reuter and Kreshchenko 2004). After architomic fissioning, the tiny body piece that is produced must regenerate a new organism. In this sense, architomy and regeneration after amputation are very similar, as they proceed through similar stages.

Mechanisms of Regeneration

As described above some triclads and macrostomids are among current model species for regeneration. In these species, regeneration requires cell proliferation and blastema formation, and they can be categorised as examples of epimorphic regeneration, together with, for example, amphibian limb regeneration. The term epimorphosis was coined by Morgan (1901) to refer to the regeneration mode 'in which proliferation of material precedes the development of the new part'. This was in contrast to morphallaxis 'in which a part is transformed directly into a new organism or part of an organism without proliferation at the cut-surfaces'. However, as Morgan already realised, these two modes of regeneration are not mutually exclusive. In fact, planarians are an example in which a mixed epimorphic/morphallactic model has been proposed (Saló and Baguñà 1984) because, in addition to cell proliferation at the cut surface, there is a remodelling of the pre-existing tissues far from the wound in order to attain the correct body proportions (Fig. 4.3). Over the years, some debate has arisen about the convenience of using these terms as originally defined by Morgan. Most flatworms regenerate through cell proliferation-dependent

blastema formation, which fits with epimorphic regeneration. The term morphallaxis, however, is more problematic because it has been mainly associated with tissue remodelling in the absence of cell proliferation. However, it has been shown that tissue remodelling in planarians also depends on the presence of stem cells (Gurley et al. 2010). Therefore, it would probably be better to use the term 'remodelling' instead of 'morphallaxis', without linking it necessarily to the presence or absence of cell proliferation.

Regeneration implies three main stages: (1) wound closure and healing, (2) cell proliferation and blastema formation, and (3) cell differentiation and morphogenesis. In triclads, wound closure is mediated by the contraction of the body

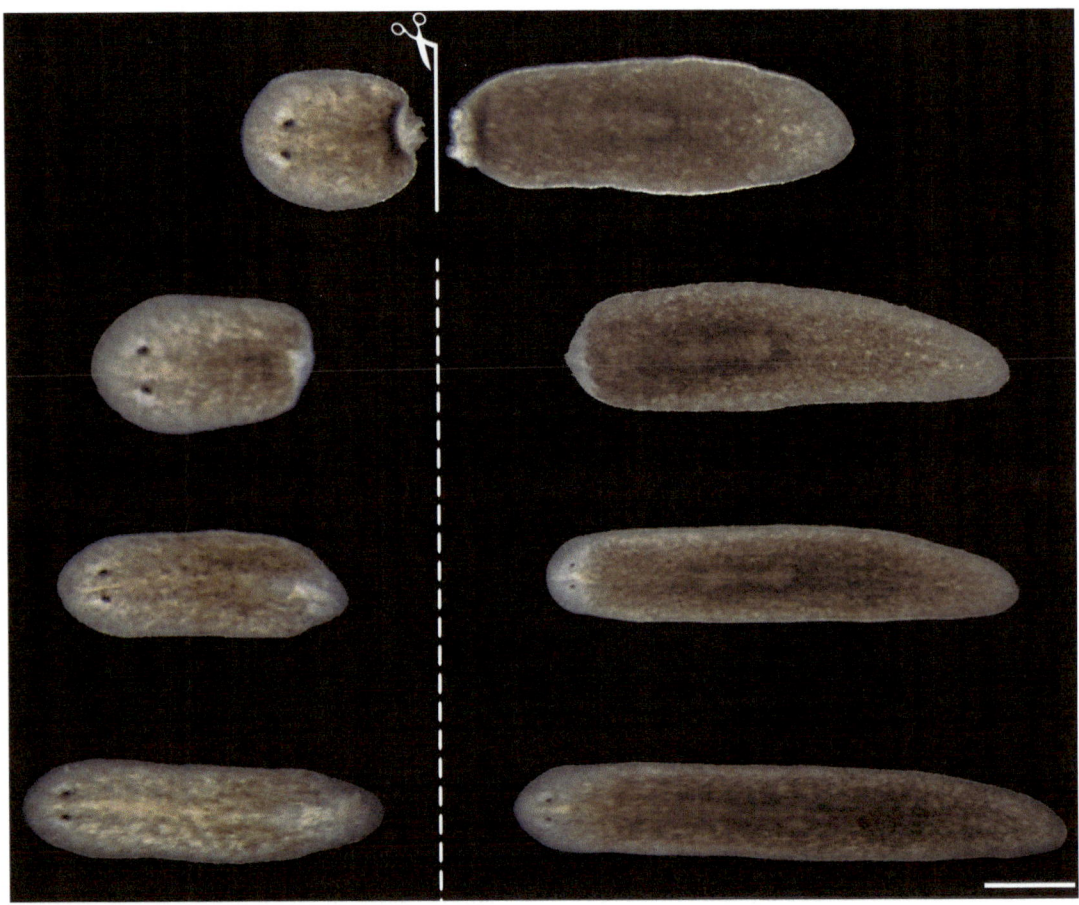

Fig. 4.3 Remodelling of pre-existing tissues during regeneration in *Schmidtea mediterranea*. During the process of regeneration while new tissue is regenerating, the pre-existing structures remodel to reach a new proportioned animal. Scale bar: 1 mm

wall muscle fibres at the cut surface, followed by the spreading of epithelial cells surrounding the amputation site (Pedersen 1976; Hori 1989). A similar process was observed in *Microstomum* (Palmberg 1986), *Catenula* (Moraczewski 1977), and *Macrostomum lignano* (Egger et al. 2009). However, in *Microstomum*, the new epidermis is formed underneath the wound epidermis, whereas in other flatworms the new epidermis forms by the insertion of new cells from the parenchyma (Palmberg 1986; Rieger et al. 1991).

In most cases, cell proliferation at the cut-edges results in the formation of a blastema, an unpigmented mass of undifferentiated cells where the missing organs and structures will differentiate. Remarkably, there are also a few examples (e.g., *Microstomum lineare* and some catenulids) in which this cell proliferation does not result in blastema formation and the new tissues and organs regenerate within the pre-existing tissues after the migration of proliferating cells from regions far away from the wound (Palmberg 1986; Rieger et al. 1991). In general, most flatworms regrow the missing parts from a regenerative blastema, similar to the process of amphibian limb regeneration. However, there are some important differences between the amphibian and flatworm blastemas. One main difference concerns the origin of the cells that form the blastema. In amphibians, following amputation, pre-existing differentiated cells of the stump dedifferentiate, re-enter the cell cycle, proliferate, and re-differentiate into new cells to restore the missing structures. In contrast, regeneration in flatworms depends upon the presence of neoblasts. Upon amputation, neoblasts proliferate at the wound region, giving rise to the blastema where they will differentiate (Saló and Baguñà 1984; Wenemoser and Reddien 2010). Thus, although some examples of transdetermination between germ and somatic cells have been observed (Gremigni et al. 1980), cell dedifferentiation does not seem to play a role in flatworm regeneration.

The classical view of a blastema as a mass of undifferentiated homogeneous cells has been challenged after the finding in amphibians and zebrafish that the dedifferentiated cells keep a memory of their origin and re-differentiate into the same cell types (Kragl et al. 2009; Knopf et al. 2011). This suggests that the blastema is in fact heterogeneous and formed by distinct populations of lineage-restricted progenitor cells (Tanaka and Reddien 2011). Remarkably, this scenario may also occur in triclads, based on the existence of distinct cell populations of progenitor-like neoblasts that express specific transcription factors required for their differentiation into eye pigment, photoreceptors, protonephridia, or different neuronal types (Lapan and Reddien 2011; Scimone et al. 2011; Cowles et al. 2013; Currie and Pearson 2013; März et al. 2013; Reddien 2013).

Another difference between vertebrate and some flatworm blastemas concerns the proliferative activity of their constituent cells that form them. Thus, in amphibians and zebrafish, cell proliferation drives blastema growth. In contrast, proliferation in triclads is mainly concentrated in the stump region adjacent to the blastema, and very few mitoses are observed within the blastema itself (Baguñà 1976; Saló and Baguñà 1984). This is more obvious in anterior blastemas than posterior ones, where more mitotic cells are observed (Wenemoser and Reddien 2010). Thus, in triclads such as *Schmidtea mediterranea*, anterior blastemas mainly grow by the migration of post-mitotic neoblasts (probably already committed) from the stump region. In contrast, the blastema of *Macrostomum lignano* and the catenulid *Paracatenula galateia* contains proliferative cells (Nimeth et al. 2007; Egger et al. 2009; Dirks et al. 2012; Verdoodt et al. 2012). Despite this difference in the distribution of the proliferating neoblasts, upon amputation, a similar biphasic mitotic and S-phase pattern is observed in triclads and *M. lignano*, respectively. An initial peak in cell proliferation that is more or less uniform throughout the animal and represents a systemic injury response is followed by a second peak restricted to the wound region (Saló and Baguñà 1984; Wenemoser and Reddien 2010; Verdoodt et al. 2012).

In summary, Platyhelminthes include some of the most popular model species in which to study regeneration. As in other phyla, there is a great variability in regenerative capabilities within Platyhelminthes, ranging from species that are incapable of any regeneration to others that can regenerate a whole animal from almost any small part of their bodies. A strong association exists between asexual reproduction and regeneration. In Platyhelminthes, regeneration depends upon the presence of adult totipotent somatic stem cells and, in most cases, proceeds through the formation of a regenerative blastema. In both Platyhelminthes and vertebrates, these blastemas appear to be formed by a heterogeneous population of undifferentiated but already committed progenitor cells. Also, in vertebrates and some Platyhelminthes, cell proliferation occurs within these blastemas. In triclads, however, very few mitotic cells are observed within them. Whereas cell dedifferentiation is the main source of cells for the forming blastemas in vertebrates, in Platyhelminthes blastemas appear by proliferation of adult pluripotent stem cells.

Schmidtea mediterranea: A Regeneration Star

Regeneration is one of the most fascinating phenomena in biology. That some animals are capable of regrowing a limb or the whole body from a tiny piece of them awakens our imagination and may even make us feel jealous! Freshwater planarians (Tricladida) have enormous regenerative power. You can cut them into many pieces and get many complete worms in just a few days. Because of the increasing interest in stem cell research, a growing number of laboratories around the world have turned their attention to these animals. The species *Schmidtea mediterranea* has emerged as a regeneration star. This tiny flatworm was originally described in the 1970s and has become a model species used in many laboratories. Remarkably, most of the clonal lines used in research nowadays derive from a population originally obtained from a small fountain in Barcelona. *S. mediterranea* exists in two forms: sexual and asexual strains. Asexual animals reproduce by fission and are the only reported case of asexuality in this genus. The difference between the two strains is a heteromorphic chromosomal translocation. Like other triclads, *S. mediterranea* is easily cultured in the laboratory; however, whereas other species with high regenerative capabilities are mixoploids or polyploids, *S. mediterranea* is a stable diploid ($2n=8$). This important feature was taken into account when searching for an appropriate species to enter the genomics era. The genome of *S. mediterranea* has been sequenced and is currently under assembly. Moreover, several transcriptomes from whole animals, specific regions (e.g., anterior or posterior blastemas), or individual cell types (e.g., neoblasts and gut cells) are available. Also, RNAi to study gene function in flatworms was originally established in *S. mediterranea* and then adapted to other species. Neoblasts were first labelled with BrdU in this flatworm and the totipotency of single neoblasts in vivo was also probed for the first time in this species. *S. mediterranea* has served as a model to understand how the anterior-posterior and dorsoventral axes are reestablished during flatworm regeneration. All these resources, tools, and studies have contributed to the establishment of *S. mediterranea* as an excellent model system to study different aspects of regeneration, such as neoblast maintenance and self-renewal, blastema formation, axial polarity, neural regeneration, growth and degrowth, and the relationship between regeneration and cancer. The results obtained so far have been relevant not only to understand regeneration in flatworms but also from a broader evolutionary perspective. Moreover, the fact that *S. mediterranea* exists in sexual and asexual forms renders it an ideal model for comparative studies on how patterning and morphogenesis occur in embryonic and postembryonic contexts.

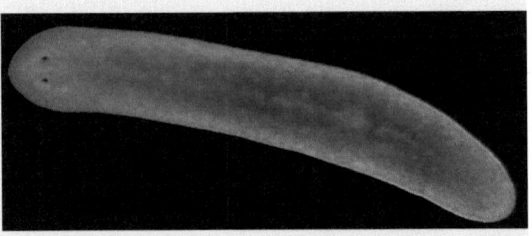

The Central Nervous System (CNS) of Platyhelminthes: Pattern and Regeneration

Structure of the CNS

Platyhelminthes have a bilaterally symmetrical CNS consisting of anterior cephalic ganglia ('brain') and one or more pairs of longitudinal nerve cords that usually run along the length of the animal (Fig. 4.4); in some catenulids, however, the nerve cords do not reach the posterior end (Rieger et al. 1991). The longitudinal cords are usually interconnected by commissures that are more or less regularly spaced in a ladder-like pattern referred to as 'orthogon' (Bullock and Horridge 1965; Reuter and Gustafsson 1995). Submuscular, subepidermal, subepithelial, and infraepithelial peripheral plexuses are interconnected with the CNS in most flatworm taxa (Baguñà 1974; Baguña and Ballester 1978; Rieger et al. 1991; Reuter and Gustafsson 1995). The pharynx contains one or two nerve rings that form a nerve net usually close to the distal part of this organ (Bullock and Horridge 1965; Baguña and Ballester 1978; Rieger et al. 1991; Reuter 1994).

The brain of platyhelminths varies substantially in form and size, as well as in the number and arrangement of neurons. In most cases (with the exception of adult Polycladida), the neuronal cell bodies are peripheral and surround a central neuropil formed by nerve fibres (Rieger et al. 1991; Reuter and Gustafsson 1995; Kotikova et al. 2002). The brain is dorsal to the ventral nerve cords, which, in some species, project into regions anterior to the brain. In many species, especially among Tricladida, the brain displays a typical spongy texture, as it is traversed by muscles and processes from secretory cells (Baguñà and Ballester 1978). Early phylogenetic offshoots usually have smaller brains with few cells lateral to a fibrillar neuropil; in later-branching Platyhelminthes the brain is formed by clearly distinguishable bilateral cephalic ganglia connected by one or more commissures. In most of the major systematic groups, there are species with or without a variably structured brain capsule (Rieger et al. 1991). An encapsulated brain has been described in Polycladida, Lecithoepitheliata, and Prolecithophora (Rieger et al. 1991). In trematodes the brain usually contains a pair of cephalic ganglia with a central neuropil connected by a commissure. Miracidium larvae already have a bilobed brain (Bullock and Horridge 1965; Leksomboon et al. 2012).

The basic orthogonal plan of the nervous system of Platyhelminthes varies mainly in the number of longitudinal cords. Whereas Bullock and Horridge (1965) described nine types of orthogons, Kotikova (1986, 1991) distinguished

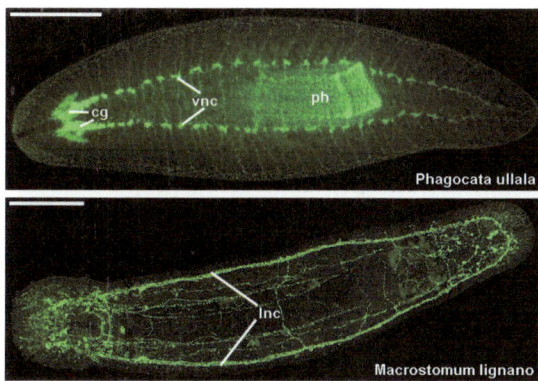
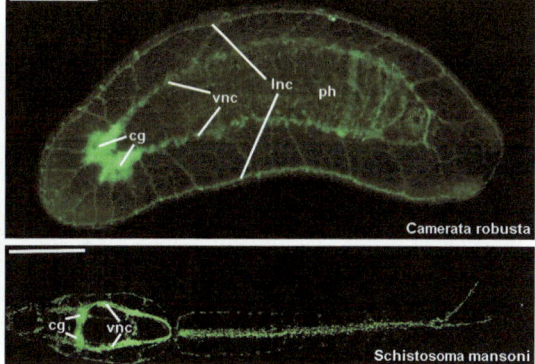

Fig. 4.4 Structure of the CNS in Platyhelminthes. Immunostaining with anti-synapsin in *Phagocata ullala*, *Camerata robusta*, and *Schistosoma mansoni*. Immunostaining with anti-GYRFamide in *Macrostomum lignano*. In *P. ullala*, *C. robusta*, and *S. mansoni*, the ventral nerve cords are the main cords. In *M. lignano* the lateral nerve cords are the main cords. In *C. robusta* (Tricladida, Maricola) the lateral nerve cords are much more prominent than in *P. ullala* (Tricladida, Continenticola) The image of *M. lignano* was originally published by Egger et al. (2007). The image of *S. mansoni* was originally published by Collins et al. (2011). *cg* cephalic ganglia, *vnc* ventral nerve cords, *lnc* lateral nerve cords, *ph* pharynx. Anterior is to the *left* in all aspects. Scale bars: *P. ullala* and *C. robusta* 0.5 mm; *M. lignano* 100 μm; *S. mansoni* 50 μm

eight different types. According to Kotikova, the body shape mainly determines the type of orthogon, which would be basically homologous and derived from the pattern observed in Catenulida. The presence of three to five pairs of cords (seen in some Rhabdocoela and Proseriata) is considered more primitive, with an evolutionary tendency towards reducing the number of cords to one prominent pair (in macrostomids, Rhabdocoela, and Tricladida) (Rieger et al. 1991). The nerve cords can be placed at different positions: dorsal, lateral, marginal, ventrolateral, and ventral. Reuter and Gustafsson (1995) proposed that, independently of their position, the term main cord should be applied to those cords that (1) start as multifibre outgrowths from the brain and (2) are thicker and possess more neurons positive for serotonin and catecholamine markers. In contrast, all other cords are thinner and have less contact with the brain.

In Catenulida, Macrostomida, and Monocelididae (Proseriata), the main cords are usually lateral (Reuter et al. 1986; Reuter 1988; Joffe and Kotikova 1991; Reiter and Wikgren 1991; Joffe and Reuter 1993). In Tricladida and Lecithoepitheliata the ventral cords are the main ones. In the Prolecithophora and Rhabdocoela, either the lateral or ventral cords dominate or the lateral and ventral ones are equally developed. For example, in the rhabdocoel *Castrella truncata*, the lateral and ventral cords are about the same thickness, although the ventral ones are associated with more serotonergic neurons and would then be considered the main cords (Kotikova et al. 2002). In Trematoda there are usually three pairs of longitudinal cords (dorsal, lateral, and ventral); the ventral cords are usually the largest (and thus constitute the so-called main cords), and they are often interconnected at the caudal tip (Bullock and Horridge 1965; Leksomboon et al. 2012). In the Cestoda, a variable number of longitudinal cords may occur, from one pair to up to 38 or 60 cords in other species; the lateral cords are often the main cords (Bullock and Horridge 1965; Joffe and Reuter 1993).

Several evolutionarily trends can be observed in the orthogon and main cords. Firstly, there seems to be a reduction in the number of longitudinal cords, which is accompanied by a strengthening of the remaining cords. There is also a trend towards concentration of plexal fibres to longitudinal cords. Whereas the main cords are localised ventrally in various lineages of Neoophora, the main cords occupy a more lateral position in the earlier-branching taxa (Catenulida and Macrostomida) (Reisinger 1972; Joffe and Reuter 1993; Reuter and Gustafsson 1995; Reuter and Halton 2001).

Traditionally, the CNS of the Platyhelminthes has been considered as primitive (Bullock and Horridge 1965) and, therefore, most similar to the urbilaterian nervous system. However, current phylogenetic analyses do not support this basal position of the Platyhelminthes within Bilateria (Riutort et al. 2012); moreover, Acoela and Nemertodermatida, previously included within the Platyhelminthes, are now placed in a separate phylum, the Acoelomorpha (Ruiz-Trillo et al. 1999). The position of the Acoelomorpha is still under debate, as they have been considered basal bilaterians or, more recently, basal deuterostomes (Philippe et al. 2011). Remarkably, although acoelomorphs and platyhelminths have been separated in evolution at least by half a billion years (Peterson and Eernisse 2001; Peterson et al. 2008), their CNS displays many similarities in the architecture of their brains and orthogons, as well as in some developmental features (Bailly et al. 2013). This has led some authors to suggest that the morphological features of the platyhelminth CNS could reflect some characteristics of their urbilaterian ancestor (Bailly et al. 2013; but see Semmler et al. 2010).

Neuronal Diversity and Molecular Complexity

In contrast to other invertebrates, platyhelminths have a large variety of neural cells, including uni-, bi-, and multipolar neurons (Rieger et al. 1991). A characteristic of flatworm neurons is their high content of vesicles, which are thought to be secretory. Examples include small clear vesicles (often regarded as cholinergic), dense-core vesicles (aminergic), and large dense vesicles (peptidergic) (Reuter and Gustafsson 1995). The neurons of some catenulids and macrostomids seem to be structurally least differentiated, whereas the vesicles of others show a greater heterogeneity (Reuter and Gustafsson 1995).

A large number of aminergic and peptidergic neuroactive substances have been detected in most flatworm taxa, both in free-living and parasitic species (Fairweather and Halton 1991; Reuter and Gustafsson 1995; Gustafsson et al. 2002; Ribeiro et al. 2005; Cebrià 2008; McVeigh et al. 2011). These markers have been used to describe the architecture of the nervous system (Fig. 4.5) and have revealed the high neuronal diversity in flatworms (see Table 4.1 for a list of Platyhelminthes in which specific neuronal populations have been identified). As Platyhelminthes do not possess true endocrine glands and a circulatory system, the nervous system can be seen as a neuroendocrine system that controls processes such as growth, development, and regeneration (Fairweather and Halton 1991). Immunohistochemistry for many of these neuroactive substances has shown that they are most often expressed in distinct, nonoverlapping neuronal populations. More recently, 51 prohormone genes and 142 peptides have been identified and characterised biochemically in the triclad *Schmidtea mediterranea*. Of these 51 genes, 85 % are expressed in the CNS and, remarkably, different prohor-

mones are often enriched in specific cell types or regions within the CNS (Collins et al. 2010), indicating a high neuronal diversity.

The complexity of the flatworm nervous system at the cellular and molecular levels is further supported by analysis of the expression of neural genes. Until now, the expression of dozens of neural genes have been reported and are used to identify distinct neuronal populations (Mineta et al. 2003; Nakazawa et al. 2003; Cebrià 2007; Cebrià et al. 2010). Most of these studies have been done in triclads such as *Schmidtea mediterranea* or *Dugesia japonica* and revealed how their brains, although quite simple at the morphological level, can be subdivided into different molecular compartments. Thus, for example, planarian homologs of the *otx/otp* homeobox genes define different domains along the medio-lateral axis of the cephalic ganglia, as *otxA* is expressed in the most medial region, *otxB* in the central neuropil, and *otp* in the brain lateral branches (Umesono et al. 1999). *Smed-netrinA* is also mainly expressed in the medial region of the cephalic ganglia (Cebrià and Newmark 2005). Other genes specifically expressed in the lateral branches are *Djnlg* (noggin-like gene, Ogawa et al. 2002a; Molina et al. 2009), *1008HH* (a glutamate receptor homolog, Cebrià et al. 2002a), *Gtsix3* (*Six/sine oculis* homolog, Pineda and Saló 2002), and *1791HH* (a G protein alpha subunit,

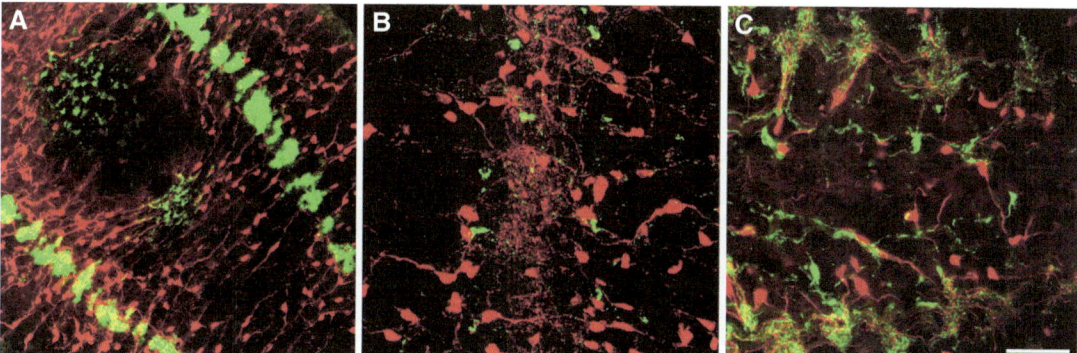

Fig. 4.5 Neuronal populations in the platyhelminth *Schmidtea mediterranea*. (**A**) Ventral nerve cords (in *green* after immunostaining with anti-synapsin) surrounded by serotonergic neurons (in *red* after immunostaining with anti-5-HT (serotonin)). Bipolar serotonergic neurons are also seen in the transverse commissures that interconnect the ventral cords. (**B**) Double immunostaining with anti-5-HT (in *red*) and anti-allatostatin (in *green*) along the ventral nerve cords. (**C**) Double immunostaining with anti-5-HT (in *red*) and anti-GYRFamide (in *green*) along the ventral nerve cords. Scale bars: (**A**) 100 µm; (**B**, **C**) 50 µm

Table 4.1 Summary of distinct neuronal populations in Platyhelminthes

Neuronal population	Species	Order	Reference
Serotonin (5-HT)	*Schmidtea mediterranea*	Tricladida	Cebrià (2008), März et al. (2013), Currie and Pearson (2013)
	Dugesia japonica	Tricladida	Nishimura et al. (2007a)
	Microstomum lineare	Macrostomida	Reuter et al. (1986)
	Girardia tigrina	Tricladida	Reuter et al. (1995b)
	Polycelis tenuis	Tricladida	Reuter et al. (1996a)
	Fasciola hepatica	Echinostomida	Gustafsson et al. (2001)
	Diphyllobothrium dendriticum	Pseudophyllidea	Lindholm et al. (1998)
	Castrella truncata	Rhabdocoela	Kotikova et al. (2002)
	Bothriomolus balticus	Seriata	Joffe and Reuter (1993)
	Stenostomum leucops	Catenulida (class)	Reuter et al. (2001)
	Echinococcus multilocularis	Cyclophyllidea	Koziol et al. (2013)
	Cryptocotyle lingua	Plagiorchiida	Pan et al. (1994)
	Schistosoma mansoni	Strigeidida	Gustafsson (1987)
	Macrostomum h. marinum	Macrostomida	Ladurner et al. (1997)
	Dendrocoelum lacteum	Tricladida	Reuter et al. (1996a)
Substance P	*Microstomum lineare*	Macrostomida	Reuter (1994)
	Stenostomum leucops	Catenulida	Reuter (1994)
	Schistosoma mansoni	Strigeidida	Gustafsson (1987)
	Diphyllobothrium dendriticum	Pseudophyllidea	Gustafsson et al. (1993)
Neuropeptide F	*Moniezia expansa*	Cyclophyllidea	Maule et al. (1992)
	Microstomum lineare	Macrostomida	Reuter et al. (1995c)
	Procerodes littoralis	Tricladida	Reuter et al. (1995a)
	Girardia tigrina	Tricladida	Reuter et al. (1995b)
	Archilopsis unipunctata	Seriata	Reuter et al. (1995c)
	Promonotus schultzei	Seriata	Reuter et al. (1995c)
	Stenostomum leucops	Catenulida	Reuter et al. (1995c)
	Schmidtea mediterranea	Tricladida	Cebrià (2008)
RFamide	*Polycelis tenuis*	Tricladida	Reuter et al. (1996a)
	Archilopsis unipunctata	Seriata	Reuter et al. (1995c)
	Promonotus schultzei	Seriata	Reuter et al. (1995c)
	Stenostomum leucops	Catenulida	Reuter et al. (1995c)
	Bothriomolus balticus	Seriata	Joffe and Reuter (1993)
	Dendrocoelum lacteum	Tricladida	Reuter et al. (1996a)
FMRFamide	*Girardia tigrina*	Tricladida	Reuter et al. (1996b)
	Diphyllobothrium dendriticum	Pseudophyllidea	Lindholm et al. (1998)
	Castrella truncata	Rhabdocoela	Kotikova et al. (2002)
	Stenostomum leucops	Catenulida	Reuter et al. (2001)
	Schmidtea mediterranea	Tricladida	Fraguas et al. (2012)
	Echinococcus multilocularis	Cyclophyllidea	Koziol et al. (2013)
	Cryptocotyle lingua	Trematoda (class)	Pan et al. (1994)
GYRFamide	*Girardia tigrina*	Tricladida	Kreshchenko et al. (1999)
	Fasciola hepatica	Echinostomida	Gustafsson et al. (2001)
	Schmidtea mediterranea	Tricladida	Cebrià (2008)
Dopamine	*Schmidtea mediterranea*	Tricladida	Fraguas et al. (2012)
	Dugesia japonica	Tricladida	Nishimura et al. (2007b)

Table 4.1 (continued)

Neuronal population	Species	Order	Reference
Octopamine	*Schmidtea mediterranea*	Tricladida	Fraguas et al. (2012)
	Dugesia japonica	Tricladida	Nishimura et al. (2008a)
GABA	*Dugesia japonica*	Tricladida	Nishimura et al. (2008b)
	Girardia tigrina	Tricladida	Eriksson and Panula (1994)

Cebrià et al. 2002a). Moreover, some genes define different domains along the anterior-posterior axis; thus, for example, *DjWntA* (a *Wnt* homolog) is mainly expressed in the posterior half of the brain, whereas *DjfzA* (a *frizzled* homolog) is expressed in the anterior half (Kobayashi et al. 2007). Among the genes expressed in the triclad CNS, there are homologs of FGF and EGFR receptors (Cebrià et al. 2002c; Ogawa et al. 2002b; Fraguas et al. 2011, 2014); *Pax6* (Pineda et al. 2002); neural cell adhesion molecules (Fusaoka et al. 2006); nicotinic acetylcholine receptors (Mineta et al. 2003); *roundabout* (Cebrià and Newmark 2007); *mushashi* (Higuchi et al. 2008); *FoxG1* (Brain factor-1) (Koinuma et al. 2003); *SNF2-like* (Rossi et al. 2003); *eye absent* (Mannini et al. 2004); *innexins* (Nogi and Levin 2005); *Smad* (Molina et al. 2007); members of the SET1/MLL family of histone methyltransferases (Hubert et al. 2013); components of the Wnt/β-catenin signalling pathway such as *β-catenin*, *apc*, and *dishevelled* (Gurley et al. 2008; Iglesias et al. 2008; Petersen and Reddien 2008); *hedgehog* (Yazawa et al. 2009); and several RNA-binding proteins (Guo et al. 2006; Rouhana et al. 2010; Wagner et al. 2012).

In 2003, Mineta et al. published the only high-throughput study so far of planarian neural genes from a comparative evolutionary perspective. From 3,101 nonredundant ESTs of a planarian head library, they identified 116 genes that showed clear homology to genes related to the nervous system in other animals. Remarkably, more than 95 % of these genes had homologs in *Drosophila*, *Caenorhabditis elegans*, and humans, indicating a high degree of conservation. However, a broad systematic comparison of the neural genes of different flatworm and acoelomorph taxa will be required to determine the extent to which the CNS of Platyhelminthes would resemble the urbilaterian nervous system.

Regeneration: Process and Genes

The amazing regenerative capabilities of some Platyhelminthes include the ability to regrow a complete and functional CNS de novo (Fig. 4.6). This remarkable feature has been best described in triclads (Cebrià 2007; Agata and Umesono 2008; Umesono et al. 2011; Fraguas et al. 2012). Table 4.2 summarises some of the genes that play relevant roles in planarian CNS regeneration. Upon head amputation, new cephalic ganglia develop within the blastema at the same time that the truncated ventral nerve cords grow to re-establish connections with the forming brain. The whole process can be divided into three main stages: (1) brain primordia formation and patterning, (2) re-establishment of connectivity, and (3) functional recovery. After 1–2 days of regeneration, two small bilateral clusters of cells corresponding to the primordia of the new cephalic ganglia are observed (Cebrià et al. 2002b; Kobayashi et al. 2007). The appearance of the brain primordia depends on the neoblasts. Until recently, it was not clear whether neural progenitors exist in planarians. However, recent studies have reported that some neoblasts already express distinct transcription factors that define different neuronal populations. Remarkably, these transcription factors correspond to conserved genes that have similar functions in neurogenesis of other invertebrate and vertebrate species. For example, a *pitx* homolog defines the serotonergic lineage during cell renewal and regeneration (Currie and Pearson 2013; März et al. 2013).

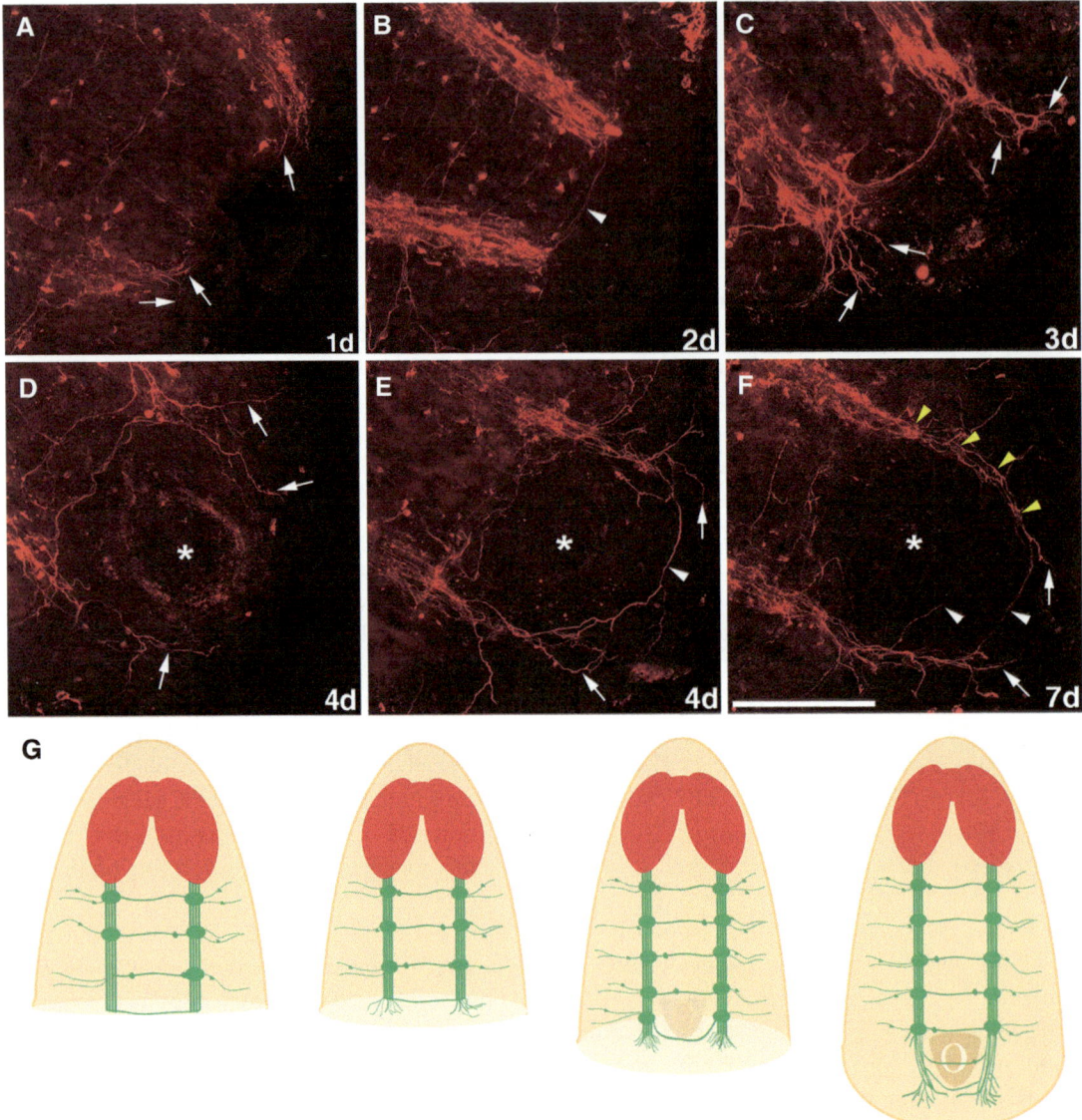

Fig. 4.6 Posterior regeneration of the CNS in *Schmidtea mediterranea*. All images show anti-neuropeptide F immunostaining in head fragments regenerating a new posterior region. (**A**) *Arrows* point to thin fibres sprouting from the truncated nerve cords. (**B**) *Arrowhead* points to a transverse commissure connecting *left* and *right* nerve cords. (**C**) Arrows point to fibres sprouting posteriorly. (**D**) Regenerating ventral nerve cords keep growing posteriorly (*arrows*). (**E**) Newly formed transverse commissures (*arrowhead*) from sprouting fibres (*arrows*). (**F**) As regeneration proceeds, new transverse commissures are formed (*white arrowheads*). *Arrows* point to thin sprouts near the posterior tip. *Yellow arrowheads* point to the new ganglia-like knots of the regenerated ventral nerve cords. (**G**) Schematic drawings of the posterior regeneration of the ventral nerve cords. Cephalic ganglia are in *red* and ventral nerve cords in *green*. Days of regeneration are indicated in (**A–F**). *Asterisks* mark the newly regenerated pharynx. Anterior to the *top left* in (**A–F**). Anterior to the *top* in (**G**). Scale bar, 200 µm (Reproduced with permission from The International Journal of Developmental Biology (Int. J. Dev. Biol. 2012 vol. 56:143–153))

Table 4.2 Selection of genes required for planarian CNS regeneration

Planarian gene	Homolog	RNAi phenotype	Reference
nou-darake	Fibroblast growth factor receptor-like	Ectopic brain differentiation in posterior regions	Cebrià et al. (2002c)
Smed-netrin2	Netrin	Disorganised meshwork of axonal projections	Cebrià and Newmark (2005)
		Mis-targeting of visual axons	
Smed-netR	Netrin receptor	Disorganised meshwork of axonal projections	Cebrià and Newmark (2005)
		Mis-targeting of visual axons	
DjCAM	N-CAM (neural cell adhesion molecule)	Defasciculation of brain axonal bundles	Fusaoka et al. (2006)
DjDSCAM	DSCAM (Down syndrome cell adhesion molecule)	Neuropil disorganised and reduced number of brain lateral branches	Fusaoka et al. (2006)
Smed-slit	Slit	CNS collapsed at the midline	Cebrià et al. (2007)
Smed-roboA	Roundabout	Abnormal reconnection of the cephalic ganglia and the ventral nerve cords	Cebrià et al. (2007)
		Mis-targeting of visual axons	
DjCHC	Clathrin heavy chain	Failed neurite extension and maintenance	Inoue et al. (2007)
DjWntA	Wnt	Brain expansion towards posterior regions	Kobayashi et al. (2007)
Smed-GSK3s[a]	Glycogen synthase kinase-3	Smaller and disconnected cephalic ganglia	Adell et al. (2008)
		Mis-targeting of visual axons	
Smed-evi/Wntless	Evenness interrupted (Wntless/Sprinter)	Disconnected and laterally displaced cephalic ganglia	Adell et al. (2009)
Smed-Wnt-5	Wnt	Disconnected and laterally displaced cephalic ganglia	Adell et al. (2009), Almuedo-Castillo et al. (2011), Gurley et al. (2010)
		Mis-targeting of visual axons	
Smed-dvl1/2	Dishevelled	Disconnected and laterally displaced cephalic ganglia	Almuedo-Castillo et al. (2011)
		Mis-targeting of visual axons	
Smed-lhx1/5-1	Lhx	Loss of serotonergic neurons	Currie and Pearson (2013)
Smed-pitx	Paired class homeobox/pituitary homeobox	Loss of serotonergic neurons	Currie and Pearson (2013), März et al. (2013)
Smed-coe	Collier/olfactory-1/early B-cell factor	Smaller and disconnected cephalic ganglia	Cowles et al. (2013)
Smed-sim	Single-minded	Reduced neuropil density	Cowles et al. (2013)
Smed-egr-4	Early growth response	Failed differentiation of the brain primordia	Fraguas et al. (2014)

Dj Dugesia japonica, Smed Schmidtea mediterranea
[a]For GSK3s, drug-mediated inhibition was performed instead of RNAi-mediated gene silencing

More recently, Cowles et al. (2013) identified 44 genes predicted to code for a bHLH (basic helix-loop-helix) domain, of which 12 are expressed in the CNS and neoblasts. Some of these genes are co-expressed with different markers of specific neuronal populations: cholinergic, GABAergic, octopaminergic, dopaminergic, and serotonergic. More importantly, two of these genes, *coe* (*collier/olfactory-1/early B-cell factor*) and *sim* (*single-minded*) are co-expressed with proliferating neoblasts and contribute to the regenerative blastema, further supporting the existence of neural progenitor cells in planarians.

The second stage in CNS regeneration involves growth of the brain primordia into mature cephalic ganglia and the extension of the amputated ventral cords into the blastema. This leads to the rewiring of the regenerating nervous system. Several families of axon guidance cues are evolutionarily and functionally conserved and have been shown to play important roles in the wiring of the developing nervous system. Among these cues, *netrin* and *slit* occupy a prominent position (Guan and Rao 2003; O'Donnell et al. 2009). Planarians have allowed analysing the function of such cues during the rewiring of the regenerative nervous system. Interestingly, these cues are not only found in planarians but are also required for proper CNS regeneration (Cebrià and Newmark 2005; Cebrià et al. 2007). For example, silencing of the homologs of either *netrin* or a *netrin receptor* disrupts neural architecture in both intact and regenerating animals. The ventral nerve cords regenerate as a completely disorganised meshwork of neural processes instead of the normal two parallel cords of compacted axonal processes. Also, the regenerated neuropil appears loosely packed after silencing of any of these genes. In addition, these genes are required for the proper targeting of the visual axons (Cebrià and Newmark 2005). These results exemplify the value of the flatworm CNS as a tool to investigate how the poor regenerative capabilities of mammals could be enhanced.

Although the development of the nervous system has been described in different groups of Platyhelminthes (Younossi-Hartenstein and Hartenstein 2000; Younossi-Hartenstein et al. 2000; Hartenstein and Jones 2003; Cardona et al. 2005; Rawlinson 2010; Bailly et al. 2013), much less is known about the genes expressed at the different embryonic stages and their putative functions during neurogenesis. It would be interesting to characterise how neurogenesis occurs at the molecular level in developing embryos and compare it with the regenerative process.

The Role of the CNS in Regeneration

It has been suggested that the nervous system might play an important role during regeneration in several models (Kumar and Brockes 2012). For example, it is well known that the amphibian limb cannot regenerate if previously denervated (Singer and Craven 1948). Also, head regeneration in *Hydra* depends on de novo neurogenesis (Miljkovic-Licina et al. 2007). However, finding the exact nature of such neurally derived molecules has been very elusive. In newts, a molecule called nAG appears to be responsible for the nerve dependence of limb regeneration (Kumar et al. 2007). However, in most cases the molecular basis for nerve-dependent regeneration remains to be characterised. In flatworms, several lines of evidence support a role for the nervous system in regeneration (Cebrià 2007). This neural influence could be in regulating different events, such as cell proliferation, differentiation, and migration, as well as patterning. Some classical studies have reported that regeneration is inhibited by removal of the nervous system in planarians such as *Leptoplana*, *Dendrocoelum*, and *Bdellocephala* (Child 1904a, b; Morgan 1905). In *Dugesia dorotocephala*, lateral pieces with no ventral nerve cord give rise to 'head-bump' regenerates: an abnormally large head with no differentiation of post-cephalic structures (Sperry et al. 1973). However, in other species, lateral pieces with no ventral nerve cord can regenerate a head (Morgan 1898, 1900). Also, it has been suggested that the brain is required to induce eye regeneration through an unknown factor of neuro-humoral nature (Wolff and Lender 1950; Lender 1955). Similarly, putative neurohormones would control neoblast migration towards the wound (Stéphan-Dubois and Lender 1956).

Other studies have analysed the role of neurotransmitters in regeneration, mainly through indirect pharmacological approaches. For example, dopamine and norepinephrine inhibitors reduce the rate of regeneration (Franquinet 1979), somatostatin might inhibit proliferation (Bautz and Schilt 1986), substance P appears to stimulate cell proliferation and differentiation in intact and regenerating planarians (Baguñà et al. 1989b), and neuropeptide F and FMRFamide stimulate pharyngeal regeneration (Sheiman et al. 2004). Serotonin also accelerates regeneration, whereas anti-serotonergic drugs delay it (Franquinet et al. 1978). More recently, Ladurner et al. (1997) proposed that the inability of *Macrostomum* to regenerate when amputated postpharyngeally could be correlated with the lack of serotonergic neurons in the posterior region. In fact, the presence of serotonergic neurons along the main nerve cords has also been associated with asexual reproduction by paratomy in *Microstomum lineare*. In these animals, the first cells of the new brain differentiate in close contact with the existing nerve cords, suggesting a neural effect on neoblast proliferation and/or differentiation (Reuter and Palmberg 1989; Reuter and Gustafsson 1996).

At the molecular level, the silencing of neural genes such as *nou-darake* (Cebrià et al. 2002c) and *roboA* (Cebrià and Newmark 2007) results in patterning and morphogenesis defects during regeneration. Thus, after *nou-darake* RNAi, brain tissues ectopically differentiate along the body, indicating that this gene is necessary to restrict the differentiation of brain tissues to the head region. In contrast, after silencing *roboA*, the newly differentiated cephalic ganglia appear disconnected from the truncated ventral nerve cords, which do not regenerate normally. Ectopic pharynges and dorsal outgrowths develop around these regions of failed connection, suggesting that, in the absence of a proper connected CNS, some neurally derived signal could alter the behaviour of the neoblasts nearby and induce the

morphogenesis defects observed (Cebrià and Newmark 2007). Other studies have suggested that ventral nerve cords could provide some long-range cue to control polarity during regeneration (Oviedo et al. 2010). A recent report suggested that the progression of head regeneration is blocked in the absence of a proper brain primordium. The proposed model hypothesises that the initial brain primordium produces a signal that induces neoblast proliferation and/or differentiation or migration to allow blastema growth (Fraguas et al. 2014).

Eye Specification and Regeneration

Due to its remarkable precision and complexity, the eye has attracted the interest of many evolutionary and developmental biologists. The animal kingdom contains a great variety of eyes, with few representative basic types. The wide variation in morphology and organisation of the eye, as well as the differences in embryological development of morphologically similar eyes, suggests that eyes have evolved independently a large number of times in animals (Salvini-Plawen and Mayr 1977). However, despite their morphological diversity, eyes share many molecular similarities, supporting the existence of conserved biochemical and developmental pathways (Gehring and Ikeo 1999; Pichaud et al. 2001). The presence of a common eye morphogenetic programme supports a monophyletic origin of an ancestral light sensory system. To reconcile such disparate observations, it has been suggested that the common ancestor had a prototypical light sensory organ. 'Eyes' then, as organs, evolved independently, in some cases by convergence, such as the camera eyes of vertebrates and cephalopods or the various eyes found in different molluscan taxa.

A key question is how to define the prototypical eye (Gehring 2002). Planarian eyes, consisting of a combination of photoreceptor cell and pigmented cell (Fig. 4.7; Hesse 1897), are considered good representatives of such a primitive visual system. This structure collects directional information by using the pigment to block

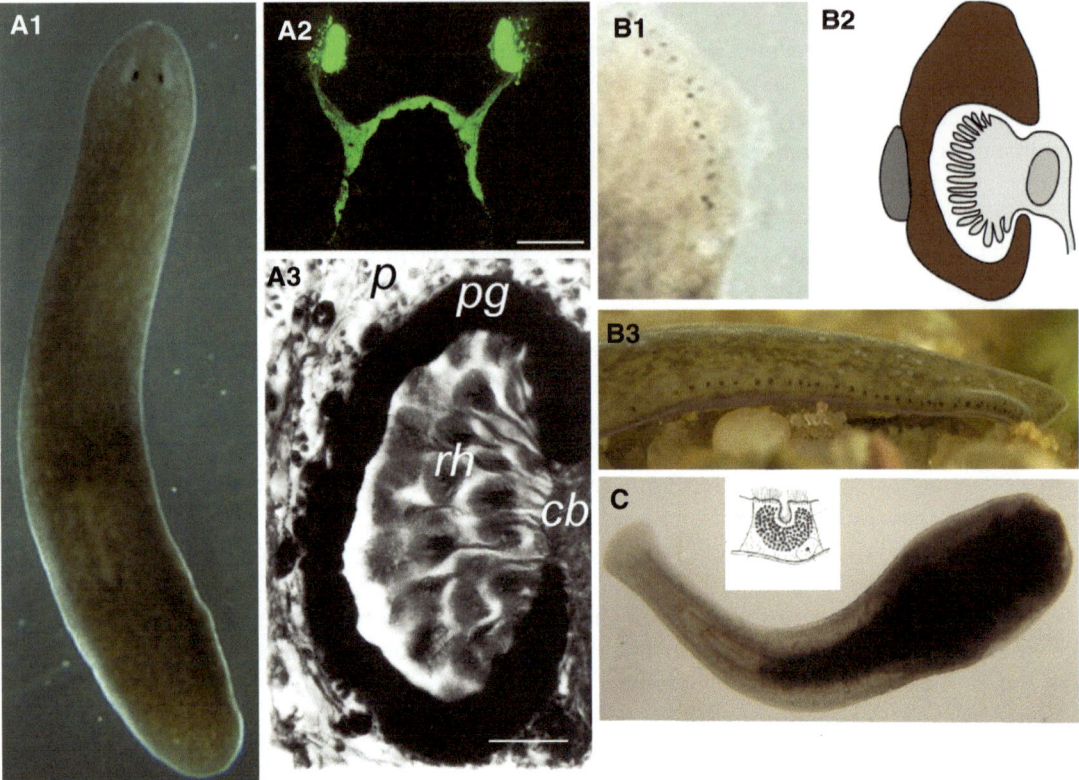

Fig. 4.7 Different types of simple eyes in Platyhelminthes. Tricladida pigment cup ocelli: (**A1**) Two eye spots located anteriorly, dorsal to the cephalic ganglia with the extensions of bipolar photoreceptors connecting to the cephalic ganglia with a partial optic chiasma (**A2**). (**A3**) The eye-cup is produced by the opposition of several pigment cells and, inside, the dendritic extremities of photoreceptor cells have a rhabdomeric structure in which opsin accumulates. (**B1**) Dorsolateral view of a planarian head with multiple eye spots formed by assembly of one photore- ceptor cell and one pigment cell (**B2**). (**B3**) Lateral view of a polycelis planarian showing long rows of eyes along the dorsoventral border. (**C**) Lecithoepitheliata epidermal eyes represent one of the simplest eyes. They are produced by a single cell and consist of ciliated sensors accumulated in a pocket that is produced by cell membrane invagination and surrounded by a pigmented area. *cb* photoreceptor cell bodies, *p* parenchyma, *pg* pigment cells, *rh* rhabdomeric structures. Scale bars in (**A1**) 0.5 mm; (**A2**, **B1**, **B3**, and **C**) 0.25 mm; (**B2** and **A3**) 0.04 mm

the light coming from specific directions. The majority of eyes among Platyhelminthes are pigment cup ocelli situated close to the dorsal basal membrane, although special epidermal eyes or rhabdomeric receptors without pigmented cells occur in some taxa. Such simple ocelli are in close association with the brain and present a large variety of shapes and morphological strategies. Since photoreceptors are bipolar neurons, their axons extend towards the dorsomedian side of the cephalic ganglia and form a partial optic chiasma, which integrates photosensory inputs on both sides (Fig. 4.7). Dendritic extremities generally have a rhabdomeric structure, a regularly ordered microvilli assembly where opsin accumulates (Orii et al. 1998). The pigmented cells group to form an eyecup that surrounds the rhabdomeres (Fig. 4.7), removing the damaged photoreceptive membranes and orientating the direction of light (Tamamaki 1990; Rieger et al. 1991). In planarians, some muscle fibres insert into the eyecup, suggesting that the diameter of the eye opening can be adapted to varying light intensities and that ocelli can rotate towards incident light direction. Finally, various light-refractive or lenticular structures have been described in planarians in the area of the eyecup opening, for example, cell extensions from the pigmented cell covering the eye opening (MacRae 1964; Carpenter et al. 1974) or accumulation of mitochondria in pigment cell extensions in front of the photosensitive cells to form 'mitochondrial lenses' (Sopott-Ehlers and Ehlers 2003). Such apparent sophistication of structures is only enough to confer phototactic behaviour with perception of variation in light intensity but not image formation (Rieger et al. 1991).

The second type of photoreceptive membrane amplifications are ciliary elaborations, which have only been found in some species of the orders Prolecithophora, Proseriata, and Macrostomida. It is also found in the larvae of polyclads, whereas the adults have rhabdomeric structures. This duality in photoreceptive membrane has been found in few groups and may indicate that this type of specialisation evolved only once. Commonly, the rhabdomeric eyes consist of one or a few covering cells, most often pigmented, and one or a few specialised bipolar nerve cells whose dendritic parts project into the covering. These eyes, described in many species among polyclads and macrostomids, measure 30–50 µm in diameter. Triclads only have a single pair of eyes (between 50 and 100 µm) containing between 35 and 70 photoreceptive cells in a cup made of many cells (Fig. 4.7). In other species, the number of pigment cup ocelli varies. Most have from one to several pairs. The first pair is dorsally located in front of the brain and the next pair occupies a more lateral position. In some triclads, such as *Polycelis*, numerous small ocelli are scattered over the head (Fig. 4.7) or from the front end of the body and in rows along the lateral body margin, even to the posterior pole of the body (Fig. 4.7).

The simplest platyhelminth eyes have been observed in the lecithoepitheliate *Gnosonesima brattstroemi*. It has a single epithelial eye cell (Karling 1968) with ciliary pits inside the cell membrane invagination which is surrounded by pigmented vacuoles (Fig. 4.7). Proseriates also have cerebral rhabdomeric ocelli with intracellular modified cilia in the same ocelli (Sopott-Ehlers 1982). Cerebral light-refracting bodies

have been described in *Stenostomum* (Catenulida). These are single nerve cells with a giant light-refracting mitochondrion (Ruppert and Schreiner 1980), which, since these animals are not phototactic, are thought to have some function related to photoperiodicity. Finally, in the larval stages of parasitic taxa, such as the free-swimming cercariae of trematodes, simple paired pigment cup ocelli contain a single pigmented cell and a photoreceptor rhabdomeric cell located in front of the posterior pole of the cephalic ganglion. In addition, three unpigmented rhabdomeric photoreceptors in front of the pigmented cup ocelli have been described. The pigmented ocelli serve to detect the direction of incoming light and to control the direction of swimming in relation to the light direction, whereas the unpigmented photoreceptors serve as monitors for light intensity (Sopott-Ehlers et al. 2001). In summary, the random distribution of eye spots or ocelli with different complexities and cell number suggests that the various types of eyes evolved multiple times independently within the Platyhelminthes.

The conserved eye genetic toolkit found in *Drosophila* and confirmed for different model systems (Gehring 2002) is also present in Platyhelminthes, at least in triclads (Saló et al. 2002). Co-expression of planarian *pax6*, *six1*, and *eya* in eye cells appears to explain a lack of requirement for the master control gene *pax6* (Pineda et al. 2000), which may be functionally substituted during planarian eye regeneration (Saló and Batistoni 2008). Table 4.3 summarises the most important genes expressed in the planarian eye; an exhaustive eye transcriptome was described in Lapan and Reddien 2012.

During eye regeneration, a stripe of migrating eye precursor cells appears once the cephalic brain primordium is determined. These cells express the conserved transcription factors *ovo*, *six1*, and *eya*, which continue to be expressed in the differentiated eye cells. Ovo is a zinc finger domain-containing protein with a conserved role in determination of eye fate (Mackay et al. 2006). The additional expression of *otxA* specifies the photoreceptor cells, whereas expression of *sp6-9* (a zinc finger-containing gene) and *dlx*

Table 4.3 Summary of genes expressed in planarian eyes and phenotypes observed after their silencing by RNAi

Planarian gene	Homolog	RNAi phenotype	Reference
Smed-otxA	*Crx/otd*	Absence of photoreceptor cells	Lapan and Reddien (2011)
Dj/Gt/Smed-Pax6A *Dj/Gt/Smed-Pax6B*	*Pax6/eyeless/twin of eyeless*	Not reported	Callaerts et al. (1999), Rossi et al. (2001), Pineda et al. (2001, 2002)
Dj/Gt/Smedsix-1	*Six1-2/Sine oculis*	No eyes	Pineda et al. (2000, 2001), Mannini et al. (2004)
Dj/Smed-eya	*Eye absent 1-4*	No eyes	Mannini et al. (2004)
Dj/Gt/Smed-opsin	*Opsin*	Loss of negative phototaxis	Sanchez-Alvarado and Newmark (1999), Pineda et al. (2000, 2001), Saló et al. (2002)
Smednos	*Nanos*	Not reported	Handberg-Thorsager and Saló (2007)
Smed-netrin2 *Smed- netR*	*Netrin* and *netrin receptor*	Defects in visual axon targeting and abnormal phototaxis	Cebrià and Newmark (2005)
Dj b-arrestin	*Arrestin*	Not reported	Agata et al. (1998), Nakazawa et al. (2003)
DjTPH	*Tryptophan hydroxylase*	Not reported	Nishimura et al. (2007a)
Smed-Ovo	*ovol1-3/shavenbaby* zinc finger	No eyes	Lapan and Reddien (2012)
Smed-soxB *Smed-foxQ2* *Smed-klf* *Smed-meis*	Transcription factors	Smaller eyes	Lapan and Reddien (2012)
Smed-SP6-9	*Sp6-9 zinc finger gene family*	Absence of eye progenitor cells	Lapan and Reddien (2011)
Smed-dlx	*Distal-less*	Absence of eye progenitor cells	Lapan and Reddien (2011)
Djeye53 *Smed-eye53-1* *Smed-eye53-2*	Unknown prohormone gene	Impaired negative phototaxis	Inoue et al. (2004), Collins et al. (2010)
Smed-npp-12	Neuropeptide precursor 12	Not reported	Collins et al. (2010)
Smed-mpl-2	*Myomodulin prohormone-like-2*	Not reported	Collins et al. (2010)
Smed-smad6/7-2	*Smad*	Reduced eyes, phototaxis and lack of anterior photoreceptor cells	Gonzalez-Sastre et al. (2012)
Smed-BMP	*BMP*	Elongated eyes with expanded anterior photoreceptor cells	Gonzalez-Sastre et al. (2012)

Dj Dugesia japonica, Gt Girardia tigrina, Smed Schmidtea mediterranea

(a homeobox-containing gene) specifies the pigment cell lineage. Finally, the transcription factors *meis*, *klf*, *foxQ2*, and *soxB* regulate eye differentiation and morphogenesis (Lapan and Reddien 2011, 2012). The eye cells aggregate in two bilaterally symmetric visual cell clusters in the dorsal blastema. The pigment cells form an eyecup that surrounds the rhabdomeres. A series of prohormone genes revealed the existence of three different subpopulations of photoreceptor cells, located in the anterior, dorsal posterior, and posterior regions of the eyecup (Collins et al. 2010). The proper photoreceptor subpopulation distribution inside the eyecup is regulated by the BMP pathway (González-Sastre et al. 2012). Interestingly, analysis of the expression pattern of eye-specific transcription factors in the planarian species *Schmidtea polychroa* revealed extensive similarities between embryonic development and regeneration of the eye (Martín-Durán et al. 2012).

SIGNALLING PATHWAYS IN REGENERATION, GROWTH, AND HOMEOSTASIS

The Establishment of the Body Axis

Platyhelminthes show bilateral symmetry, with an anterior-posterior (AP) axis (in which the anterior pole corresponds to the brain region) and a very short dorsoventral (DV) axis. Almost nothing is known about the molecular mechanisms that specify axial identity during embryogenesis in any platyhelminth (Chapter 3). However, the molecular mechanisms of axial re-establishment during regeneration have been studied in detail in the planarian species *Schmidtea mediterranea* and, to a lesser extent, *Dugesia japonica*.

Re-establishment of the body axis after amputation has fascinated biologists since the nineteenth century, and planarians have remained a favourite model system in which to investigate it (Morgan 1904). Classical amputation experiments in different planarian species demonstrated that planarians are able to regenerate the missing organs with the proper polarity in relation to the pre-existing tissues after amputation. However, it was observed that very narrow bipolar regenerating pieces occasionally generated bi-headed planarians ('Janus heads') (Morgan 1898). Although the molecules involved in this process were completely unknown at that time, these experiments led to the concept of 'gradient activity'. It was hypothesised that adult planarians should have an intrinsic gradient that, when broken through any amputation, tends to be restored, thus inducing

the re-establishment of polarity in the missing parts. According to this hypothesis, the appearance of Janus heads would indicate that a minimal AP distance is required in a region of tissue in order to be interpreted as a gradient and to specify polarity (Morgan 1905; Child 1911). The study of planarian regeneration was one of the mainstays of the 'morphogenetic gradient' and 'organising centre' theories (Lewis et al. 1977; Meinhardt 1978), which continue to be central to developmental biology today.

In the last 10 years, with the sequencing of the genome of the planarian *Schmidtea mediterranea* and the introduction of RNAi techniques, the molecular nature of the gradients that govern axial patterning has been elucidated (reviewed in Reddien 2011; Almuedo-Castillo et al. 2012). Thus, it was demonstrated that Wnts and BMPs are the morphogens that specify AP and DV axial identities, respectively, in adult planarians. The following section discusses the evidence for a conserved role of these signalling pathways in planarians and across metazoans.

The Wnt/β-catenin Pathway Establishes the Planarian AP Axis

The Wnt/β-catenin signalling pathway has an evolutionarily conserved role in establishing polarity during embryonic development. It specifies the main axis in cnidarians (Wikramanayake et al. 2003) and echinoderms (Logan et al. 1999) and the AP axis in most bilaterians (Holland 2002; Croce and McClay 2006; Petersen and Reddien 2009). The binding of Wnts, the secreted elements of the pathway, to the receptors frizzled and coreceptors LRP leads to the disruption of the β-catenin 'degradation complex', composed of axin, glycogen synthase kinase 3 (GSK3), casein kinase 1 (CKI), and adenomatous polyposis coli (APC). Afterwards, β-catenin, the key intracellular element of the pathway, accumulates in the cytoplasm, enters the nucleus, and activates TCF transcription factors, which regulate the expression of multiple genes.

Several elements of the Wnt/β-catenin signalling pathway that have been characterised in *Schmidtea mediterranea* reveal a functional conservation of this pathway in the specification of the AP axial identities (reviewed in Almuedo-Castillo et al. 2012). Thus, silencing of *Smed-βcatenin1* leads to an extreme anteriorised phenotype in which 'radial-like hypercephalised' planarians develop large circular cephalic ganglia together with several ectopic eyes all around the planarian body (Fig. 4.8; Iglesias et al. 2008). In amputated planarians that must regenerate the head and the tail, inhibition of *Smed-βcatenin1* induces the regeneration of bi-headed planarians that gradually anteriorise completely. Interestingly, inhibition of *Smed-βcatenin1* in intact planarians also leads to anteriorisation of the animals, dem-

onstrating that adult animals require a continuously active signalling pathway to maintain the correct axial organisation (Iglesias et al. 2008). The anteriorisation of the RNAi animals can be followed by the appearance of external morphological traits, such as the eyes, as well as by the ectopic appearance of anterior (brain and eyes) markers and the disappearance of posterior ones (posterior Hox genes) (Gurley et al. 2008; Iglesias et al. 2008; Petersen and Reddien 2008). In the most extreme phenotypes, not only are posterior regions absent, but the pharynx, which is located in the central region, also eventually disappears

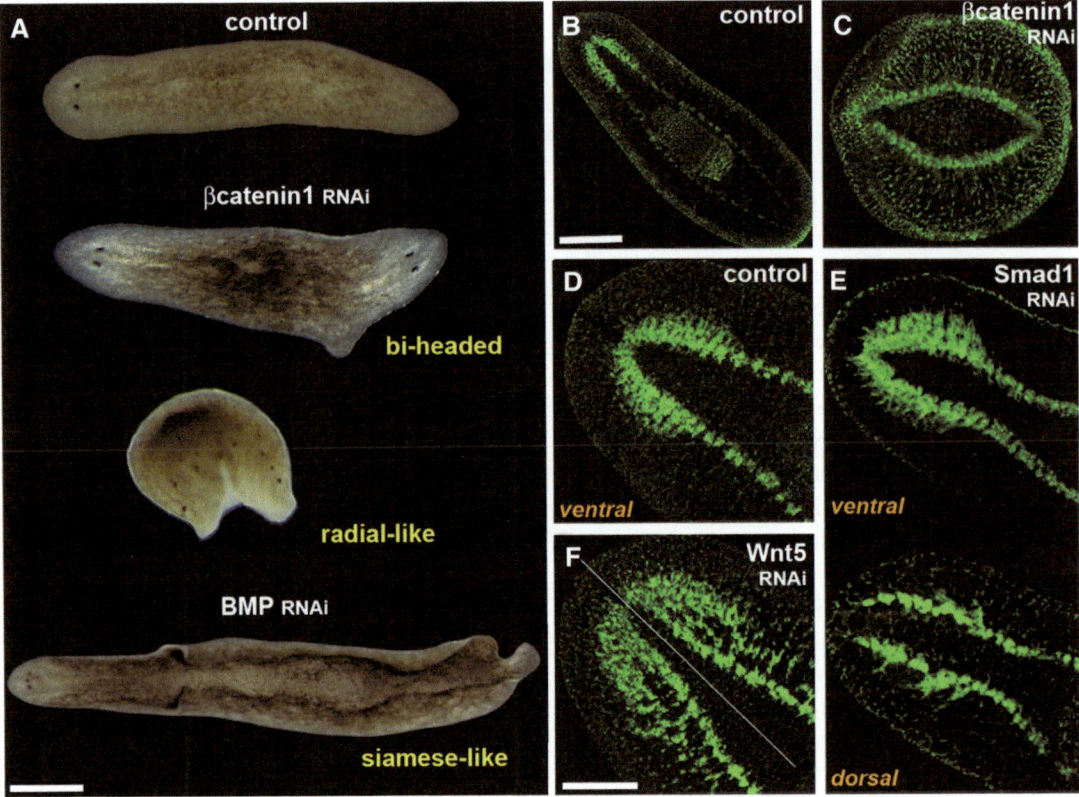

Fig. 4.8 Phenotypes generated after silencing Wnt or BMP signalling pathways. (**A**) In vivo images of planarians in which *βcatenin1* or *BMP* has been silenced by RNAi. Moderate silencing of *βcatenin1* induces the regeneration of a head in the posterior part (bi-headed planarians) and higher inhibition leads to a fully anteriorised phenotype (radial-like planarians; note the appearance of ectopic eyes around the animal). Silencing of *BMP* leads to a ventralisation and duplication of the dorsoventral border (Siamese-like planarians). (**B–F**) Anti-synapsin staining of the planarian central nervous system showing its complete anteriorisation after *βcatenin1* silencing (**B, C**), its duplication after *Smad1* inhibition (**E**) (note the appearance of an ectopic nervous system in the dorsal side of planarians), and its expansion along the medio-lateral axis after the inhibition of *Wnt5* (**F**) (*white line* indicates the midline), compared to controls (**D**). Anterior is to the *left* (**A, C**) or the *top left* (**B, D–F**). Scale bars: (**A**) 1 mm; (**B–C**) 400 μm; (**D–F**) 200 μm

(Iglesias et al. 2008). The timing of appearance and disappearance of these molecular markers demonstrates that anteriorisation is a gradual and dose-dependent process. This is consistent with a model in which a morphogenetic gradient activity from a posterior organiser is responsible for patterning the different identities along the AP axis (Adell et al. 2010). The involvement of the *βcatenin-dependent* Wnt signalling in this process is further supported by the finding that RNAi inhibition of elements of the β-catenin destruction complex (*Smed-APC* and *Smed-axin*) leads to the opposite phenotype, that is, posteriorisation of the animals (Gurley et al. 2008; Iglesias et al. 2011). In agreement with the existence of a posterior organising region as a source of morphogens, four of the nine Wnts that comprise the Wnt family in planarians (*Smed-Wnt1/11-1/11-2/11-5*) are expressed in nested domains in the most posterior part of planarians, and RNAi silencing of one of them (*Smed-Wnt1*) generates bi-headed planarians (Adell et al. 2009). Moreover, a Wnt inhibitor (*Smed-notum*) is expressed in the most anterior tip, and RNAi for *Smed-notum* leads to the opposite phenotype, that is, to 'bi-tailed' planarians (Petersen and Reddien 2011). Taken together, these results strongly support a conserved role for Wnt signalling in the specification of the AP axis. Furthermore, the role seems also to be conserved in different developmental contexts such as embryonic development and adult regeneration.

Interestingly, this situation resembles the one described for non-bilaterians, since the nested expression of Wnts in the aboral region and Wnt inhibitors (*dkk*) in the oral region defines the main body axis of cnidarians (Fig. 4.9). Thus, β-catenin signalling appears as a general mechanism to define axial polarity.

It is still not known whether the same signalling networks pattern the embryonic body axis in planarians. The core components of the canonical Wnt pathway are expressed in planarian embryos at the stage where the yolk has been completely ingested, that is, when the embryo differentiates the definitive structures that will replace the embryonic ones (Martín-Durán et al. 2010; see Chapter 3). The expression pattern resembles that of adults, suggesting that the Wnt signalling pathway also controls the establishment of definitive axial identities during embryogenesis. However, the molecular control of early polarity requires further attention. *Smed-βcatenin1* is expressed earlier, mostly associated with the development of the transient embryonic pharynx. At this stage, genes involved in gastrulation (*snail*) and germ layer determination (*foxA* and *twist*) are specifically expressed in migrating blastomeres giving rise to the temporary pharynx. Thus, β-catenin could have a primary role in endomesoderm specification, which seems to be an ancestral role of the Wnt signalling, linked to its function in the specification of the main body

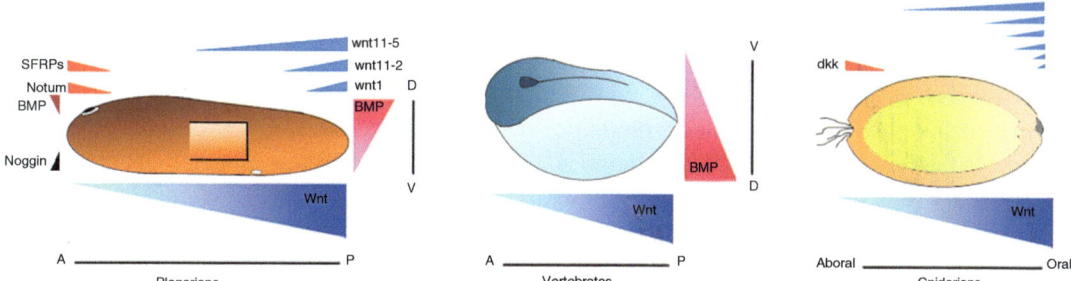

Fig. 4.9 Evolutionary conservation of the Wnt and BMP signalling pathways in metazoan anterior-posterior and dorsoventral axis establishment. In adult planarians, the Wnt and the BMP signals pattern the anterior-posterior (AP) and dorsoventral (DV) axes both during regeneration and homeostasis. Patterning is similar to *Xenopus* embryo-genesis, although in vertebrates a shift in the DV axis has occurred. In prebilaterian animals such as cnidarians, Wnt signalling also patterns the main (oral-aboral) axis. The expression of Wnts (in *blue*) and Wnt inhibitors (in *red*: *sFRP*, *notum* and *dkk*) in opposite poles is shown in planarians and cnidarians

axis (Martín-Durán and Romero 2011). Moreover, expression analysis indicates that β-catenin could be a maternal gene (Martín-Durán et al. 2010), as found in several animals across the phylogenetic tree.

From an evolutionary point of view, it is also interesting to note that a duplication and functional specialisation of β-catenin has been reported in *Schmidtea mediterranea*. In most animals, β-catenin is a bifunctional protein that has a transcriptional role when activated by the interaction of Wnts with their receptors and also a role in cell adhesion, as a component of adherens junctions (Schneider et al. 2003). In planarians, while *Smed-βcatenin1* is exclusively involved in transducing Wnt signalling, a second *β-catenin* (*Smed-βcatenin2*) is specifically involved in cell adhesion (Chai et al. 2010). Duplications of β-catenin have also been described in the nematode *Caenorhabditis elegans* and in vertebrates, since plakoglobin is a duplication of β-catenin. However, phylogenetic analyses demonstrate that they originated from independent duplication events (Korswagen et al. 2000). Interestingly, two β-catenins, which probably share an evolutionary origin with the planarian ones, are also found in *Schistosoma* (Chai et al. 2010). Studies on representatives from other platyhelminth subclades will clarify the evolutionary relationship of these duplications and to which extent the separation of the two roles could have implications for the plasticity (regeneration and growth/degrowth) found in the phylum.

The BMP Pathway Establishes the Planarian DV Axis

BMP proteins (bone morphogenetic proteins; Decapentaplegic [Dpp] in *Drosophila*) comprise a subfamily within the TGF beta superfamily of extracellular ligands, which have essential roles in key developmental processes in all metazoans. Specification of the DV axis is an evolutionarily conserved role of the BMP pathway (De Robertis and Kuroda 2004; Ashe and Briscoe 2006), although in vertebrates BMP signalling promotes ventral fates and in invertebrates it promotes dorsal fates, linked to

the shift in the positioning of the nervous system in chordates (De Robertis and Kuroda 2004). During *Xenopus* development, both dorsal and ventral signalling centres serve as sources of BMPs and their modulators (De Robertis and Kuroda 2004; De Robertis 2009). BMPs and BMP antagonists are secreted ventrally and ADMP (anti-dorsalising morphogenetic protein, a member of the BMP family) and other BMP antagonists (chordin and noggin) are secreted by the dorsal centre. They together configure a complex self-regulatory circuit that finally restricts BMP signalling to the ventral region of the embryo.

Homologs of BMPs and their inhibitors have been identified in *Schmidtea mediterranea*. Their expression patterns and functional characterisation by RNAi support the existence of an equivalent self-regulatory circuit in adult planarians that patterns their DV axis during regeneration and also during normal homeostasis (Molina et al. 2011a). RNAi silencing of planarian *BMP*, *ADMP*, and *Smads*, which are the intracellular effectors of BMP signalling, results in animals in which the dorsal side is transformed into a ventral one (Molina et al. 2007; 2011b; Orii and Watanabe 2007; Reddien et al. 2007; Gaviño and Reddien 2011). This ventralisation is shown by the disappearance of dorsal molecular markers, together with the ectopic appearance of ventral ones on the dorsal sides and the differentiation of ectopic structures (Fig. 4.8; Molina et al. 2007; Reddien et al. 2007). In severe phenotypes, there is a duplication of the body margin and an almost complete ectopic CNS develops on the ventralised dorsal side, resulting in 'Siamese twin-like' planarians (Molina et al. 2007). As in *Xenopus*, planarian *BMP* and *ADMP* show complementary expression patterns along the dorsal and ventral midlines, respectively (Molina et al. 2007, 2011b; Reddien et al. 2007; Gaviño and Reddien 2011), and in accordance with the DV inversion that occurred in chordates. Furthermore, *ADMP* promotes *BMP* expression and *BMP* inhibits *ADMP* expression, in agreement with its role as a regulatory circuit that buffers against perturbations of BMP signalling. Moreover, the planarian

BMP/ADMP circuit seems to be regulated by canonical antagonists of the *noggin* family, since RNAi silencing of planarian *noggin* genes (*noggin1-2*) produces a dorsalisation of the animals (Molina et al. 2011b).

Very recently, an expanded *noggin* family (up to ten members) was found to be present in planarians. Interestingly, its characterisation allowed the discovery of a new type of *noggin* genes that were called *noggin-like* genes (*nlg*), which carry an insertion within the noggin domain (Molina et al. 2011b). Functional analysis of these genes in planarians shows that *Smed-nlg-8* does not function as a canonical BMP inhibitor. Instead, silencing induces ventralisation and enhances *BMP* RNAi phenotypes. Importantly, *nlg* homologs are found to be present in the genome of all metazoans, from sponges to chordates (Molina et al. 2011b). The unexpected activity of this family of novel regulatory elements adds a new step in the complex regulation of DV axis establishment that should be further investigated in different developmental models.

To conclude, despite the shift in the positioning of the nervous system, the comparative results available demonstrate conservation of BMP/ADMP signalling in establishment of the DV axis in protostomes and deuterostomes. They also support the notion of an ancestral role for the pathway in the specification of neural fates within ectodermal derivatives, since neural territories are specified in the region of lowest BMP signalling (Reversade and De Robertis 2005).

Wnt5 and Slit Establish the Planarian Medio-lateral Axis

Three axes are required to define the positional identity of a three-dimensional body. Thus, besides the AP and the DV axis, a third perpendicular axis from the midline to the edge of the animal, the medio-lateral (M-L) axis, is defined in bilaterians. This axis can be extended to the left-right axis, when left-right asymmetries are observed. The molecular signals that pattern this axis seem not to be so general and broadly conserved as the ones that pattern the AP and

DV identities. In planarians, which are the only Platyhelminthes in which this issue has been studied, two signals are known to be implemented, the Wnt5- and the Slit-secreted factors. Neither of these signalling inputs controls the cellular transcriptional status, as with Wnt or BMP signals in AP and DV patterning. Instead, they control the assembly of the cellular cytoskeleton in neural cells, acting as evolutionarily conserved axon-guidance cues (Ciani and Salinas 2005).

The *Wnt5* homologs of the planarian species *Schmidtea mediterranea* and *Girardia tigrina* have been identified (*Smed-Wnt5* and *Gt-Wnt5*) (Marsal et al. 2003; Adell et al. 2009). *Smed-Wnt5* is expressed from the most external part of the CNS to the lateral edges and RNAi leads to a lateral displacement of the CNS (Fig. 4.8; Adell et al. 2009; Gurley et al. 2010). This phenotype suggests that *Smed-Wnt5* functions in restricting the location of the CNS along the ML axis, possibly through its role as a repulsive cue for growing axons. This would be consistent with the role of *Wnt5* in other models, such as *Drosophila* and vertebrates, where *Wnt5* acts through *Slit* or *ROR* receptors to control the decision of growing axons to cross the midline or remain on the same side (Yoshikawa et al. 2003; Ypsilanti et al. 2010). Interestingly, the planarian homolog of slit (*Smed-slit*), which is also a conserved signal involved in repelling growing axons, shows a complementary expression pattern with respect to *Smed-Wnt5*, since it is expressed from the internal part of the CNS to the midline (Cebriá et al. 2007). Moreover, silencing of *Smed-slit* in planarians leads to collapse of the CNS at the midline (Cebrià et al. 2007), a phenotype that could be considered opposite to the one generated after silencing *Smed-Wnt5*, at least in relation to the ML positioning of the nervous system. All this may indicate that *Smed-Wnt5* and *Smed-slit* could establish a signalling network that restricts the positioning of the nervous tissues along the ML axis (Gurley et al. 2010; Almuedo-Castillo et al. 2011). Although a cooperative role of these systems has not been described before, their role as repulsive cues for axons is conserved among metazoans. However, two immediate questions related to the role of *Wnt5* and *slit* in planarians remain open, that is, the specific molecular relationship between *Smed-Wnt5* and *Smed-slit* in the control of CNS positioning and the nature of their receptors. From data in other models, a *Derailed* or *ROR* homolog could be the *Smed-Wnt5* receptors, and a *Robo* homolog should be the *Smed-slit* receptor. In fact, a *Smed-RoboA* gene has been identified, but its RNAi phenotype suggests that it is not the *Smed-slit* receptor (Cebrià and Newmark 2007).

Altogether, the results found in planarians demonstrate that the Wnt and the BMP signalling pathways are broadly used for the specification of the AP and DV axis across metazoans and in all developmental contexts, from embryonic development to the re-establishment of the axial identities during adult regeneration and also for its maintenance during homeostasis. Remarkably, very recently it has been shown that in the acoel *Hofstenia miamia*, which also shows whole-body regeneration properties based on adult stem cells, the Wnt/β-catenin and BMP/ADMP signalling pathways control the re-establishment of AP and DV axes, respectively, during regeneration. These results suggest that animals such as acoels and planarians, separated by more than 550 million years of independent evolution, would share similar molecular regeneration mechanisms (Srivastava et al. 2014).

The conservation of these mechanisms during the process of embryogenesis in planarians and Platyhelminthes in general is a question that still requires further studies, since it would allow for direct comparison of the patterning mechanisms in embryonic and adult stages in the same species. In addition, understanding the mechanisms that control axial polarity during such a divergent cleavage process would help to understand this mode of development, which currently remains very poorly understood.

Regarding the evolutionary relationship of axial specification within Platyhelminthes, the existence of Wnt and BMP elements has been reported in the planarian *Dugesia japonica* (Orii et al. 1998) and a *BMP* homolog has also been found in *Schistosoma* (Liu et al. 2013). Their role has only been characterised in *D. japonica*, where

it appears completely conserved with respect to *Schmidtea mediterranea* (Yazawa et al. 2009). As mentioned above, RNAi for *β-catenin* in planarian species with restricted regenerative abilities (*Dendrocoelum lacteum*, *Procotyla fluviatilis*, *and Phagocata kawakatsui*) restores their ability to regenerate the head. This demonstrates the conservation of Wnt signalling in specifying AP axis identities. Moreover, it highlights the essential link between the activation of the intercellular signalling pathway and the ability to regenerate. These experiments highlight the relevance of increasing the number of model species to advance our knowledge on animal development, regeneration, and evolution.

Growth Control in Platyhelminthes

A long-standing question in developmental biology is how the size of an organ or a whole organism is determined. Classical embryological studies indicate that developing organs possess intrinsic information about their final size. In the same manner, regenerating tissues also have mechanisms to control their growth and reach the correct size. For example, when a piece of a liver is removed, the remaining part regenerates and stops growing when the liver reaches the original size. In the same manner, when first-instar larvae or amputated imaginal discs are transplanted into adult flies, they grow or regenerate, respectively, to reach the same final size as unmanipulated ones. The molecular mechanisms that determine this 'size checkpoint' in order to stop organ growth at the appropriate point during development or regeneration are just beginning to be elucidated, mainly with the discovery of the intercellular communication mechanism of the Hippo pathway.

The Hippo Signalling Pathway

The Hippo cascade appears as an evolutionarily conserved growth suppressive signalling pathway. It was initially discovered in the fruit fly, where mutations in components of the pathway resulted in dramatic overgrowth of tissues, and later on it was found to be conserved in mammals (reviewed in Johnson and Halder 2014). The main function of the Hippo pathway is to nega-tively regulate the activity of Yorkie (YAP/TAZ in mammals), a transcriptional co-activator that is the main downstream mediator of Hippo (Fig. 4.10). Hippo kinase, the protein that gives the name to the pathway, is responsible for phos-phorylating Warts and Mats, which, in turn, phosphorylate Yorkie. The phosphorylated form of Yorkie is inactive, since it is retained in the cytoplasm and ubiquitinated. When Hippo is inactive, dephosphorylated Yorkie can enter the nucleus and transcriptionally control its targets, which generally promote cell proliferation and inhibit cell death. Although several direct down-stream target genes of the Hippo pathway have been identified, including cyclins, growth factors and inhibitors of apoptosis, and genes involved in cell proliferation, cell survival, and stem cell functions, the mechanism through which Yorkie drives tissue growth is not understood.

Hippo signalling is similar to other signalling networks in that it depends on a cascade of phos-phorylation events. However, in contrast to other pathways, it does not appear to have dedicated extracellular signalling peptides and receptors and is instead regulated by a network of upstream components, many of which are involved in regulating cell adhesion and cell polarity (Fig. 4.10). Hippo signalling is currently thought to function as a sensor for the physical organisa-tion of cells in tissues (cell-cell contact, cell polarity, cell adhesion, etc.) to control cell prolif-eration and cell death. Thus, the pathway would coordinate the physical cues with the classic growth factor-mediated signalling pathways (Gumbiner and Kim 2014). Therefore, the Hippo pathway has a fundamental role not only in organ growth control during embryonic development but also during regeneration and may be a hub in the control of stem cell and tumour growth.

Hippo signalling has been most extensively studied in mouse liver and heart, and in *Drosophila* embryos, where *YAP/Yorkie* overexpression or loss of *lats/Warts* kinase activities increases size by increasing cell numbers (more proliferation and less cell death). In general, the activation of the Hippo pathway limits tissue growth by restricting proliferation and promoting apoptosis, and inhibi-tion of the pathway (by activation of *Yorkie*) is associated with stem cell expansion. Therefore, Hippo pathway inhibition is correlated with the ability to regenerate missing tissues. Consistent with this, several studies have demonstrated that *YAP* is a crucial regulator of cardiomyocyte prolif-eration and cardiac morphogenesis. Cardiac deletion of *YAP* impedes neonatal heart

Fig. 4.10 The Hippo signalling pathway. Signals from the components of the cellular membrane modulate the phosphorylat-ing activity of Hippo and warts/lats kinases. When Hippo is active, it phosphorylates Yorkie/ YAP/TAZ inhibiting its entrance to the nucleus. When Hippo is inactive, Yorkie/YAP/TAZ can enter the nucleus and modulate the transcription of target genes, mainly associated with proliferation and apoptotic responses. First the *Drosophila* and then the mammalian name for each protein are indicated

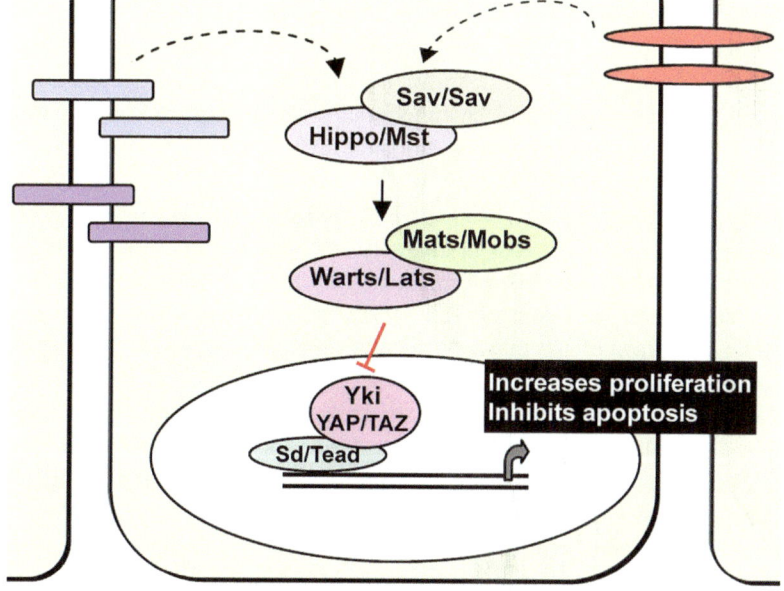

regeneration, whereas its forced expression in the adult heart stimulates cardiac regeneration after myocardial infarction (Xin et al. 2013). In the same manner, during limb bud regeneration in the amphibian *Xenopus laevis*, YAP is upregulated, and inhibition causes limb bud regeneration defects (Hayashi et al. 2014). However, the relationship between Hippo signalling, regeneration, and organ size is not absolute. The role of the Hippo signalling pathway in the intestine, for example, remains unclear, since overexpression of *YAP* causes an enlargement of the stem cell compartment but does not lead to an overall increase in organ size (Li and Clevers 2013).

Role of the Hippo Pathway in Platyhelminthes

Very recently the Hippo pathway has been characterised in two Platyhelminthes: *Macrostomum lignano* (Macrostomida) and the planarian *Schmidtea mediterranea* (Tricladida) (Demircan and Berezikov 2013; Lin and Pearson 2014). Knockdown of the Hippo pathway core genes in *M. lignano* during regeneration causes hyperproliferation of neoblasts, which leads to the formation of outgrowths and to the disruption of allometric scaling, since regenerated parts appear bigger than the original ones (Demircan and Berezikov 2013). In contrast, *Yorkie* is essential for neoblast self-renewal, since *Yorkie* silencing leads to the same 'tissue regression' phenotype observed after the inhibition of neoblast genes (Reddien et al. 2005). Thus, in *M. lignano*, the Hippo pathway appears to be functionally conserved in relation to *Drosophila* and mammals.

Although the homologs of the core Hippo pathway elements have been identified in *Schmidtea mediterranea*, only the role of the transcriptional effector, *Yorkie*, has been reported (Lin and Pearson 2014). Inhibition of *Smed-Yorkie* generates a plethora of effects that are difficult to associate with a single function. However, it seems to be directly involved in the maintenance and regeneration of the planarian excretory system. A role for *YAP*, the mammalian *Yorkie*, in nephrogenesis during mouse kidney development has also been reported (Reginensi et al. 2013). Regarding its role in growth control, as expected, *Yorkie* RNAi pla-

narians do not regenerate properly, but the cause seems not to be reduced proliferation or increased apoptosis, as in other systems. Instead, it seems to misregulate patterning, acting through the Wnt pathway. This result is not surprising, since crosstalk with other intercellular signalling pathways emerges as an intrinsic feature of the Hippo pathway (Konsavage and Yochum 2013). Thus, although functional analysis of the cytoplasmic Hippo kinases is lacking, Hippo signalling does not appear to directly regulate apoptotic and proliferative rates in planarians.

An important consideration regarding the differences found between *Macrostomum* and *Schmidtea* is the expression pattern of the Hippo elements. In *Macrostomum*, as in the mammalian heart and liver, *Yorkie* is expressed in the stem cell compartment, that is, in neoblasts. In contrast, in planarians it is expressed in post-mitotic cells (Demircan and Berezikov 2013; Lin and Pearson 2014). The situation in planarians strikingly resembles the one found in the mammalian and *Drosophila* gut system. In the mouse intestine, *Yorkie/YAP* is found in intestinal stem cells but also in enterocytes. Its overexpression first activates proliferation and suppresses differentiation, in agreement with its generic growth-promoting effect, but it eventually causes the loss of Paneth cells (specialised epithelial cells of the small intestine located at the base of the crypts) and of the intestinal crypts. Unexpectedly, inhibition causes an increase in Paneth cells and intestinal stem cells (Li and Clevers 2013). These observations may be explained by the functional relationship between the Hippo and the Wnt signalling pathways. Since, in mouse, Paneth cells are an essential source of growth factors, such as Wnts, they appear to be critical for the maintenance of the intestinal stem cell niche (Li and Clevers 2013). In planarians, like in the gut, inactivation of *Yorkie* also leads to activation of proliferation, and moreover, a relationship between Hippo and Wnt signalling has been demonstrated. Thus, it could be proposed that the precise role of *Yorkie* depends on the cellular composition of the tissues. In stem cell-based systems, like planarians and the intestine, which continuously undergo a stem cell-driven renewal programme, Hippo and Yorkie elements would not be directly required for stem cell proliferation but rather for the maintenance of the stem cell niche. In other systems, such as the vertebrate liver and heart, which are composed of mostly quiescent cells and only enter the cell cycle in extreme and rare conditions, Yorkie would directly control entry into the stem cell cycle. Supporting this view, *Macrostomum* does not show such broad regenerative capability as do planarians. *Macrostomum* normally reproduces sexually, and when forced, they are only able to regenerate the posterior part but never the head region. Thus, Yorkie behaviour resembles that described in mammals and vertebrate heart tissues.

The different functions found for *Yorkie*, for instance, in *Macrostomum* versus planarians, highlight the existence of context-specific behaviours that need to be addressed. The study of metazoan species with different stem cell natures (regenerating versus non-regenerating species) will help to clarify the extent to which *Yorkie* has a conserved role specifically in the stem cell niche. The broader conserved role of Hippo signalling in sensing cell contacts to control the final cell density and organ size suggests that its role is not restricted to the stem cell population but rather that it functions as a hub in coupling proliferative and apoptotic responses. Based on limited data from unicellular organisms, at least some elements of the pathway appear to have been present before the origin of multicellularity (Sebé-Pedrós et al. 2011; Artemenko and Devreotes 2013; Rock et al. 2013). Functional and comparative studies in uni- and multicellular models should provide insights into the ancient role of those kinases and the evolution of their linkage to extracellular signals in multicellular organisms.

OPEN QUESTIONS

Many issues concerning regeneration in Platyhelminthes remain open and answering them will require improved molecular tools such as transgenesis for the study of model species. In particular, it will be necessary to use species with different regenerative capacities and modes of development.

- Why is the ability to regenerate all missing tissues restricted to certain Platyhelminthes? Why can only some triclads regenerate the head?
- What is the evolutionary origin of neoblasts?
- What is the percentage of true totipotent stem cells in Platyhelminthes with high regenerative capabilities?
- How plastic are the progenitor cells to change their fate after amputation?
- Does the maintenance of adult stem cells and continuously active intercellular signalling play an equal role in regenerative capacity or is one more relevant than the other?
- Is there a neoblast niche?
- Are similar molecular mechanisms involved in the control of embryogenesis and regeneration?

Acknowledgements We thank Bernhard Egger and Jim Collins and Phil Newmark for providing the images of *Macrostomum lignano* and *Schistosoma mansoni*, respectively, used in Fig. 4.4. We thank Miquel Vila-Farré for providing the specimens of *Phagocata ullala* and *Camerata robusta* used for the immunostainings shown in Fig. 4.4. We thank Maria Almuedo-Castillo for providing planarian images in Fig. 4.3. We thank Iain Patten for advice on the English. This work was supported by grant BFU2012-31701 (Ministerio de Economía y Competitividad, Spain) to F.C, grant BFU2008-01544 (Ministerio de Economía y Competitividad, Spain) to ES, grant 2009SGR1018 (Agència de Gestió d'Ajuts Universitaris i de Recerca) to ES and FC, and grant AIB2010DE-00402 (Ministerio de Economia y Competitividad Accion Integrada) to ES.

References

Adell T, Marsal M, Saló E (2008) Planarian *GSK3s* are involved in neural regeneration. Dev Genes Evol 218:105–106

Adell T, Saló E, Boutros M, Bartscherer K (2009) *Smed-Evi/Wntless* is required for *b-catenin*-dependent and -independent processes during planarian regeneration. Development 136:905–910

Adell T, Cebrià F, Saló E (2010) Gradients in planarian regeneration and homeostasis. Cold Spring Harb Perspect Biol 2(1):a000505

Adell T, Cebrià F, Saló F (2014) Planarian totipotent stem cells. In: Calegari F, Waskow C (eds) Stem cells from basic research to therapy, vol 1, Basic stem cell biology, tissue formation during development, and model organisms. CRC Press, Boca Raton, pp 433–472

Agata K (2003) Regeneration and gene regulation in planarians. Curr Opin Genet Dev 13:492–496

Agata K, Umesono Y (2008) Brain regeneration from pluripotent stem cells in planarian. Philos Trans R Soc Lond B Biol Sci 363:2071–2078

Agata K, Soejima Y, Kato K, Kobayashi C, Umesono Y, Watanabe K (1998) Structure of the planarian central nervous system (CNS) revealed by neuronal cell markers. Zool Sci 15:433–440

Almuedo-Castillo M, Saló E, Adell T (2011) *Dishevelled* is essential for neural connectivity and planar cell polarity in planarians. Proc Natl Acad Sci U S A 108:2813–2818

Almuedo-Castillo M, Sureda-Gómez M, Adell T (2012) Wnt signaling in planarians: new answers to old questions. Int J Dev Biol 56:53–65

Artemenko Y, Devreotes PN (2013) Hippo on the move: tumor suppressor regulates adhesion and migration. Cell Cycle 12:535–536

Ashe HL, Briscoe J (2006) The interpretation of morphogen gradients. Development 133:385–394

Auladell C, García-Valero J, Baguñà J (1993) Ultrastructural localization of RNA in the chromatoid bodies of undifferentiated cells (neoblasts) in planarians by RNase gold complex technique. J Morphol 216:319–326

Baguñà J (1974) A demonstration of a peripheral and a gastrodermal nervous plexus in planarians. Zool Anz 193:240–244

Baguñà J (1976) Mitosis in the intact and regenerating planarian *Dugesia mediterranea n. sp.* I. Mitotic studies during growth, feeding and starvation. J Exp Zool 195:65–80

Baguñà J (2012) The planarian neoblast: the rambling history of its origin and some current black boxes. Int J Dev Biol 56:19–37

Baguñà J, Ballester R (1978) The nervous system in planarians: peripheral and gastrodermal plexuses, pharynx innervation, and the relationship between central nervous system structure and the acoelomate organization. J Morphol 155:237–252

Baguñà J, Saló E, Auladell C (1989a) Regeneration and pattern formation in planarians. III. Evidence that neoblasts are totipotent stem-cells and the source of blastema cells. Development 107:77–86

Baguñà J, Saló E, Romero R (1989b) Effects of activators and antagonists of the neuropeptides substance P and substance K on cell proliferation in planarians. Int J Dev Biol 33:261–266

Bailly X, Reichert H, Hartenstein V (2013) The urbilaterian brain revisited: novel insights into old questions from new flatworm clades. Dev Genes Evol 223:149–157

Bautz A, Schilt J (1986) Somatostatin-like peptide and regeneration capacities in planarians. Gen Comp Endocrinol 64:267–272

Bely AE (2010) Evolutionary loss of animal regeneration: pattern and process. Integr Comp Biol 50:515–527

Bely AE, Nyberg KG (2010) Evolution of animal regeneration: re-emergence of a field. Trends Ecol Evol 25:161–170

Birnbaum KD, Sánchez-Alvarado A (2008) Slicing across kingdoms: regeneration in plants and animals. Cell 132:697–710

Brehm K (2010) *Echinococcus multilocularis* as an experimental model in stem cell research and molecular host-parasite interaction. Parasitology 137:537–555

Brøndsted HV (1969) Planarian regeneration. Pergamon Press, Oxford

Bueno D, Fernàndez-Rodríguez J, Cardona A, Hernàndez-Hernàndez V, Romero R (2002) A novel invertebrate trophic factor related to invertebrate neurotrophins is involved in planarian body regional survival and asexual reproduction. Dev Biol 252:188–201

Bullock TH, Horridge GA (1965) Structure and function in the nervous systems of invertebrates. Freeman, San Francisco

Callaerts P, Muñoz-Mármol AM, Glardon S, Castillo E, Sun H, Li WH, Gehring WJ, Saló E (1999) Isolation and expression of a *Pax-6* gene in the regenerating and intact planarian *Dugesia(G) tigrina*. Proc Natl Acad Sci U S A 96:558–563

Cardona A, Hartenstein V, Romero R (2005) The embryonic development of the triclad *Schmidtea polychroa*. Dev Genes Evol 215:109–131

Carpenter KS, Morita M, Best JB (1974) Ultrastructure of the photoreceptor of the planarian *Dugesia dorotocephala*. I. Normal eye. Cell Tissue Res 148:143–158

Cebrià F (2007) Regenerating the central nervous system: how easy for planarians! Dev Genes Evol 217:733–748

Cebrià F (2008) Organization of the nervous system in the model planarian *Schmidtea mediterranea*: an immunocytochemical study. Neurosci Res 61:375–384

Cebrià F, Newmark PA (2005) Planarian homologs of *netrin* and *netrin receptor* are required for proper regeneration of the central nervous system and the maintenance of nervous system architecture. Development 132:3691–3703

Cebrià F, Newmark PA (2007) Morphogenesis defects are associated with abnormal nervous system regeneration after *roboA* RNAi in planarians. Development 134:833–837

Cebrià F, Kudome T, Nakazawa M, Mineta K, Ikeo K, Gojobori T, Agata K (2002a) The expression of neural-specific genes reveals the structural and molecular complexity of the planarian central nervous system. Mech Dev 116:199–204

Cebrià F, Nakazawa M, Mineta K, Ikeo K, Gojobori T, Agata K (2002b) Dissecting planarian central nervous system regeneration by the expression of neural-specific genes. Dev Growth Differ 44:135–146

Cebrià F, Kobayashi C, Umesono Y, Nakazawa M, Mineta K, Ikeo K, Gojobori T, Itoh M, Taira M, Sánchez-Alvarado A, Agata K (2002c) FGFR-related gene *nou-darake* restricts brain tissues to the head region of planarians. Nature 419:620–624

Cebrià F, Guo T, Jopek J, Newmark PA (2007) Regeneration and maintenance of the planarian midline is regulated by a *slit* ortholog. Dev Biol 307:394–406

Cebrià F, Adell T, Saló E (2010) Regenerative medicine: lessons from planarians. In: Singh SR (ed) Stem cell, regenerative medicine and cancer. Nova Science Publisher, Hauppauge, NY, pp 29–68

Chai G, Ma C, Bao K, Zheng L, Wang X, Sun Z, Salò E, Adell T, Wu W (2010) Complete functional segregation of planarian *beta-catenin-1* and *-2* in mediating Wnt signaling and cell adhesion. J Biol Chem 285(31):24120–24130

Child CM (1904a) Studies on regulation. V. The relation between the central nervous system and regeneration in Leptoplana: posterior regeneration. J Exp Zool 1:463–512

Child CM (1904b) Studies on regulation. VI. The relation between the central nervous system and regeneration in Leptoplana: anterior and lateral regeneration. J Exp Zool 1:513–558

Child CM (1911) Studies on the dynamics of morphogenesis and inheritance in experimental reproduction. I The axial gradient in Planaria dorotocephala as a limiting factor in regulation. J Exp Zool 10:265–320

Ciani L, Salinas PC (2005) WNTs in the vertebrate nervous system: from patterning to neuronal connectivity. Nat Rev Neurosci 6:351–362

Collins JJ 3rd, Hou X, Romanova EV, Lambrus BG, Miller CM, Saberi A, Sweedler JV, Newmark PA (2010) Genome-wide analyses reveal a role for peptide hormones in planarian germline development. PLoS Biol 8:e1000509

Collins JJ 3rd, King RS, Cogswell A, Williams DL, Newmark PA (2011) An atlas for *Schistosoma mansoni* organs and life-cycle stages using cell type-specific markers and confocal microscopy. PLoS Negl Trop Dis 5:e1009

Collins JJ 3rd, Wang B, Lambrus BG, Tharp ME, Iyer H, Nemwark PA (2013) Adult somatic stem cells in the human parasite *Schistosoma mansoni*. Nature 494:476–479

Coultas KA, Zhang SM (2012) In vitro cercariae transformation: comparison of mechanical and nonmechanical methods and observation of morphological changes of detached cercariae tails. J Parasitol 98:1257–1261

Cowles MW, Brown DD, Nisperos SV, Stanley BN, Pearson BJ, Zayas RM (2013) Genome-wide analysis of the bHLH gene family in planarians identifies factors required for adult neurogenesis and neuronal regeneration. Development 140:4691–4702

Croce JC, McClay DR (2006) The canonical Wnt pathway in embryonic axis polarity. Semin Cell Dev Biol 2:168–174

Currie KW, Pearson BJ (2013) Transcription factors *lhx1/5-1* and *pitx* are required for the maintenance and regeneration of serotonergic neurons in planarians. Development 140:3577–3588

Dalyell JG (1814) Observations on some interesting phenomena in animal physiology exhibited by several species of planariae. Archibald Constable & Co, Edinburgh

De Robertis EM (2009) Spemann's organizer and the self-regulation of embryonic fields. Mech Dev 126(11–12): 925–941

De Robertis EM, Kuroda H (2004) Dorsal-ventral patterning and neural induction in *Xenopus* embryos. Annu Rev Cell Dev Biol 20:285–308

Demircan T, Berezikov E (2013) The Hippo pathway regulates stem cells during homeostasis and regeneration of the flatworm *Macrostomum lignano*. Stem Cells Dev 22:2174–2185

Dirks U, Gruber-Vodicka HR, Egger B, Ott JA (2012) Proliferation pattern during rostrum regeneration of the symbiotic flatworm *Paracatenula galateia*: a pulse-chase-pulse analysis. Cell Tissue Res 349:517–525

Egger B, Ladurner P, Nimeth K, Gschwentner R, Rieger R (2006) The regeneration capacity of the flatworm *Macrostomum lignano*—on repeated regeneration, rejuvenization, and the minimal size needed for regeneration. Dev Genes Evol 216:565–577

Egger B, Gschwentner R, Rieger R (2007) Free-living flatworms under the knife: past and present. Dev Genes Evol 217:89–104

Egger B, Gschwentner R, Hess MW, Nimeth KT, Adamski Z, Willems M, Rieger R, Salvenmoser W (2009) The caudal regeneration blastema is an accumulation of rapidly proliferating stem cells in the flatworm *Macrostomum lignano*. BMC Dev Biol 9:41

Eriksson KS, Panula P (1994) Gamma-aminobutyric acid in the nervous system of a planarian. J Comp Neurol 345:528–536

Extravour CG, Akam M (2003) Mechanisms of germ cell specification across the metazoans: epigenesis and preformation. Development 130:5869–5884

Fairweather I, Halton DW (1991) Neuropeptides in platyhelminths. Parasitology 102:S77–S92

Fernández-Taboada E, Moritz S, Zeuschner D, Stehling M, Schöler HR, Saló E, Gentile L (2010) *Smed-SmB*, a member of the LSm protein superfamily, is essential for chromatoid body organization and planarian stem cell proliferation. Development 137: 1055–1065

Fraguas S, Barberán S, Cebrià F (2011) EGFR signalling regulates cell proliferation, differentiation and morphogenesis during planarian regeneration and homeostasis. Dev Biol 56:143–153

Fraguas S, Barberán S, Ibarra B, Stöger L, Cebrià F (2012) Regeneration of neuronal cell types in *Schmidtea mediterranea*: an immunohistochemical and expression study. Int J Dev Biol 56:143–153

Fraguas S, Barberán S, Iglesias M, Rodríguez-Esteban G, Cebrià F (2014) *egr-4*, a target of EGFR signalling is required for the formation of the brain primordia and head regeneration in planarians. Development 141:1835–1847

Franquinet R (1979) The role of serotonin and catecholamines in the regeneration of the planaria *Polycelis tenuis*. J Embryol Exp Morphol 51:85–95

Franquinet R, Le Moigne A, Hanoune J (1978) The adenylate cyclase system of planarian *Polycelis tenuis*. Activation by serotonin and guanine nucleotides. Biochim Biophys Acta 539:88–97

Fusaoka E, Inoue T, Mineta K, Agata K, Takeuchi K (2006) Structure and function of primitive immunoglobulin superfamily neural cell adhesion molecules: a lesson from studies on planarian. Genes Cells 11:541–555

Garza-Garcia AA, Driscoll PC, Brockes JP (2010) Evidence for the local evolution of mechanisms underlying limb regeneration in salamanders. Integr Comp Biol 50:528–535

Gaviño MA, Reddien P (2011) A *Bmp/Admp* regulatory circuit controls maintenance and regeneration of dorsal-ventral polarity in planarians. Curr Biol 21:294–299

Gehring WJ (2002) The genetic control of eye development and its implications for the evolution of the various eye-types. Int J Dev Biol 46:65–73

Gehring WJ, Ikeo K (1999) *Pax 6*: mastering eye morphogenesis and eye evolution. Trends Genet 15:371–377

Gentile L, Cebrià F, Bartscherer K (2011) The planarian flatworm: an in vivo model for stem cell biology and nervous system regeneration. Dis Model Mech 4:12–19

González-Sastre A, Molina MD, Saló E (2012) *Inhibitory Smads* and *bone morphogenetic protein (BMP)* modulate anterior photoreceptor cell number during planarian eye regeneration. Int J Dev Biol 56:155–163

Gremigni V, Miceli C, Puccinelli I (1980) On the role of germ cells in planarian regeneration. A karyological investigation. J Embryol Exp Morpholog 55:53–63

Guan KL, Rao Y (2003) Signalling mechanisms mediating neuronal responses to guidance cues. Nat Rev Neurosci 4:941–956

Gumbiner BM, Kim NG (2014) The Hippo-YAP signaling pathway and contact inhibition of growth. J Cell Sci 127:709–717

Guo T, Peters AH, Newmark PA (2006) A *Bruno*-like gene is required for stem cell maintenance in planarians. Dev Cell 11:159–169

Gurley KA, Rink JC, Sánchez-Alvarado A (2008) *b-catenin* defines head versus tail identity during planarian regeneration and homeostasis. Science 319:323–327

Gurley KA, Elliott SA, Simakov O, Schmidt HA, Holstein TW, Sánchez-Alvarado A (2010) Expression of secreted Wnt pathway components reveals unexpected complexity of the planarian amputation response. Dev Biol 347:24–39

Gustafsson MKS (1987) Immunocytochemical demonstration of neuropeptides and serotonin in the nervous system of adult *Schistosoma mansoni*. Parasitol Res 74:168–174

Gustafsson MKS, Nässel D, Kuusisto A (1993) Immunocytochemical evidence for the presence of substance P-like peptide in *Diphyllobothrium dendriticum*. Parasitology 106:83–89

Gustafsson MKS, Terenina NB, Kreshchenko ND, Reuter M, Maule AG, Halton DW (2001) Comparative study of the spatial relationship between nicotinamide adenine dinucleotide phosphate-diaphorase activity, serotonin immunoreactivity, and GYRFamide immunoreactivity and the musculature of the adult liver fluke, *Fasciola hepatica* (Digenea, Fasciolidae). J Comp Neurol 429: 71–79

Gustafsson MKS, Halton DW, Kreshchencko ND, Movsessian SO, Raikova OI, Reuter M, Terenina NB (2002) Neuropeptides in flatworms. Peptides 23:2053–2061

Handberg-Thorsager M, Saló E (2007) The planarian *nanos*-like gene *Smednos* is expressed in germline and eye precursor cells during development and regeneration. Dev Genes Evol 217:403–411

Hartenstein V, Jones M (2003) The embryonic development of the bodywall and nervous system of the cestode flatworm *Hymenolepis diminuta*. Cell Tissue Res 311:427–435

Hayashi S, Tamura K, Yokoyama H (2014) Yap1, transcription regulator in the Hippo signaling pathway, is required for *Xenopus* limb bud regeneration. Dev Biol 388:57–67

Hesse R (1897) Untersuchungen über die Organe der Lichtempfindung bei niederen Thieren. II. Die Augen der Plathelminthen. Z Wiss Zool 62:527–582

Higuchi S, Hayashi T, Tarui H, Nishimura O, Nishimura K, Shibata N, Sakamoto H, Agata K (2008) Expression and functional analysis of *musashi*-like genes in planarian CNS regeneration. Mech Dev 125:631–645

Holland LZ (2002) Heads or tails? Amphioxus and the evolution of anterior-posterior patterning in deuterostomes. Dev Biol 24:209–228

Hori I (1989) Observations on planarian epithelization after wounding. J Submicrosc Cytol Pathol 21:307–315

Hubert A, Henderson JM, Ross KG, Cowles MW, Torres J, Zayas RM (2013) Epigenetic regulation of planarian stem cells by the SET1/MLL family of histone methyltransferases. Epigenetics 8:79–91

Hyman LH (1951) The invertebrates. II. Platyhelminthes and rhynchocoela. The acoelomate bilateria. McGraw-Hill, New York

Iglesias M, Gomez-Skarmeta JL, Saló E, Adell T (2008) Silencing of *Smed-betacatenin1* generates radial-like hypercephalized planarians. Development 135:1215–1221

Iglesias M, Almuedo-Castillo M, Aboobaker AA, Saló E (2011) Early planarian brain regeneration is independent of blastema polarity mediated by the Wnt/b-catenin pathway. Dev Biol 358:68–78

Inoue T, Kumamoto H, Okamoto K, Umesono Y, Sakai M, Sanchez Alvarado A, Agata K (2004) Morphological and functional recovery of the planarian photosensing system during head regeneration. Zool Sci 21:275–283

Inoue T, Hayashi T, Takechi K, Agata K (2007) Clathrin-mediated endocytic signals are required for the regeneration of, as well as homeostasis in, the planarian CNS. Development 134:1679–1689

Joffe BI, Kotikova EA (1991) Distribution of catecholamines in turbellarians (with discussion of neuronal homologies in the Platyhelminthes). Stud Neurosci 13:77–113

Joffe BI, Reuter M (1993) The nervous system of *Bothriomolus balticus* (Proseriata) – a contribution to the knowledge of the orthogon in the Plathelminthes. Zoomorphology 113:113–127

Johnson R, Halder G (2014) The two faces of Hippo: targeting the Hippo pathway for regenerative medicine and cancer treatment. Nat Rev Drug Discov 13:63–79

Karling TG (1968) On the genus gnosonesima teisinger (Turbellaria). Sarsia 33:81–108

Knopf F, Hammond C, Chekuru A, Kurth T, Hans S, Weber CW, Mahatma G, Fisher S, Brand M, Schulte-Merker S, Weidinger G (2011) Bone regenerates via dedifferentiation of osteoblasts in the zebrafish fin. Dev Cell 20:713–724

Kobayashi C, Saito Y, Ogawa K, Agata K (2007) Wnt signalling is required for antero-posterior patterning of the planarian brain. Dev Biol 306:714–724

Koinuma S, Umesono Y, Watanabe K, Agata K (2003) The expression of planarian brain factor homologs *DjFoxG* and *DjFoxD*. Gene Expr Patterns 3:21–27

Konsavage WM, Yochum GS (2013) Intersection of Hippo/YAP and Wnt/B-catenin signaling pathways. Acta Biochim Biophys Sin (Shanghai) 45:71–79

Korswagen HC, Herman MA, Clevers HC (2000) Distinct beta-catenins mediate adhesion and signalling functions in *C. elegans*. Nature 406:527–532

Kotikova EA (1986) Comparative characterization of the nervous system of the Turbellaria. Hydrobiologia 132:89–92

Kotikova EA (1991) The orthogon of the plathelminthes and main trends of its evolution. Proc Zool Inst St Petersburg 241:88–111

Kotikova EA, Raikova OI, Reuter M, Gustafsson MKS (2002) The nervous and muscular systems in the free-living flatworm *Castrella truncata* (Rhabdocoela): an immunocytochemical and phalloidin fluorescence study. Tissue Cell 34:365–374

Koziol U, Krohne G, Brehm K (2013) Anatomy and development of the larval nervous system in *Echinococcus multilocularis*. Front Zool 10:24

Koziol U, Rauschendorfer T, Zanon Rodríguez L, Krhone G, Brehm K (2014) The unique stem cell system of the immortal larva of the human parasite *Echinococcus multilocularis*. Evodevo 5:10

Kragl M, Knapp D, Nacu E, Khatta S, Maden M, Epperlein HH, Tanaka EM (2009) Cells keep a memory of their tissue origin during axolotl limb regeneration. Nature 460:60–65

Kreshchenko ND, Reuter M, Sheiman IM, Halton DW, Johnston RN, Shaw C, Gustafsson MKS (1999) Relationship between musculature and nervous system in the regenerating pharynx in *Girardia tigrina* (Plathelminthes). Invertebr Reprod Dev 35:109–125

Kumar A, Brockes JP (2012) Nerve dependence in tissue, organ, and appendage regeneration. Trends Neurosci 35:691–699

Kumar A, Godwin JW, Gates PB, Garza-Garcia AA, Brockes JP (2007) Molecular basis for the nerve dependence of limb regeneration in an adult vertebrate. Science 318:772–777

Labbé RM, Irimia M, Currie KW, Lin A, Zhu SJ, Brown DD, Ross EJ, Voisin V, Bader GD, Blencowe BJ, Pearson BJ (2012) A comparative transcriptomic analysis reveals conserved features of stem cell pluripotency in planarians and mammals. Stem Cells 30:1734–1745

Ladurner P, Mair GR, Reiter D, Salvenmoser W, Rieger RM (1997) Serotonergic nervous system of two macrostomid species: recent or ancient divergence? Invert Biol 116:178–191

Ladurner P, Rieger R, Baguñà J (2000) Spatial distribution and differentiation potential of stem cells in hatchlings and adults in the marine platyhelminth *Macrostomum sp*: a bromodeoxyuridine analysis. Dev Biol 226:231–241

Lapan SW, Reddien PW (2011) *dlx* and *sp6-9* control optic cup regeneration in a prototypic eye. PLoS Genet 7:e1002226

Lapan SW, Reddien PW (2012) Transcriptome analysis of the planarian eye identifies *ovo* as a specific regulator of eye regeneration. Cell Rep 2:294–307

Laumer CE, Giribet G (2014) Inclusive taxon sampling suggest a single, stepwise origin of ectolecithality in Platyhelminthes. Biol J Linn Soc 111:570–588

Leksomboon R, Chaijaroonkhanarak W, Arunyanart C, Umka J, Jones MK, Sripa B (2012) Organization of the nervous system in *Opisthorchis viverrini* investigated by histochemical and immunohistochemical study. Parasitol Int 61:107–111

Lender T (1955) Some properties of the organisine of eye regeneration in the planaria *Polycelis nigra*. C R Heabd Seances Acad Sci 240:1726–1728

Lewis J, Slack JM, Wolpert L (1977) Thresholds in development. J Theor Biol 65:579–590

Li VS, Clevers H (2013) Intestinal regeneration: YAP-tumor suppressor and oncoprotein? Curr Biol 23:R110–R112

Lin AY, Pearson BJ (2014) Planarian *yorkie/YAP* functions to integrate adult stem cell proliferation, organ homeostasis and maintenance of axial patterning. Development 141:1197–1208

Lindholm AM, Reuter M, Gustafsson MKS (1998) The NADPH-diaphorase staining reaction in relation to the aminergic and peptidergic nervous system and the musculature of adult *Diphyllobothrium dendriticum*. Parasitology 117:283–292

Liu SY, Selck C, Friedrich B, Lutz R, Vila-Farré M, Dahl A, Brandl H, Lakshmanaperumal N, Henry I, Rink JC (2013) Reactivating head regrowth in a regeneration-deficient planarian species. Nature 500:81–84

Logan CY, Miller JR, Ferkowicz MJ, McClay DR (1999) Nuclear beta-catenin is required to specify vegetal cell fates in the sea urchin embryo. Development 126:345–357

Mackay DR, Hu M, Li B et al (2006) The mouse *Ovol2* gene is required for cranial neural tube development. Dev Biol 291:38–52

MacRae EK (1964) Observations on the fine structure of photoreceptor cells in the planarian *Dugesia tigrina*. J Ultrastruct Res 10:334–349

Mannini L, Rossi L, Deri P, Gremigni V, Salvetti A, Salo E, Batistoni R (2004) *Djeyes absent* (*Djeya*) controls prototypic planarian eye regeneration by cooperating with the transcription factor *Djsix-1*. Dev Biol 269:346–359

Marsal M, Pineda D, Saló E (2003) *Gtwnt-5* a member of the wnt family expressed in a subpopulation of the nervous system of the planarian *Girardia tigrina*. Gene Expr Patterns 3:489–495

Martín-Durán JM, Romero R (2011) Evolutionary implications of morphogenesis and molecular patterning of the blind gut in the planarian *Schmidtea polychroa*. Dev Biol 352:164–176

Martín-Durán JM, Amaya E, Romero R (2010) Germ layer specification and axial patterning in the embryonic development of the freshwater planarian *Schmidtea polychroa*. Dev Biol 340:145–158

Martín-Durán JM, Monjo F, Romero R (2012) Morphological and molecular development of the eyes during embryogenesis of the freshwater planarian *Schmidtea polychroa*. Dev Genes Evol 222:45–54

März M, Seebeck F, Bartscherer K (2013) A *Pitx* transcription factor controls the establishment and maintenance of the serotonergic lineage in planarians. Development 140:4499–4509

Maule AG, Shaw C, Halton DW, Brennan GP, Johnston CF, Moore S (1992) Neuropeptide F (*Moniezia expansa*): localization and characterization using specific antisera. Parasitology 105:505–512

McVeigh P, Mair GR, Novozhilova E, Day A, Zamanian M, Marks NJ, KImber MJ, Day TA, Maule AG (2011) Schistosome I/Lamides—a new family of bioactive helminth neuropeptides. Int J Parasitol 41:905–913

Meinhardt H (1978) Space-dependent cell determination under the control of morphogen gradient. J Theor Biol 74:307–321

Merchant MT, Corella C, Willms K (1997) Autoradiographic analysis of the germinative tissue in evaginated *Taenia solium* metacestodes. J Parasitol 83:363–367

Miljkovic-Licina M, Chera S, Ghila L, Galliot B (2007) Head regeneration in wild-type hydra requires de novo neurogenesis. Development 134:1191–1201

Mineta K, Nakazawa M, Cebrià F, Ikeo K, Agata K, Gojobori T (2003) Origin and evolutionary process of the CNS elucidated by comparative genomics analysis of planarian ESZTs. Proc Natl Acad Sci U S A 100:7666–7671

Molina MD, Saló E, Cebrià F (2007) The BMP pathway is essential for re-specification and maintenance of the dorsoventral axis in regenerating and intact planarians. Dev Biol 311:79–94

Molina MD, Saló E, Cebrià F (2009) Expression pattern of the expanded *noggin* gene family in the planarian *Schmidtea mediterranea*. Gene Expr Patterns 9:246–253

Molina MD, Saló M, Cebrià F (2011a) Organizing the DV axis during planarian regeneration. Commun Integr Biol 4:498–500

Molina MD, Neto A, Maeso I, Gómez-Skarmeta JL, Saló E, Cebrià F (2011b) *Noggin* and *noggin*-like genes control dorsoventral axis regeneration in planarians. Curr Biol 21:300–305

Moraczewski J (1977) Asexual reproduction and regeneration of Catenula (Turbellaria, Archoophora). Zoomorphology 88:65–80

Morgan TH (1898) Experimental studies of the regeneration of *Planaria maculata*. Arch Entwickelungsmech Org 7:364–397

Morgan TH (1900) Regeneration in planarians. Arch Entwicklungsmech Org 10:58–119

Morgan TH (1901) Regeneration. Macmillan, New York

Morgan TH (1904) Polarity and axial heteromorphosis. Am Nat 38:502–505

Morgan TH (1905) "Polarity" considered as a phenomenon of gradation of materials. J Exp Zool 2:495–506

Morris J, Ladurner P, Rieger R, Pfister D, Del Mar De Miguel-Bonet M, Jacobs D, Hartenstein V (2006) The *Macrostomum lignano* EST database as a molecular resource for studying platyhelminth development and phylogeny. Dev Genes Evol 216:695–707

Nakazawa M, Cebria F, Mineta K, Ikeo K, Agata K, Gojobori T (2003) Search for the evolutionary origin of a brain: planarian brain characterized by microarray. Mol Biol Evol 20:784–791

Newmark PA, Sánchez-Alvarado A (2000) Bromodeoxyuridine specifically labels the regenerative stem cells of planarians. Dev Biol 220:142–153

Newmark PA, Sánchez-Alvarado A (2002) Not your father's planarian: a classic model enters the era of functional genomics. Nat Rev Genet 3:210–219

Nimeth KT, Egger B, Rieger R, Salvenmoser W, Peter R, Gschwentner R (2007) Regeneration in *Macrostomum lignano* (Platyhelminthes): cellular dynamics in the neoblast stem cell system. Cell Tissue Res 327:637–646

Nishimura K, Kitamura Y, Inoue T, Umesono Y, Yoshimoto K, Takeuchi K, Taniguchi T, Agata K (2007a) Identification and distribution of *tryptophan hydroxylase* (*TPH*)-positive neurons in the planarian *Dugesia japonica*. Neurosci Res 59:101–106

Nishimura K, Kitamura Y, Inoue T, Umesono Y, Sano S, Yoshimoto K, Inden M, Takata K, Taniguchi T, Shimohama S, Agata K (2007b) Reconstruction of dopaminergic neural network and locomotion function in planarian regenerates. Dev Neurobiol 67: 1059–1078

Nishimura K, Kitamura Y, Inoue T, Umesono Y, Yoshimoto K, Taniguchi T, Agata K (2008a) Characterization of *tyramine beta-hydroxylase* in planarian *Dugesia japonica*: cloning and expression. Neurochem Int 53:184–192

Nishimura K, Kitamura Y, Umesono Y, Takeuchi K, Takata K, Taniguchi T, Agata K (2008b) Identification of *glutamic acid decarboxylase* gene and distribution of GABAergic nervous system in the planarian *Dugesia japonica*. Neuroscience 153:1103–1114

Nogi T, Levin M (2005) Characterization of innexin gene expression and functional roles of gap-junctional communication in planarian regeneration. Dev Biol 287:314–335

O'Donnell M, Chance RK, Bashwa GJ (2009) Axon growth and guidance: receptor regulation and signal transduction. Ann Rev Neurosci 32:383–412

Ogawa K, Ishihara S, Saito Y, Mineta K, Nakazawa M, Ikeo K, Gojobori T, Watanbe K, Agata K (2002a) Induction of a *noggin*-like gene by ectopic DV interaction during planarian regeneration. Dev Biol 250:59–70

Ogawa K, Kobayashi C, Hayashi T, Orii H, Watanabe K, Agata K (2002b) Planarian *fibroblasts growth factor receptor* homologs expressed in stem cells and cephalic ganglions. Dev Growth Differ 44:191–204

Onal P, Grün D, Adamidi C, Rybak A, Solana J, Mastrobuoni G, Wang Y, Rahn HP, Chen W, Kempa S, Ziebold U, Rajewsky N (2012) Gene expression of pluripotency determinants is conserved between mammalian and planarian stem cells. EMBO J 31:2755–2769

Orii H, Watanabe K (2007) *Bone morphogenetic protein* is required for dorso-ventral patterning in the planarian *Dugesia japonica*. Dev Growth Differ 49:345–349

Orii H, Katayama T, Sakurai T, Agata K, Watanabe K (1998) Immunohistochemical detection of opsins in turbellarians. Hydrobiologia 383:183–187

Oviedo NJ, Morokuma J, Walentek P, Kema IP, Gu MB, Ahn JM, Hwang JS, Gojobori T, Levin M (2010) Long-range neural and gap junction protein-mediated cues control polarity during planarian regeneration. Dev Biol 339:188–199

Pallas PS (1774) Spicilegia zoological quibus novae imprimis et obscurae animaliu. Speciosiconibus atque conamentariis illustrator. Fasc. X, Berolini

Palmberg I (1986) Cell migration and differentiation during wound healing and regeneration in *Microstomum lineare* (Turbellaria). Hydrobiologia 132:181–188

Pan JZ, Halton DW, Shaw C, Maule AG, Johnston CF (1994) Serotonin and neuropeptide immunoreactivities in the intramolluscan stages of three marine trematode parasites. Parasitol Res 80:388–395

Pedersen KJ (1976) Scanning electron microscopical observations on epidermal wound healing in the planarian *Dugesia tigrina*. Wilhelm Rouxs Arch Dev Biol 179:251–273

Petersen CP, Reddien PW (2008) *Smed-bcatenin-1* is required for anteroposterior blastema polarity in planarian regeneration. Science 319:327–330

Petersen CP, Reddien PW (2009) A wound-induced Wnt expression program controls planarian regeneration polarity. Proc Natl Acad Sci U S A 106:17061–17066

Petersen CP, Reddien PW (2011) Polarized *notum* activation at wounds inhibits *Wnt* function to promote planarian head regeneration. Science 332:852–855

Peterson KJ, Eernisse DJ (2001) Animal phylogeny and the ancestry of bilaterians: inferences from morphology and 18S rDNA gene sequences. Evol Dev 3:170–205

Peterson KJ, Cotton JA, Gehling JG, Pisani D (2008) The Ediacaran emergence of bilaterians: congruence between the genetic and the geological fossil records. Phil Trans Soc B 363:1435–1443

Philippe H, Brinkmann H, Copley RR, Moroz LL, Nakano H, Poustka AJ, Wallberg A, Peterson KJ, Telford MJ (2011) Acoelomorph flatworms are deuterostomes related to *Xenoturbella*. Nature 470:255–258

Pichaud F, Treisman J, Desplan C (2001) Reinventing a common strategy for patterning the eye. Cell 105:9–12

Pineda D, Saló E (2002) Planarian *Gtsix3*, a member of the *Six/so* gene family, is expressed in brain branches but not in eye cells. Mech Dev 119(suppl1):S161–S171

Pineda D, Gonzalez J, Callaerts P, Ikeo K, Gehring WJ, Saló E (2000) Searching for the prototypic eye genetic network: *Sine oculis* is essential for eye regeneration in planarians. Proc Natl Acad Sci U S A 97:4525–4529

Pineda D, Gonzalez J, Marsal M, Saló E (2001) Evolutionary conservation of the initial eye genetic pathway in planarians. Belg J Zool 131:77–82

Pineda D, Rossi L, Batistoni R, Salvetti A, Marsal M, Gremigni V, Falleni A, Gonzalez-Linares J, Deri P, Saló E (2002) The genetic network of prototypic planarian eye regeneration is *Pax6* independent. Development 129:1423–1434

Poss KD (2010) Advances in understanding tissue regenerative capacity and mechanisms in animals. Nat Rev Genet 11:710–722

Randolph H (1892) The regeneration of the tail in *Lumbriculus*. J Morphol 7:317–344

Randolph H (1897) Observations and experiments on regeneration in planarians. Arch EntwMech Org 5:352–372

Rawlinson KA (2010) Embryonic and post-embryonic development of the polyclad flatworm Maritigrella crozieri; implications for the evolution of spiralian life history traits. Front Zool 7:12

Reddien PW (2011) Constitutive gene expression and the specification of tissue identity in adult planarian biology. Trends Genet 27:277–285

Reddien PW (2013) Specialized progenitors and regeneration. Development 140:951–957

Reddien PW, Oviedo NJ, Jennings JR, Jenkin JC, Sánchez Alvarado A (2005) SMEDWI-2 is a PIWI-like protein that regulates planarian stem cells. Science 310:1327–1330

Reddien PW, Bermange AL, Kicza AM, Sanchez Alvarado A (2007) BMP signaling regulates the dorsal planarian midline and is needed for asymmetric regeneration. Development 134:4043–4051

Reginensi A, Scott RP, Gregorieff A, Bagherie-Lachidan M, Chung C, Lim DS, Pawson T, Wrana J, McNeill H (2013) *Yap-* and *Cdc42*-dependent nephrogenesis and morphogenesis during mouse kidney development. PLoS Genet 9:e1003380

Reisinger E (1972) Die Evolution des Orthogons der Spiralier und das Archicoelomatenproblem. Z Zool Syst Evolutionforsch 10:1–43

Reiter D, Wikgren M (1991) Immunoreactivity to a specific echinoderm neuropeptide in the nervous system of the flatworm *Macrostomum hystricinum marinum* (Turbellaria, Macrostomida). Hydrobiologia 227:229

Reuter M (1988) Development and organization of nervous system visualized by immunocytochemistry in three flatworm species. Fortschr Zool 36:181–184

Reuter M (1994) Substance P immunoreactivity in sensory structures and the central and pharyngeal nervous system of *Stenostomum leucops* (Catenulida) and *Microstomum lineare* (Macrostomida). Cell Tissue Res 276:173–180

Reuter M, Gustafsson MKS (1995) The flatworm nervous system: pattern and phylogeny. In: Breidbach O, Kutsch W (eds) The nervous systems of invertebrates: an evolutionary and comparative approach. Birkhäuser Verlag, Basel, pp 25–59

Reuter M, Gustafsson MKS (1996) Neuronal signal substances in asexual multiplication and development in flatworms. Cell Mol Neurobiol 16:591–616

Reuter M, Halton DW (2001) Comparative neurobiology of Platyhelminthes. In: Littlewood TJ, Bray RA (eds) The interrelationships of Platyhelminthes. Academic, London, pp 239–259

Reuter M, Kreshchenko N (2004) Flatworm asexual multiplication implicates stem cells and regeneration. Can J Zool 82:334–356

Reuter M, Palmberg I (1989) Development and differentiation of neuronal subsets in asexually reproducing *Microstomum lineare*. Immunocytochemistry of 5-HT, RF-amide and SCPv. Histochemistry 91:123–131

Reuter M, Wikgren M, Lehtonen M (1986) Immunocytochemical demonstration of 5-HT-like and FMRF-amide-like substances in whole mounts of *Microstomum lineare* (Turbellaria). Cell Tissue Res 246:7–12

Reuter M, Gustafsson MKS, Sheiman IM, Terenina N, Halton DW, Maule AG, Shaw C (1995a) The nervous system of Tricladida. II. Neuroanatomy of *Dugesia tigrina* (Paludicola, Dugesiidae): an immunocytochemical study. Invertebr Neurosci 1:133–143

Reuter M, Gustafsson MKS, Sahlgren C, Halton DW, Maule AG, Shaw C (1995b) The nervous system of Tricladida. I. Neuroanatomy of *Procerodes littoralis* (Maricola, Procerodidae): an immunocytochemical study. Invertebr Neurosci 1:113–122

Reuter M, Maule AG, Halton DW, Gustafsson MS, Shaw C (1995c) The organization of the nervous system in Plathelminthes. The neuropeptide F-immunoreactivity pattern in Catenulida, Macrostomida, Proseriata. Zoomorphology 115:83–97

Reuter M, Gustafsson MKS, Mäntylä K, Grimmelikhuijzen CJP (1996a) The nervous system of Tricladida. III. Neuroanatomy of *Dendrocoelum lacteum* and *Polycelis tenuis* (Plathelminthes, Paludicola): an immunocytochemical study. Zoomorphology 116:111–122

Reuter M, Sheiman IM, Gustafsson MKS, Halton DW, Maule AG, Shaw C (1996b) Development of the nervous system in *Dugesia tigrina* during regeneration after fission and decapitation. Invertebr Reprod Dev 29:199–211

Reuter M, Raikova OI, Gustafsson MKS (2001) Patterns in the nervous and muscle systems in lower flatworms. Belg J Zool 131:47–53

Reversade B, De Robertis EM (2005) Regulation of *ADMP* and *BMP2/4/7* at opposite embryonic poles generates a self-regulating morphogenetic field. Cell 123:1147–1160

Ribeiro P, El-Shehabi F, Patocka N (2005) Classical transmitters and their receptors in flatworms. Parasitology 131:S19–S40

Rieger R, Tyler S, Smith JPS III, Rieger G (1991) Platyhelminthes: Turbellaria. In: Harrison FW, Bogitsh BJ (eds) Microscopic anatomy of invertebrates, vol 3. Wiley-Liss, New York, pp 7–140

Rink J (2013) Stem cell systems and regeneration in planaria. Dev Genes Evol 223:67–84

Riutort M, Álvarez-Presas M, Lázaro E, Solà E, Paps J (2012) Evolutionary history of the Tricladida and the Platyhelminthes: an up-to-date phylogenetic and systematic account. Int J Dev Biol 56:5–17

Rock JM, Lim D, Stach L, Ogrodowicz RW, Keck JM, Jones MH, Wong CC, Yates JR 3rd, Winey M, Smerdon SJ, Yaffe MB, Amon A (2013) Activation of the yeast Hippo pathway by phosphorylation-dependent assembly of signaling complexes. Science 340:871–875

Rossi L, Batistoni R, Salvetti A, Deri P, Bernini F, Andreoli I, Falleni A, Gremigni V (2001) Molecular aspects of cell proliferation and neurogenesis in planarians. Belg J Zool 131:83–87

Rossi L, Deri P, Andreoli I, Gremigni V, Salvetti A, Batistoni R (2003) Expression of *DjXnp*, a novel member of the *SNF2*-like ATP-dependent chromatin remodelling genes, in intact and regenerating planarians. Int J Dev Biol 47:293–298

Rossi L, Salvetti A, Marincola FM, Lena A, Deri P, Mannini L, Batistoni R, Wang E, Gremigni V (2007) Deciphering the molecular machinery of stem cells: a look at the neoblast gene expression profile. Genome Biol 8:R62

Rouhana L, Shibata N, Nishimura O, Agata K (2010) Different requirements for conserved post-transcriptional regulators in planarian regeneration and stem cell maintenance. Dev Biol 341:429–443

Ruiz-Trillo I, Riutort M, Littlewood DTJ, Herniou EA, Baguñà J (1999) Acoel flatworms: earliest extant bilaterian metazoans, not members of Platyhelminthes. Science 283:1919–1923

Ruppert EE, Schreiner SP (1980) Ultrastructure and potential significance of cerebral light refracting bodies of *Stenostomum virginianum* (Turbellaria, Catenulida). Zoomorphology 96:21–31

Saló E (2006) The power of regeneration and the stem-cell kingdom: freshwater planarians (Platyhelminthes). Bioessays 28:546–559

Saló E, Baguñà J (1984) Regeneration and pattern formation in planarians. I. The pattern of mitosis in anterior and posterior regeneration in *Dugesia* (*G*) *tigrina*, and a new proposal for blastema formation. J Embryol Exp Morphol 83:63–80

Saló E, Batistoni R (2008) Planarian eye, a simple and plastic system with great regenerative capacity. In: Tsonis PA (ed) Animal models in eye research. Elsevier, Amsterdam

Saló E, Pineda D, Marsal M, González J, Gremigni V, Batistoni R (2002) Genetic network of the eye in Platyhelminthes: expression and functional analysis of some players during planarian regeneration. Gene 287:67–74

Salvini-Plawen LV, Mayr E (1977) On the evolution of photoreceptors and eyes. Evol Biol 10:207–263

Sánchez Alvarado A, Newmark PA (1999) Double-stranded RNA specifically disrupts gene expression during planarian regeneration. Proc Natl Acad Sci U S A 96:5049–5054

Schneider SQ, Finnerty JR, Martindale MQ (2003) Protein evolution: structure-function relationships of the oncogene beta-catenin in the evolution of multicellular animals. J Exp Zool B Mol Dev Evol 295:25–44

Scimone ML, Srivastava M, Bell GW, Reddien PW (2011) A regulatory program for excretory system regeneration in planarians. Development 38:4387–4398

Sebé-Pedrós A, de Mendoza A, Lang BF, Degnan BM, Ruiz-Trillo I (2011) Unexpected repertoire of metazoan transcription factors in the unicellular holozoan *Capsaspora owczarzaki*. Mol Biol Evol 28: 1241–1254

Semmler H, Chiodin M, Bailly X, Martinez P, Wanninger A (2010) Steps towards a centralized nervous system in basal bilaterians: insights from neurogenesis of the acoel *Symsagittifera roscoffensis*. Dev Growth Differ 52:701–713

Senft AW, Weller TH (1956) Growth and regeneration of *Schistosoma mansoni* in vitro. Proc Soc Exp Biol Med 93:16–19

Sheiman IM, Kreshchenko ND, Sedelnikov ZV, Groznyi AV (2004) Morphogenesis in planarians *Dugesia tigrina*. Ontogenez 35:285–290

Sickes JM, Newmark PA (2013) Restoration of anterior regeneration in a planarian with limited regenerative ability. Nature 500:77–80

Singer M, Craven L (1948) The growth and morphogenesis of the regenerating forelimb of adult *Triturus* following denervation at various stages of development. J Exp Zool 108:279–308

Sivickis PB (1930) A quantitative study of regeneration along the main axis of the triclad body. Arch Zool Ital 16:430–449

Solana J (2013) Closing the circle of germline and stem cells: the primordial stem cell hypothesis. Evodevo 4:2

Sopott-Ehlers B (1982) Ultrastruktur potentiell photorezeptorischer Zellen unterschiedlicher Organisation bei einem Proseriat (Platyhelminthes). Zoomorphology 101:165–176

Sopott-Ehlers B, Ehlers U (2003) Eyes covered by mitochondrial lenses in *Petaliella spiracauda* and *Ptychopera purasjokii* (Plathelminthes, Rhabdocoela, Trigonostominae). Ultrastructural features and phylogenetic implications. J Submicrosc Cytol Pathol 35:415–421

Sopott-Ehlers B, Kearn GC, Ehlers U (2001) Evidence for the mitochondrial origin of the eye lenses in embryos of *Entobdella soleae* (Plathelminthes, Monogenea). Parasitol Res 87:421–427

Sperry PJ, Ansevin KD, Tittel FK (1973) The inductive role of the nerve cord in regeneration of isolated postpharyngeal body sections of *Dugesia dorotocephala*. J Exp Zool 186:159–174

Spiliotis M, Lechner S, Tappe D, Scheller C, Krohne G, Brehm K (2008) Transient transfection of *Echinococcus multilocularis* primary cells and complete in vitro regeneration of metacestode vesicles. Int J Parasitol 38:1025–1039

Srivastava M, Mazza-Curll KL, van Wolfswinkel JC, Reddien PW (2014) Whole-body acoel regeneration

is controlled by Wnt and Bmp-Admp signalling. Curr Biol 24:1107–1113

Stéphan-Dubois F, Lender Th (1956) Corrélation humorales dans le regeneration des planaires paludicoles. Ann Sci Nat Zool 11 ser

Tamamaki N (1990) Evidence for the phagocytotic removal of photoreceptive membrane by pigment cells in the eye of the planarian, *Dugesia japonica*. Zool Sci 7:385–393

Tanaka EM, Reddien PW (2011) The cellular basis for animal regeneration. Dev Cell 21:172–185

Teshirogi W, Ishida S, Yamazaki H (1977) Regenerative capacities of transverse pieces in the two species of freshwater planarian, *Dendrocoelopsis lactea* and *Polycelis sapporo*. Sci Rep Hirosaki Univ 24: 55–72

Umesono Y, Watanabe K, Agata K (1999) Distinct structural domains in the planarian brain defined by the expression of evolutionarily conserved homeobox genes. Dev Genes Evol 209:31–39

Umesono Y, Tasaki J, Nishimura K, Inoue T, Agata K (2011) Regeneration in an evolutionarily primitive brain—the planarian *Dugesia japonica* model. Eur J Neurosci 34:863–869

Umesono Y, Tasaki J, Nishimura Y, Hrouda M, Kawaguchi E, Yazawa S, Nishimura O, Hosoda K, Inoue T, Agata K (2013) The molecular logic for planarian regeneration along the anterior-posterior axis. Nature 500:73–76

Verdoodt F, Bert W, Couvreur M, De Mulder K, Willems M (2012) Proliferative response of the stem cell system during regeneration of the rostrum in *Macrostomum lignano* (Platyhelminthes). Cell Tissue Res 347:397–406

Vorontsova MA, Liosner LD (1960) Asexual propagation and regeneration. Pergamon Press, London

Wagner DE, Wang IE, Reddien PW (2011) Clonogenic neoblasts are pluripotent adult stem cells that underlie planarian regeneration. Science 332:811–816

Wagner DE, Ho JJ, Reddien PW (2012) Genetic regulators of a pluripotent adult stem cell system in planarians identified by RNAi and clonal analysis. Cell Stem Cell 10:299–311

Wang B, Collins JJ 3rd, Newmark PA (2013) Functional characterization of neoblast-like stem cells in larval *Schistosoma mansoni*. Elife 2:e00768

Wenemoser D, Reddien PW (2010) Planarian regeneration involves distinct stem cell responses to wounds and tissue absence. Dev Biol 344:979–991

Wikramanayake AH, Hong M, Lee PN, Pang K, Byrum CA, Bince JM, Xu R, Martindale MQ (2003) An ancient role for nuclear beta-catenin in the evolution of axial polarity and germ layer segregation. Nature 426:446–450

Wolff E, Lender T (1950) Sur le role organisateur du cerveau dans la regeneration des yeux chez une planaire d'eau douce. C R Acad Sci 230:2238–2239

Xin M, Kim Y, Sutherland LB, Murakami M, Qi X, McAnally J, Porrello ER, Mahmoud AI, Tan W,

Shelton JM, Richardson JA, Sadek HA, Bassel-Duby R, Olson EN (2013) Hippo pathway effector Yap promotes cardiac regeneration. Proc Natl Acad Sci U S A 110:13839–13844

Yazawa S, Umesono Y, Hayashi T, Tarui H, Agata K (2009) Planarian *Hedgehog/Patched* establishes anterior-posterior polarity by regulating Wnt signaling. Proc Natl Acad Sci U S A 106:22329–22334

Yoshikawa S, McKinnon RD, Kokel M, Thomas JB (2003) Wnt-mediated axon guidance via the *Drosophila* Derailed receptor. Nature 422:583–588

Younossi-Hartenstein A, Hartenstein V (2000) The embryonic development of the polyclad flatworm *Imogine mcgrathi*. Dev Genes Evol 210:383–398

Younossi-Hartenstein A, Ehlers U, Hartenstein V (2000) Embryonic development of the nervous system of the rhabdocoel flatworm *Mesostoma lingua* (Abilgaard 1789). J Comp Neurol 16:461–474

Ypsilanti AR, Zagar Y, Chédotal A (2010) Moving away from the midline: new developments for *Slit* and *Robo*. Development 137:1939–1952

Cycliophora

5

Andreas Wanninger and Ricardo Neves

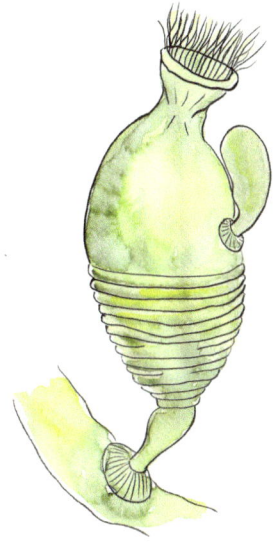

Chapter vignette artwork by Brigitte Baldrian.
© Brigitte Baldrian and Andreas Wanninger

A. Wanninger (✉)
Department of Integrative Zoology, University
of Vienna, Althanstrasse 14, Vienna 1090, Austria
e-mail: andreas.wanninger@univie.ac.at

R. Neves
Biozentrum – Molecular Zoology, University of Basel,
Klingelbergstrasse 50, Basel 4056, Switzerland

A. Wanninger (ed.), *Evolutionary Developmental Biology of Invertebrates 2: Lophotrochozoa (Spiralia)*
DOI 10.1007/978-3-7091-1871-9_5, © Springer-Verlag Wien 2015

INTRODUCTION

Cycliophora is an acoelomate, monogeneric phylum that includes two described species, *Symbion pandora* and *Symbion americanus*, but population genetics and phylogeographic studies suggest that there are additional species awaiting formal description (Nedvěd 2004; Obst et al. 2005, 2006; Baker et al. 2007; Baker and Giribet 2007). Cycliophorans are found epizoically on setae of the mouthparts of lobsters, including the Norway lobster, *Nephrops norvegicus*, the European lobster, *Homarus gammarus*, and the American lobster, *Homarus americanus*. The minute animals are characterized by a highly complex metagenetic life cycle that includes an asexual, polyp-like feeding stage that is firmly attached to the host lobster (Figs. 5.1 and 5.2A).

The feeding stage is also the easiest recognizable life cycle stage. Its three-partite body is covered by a thick cuticle and is subdivided into a buccal funnel that is surrounded by compound cilia and houses the mouth and esophagus, an oval-shaped trunk with all internal organs, and a short, acellular stalk with attachment disc (Funch and Kristensen 1995). It is the only stage in the cycliophoran life cycle known to feed, and its buccal funnel and gut are recurrently regenerated and replaced by the next generation of the internally preformed corresponding structures. As a consequence, older feeding stages have a series of cuticular wrinkles (scars) on their trunk (Fig. 5.2A). The gut of the feeding stage is U-shaped, with an anal opening located close to the transition zone between the trunk and the buccal funnel (Funch and Kristensen 1995). Feeding stage individuals give rise to a number of other important stages in the cycliophoran life cycle by internal clonal reproduction. These include the so-called Pandora and Prometheus larva and the sexual female (Fig. 5.2B, C, E). These stages, as well as the male that develops from an encysted Prometheus larva and the sexually produced chordoid larva, are free-swimming for at least some time and bear characteristic ciliary fields, usually in anteroventral and posterior position (Fig. 5.2B–D). The chordoid larva is probably the longest-lived cycliophoran larval

stage and colonizes a new host (Fig. 5.2G). It has been interpreted as a derived trochophore-like larva, although characteristic features such as an apical organ or a distinct prototroch are lacking (see Chapters 6, 7, and 9; Funch and Kristensen 1995).

In the original description of the first cycliophoran species, *Symbion pandora*, Entoprocta and Ectoprocta had been considered its closest extant relatives, mainly because all three phyla show clonal reproduction by budding (cf, Chapters 6 and 11; Funch and Kristensen 1995). This assumption has received support by some molecular phylogenetic studies that have united Ectoprocta, Entoprocta, and Cycliophora in a monophyletic clade, the Polyzoa (Hejnol et al. 2009; Paps et al. 2009). Other analyses, however, have proposed close rotifer/syndermate affinities of Cycliophora (Winnepenninckx et al. 1998; Giribet et al. 2000). Since cycliophoran nucleotide sequence data have so far only been rarely included in large-scale phylogenomic studies, the definite position of Cycliophora within the bilaterian tree remains contentious.

LIFE CYCLE STAGES AND DEVELOPMENT

Cycliophorans exhibit one of the most complex animal life cycles known to date (Fig. 5.1), which is characterized by an alternation of asexual and sexual reproductive cycles (metagenesis) (Funch and Kristensen 1999). With a size of 300–400 µm the so-called feeding stage is the largest and most prominent stage in the life cycle. It has a simple muscular body plan that includes a myoepithelial ring musculature surrounding the mouth, an anal sphincter, and a few longitudinal muscles in the buccal funnel and trunk (Fig. 5.3). Circular body wall muscles are absent, and the stalk and attachment disc are devoid of musculature altogether (Neves et al. 2009a, 2010a).

The neuroanatomy of the feeding stage is poorly known. Two anterior ganglia, one around the esophagus and one in the buccal cavity, were described (Funch and Kristensen 1997), but immunocytochemical studies using a suite of commercially

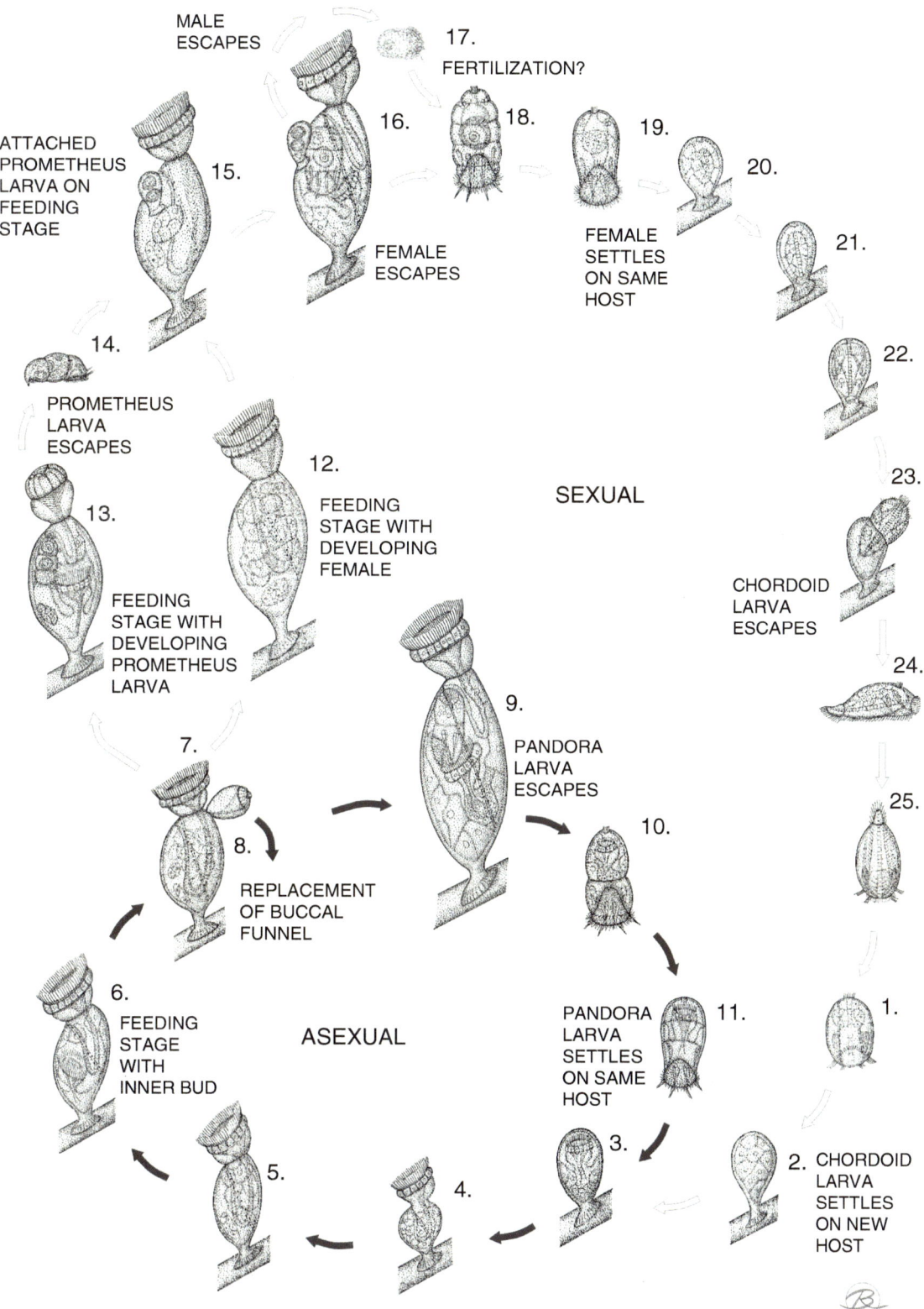

Fig. 5.1 The hypothetical life cycle of *Symbion pandora* (From Neves et al. (2012), modified after Obst and Funch (2003))

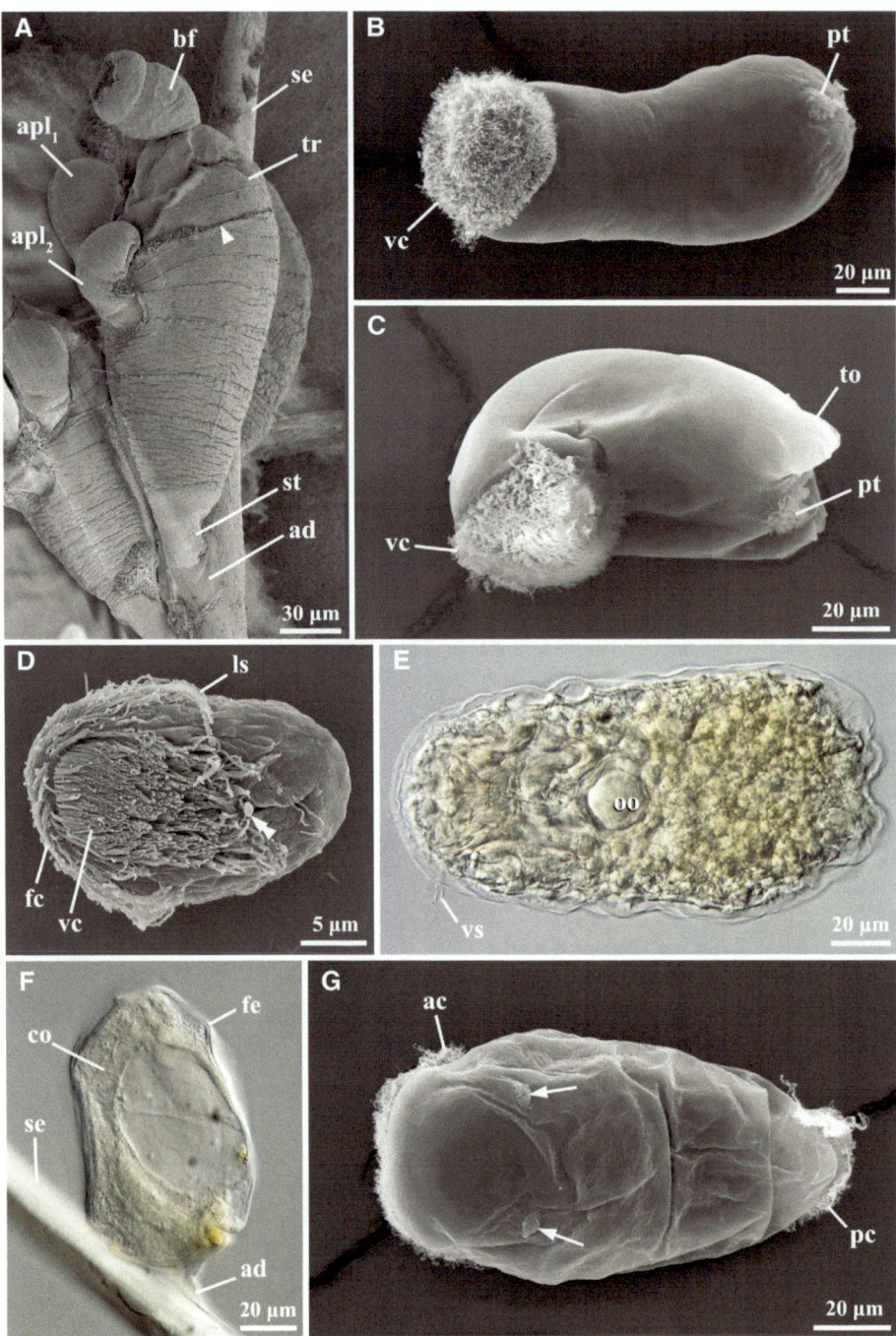

Fig. 5.2 Representative life cycle stages of the cycliophoran species *Symbion pandora* (**A**, **D**–**G**) and *S. americanus* (**B**, **C**). Scanning electron (**A**–**D**, **G**) and light micrographs (**E**, **F**). Anterior faces left in all aspects except for **A**, in which the buccal funnel faces up. (**A**) Feeding stage with two Prometheus larvae attached (*apl1*–*2*). Note the wrinkle in the cuticle of the trunk (*arrowhead*). (**B**) Pandora larva with anterior ciliated field and posterior ciliated tuft; ventral view. (**C**) Prometheus larva with a posterior pair of toes (*to*); lateral view. Note that the toes are absent in the attached Prometheus larvae of *S.* *pandora* shown in **A**. (**D**) Dwarf male with penis (*double arrowhead*) located inside a pouch-like structure; ventral view. (**E**) Free-living female with large oocyte (*oo*). (**F**) Chordoid cyst enclosing a chordoid larva. Note the chordoid organ (*co*). (**G**) Chordoid larva with paired dorsal ciliated organs (*arrows*). Abbreviations: *ac* anterior ciliated field, *ad* attachment disc, *bf* buccal funnel, *fc* frontal ciliated field, *fe* female exuvium, *ls* lateral sensorial organ, *pc* posterior ciliated field, *pt* posterior ciliated tuft, *se* seta of the mouthparts of the host lobster, *st* stalk, *tr* trunk, *vs* ventral sensory organ (After Neves et al. (2010a, b; 2012))

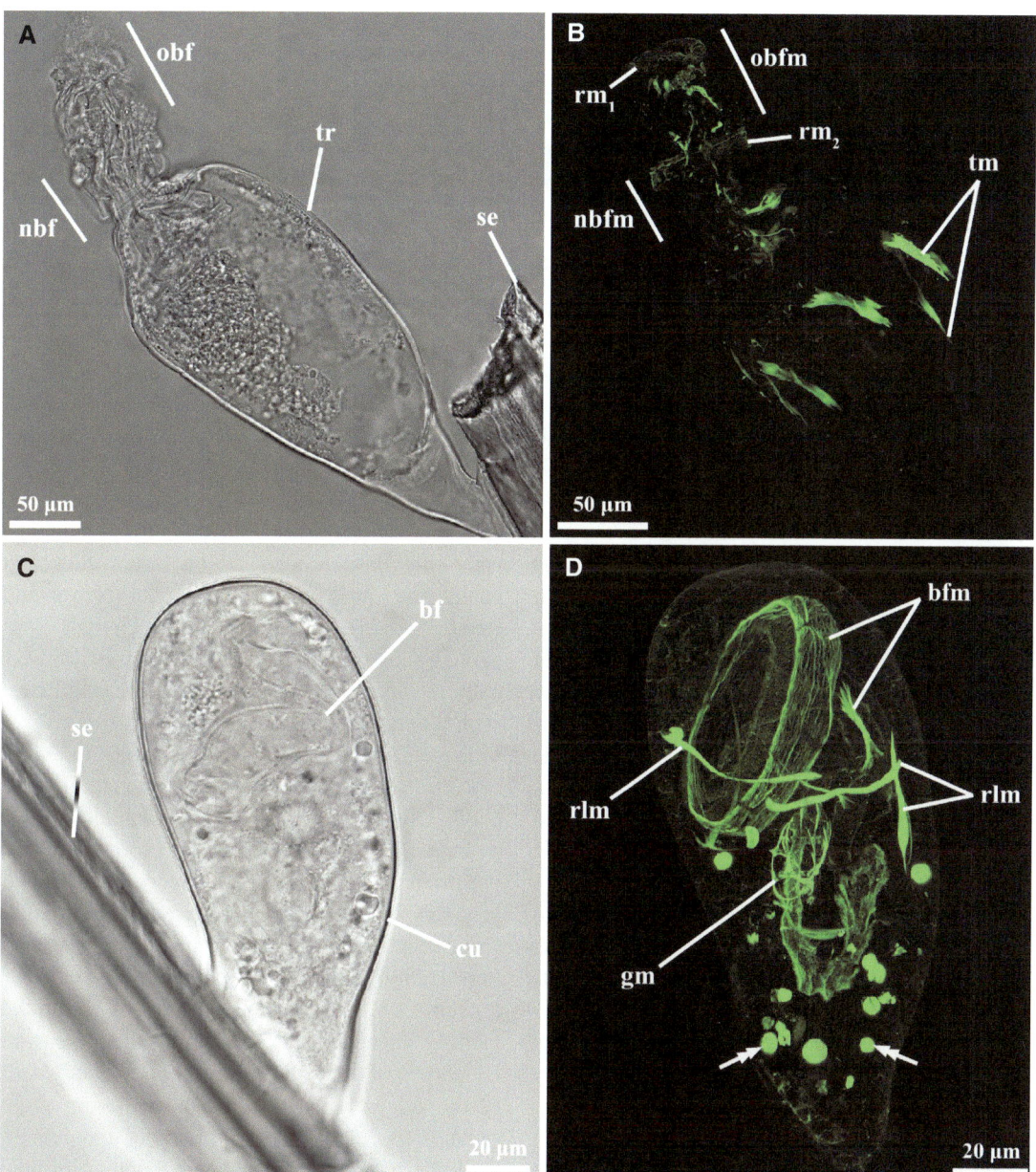

Fig. 5.3 Feeding stage individual of *Symbion* sp. and settled Pandora larva with developing feeding stage inside. Distal is up in all aspects. (**A, C**) are light micrographs, (**B, D**) show musculature visualized by F-actin staining using fluorochrome-coupled phalloidin and confocal microscopy. (**A**) Feeding stage where the new buccal funnel (*nbf*) emerges and is about to replace the old buccal funnel (*obf*). (**B**) Musculature of the feeding stage shown in **A**. (**C**) Settled Pandora larva with feeding stage developing inside. (**D**) Musculature of the feeding stage developing inside the Pandora larva shown in **C**. Abbreviations: *bf* buccal funnel, *bfm* musculature of the buccal funnel, *cu* cuticle, *nbfm* musculature of the new buccal funnel, *gm* musculature of the gut, *obfm* musculature of the old buccal funnel, *rlm* remnants of the musculature of the Pandora larva, *rm1* ring muscle of the old buccal funnel, *rm2* ring muscle of the new buccal funnel, *se* seta of the mouthparts of the host lobster, *tlm* longitudinal muscles of the trunk, *tr* trunk, *double-headed arrows* F-actin signal of degenerating musculature of the Pandora larva

available antibodies directed against a number of neural markers (e.g., serotonin) only showed scattered and weak signal in the buccal funnel of *Symbion pandora* and did not recognize any ganglionic structures. Transmission electron microscopy analyses identified nervous tissue in the buccal funnel, but whether this is ganglionic in nature could not be clarified (Neves et al. 2010b).

According to the data available, all other life cycle stages are at least for some time free-swimming (Wanninger 2005; Neves et al. 2010b, c). They have varying sets of circular, longitudinal, dorsoventral, and other muscles that support the body (Wanninger 2005; Neves et al. 2009a, 2010a). In these stages, the nervous system comprises a relatively large, bilobed cerebral ganglion and two ventral longitudinal nerves. These elements contain serotonin (Wanninger 2005; Neves et al. 2010b, d). The chordoid larva alone has four ventral neurite bundles which fuse in the posterior region (Wanninger 2005). In addition to the serotonergic signal, these nerves showed synapsin as well as FMRFamide immunoreactivity, with the latter being confined to the two outer neurites (Neves et al. 2010d). Posterolateral sensory organs ("lateral ciliated pits") and a dorsoanterior ciliated sensory organ are present in the chordoid larva (Funch 1996), but the latter most likely does not correspond to the apical organ of other lophotrochozoan larvae (Neves et al. 2010d). Distinct neural subsets underlying (and potentially innervating) the ciliary fields, as present in most lophotrochozoan larvae, were not found. Protonephridia are only known from the chordoid larva. The name-giving "chordoid organ" of this larva is composed of a series of muscular subunits that form a bow-like structure that extends from the posterior pole in ventral direction from where it continues into the anterior region (Wanninger 2005). Light sensory organs are absent in all stages.

As part of the asexual life cycle, the *Pandora larva* emerges from the feeding stage and, after a supposedly brief free-swimming stage, settles on setae of the mouth parts of the same host lobster. It attaches with the apical region and gives rise to a new feeding stage, whose anlagen are already formed in Pandora larvae liberated from the maternal feeding individual (Funch and Kristensen 1995). The cerebral ganglion of the Pandora larva and all other free-swimming stages degenerates during settlement.

In order to enter the sexual reproduction phase, feeding stages may either develop, one at a time, a female or a Prometheus larva, both internally (Figs. 5.1 and 5.2C, E). While still within the feeding stage, the female develops one single oocyte (Fig. 5.2E). The Prometheus larva settles on the trunk of a feeding individual after release (Fig. 5.2A), encysts, and generates internally one to three mature dwarf males with a cuticular penile structure (Fig. 5.2D; Neves et al. 2010c). Electron microscopy studies showed that younger males are larger and possess more nucleated somatic cells (aproximately 200) than mature individuals, who only have around 50 nucleated cells and nuclei-free muscles and epidermis (Neves and Reichert 2015). With 30–40 µm in length these animals range among the smallest free-living, sexually mature metazoans. Despite their small size, cycliophoran dwarf males have a distinct body wall musculature, a brain that occupies the anterior third of the body, as well as two ventral nerve cords (Obst and Funch 2003; Neves et al. 2010b).

Feeding stages with encysted Prometheus larvae often bear females inside the trunk. The male probably hatches from the cyst, and the mature female is released from the feeding stage. Fertilization of the oocyte is internal, but whether it occurs while the female still resides within the feeding stage (through the wall of the male cyst, the cuticle of the feeding stage, and the body wall of the female!) or during the free-living stage remains speculative. In any case, the female with the fertilized oocyte settles with the anterior pole on the mouthparts of the same host, degenerates, and forms the so-called chordoid cyst (Fig. 5.2F). Therein, the sexually produced chordoid larva develops from the fertilized oocyte.

Isolated observations found a female with an uncleaved oocyte inside a feeding stage and several free-swimming oocyte-bearing females. This may indicate that fertilization takes place after release of the female from the feeding stage. Indeed, the only account on embryogenesis stems from settled females (Neves et al. 2012). Cleavage

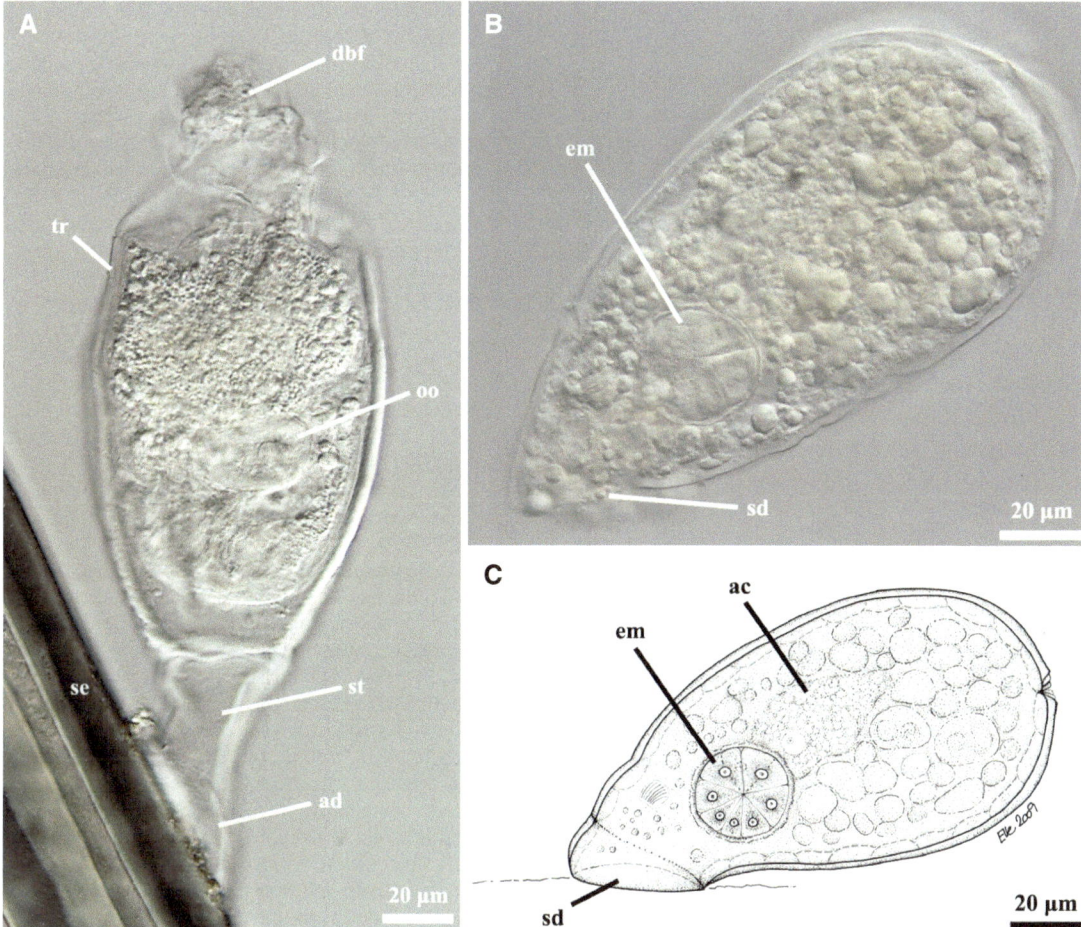

Fig. 5.4 *Symbion pandora.* (**A**) Light micrograph of a feeding stage with a developing female inside. Note the large oocyte (*oo*). (**B**) Light micrograph of a settled female with an embryo (*em*) at the eight cell stage. Anterior faces down. (**C**) *Line drawing* of a settled female based on the specimen shown in (**B**). Abbreviations: *ac* auxiliary cells, *ad* attachment disc, *dbf* degenerated buccal funnel, *sd* settlement disc, *se* seta of the mouthparts of the host lobster, *st* stalk, *tr* trunk (After Neves et al. (2010b, 2012))

appears to be holoblastic with four micromeres and four macromeres at the eight cell stage, but blastomere arrangement bears no similarities to a spiral cleavage pattern (Fig. 5.4B, C). Polar bodies were not observed. Hatched chordoid larvae probably seek a new lobster specimen, settle, encyst, and metamorphose into new feeding individuals. Accordingly, the chordoid larva is regarded as the cycliophoran dispersal stage.

While this reads like a well-founded reconstruction of the cycliophoran life cycle, most of its dynamics are only inferred from studies on the morphology of the various stages and their distribution on the host animals (Obst and Funch 2006).

Crucial events such as fertilization, settlement of the chordoid larva, and details of the embryology have never been directly observed, and details on the development of the various asexually produced stages are unknown. Needless to say that no data on gene expression patterns are currently available, leaving the field wide open for future studies into literally all directions of cycliophoran developmental biology.

A recent observation found cycliophoran individuals on copepods that live on the mouthparts of the European lobster, casting doubt on the proposed cycliophoran host specificity being confined to lobster crustaceans (Neves et al. 2014). This finding may be of utmost importance for the understating of the cycliophoran life cycle and questions concerned with the dispersal and colonization of new host lobsters, illustrating that we have only begun to understand the basic mechanisms that underlie the biology of this enigmatic phylum.

OPEN QUESTIONS

- Virtually all aspects of cycliophoran development, including fertilization, embryology, cleavage, organogenesis, and gene expression.
- Emergence of the asexually produced life cycle stages after settlement and encystation, in particular: Does the chordoid larva indeed metamorphose into a feeding stage?
- Is the chordoid larva a modified trochophore?
- What induces sexual reproduction in the *Symbion* life cycle?

Acknowledgements AW thanks the Faculty of Life Sciences, University of Vienna, for generous support while establishing his group during his first years in Vienna. The Danish Research Agency (FNU) and the Carlsberg Foundation as well as the European Commission are thanked for the support of his research during the Copenhagen years. Marion Hüffel provided invaluable support during this entire project and beyond. AW particularly thanks his coauthor, Ricardo Neves, for utmost productive years of cooperation into cycliophoran research.

References

Baker JM, Giribet G (2007) A molecular phylogenetic approach to the phylum Cycliophora provides further evidence for cryptic speciation in *Symbion americanus*. Zool Scr 36:353–359

Baker JM, Funch P, Giribet G (2007) Cryptic speciation in the recently discovered American cycliophoran *Symbion americanus*; genetic structure and population expansion. Mar Biol 151:2183–2193

Funch P (1996) The chordoid larva of *Symbion pandora* (Cycliophora) is a modified trochophore. J Morphol 230:231–263

Funch P, Kristensen RM (1995) Cycliophora is a new phylum with affinities to Entoprocta and Ectoprocta. Nature 378:711–714

Funch P, Kristensen RM (1997) Cycliophora. In: Harrison FW, Woollacott RM (eds) Microscopic anatomy of invertebrates, vol 13, Lophophorates, Entoprocta and Cycliophora. Wiley-Liss, New York, pp 409–474

Funch P, Kristensen RM (1999) Cycliophora. In: Knobil E, Neill JD (eds) Encyclopaedia of reproduction, vol 1. Academic, New York, pp 800–808

Giribet G, Distel DL, Polz M, Sterrer W, Wheeler WC (2000) Triploblastic relationships with emphasis on the acoelomates and the position of Gnathostomulida, Cycliophora, Plathelminthes, and Chaetognatha: a combined approach of 18S rDNA sequences and morphology. Syst Biol 49:539–562

Hejnol A, Obst M, Stamatakis A, Ott M, Rouse GW, Edgecombe GD, Martinez P, Baguñà J, Bailly X, Jondelius U, Wiens M, Müller WEG, Seaver E, Wheeler WC, Martindale MQ, Giribet G, Dunn CW (2009) Assessing the root of bilaterian animals with scalable phylogenomic methods. Proc R Soc B 276:4261–4270

Nedvěd O (2004) Occurrence of the phylum Cycliophora in the Mediterranean. Mar Ecol Prog Ser 277:297–299

Neves RC, Kristensen RM, Wanninger A (2009a) Three-dimensional reconstruction of the musculature of various life cycle stages of the cycliophoran *Symbion americanus*. J Morphol 270:257–270

Neves RC, Sørensen KJK, Kristensen RM, Wanninger A (2009b) Cycliophoran dwarf males break the rule: high complexity with low cell-numbers. Biol Bull 217:2–5

Neves RC, Cunha MR, Kristensen RM, Wanninger A (2010a) Comparative myoanatomy of cycliophoran life cycle stages. J Morphol 271:596–611

Neves RC, Kristensen RM, Wanninger A (2010b) Serotonin immunoreactivity in the nervous system of the Pandora larva, the Prometheus larva and the dwarf male of *Symbion americanus* (Cycliophora). Zool Anz 249:1–12

Neves RC, Cunha MR, Funch P, Wanninger A, Kristensen RM (2010c) External morphology of the cycliophoran dwarf male: a comparative study of *Symbion pandora* and *S. americanus*. Helgol Mar Res 64:257–262

Neves RC, Cunha MR, Kristensen RM, Wanninger A (2010d) Expression of synapsin and co-localization with serotonin and RFamide-like immunoreactivity in the nervous system of the chordoid larva of *Symbion pandora* (Cycliophora). Invertebr Biol 129:17–26

Neves RC, Kristensen RM, Funch P (2012) Ultrastructure and morphology of the cycliophoran female. J Morphol 273:850–869

Neves RC, Bailly X, Reichert R (2014) Are copepods secondary hosts of Cycliophora? Org Divers Evol 14:363–367

Neves RC, Reichert H (2015) Microanatomy and development of the dwarf male of Symbion pandora (Phylum Cycliophora): new insights from ultrastructural investigation based on serial section electron microscopy. PLoS ONE 10: e0122364

Obst M, Funch P (2003) Dwarf male of *Symbion pandora* (Cycliophora). J Morphol 255:261–278

Obst M, Funch P, Kristensen RM (2006) A new species of Cycliophora from the mouthparts of the American lobster, *Homarus americanus* (Nephropidae, Decapoda). Org Div Evol 6:83–97

Obst M, Funch P (2006) The microhabitat of *Symbion pandora* (Cycliophora) on the mouthparts of its host *Nephrops norvegicus* (Decapoda: Nephropidae). Mar Biol 148:945–951

Obst M, Funch P, Giribet G (2005) Hidden diversity and host specificity in cycliophorans: a phylogeographic analysis along the North Atlantic and Mediterranean Sea. Mol Ecol 14:4427–4440

Paps J, Baguña J, Riutort M (2009) Lophotrochozoan internal phylogeny: new insights from an up-to-date analysis of nuclear ribosomal genes. Proc R Soc B 276:1245–1254

Wanninger A (2005) Immunocytochemistry of the nervous system and the musculature of the chordoid larva of *Symbion pandora* (Cycliophora). J Morphol 265:237–243

Winnepenninckx B, Backeljau T, Kristensen RM (1998) Relations of the new phylum Cycliophora. Nature 393:636–638

Entoprocta

6

Andreas Wanninger

Chapter vignette artwork by Brigitte Baldrian.
© Brigitte Baldrian and Andreas Wanninger.

A. Wanninger
Department of Integrative Zoology, University
of Vienna, Althanstrasse 14, Vienna 1090, Austria
e-mail: andreas.wanninger@univie.ac.at

A. Wanninger (ed.), *Evolutionary Developmental Biology of Invertebrates 2: Lophotrochozoa (Spiralia)*
DOI 10.1007/978-3-7091-1871-9_6, © Springer-Verlag Wien 2015

INTRODUCTION

Entoprocts or kamptozoans are solitary or colonial aquatic filter feeders with a characteristic ciliated tentacle apparatus. They are acoelomate, but non-mesothelially lined lacunae form schizocoelic spaces that pervade the animals. About 200 recent species have been described, with a total body size roughly ranging between 100 μm and several millimeters. The phylum is largely confined to marine (including brackish) habitats, with only two freshwater species known to date.

The entoproct body consists of a slender, muscular stalk with terminal secretory cells that allow for reversible attachment to the substrate. An adhesive disc is often present in solitary species. The stalk is continuous with the calyx, which constitutes the major body region (Fig. 6.1). It terminates with the atrium, which houses the slit-like mouth opening and the elevated anus sitting on the anal cone. Accordingly, the "upper" side of the animal marks its morphological ventral side. The entire atrial region is surrounded by the tentacle crown (Fig. 6.1). The monolayered epidermis is covered by a cuticle which may be reduced or entirely lacking on the inner side of the tentacles and the atrium; these areas have been proposed to be the major sites of gas exchange (specific respiratory organs are generally absent in entoprocts). Within the calyx reside the major inner organs including nerves, muscles, gonads, brood pouches, protonephridia, and digestive tract (Nielsen and Jespersen 1997). The stalk (including attachment disc, if present) is highly muscular and functions as an internal skeleton, thereby allowing for flexible body movements (Wanninger 2004). The musculature of the calyx is only weakly developed. Its most prominent feature is a ring muscle at the base of the tentacles that allows for closure of the atrial region and protection of the tentacles upon contraction. A three-layered body wall musculature comprising ring, diagonal, and longitudinal muscles, typical for many (vermiform) lophotrochozoans, is lacking. The nervous system is simple and includes a dumbbell-shaped cerebral ganglion at the oral side of the atrium, from which nerves emanate and project into the stalk and tentacles (Fuchs et al. 2006; Schwaha et al. 2010). Intermediate tentacle ganglia, as found in ectoprocts, are lacking. Ciliated sensory organs are often found along the stalk and calyx (Fuchs et al. 2006). These may serve as geosensory organs, but detailed studies are lacking. Light sensory organs have so far not been found in adult entoprocts. The U-shaped digestive tract lies within the calyx and consists of a short esophagus and hindgut and a relatively prominent stomach. Usually one pair of protonephridia is found in adults as well as larvae (Nielsen and Jespersen 1997); a complex arrangement of numerous protonephridia, probably serving osmoregulatory functions, has been found in the two freshwater species, *Urnatella gracilis* and *Loxosomatoides sirindhornae* (Emschermann 1965; Schwaha et al. 2010). For further details on the morphology of adult entoprocts, see Emschermann (1982), Nielsen and Jespersen (1997), Fuchs et al. (2006), and Nielsen (2012).

Morphological and molecular analyses agree that colonial and solitary entoprocts each form monophyletic assemblages, thus subdividing Entoprota into the Coloniales (comprising the Barentsiidae, Pedicellinidae, and Loxocalypodidae) and the Solitaria (with Loxosomatidae as the sole subclade) (Emschermann 1972; Fuchs et al. 2010). It is generally assumed that the (solitary) loxosomatids have largely retained major basal features of Entoprocta, including a lecithotrophic, creeping-type larva (see below).

Traditionally, Entoprocta has been thought to cluster with the Ectoprocta to form the phylum Bryozoa (Nielsen 1971), but this view is challenged by differences in adult morphology (e.g., coelom in ectoprocts versus acoelomate condition in entoprocts) and developmental characters such as mode of cleavage (spiral in entoprocts versus radial in ectoprocts; see Chapter 11). Instead, similarities in the non-coelomic, lacuna-like circulatory system and in a number of larval features including the presence of a ciliated foot, the morphology of the apical organ, and a tetraneurous nervous system suggest entoproct-molluscan monophyly (Sinusoida, Lacunifera, or

Fig. 6.1 Gross morphology of a solitary entoproct, the loxosomatid *Loxosoma nielseni*, with major body plan features indicated. Scanning electron micrograph, the total size of the animal from foot plate to anus is 175 μm

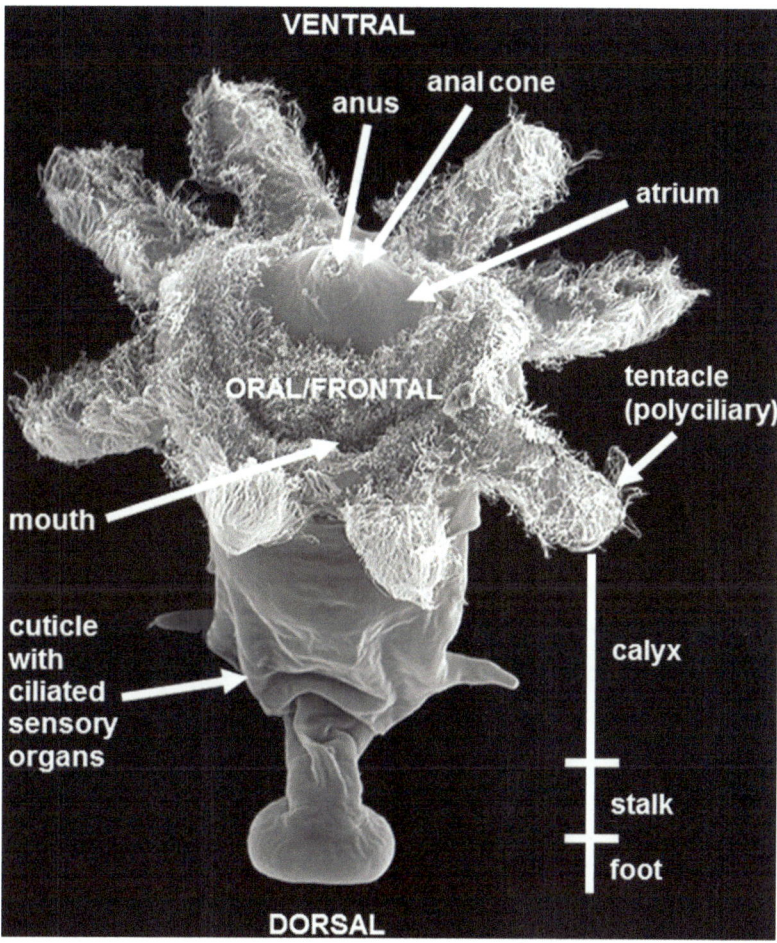

Tetraneuralia concept; Chapter 7) (Bartolomaeus 1993; Ax 1999; Wanninger 2009). Some molecular phylogenies recovered a superclade Polyzoa, with Entoprocta and Cycliophora monophyletic and the sister to Ectoprocta (e.g., Hejnol et al. 2009), but this was heavily refuted by others, who suggested that this clade resulted from a systematic bias (Nesnidal et al. 2013). Additional alternative scenarios have been proposed, including one that found Entoprocta in a basal position of a lophotrochozoan subclade that also includes the mollusks, annelids, nemerteans, brachiopods, and phoronids, but notably not the ectoprocts (Dunn et al. 2008). The ongoing vivid discussions demonstrate that the entoprocts have not yet found their final resting place within the lophotrochozoan tree of life.

EARLY DEVELOPMENT

Many solitary entoprocts are protandric hermaphrodites. From the colonial genus *Barentsia*, species are known that form either male or female colonies only or have male and female zooids in a single colony, while other species have hermaphroditic zooids, thus indicating the high plasticity of entoproct reproductive modes (Wasson 1997; see Nielsen 2012). The male gametes are released into the water column. Fertilization is internal and the embryos are maintained in the calyx. Matrotrophy has been described for some species (Harmer 1885; Nickerson 1901).

Accounts on entoproct early embryology and cleavage are scarce and not very detailed. For a long time, the main information had been on two colonial representatives, *Pedicellina* (Fig. 6.2; Hatschek 1877; Marcus 1939) and *Barentsia* (Malakhov 1990), supplemented by a classical work on solitary loxosomatids, mostly *Loxosomella leptoclini* (Harmer 1885). Recently, new data have become available on a yet undescribed *Loxosomella* species employing fluorescence staining of nuclei and 3D reconstruction of embryogenesis (Figs. 6.3 and 6.4; Merkel et al. 2012). Most authors noted a holoblastic, more or less equal, spiral cleavage pattern, comparable to that of other Spiralia. For *Pedicillina*, synchronous as well as asynchronous cell divisions have been reported during early cleavage cycles (Hatschek 1877; Marcus 1939).

In *Loxosomella* sp., cleavage is highly asynchronous. An apical rosette was found in *Pedicillina* and *Loxosomella*, together with two

spiralian cross-like patterns resembling those of annelids and mollusks at around 43 cells in *Loxosomella* (Fig. 6.5; Merkel et al. 2012). Gastrulation is preceded by flattening of the embryo, occurs when the embryo has slightly more than 100 cells, and eventually results in a coeloblastula (Merkel et al. 2012). This is in accordance with the data provided by Marcus (1939) on *Pedicillina cernua*, who found the embryos to gastrulate at around 120 cells.

For *Loxosomella leptoclini*, Harmer (1885) mentioned subsequent closure of the blastopore and the formation of two lateral mesodermal bands that give rise to the future mesoderm, as well as de novo generation of the mouth and emergence of the anus in the region of the former blastopore, but these data need careful reinvestigation using modern methods.

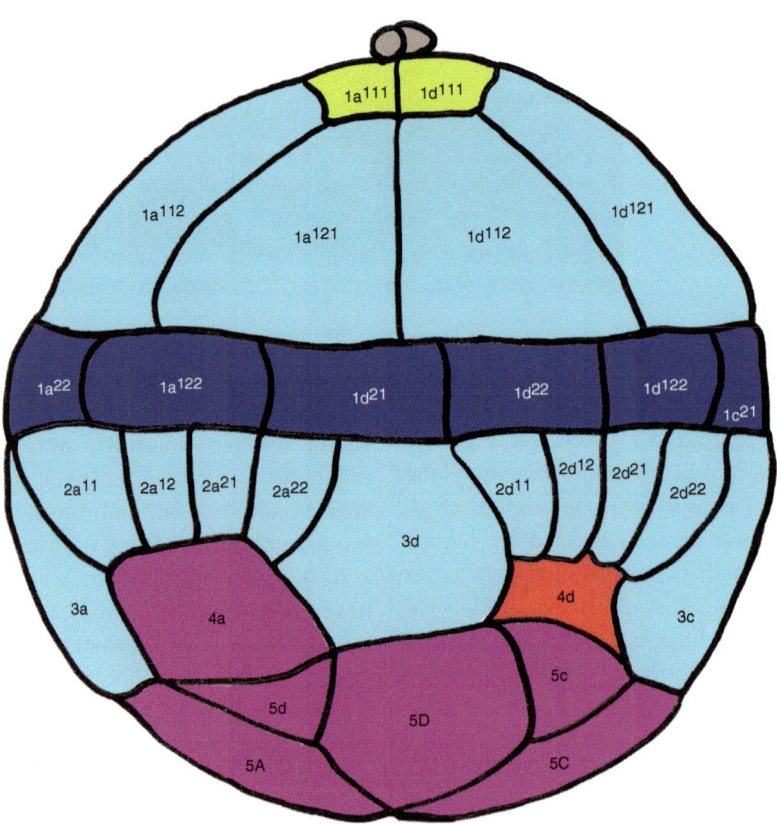

Fig. 6.2 Cell lineage of the colonial entoproct *Pedicillina cernua*. Cells of the apical rosette are in *yellow*, trochoblasts are in *dark blue*, and the remaining blastomeres that contribute to the ectoderm are in *light blue*. The mesoderm founder cell (mesendoblast) 4d is in *red* and blastomeres that give rise to endodermal cells are in *magenta*. Polar bodies are depicted in *gray*. (Redrawn and modified after Marcus (1939))

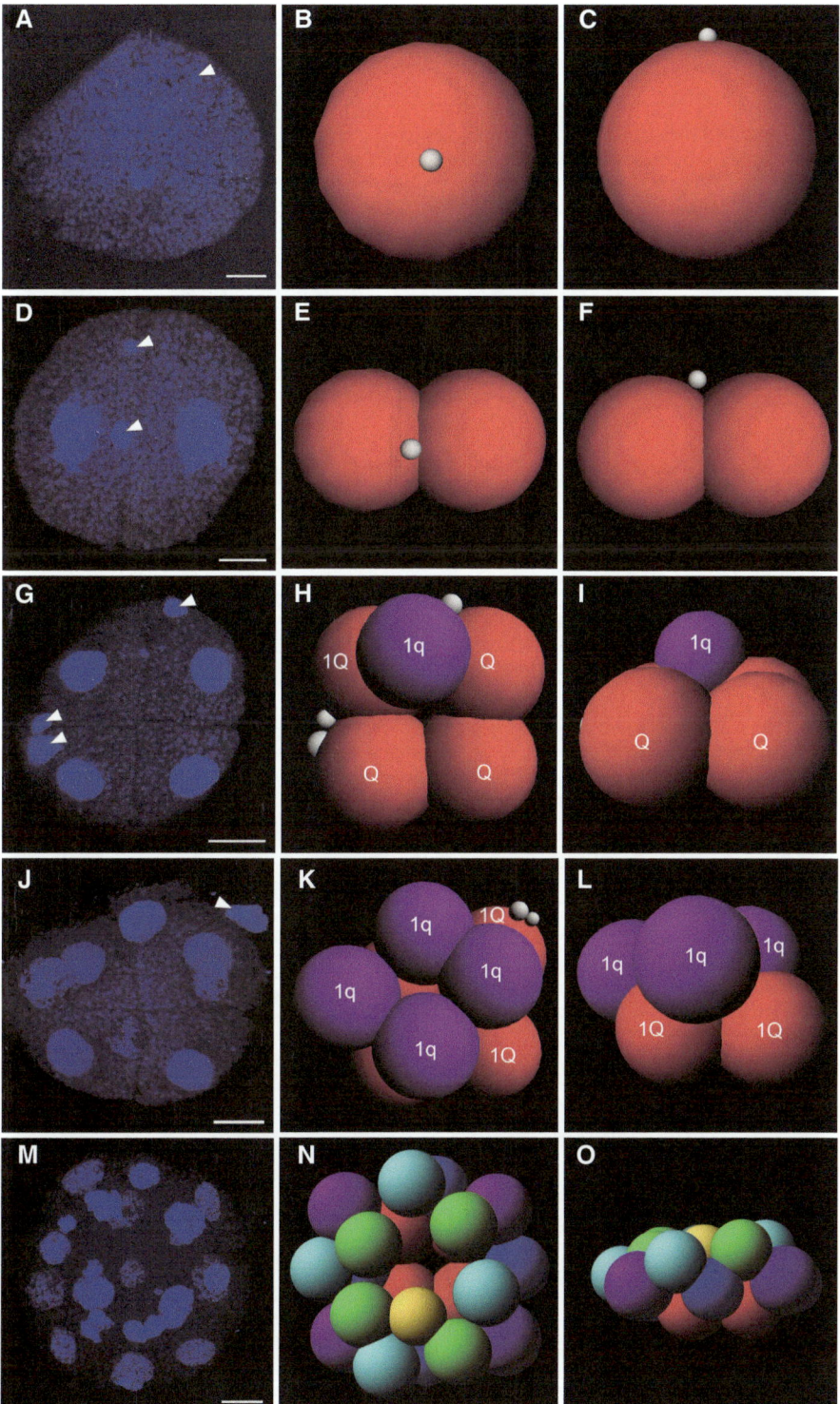

Fig. 6.3 Early entoproct embryology of *Loxosomella* sp. showing spiral cleavage. Images in the *left column* are confocal micrographs of nuclei stainings; *middle* and *right columns* are 3D reconstructions. *Middle column*, animal view; *right column*, lateral view. Cells resulting from a distinct cleavage cycle are labeled with the same color in the 3D reconstructions. Polar bodies are in *gray* in the reconstructions and indicated by *arrowheads* in the confocal images. Note the almost similar size of macromeres (*Q*) and micromeres (*q*). Scale bars represent 10 μm (From Merkel et al. (2012))

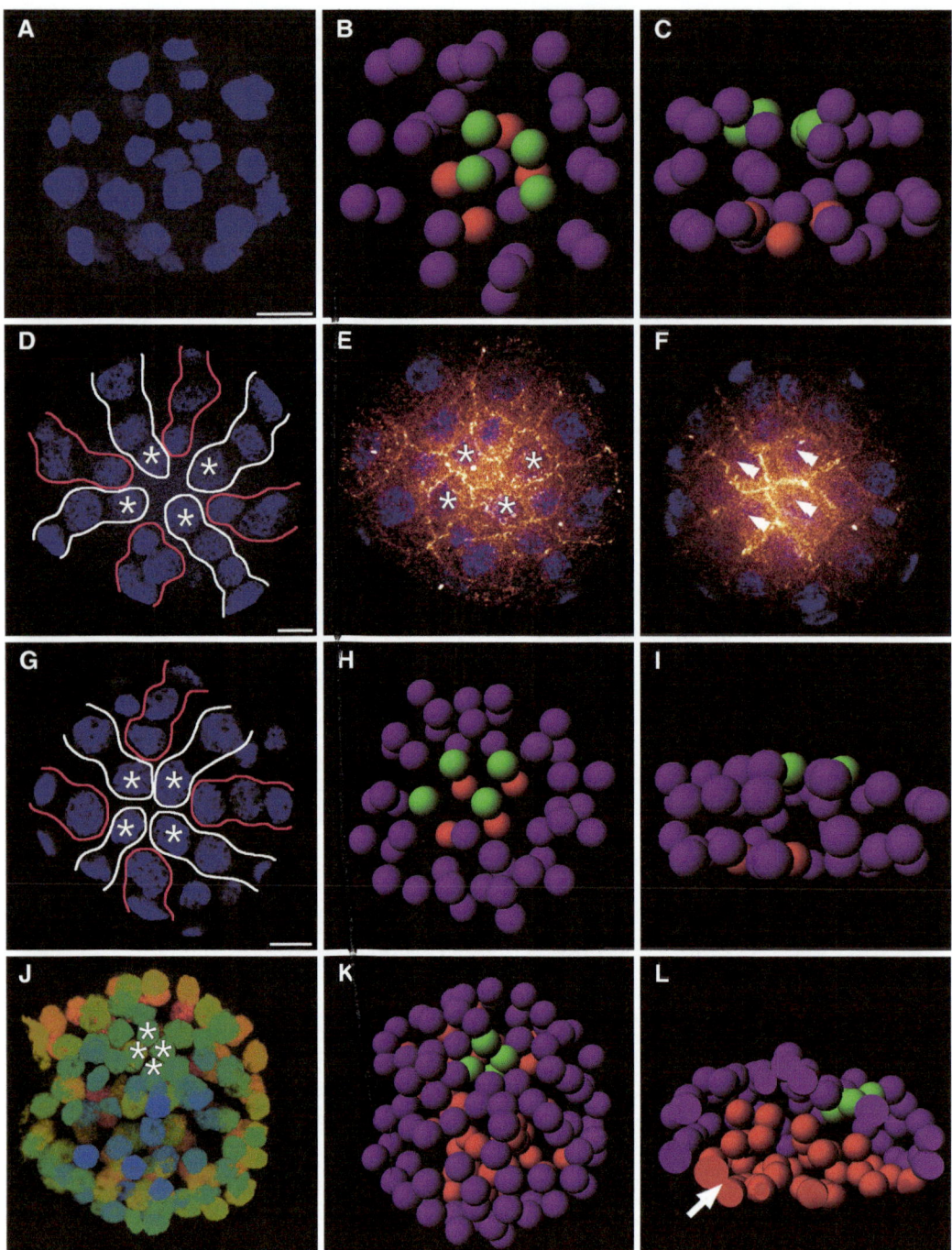

Fig. 6.4 Confocal micrographs and 3D reconstructions of blastula and gastrula stages of *Loxosomella* sp. "Apical cross patterns" are indicated by *white* and *red lines* in **D**, **G**. Apical rosette cell nuclei are marked with *asterisks* in **D, E, G, J** and vegetal "macromere" quartet cells by *double arrowheads* in **F**. *Scale bars* represent 10 μm. (**A, D, G**) Nucleic acid staining (*blue*). (**E, F**) Nucleic acid (*blue*) and F-actin staining (*red*). (**J**) Nucleic acid staining shown as depth-coded confocal projection. (**B, C, H, I, K, L**) 3D reconstructions. (**A, F**) Vegetal view. (**B, D, E, G, H, J, K**) animal view. (**C, I, L**) Lateral view. *Red*: "macromere" quartet (i.e., vegetal) cell nuclei; *green*, (derivatives of) apical rosette cell nuclei; *purple*, other cell nuclei. (**A–C**) 36-cell stage. (**D–F**) 43-cell stage. Cells surrounding the apical rosette (*asterisks*) show both a molluscan- and an annelid-like cross pattern (**D**). (**G–I**) 51-cell stage. (**J–L**) Gastrula stage (107 cells). Derivatives of apical rosette cells lie in a lower plane than the surrounding cells (**J**). (**K, L**) Gastrulation. Vegetal cells (*red*), blastocoel (*arrow*) (From Merkel et al. (2012))

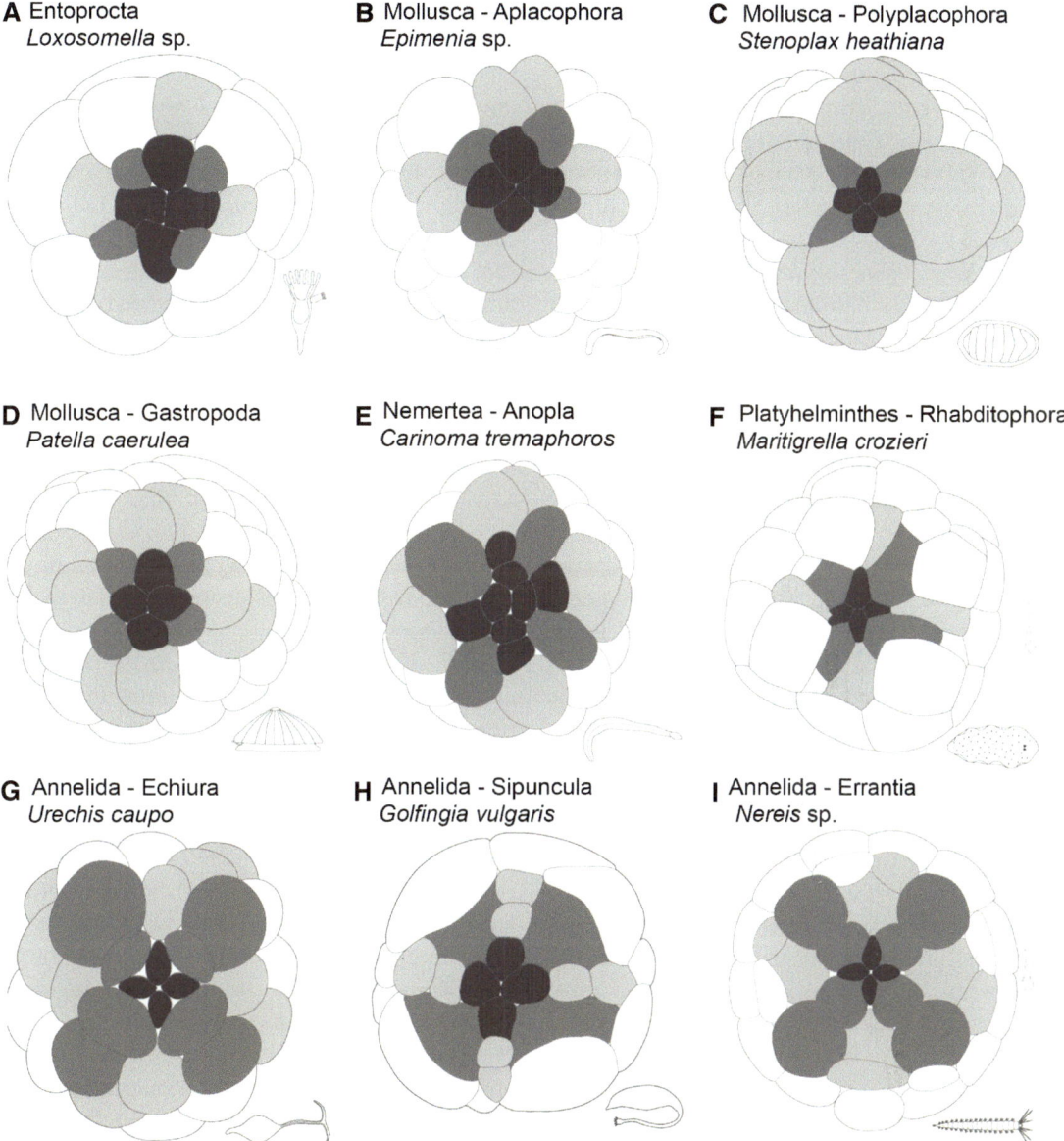

Fig. 6.5 Spiralian apical cross cell patterns based on several authors. *Dark gray*, apical rosette cells; *middle gray*, periphere rosette/"annelid cross" cells; *light gray*, "molluscan cross" cells. (**A**) 43-cell stage of *Loxosomella* sp. (**B**) Approximately 64-cell stage of the aplacophoran mollusk *Epimenia* sp.; after Baba (1951). (**C**) Approximately 62-cell stage of the polyplacophoran mollusk *Stenoplax heathiana* (= *Ischnochiton magdalenensis*); after Heath (1898). (**D**) 58-cell stage of the gastropod mollusk *Patella caerulea*; after Wilson (1904). (**E**) 64-cell stage of the nemertean *Carinoma tremaphoros*; after Maslakova et al. (2004). (**F**) Approximately 64-cell stage of the polyclad flatworm *Maritigrella crozieri*; after Rawlinson (2010). (**G**) 64-cell stage of the echiurid *Urechis caupo*; after Newby (1932). (**H**) 48-cell stage of the sipunculan *Golfingia vulgaris* (= *Phascolosoma vulgare*); after Gerould (1903). (**I**) Approximately 58-cell stage of the polychaete *Nereis* sp.; after Wilson (1892) (From Merkel et al. (2012))

Following cell fates is particularly difficult in entoprocts due to the brooded embryos and because the micromeres and macromeres are of almost the same size (Merkel et al. 2012). The only data on entoproct cell lineage are available for *Pedicellina*, where progenies of the 1a-d lineages form the apical rosette and the trochoblasts. Second quartet micromeres, as, e.g., in mollusks, were not identified to contribute to the prototroch (Fig. 6.2; Marcus 1939). As in other spiralians (in particular mollusks; see Chapter 7: Table 1), the second and third quartet micromeres contribute to ectodermal domains, the 4a–c and 5a–d lineages to the endoderm, and the mesendoblast 4d was found to constitute the endomesoderm mother cell (Fig. 6.2; Marcus 1939), but more detailed data using modern methods are highly desired to further assess entoproct cell genealogy. Data on organogenesis and the precise contribution of individual blastomeres to larval organ systems are virtually unknown.

LATE DEVELOPMENT

Larval Morphology

Entoprocts exhibit a wide variety of larvae which can be subdivided into two major categories, a swimming- and a creeping-type larva, with the latter most likely constituting the basal type (Figs. 6.6 and 6.7; Nielsen 1971; Fuchs et al. 2010). The swimming-type larva has a predominant episphere. The hyposphere is often reduced and obscured by the compound cilia of the prototroch (Fig. 6.7A), although a short, ciliated protrusion, the larval foot, is at least rudimentarily present in most species. Swimming-type larvae often have a long planktonic and planktotrophic phase which can last for several weeks. The apical organ is simple and comprises a ciliated tuft and three or four serotonin-positive flask-shaped cells (Fig. 6.7D, G; Fuchs and Wanninger 2008). From there, a paired nerve projects to the ganglion of the frontal organ, a sensory organ characteristic for entoproct larvae, and further to the prototroch nerve ring (Fig. 6.7G). The musculature of the swimming-type larva is fairly complex. It includes a dense set of ring muscles that engulf the episphere including the apical organ and a prominent ring muscle underlying the prototroch, as well as numerous supporting muscles. Several apical organ retractor muscles project from the episphere toward the prototroch (Fuchs and Wanninger 2008).

In contrast to the swimming-type larvae, creeping-type larvae are released from the mother animal as (almost?) metamorphic competent individuals. Although they do have a mouth and gut, they do not seem to feed. They immediately start a semi-benthic life during which they creep over the substrate with their distinct foot. This behavior is only briefly interrupted by short excursions into the water column. As in the swimming-type larvae, the foot is situated posttrochally but, in contrast to the former, is significantly pronounced (Figs. 6.6 and 6.7C). In the creeping-type larva, both swimming and creeping is directed with the frontal organ forward (not with the apical organ as in other lophotrochozoans) and thus reflects a 90° shift of the (functional) anterior-posterior axis of the animal compared to that of all other lophotrochozoan larvae. Attachment prior to metamorphosis is either with the frontal organ or the foot facing the substrate, similar to the swimming-type larvae.

The entoproct creeping-type larva exhibits a number of morphological features that have been homologized with corresponding structures of mollusks and thus entoproct-mollusk monophyly has been proposed (see above). Shared characters include the ciliated creeping foot with distinct foot glands and compound cirri in the anterior region of the foot, which have been homologized with those of adult neomeniomorph mollusks (Fig. 6.6; Haszprunar and Wanninger 2008). Both the creeping-type larva and polyplacophoran larvae have an apical organ with a complex arrangement of eight to ten serotonin-containing flask-shaped (ampullary) and surrounding, non-flask-shaped, peripheral cells (Fig. 6.7E, F, H; Wanninger et al. 2007). Further distinct features shared by the entoproct creeping larva and adult mollusks are a tetraneurous nervous system comprising two ventral (pedal) nerve cords (interconnected by commissures) and two lateral, more dorsally positioned (visceral) cords (Fig. 6.7E, F), as well as a ventrally intercrossing dorsoventral musculature (Haszprunar and Wanninger 2008). Additional neural features of the creeping-type larva include a buccal system and iterated sets of perikarya associated with the pedal nerve cords (Wanninger et al. 2007).

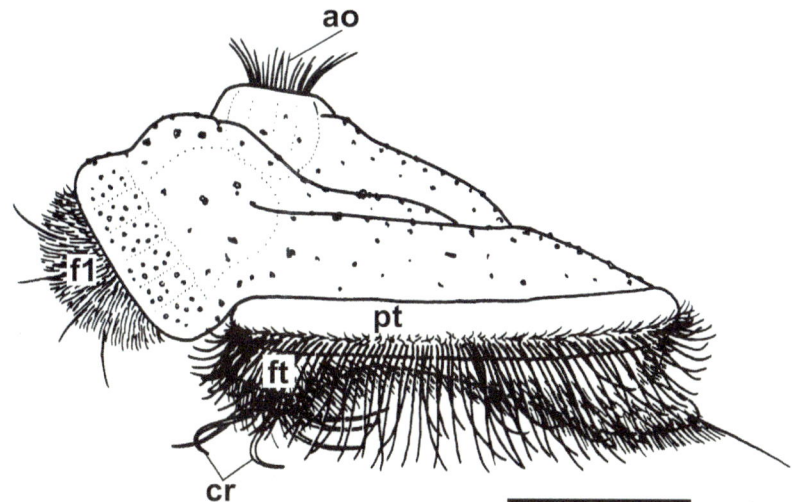

Fig. 6.6 Schematic representation of the creeping-type larva of the solitary entoproct *Loxosomella murmanica*. Swimming/creeping direction is with the frontal organ (*fl*) forward and not with the apical organ (*ao*) as in other trochozoan larvae. Scale bar is 50 μm. The prototroch (*pt*) encircles the larval foot (*ft*) with characteristic anterior cirri (*cr*) (From Haszprunar and Wanninger (2008), based on the original drawing by Nielsen (1971))

Entoproct creeping-type larvae are highly muscular and exhibit an extremely complex arrangement of distinct muscle units that is unmatched among lophotrochozoan larvae (Merkel and Wanninger, unpublished). These include ring muscles underlying the prototroch, the frontal and the apical organ, as well as various longitudinal muscles including frontal organ and apical organ retractors, body wall muscles, and dorsoventral muscles, allowing for the high flexibility of the body observed in live larvae.

Entoproct larvae have one pair of protonephridia and eyes are common. They are usually situated in the region of the frontal organ and consist of a photoreceptor cell, a pigment cell, and a lens cell; they are lost at metamorphosis (Woollacott and Eakin 1973).

Metamorphosis

Our current knowledge on entoproct metamorphosis largely relies on a few decades- to century-old works (e.g., Harmer 1885; Marcus 1939; Nielsen 1967, 1971) and thus requires reassessment using modern methodology. The data available indicate a high degree of variability in the dynamics of this process. In many *Loxosomella* species, the metamorphic competent larva attaches to the substrate with the frontal organ (Nielsen 1971). In other entoprocts, e.g., *Pedicellina* and *Barentsia*, the frontal organ is used for probing the substrate, while fixation to the substrate takes place by secretions of specific gland cells of the foot (Fig. 6.8). Subsequently, the larval retractors and the prototroch ring muscle contract, thus enclosing the entire larval body. As typical for lophotrochozoan larvae, the apical organ and the prototroch are lost. This is also true for the frontal organ, although the gland cells used for adhesion to the substrate in some *Loxosomella* species may persist in the adult as a distinct foot gland (Nielsen 1971). The fate of larval key features such as neural elements, muscles, protonephridia, and the subsequent ontogeny of the corresponding adult systems remain unknown, although loss of major parts of the larval body plan features is commonly assumed, with de novo generation of adult organ systems (catastrophic metamorphosis).

Following adhesion of the larva to the substrate, the digestive tract rotates such that the

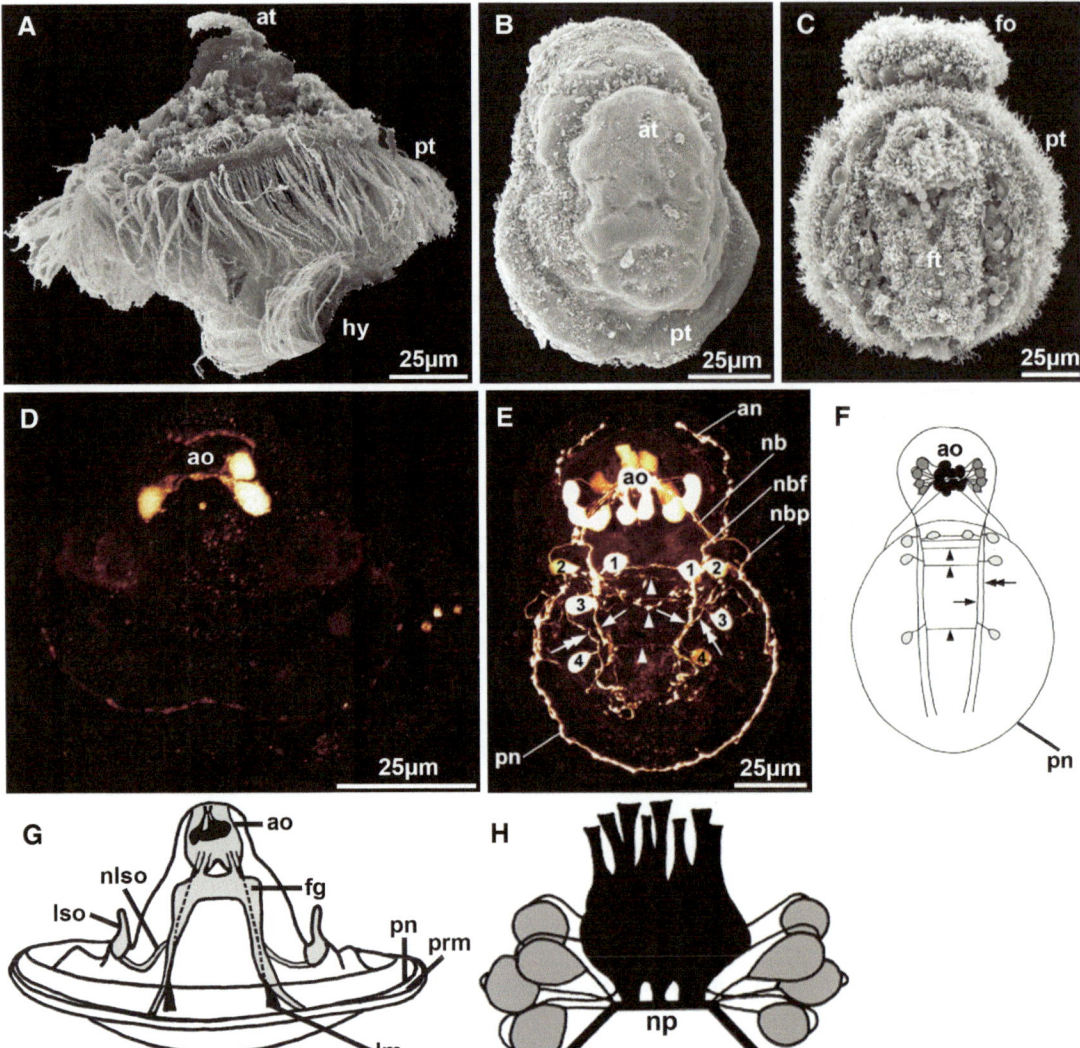

Fig. 6.7 Entoproct larvae and larval nervous systems. **A–C** are scanning electron micrographs; **D** and **E** are confocal projections of the serotonin-like nervous system. Line drawings **F–H** are based on various confocal analyses of the respective larval types. The apical organ is up in **A, D, G,** and **H**. Frontal organ is up in **B, C, E,** and **F**. (**A**) Swimming-type larva with apical tuft (*at*), prototroch (*pt*), and small hyposphere (*hy*). (**B**) Creeping-type larva with view onto the retracted apical organ (*ao*). (**C**) Ventral view of a creeping-type larva with ciliated frontal organ (*fo*), prototroch (*pt*), and foot (*ft*). (**D**) Three serotonin-positive flask cells in the apical organ (*ao*) of a swimming-type larva, antero-ventral view. (**E**) Serotonin-like nervous system of a creeping-type larva in ventral view. Note the much higher complexity of the nervous system than that of the swimming-type larva in **D**, including a complex apical organ (*ao*), anterior nerve loop (*an*), prototroch neurites (*pn*), pedal neurite bundles (*arrows*) with commissures (*arrowheads*), lateral neurite bundles (*double arrows*), and paired perikarya (1–4). Note also the split of the neurite bundles (*nb*) that arise from the apical neuropil and give rise to neurites that run toward the pedal neurite bundles (*nbf*) and to the prototroch neurites (*nbp*). (**F**) Schematic representation of the serotonin-like nervous system of the creeping-type larva depicted in **E**. (**G**) Schematic representation of the nervous system based on semithin section (*light gray*) data by Nielsen (1971) with the flask cells of the apical organ (*ao*) included (*black*). Other components include the frontal ganglion (*fg*), prototroch ring nerve (*pn*), and lateral sense organ (*lso*) with connecting nerve (*nlso*). *prm*, prototroch ring muscle, *lm* longitudinal muscle. (**H**) Serotonin-like apical organ of the creeping-type larva shown in **E** and **F** shifted by 90° to visualize the eight flask cells (*black*) and the surrounding bipolar peripheral cells (*gray*), both connected to the underlying neuropil (*np*). (**B, C, E, F, H** modified from Wanninger et al. (2007); **D, G** modified from Fuchs and Wanninger (2008))

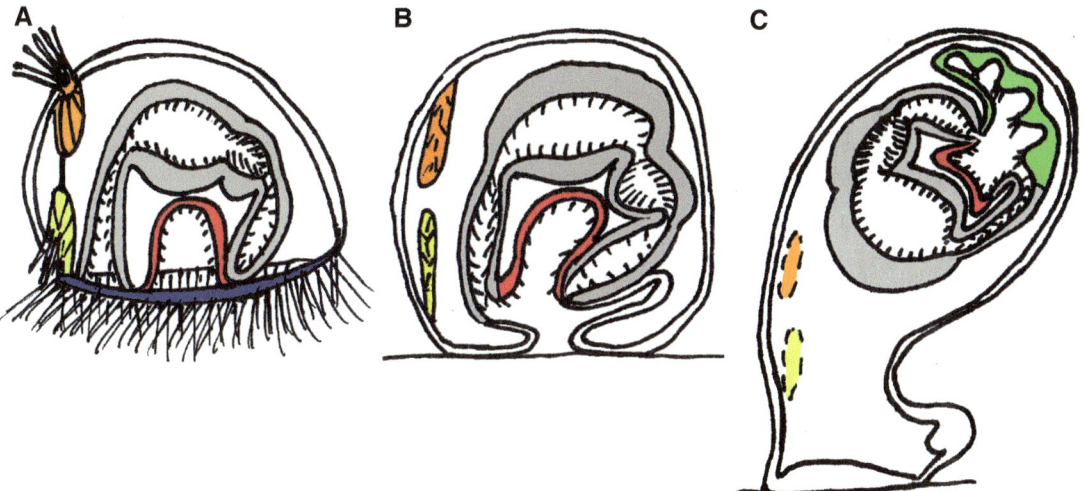

Fig. 6.8 Entoproct metamorphosis. (**A**) Metamorphic competent larva with ciliated prototroch (*blue*), contracted foot (*red*), as well as apical organ (*orange*), frontal organ (*yellow*), and rudiment of the digestive tract (*gray*). (**B**) Settlement and adhesion is followed by loss of the prototroch and subsequent degeneration of the apical and frontal organ. Note that some entoprocts seem to settle and attach with the frontal organ forward and not with the foot as indicated here. (**C**) Metamorphosis involves a 90° rotation of the digestive tract and de novo formation of the tentacles (*green*). The region of the larval foot becomes part of the atrium of the adult (Redrawn and modified after Marcus (1939))

mouth and anal opening eventually come to lie on the opposite side of the substrate (Fig. 6.8A, B). The stalk grows and the adult tentacles start to form as epithelial outgrowths, eventually surrounding the mouth, the anus, and the newly formed atrium (Fig. 6.8C). Significant deviations from this process have been reported in species where the juvenile emerges from the larval body after settlement by budding or is preformed in the live larva (Nielsen 1971). Details on the molecular, cellular, and morphogenetic mechanisms underlying these processes are unknown.

tive tract and the protonephridia are folded off from the atrium (Emschermann 1972). The scarce data on the exact mechanisms of the budding process (see Nielsen 1971 for a brief review of the classical works on the topic) suggest that no distinct endodermal or mesodermal tissues and only few individual mesodermal cells of the mother animal appear to be involved in the formation of the new bud. Accordingly, muscles, gonads, and other organs are most likely formed from immigrating (mesenchymal? stem?) cells, but these processes are not yet understood.

Asexual Reproduction

Asexual reproduction is common in colonial and solitary entoprocts. Buds form as ectodermal protrusions, usually on the latero-oral side of the calyx in solitary species. In colonial representatives, they form at the base of the stalk and on the stolons that interconnect individual zooids (Emschermann 1972; Emschermann and Wanninger 2013). The atrium is formed by invagination from the juvenile bud, which is followed by growth of the tentacles. The diges-

GENE EXPRESSION

No gene expression data on any entoproct species exist until today. The few transcriptomes that have been generated have not yet resulted in published sequences of important developmental patterning genes such Hox or other homeobox genes, but are likely to be underway. Accordingly, developmental genetics of Entoprocta is still a field wide open to the ambitious developmental biologist, but methods for culturing and yield of significant amounts of embryonic, larval, and metamorphosing stages remain to be established.

OPEN QUESTIONS

- Early developmental events following gastrulation, including gene expression, cell lineage, and organogenesis.
- Developmental gene expression (especially Hox, ParaHox genes) during larval development and metamorphosis.
- Gene expression profiles of the prototroch and the apical organ.
- Body plan remodeling during metamorphosis: Which, if any, larval elements contribute to the adult body plan?
- Fate of the apical organ after metamorphosis.
- Cellular and molecular mechanisms of clonal reproduction (budding).

Acknowledgments I thank Marion Hüffel for help with the graphical representations used in this chapter as well as for invaluable support on many aspects connected to this treatise project. The University of Vienna, especially the Faculty of Life Sciences, is thanked for the generous support in establishing the Wanninger lab during the past years. I am also grateful for the previous support from the Danish Research Council (FNU), the Carlsberg Foundation, and the European Commission for funding my research during my years in Copenhagen.

References

Ax P (1999) Das System der Metazoa, 2nd edn. Gustav Fischer, Stuttgart

Baba K (1951) General sketch of the development in a solenogastre, *Epimenia verrucosa* (Nierstrasz). Misc Rep Res Inst Nat Res (Tokyo) 19–21:38–46

Bartolomaeus T (1993) Die Leibeshöhlenverhältnisse und Nephridialorgane der Bilateria – Ultrastruktur, Entwicklung und Evolution. University of Göttingen, Göttingen

Dunn CW, Hejnol A, Matus DQ, Pang K, Browne WE, Smith SA, Seaver E, Rouse GW, Obst M, Edgecombe GD, Sørensen MV, Haddock SHD, Schmidt-Rhaesa A, Okusu A, Kristensen RM, Wheeler WC, Martindale MQ, Giribet G (2008) Broad phylogenomic sampling improves resolution of the animal tree of life. Nature 452:745–749

Emschermann P (1965) Das Protonephridiensystem von *Urnatella gracilis* Leidy (Kamptozoa). Bau, Entwicklung und Funktion. Z Morphol Ökol Tiere 55:859–914

Emschermann P (1972) *Loxokalypus socialis* gen. et spec. nov. (Kamptozoa, Loxokalypodidae fam. nov.), ein neuer Kamptozoentyp aus dem nördlichen Pazifischen Ozean. Ein Vorschlag zur Neufassung der Kamptozoensystematik. Mar Biol 12:237–254

Emschermann P (1982) Les Kamptozoaires. État actuel de nos connaissances sur leur anatomie, leur développement, leur biologie et leur position phylogénétique. Bull Soc Zool Fr 107:317–344

Emschermann P, Wanninger A (2013) Kamptozoa. In: Rieger G, Westheide W (eds) Spezielle Zoologie. Springer Spektrum, Heidelberg

Fuchs J, Wanninger A (2008) Reconstruction of the neuromuscular system of the swimming-type larva of *Loxosomella atkinsae* (Entoprocta) as inferred by fluorescence labelling and confocal microscopy. Org Divers Evol 8:325–335

Fuchs J, Bright M, Funch P, Wanninger A (2006) Immunocytochemistry of the neuromuscular systems of *Loxosomella vivipara* and *L. parguerensis* (Entoprocta: Loxosomatidae). J Morphol 267:866–883

Fuchs J, Iseto T, Hirose M, Sundberg P, Obst M (2010) The first internal molecular phylogeny of the animal phylum Entoprocta (Kamptozoa). Mol Phylogenet Evol 56:370–379

Gerould JH (1903) Studies on the embryology of the Sipunculidae. I. The embryonal envelope and its homologue. Mark Anniversary Volume. pp 437–452

Harmer SF (1885) On the structure and life history of *Loxosoma*. Q J Microsc Sci 25:261–337, pls. 19–21

Haszprunar G, Wanninger A (2008) On the fine structure of the creeping larva of *Loxosomella murmanica*: additional evidence for a clade of Kamptozoa (Entoprocta) and Mollusca. Acta Zool (Stockholm) 89:137–148

Hatschek B (1877) Embryonalentwicklung und Knospung der *Pedicellina echinata*. Z Wiss Zool 29:502–549

Heath H (1898) The development of *Ischnochiton*. Zool Jahrb Abt Anat Ontog Tiere 12:567–656

Hejnol A, Obst M, Stamatakis A, Ott M, Rouse GW, Edgecombe GD, Martinez P, Baguñà J, Bailly X, Jondelius U, Wiens M, Müller WEG, Seaver E, Wheeler WC, Martindale MQ, Giribet G, Dunn CW (2009) Assessing the root of bilaterian animals with scalable phylogenomic methods. Proc R Soc Lond B 276:4261–4270

Malakhov VV (1990) Description of the development of *Ascopodaria discreta* (Coloniales, Barentsiidae) and discussion of the Kamptozoa status in the animal kingdom. Zool Zh 69:20–30

Marcus E (1939) Bryozoários marinhos brasileiros III. Bol Fac Fil Ciên Letr Univ S Paulo, XIII. Zoologica 3:111–354

Maslakova SA, Martindale MQ, Norenburg JL (2004) Fundamental properties of the spiralian developmental program are displayed by the basal nemertean

Carinoma tremaphoros (Palaeonemertea, Nemertea). Dev Biol 267:342–360

Merkel J, Wollesen T, Lieb B, Wanninger A (2012) Spiral cleavage and early embryology of a loxosomatid entoproct and the usefulness of spiralian apical cross patterns for phylogenetic inferences. BMC Dev Biol 12:11

Nesnidal MP, Helmkampf M, Meyer A, Witek A, Bruchhaus I, Ebersberger I, Hankeln T, Lieb B, Struck TH, Hausdorf B (2013) New phylogenomic data support the monophyly of Lophophorata and an ectoproct-phoronid clade and indicate that Polyzoa and Kryptrochozoa are caused by systematic bias. BMC Evol Biol 13:253

Newby WW (1932) The early embryology of the echiuroid, *Urechis*. Biol Bull 63:387–399

Nickerson WS (1901) On *Loxosoma davenporti* sp. nov. an endoproct from the New England coast. J Morphol 17:351–380, pls. 32–33

Nielsen C (1967) Metamorphosis of the larva of *Loxosomella murmanica* (Nilus) (Entoprocta). Ophelia 4:85–89

Nielsen C (1971) Entoproct life-cycles and the entoproct/ectoproct relationship. Ophelia 9:209–341

Nielsen C (2012) Animal evolution: interrelationships of the living phyla. Oxford University Press, Oxford

Nielsen C, Jespersen A (1997) Entoprocta. In: Harrison FW (ed) Microscopic anatomy of invertebrates, vol 13. Wiley-Liss, New York

Rawlinson KA (2010) Embryonic and post-embryonic development of the polyclad flatworm *Maritigrella*

crozieri; implications for the evolution of spiralian life history traits. Front Zool 7:12

Schwaha T, Wood TS, Wanninger A (2010) Trapped in freshwater: the internal anatomy of the entoproct *Loxosomatoides sirindhornae*. Front Zool 7:7

Wanninger A (2004) Myo-anatomy of juvenile and adult loxosomatid Entoprocta and the use of muscular body plans for phylogenetic inferences. J Morphol 261:249–257

Wanninger A (2009) Shaping the things to come: ontogeny of lophotrochozoan neuromuscular systems and the Tetraneuralia concept. Biol Bull 216:293–306

Wanninger A, Fuchs J, Haszprunar G (2007) The anatomy of the serotonergic nervous system of an entoproct creeping-type larva and its phylogenetic implications. Invertebr Biol 126:268–278

Wasson K (1997) Sexual modes in the colonial kamptozoan genus *Barentsia*. Biol Bull 193:163–170

Wilson EB (1892) The cell-lineage of *Nereis*. A contribution to the cytogeny of the annelid body. J Morphol 6:361–480

Wilson EB (1904) Experimental studies in germinal localization. II. Experiments on the cleavage-mosaic in *Patella* and *Dentalium*. J Exp Zool 1:197–268

Woollacott RM, Eakin RM (1973) Ultrastructure of a potential photoreceptoral organ in the larva of an entoproct. J Ultrastruct Res 43:412–425

Mollusca

Andreas Wanninger and Tim Wollesen

Chapter vignette artwork by Brigitte Baldrian.
© Brigitte Baldrian and Andreas Wanninger.

A. Wanninger (✉) • T. Wollesen
Department of Integrative Zoology,
University of Vienna, Althanstrasse 14,
Vienna 1090, Austria
e-mail: andreas.wanninger@univie.ac.at

A. Wanninger (ed.), *Evolutionary Developmental Biology of Invertebrates 2: Lophotrochozoa (Spiralia)*
DOI 10.1007/978-3-7091-1871-9_7, © Springer-Verlag Wien 2015

INTRODUCTION

With probably around 200,000 extant species, Mollusca is the second-most speciose phylum after Hexapoda. However, what makes mollusks particularly interesting from an evolutionary perspective is not their richness in species as such, but rather the huge variety of body plan phenotypes exhibited by its representatives. These include cylindrical, shell-less, spicule-bearing, wormlike, crawling, and burrowing creatures (Neomeniomorpha or Solenogastres and Chaetodermomorpha or Caudofoveata), eight-shelled grazers (Polyplacophora or chitons),

two-valved filter feeders (Bivalvia such as mussels and clams), as well as the single-shelled Monoplacophora (Tryblidia), Gastropoda (snails, slugs), Scaphopoda (tusk shells), and Cephalopoda (octopuses, squids, nautiluses) (Fig. 7.1; for a comprehensive recent account on various aspects on molluscan phylogeny and evolution, see Ponder and Lindberg 2008).

With such a diverse morphological toolkit at hand, mollusks have managed to conquer almost all natural aquatic and terrestrial realms, and the spectacular "flying squid" (*Todarodes pacificus*) is even able to get partly airborne by using its fins, arms, and tentacles as wings. The origin of

Fig. 7.1 Molluscan diversity. Anterior faces to the right in all aspects. Polyplacophora (*Acanthochitona crinita*, dorsal view). Neomeniomorpha (*Helluoherpia aegiri*, lateral view; courtesy of Maik Scherholz). Chaetodermomorpha (*Falcidens crossotus*, lateral view). Monoplacophora (*Laevipilina theresae*, dorsal view; © Michael Schrödl, 2015. All Rights Reserved). Bivalvia (*Nucula tumidula*, lateral view). Gastropoda (*Patella vulgata*, dorsal view). Scaphopoda (*Antalis entalis*, lateral view). Cephalopoda (*Idiosepius notoides*, lateral view). Images not to scale

the phylum dates back to at least the Early Cambrian some 540 million years ago, but is most likely rooted deep in the Precambrian (Vendian) era (Parkhaev 2008).

Man and Mollusk

Humans have had long-standing ambivalent relationships with mollusks. Oysters, scallops, cockles, mussels, octopods, cuttlefish, and certain snails provide a welcome protein source, and gastropod, bivalve, and scaphopod shells have been widely used for decorative and monetary purposes for centuries (including, e.g., the commercial production of pearls). On the other hand, the attractive cone snails (*Conus*) cause regular, fatal accidents among shell collectors, but at the same time high hopes are put in research on the powerful neurotoxins produced in their salivary glands for pain relief and the treatment of chronic neural diseases including Alzheimer's, Parkinson's, or depression (Anderson and Bokor 2012). These current approaches may be seen as a continuation of a long tradition of mollusks as contributors to neurobiological research which, for example, has led to detailed reconstructions of the immensely complex cephalopod brain (reviewed in Nixon and Young 2003) and the discovery and distribution of neuroactive substances (Messenger 1996).

A number of mollusks are viewed as notorious pest species, such as several freshwater gastropods that house intermediate developmental stages of important human pathogenic parasites, including flukes and nematodes, which may cause severe medical problems, particularly in sub-Saharan Africa and Southeast Asia. Semisessile bivalves such as the blue mussel (*Mytilus*) or the zebra mussel (*Dreissena polymorpha*) may inflict economic damage as biofouling organisms that attach to plumbing and ship bodies by their byssus threads. *Dreissena* represents one of the several highly competitive neozoan mollusks with a particularly broad ecological tolerance that has resulted in rapid colonialization of formerly non-native habitats, with in parts dramatic consequences for the native fauna. This is due to its high efficiency as adult filter feeder (which has

significant impact on the plankton communities in the respective habitats, for example, in freshwater streams and lakes throughout Europe and North America), in combination with fast distribution rates by free-swimming veliger larvae that result from seasonal mass spawning events. Other important invasive species include the terrestrial pulmonate gastropod *Arion vulgaris* (Spanish slug) that has spread throughout Europe during the last half-century, probably from Portugal or France. Originally constrained in reproductive success by dry climates, the moister and cooler conditions in Central Europe have led to regional explosions in individual numbers of this species, thereby causing severe horticultural and economic damage.

These examples illustrate that molluscan evolutionary and ecological success is not merely historical, but instead an ongoing process that is intimately linked to the reproductive and developmental biology of the phylum's protagonists. Considering the numerous aspects in which molluscan developmental biology plays a vital role, ranging from applied approaches such as pest control and protein production to evolutionary-oriented research, the importance of comparative developmental data becomes immediately obvious. This potential has, however, only been scarcely exploited to date, which is at least partly due to the lack of a solid pool of model system species from the various class-level taxa from which ontogenetic material can readily be obtained (see boxed text). Thus, today's comparative developmental malacologists still need to travel the world to collect adult specimens from natural habitats and experiment with rearing conditions including induction of spawning and (in many aquatic species) metamorphosis, often with erratic success. In times of increased pressure on high publication output, these obstacles still render mollusks relatively unattractive models for many developmental biologists. Thus, what we *do* know on the various aspects of molluscan comparative evolutionary developmental biology sadly lags behind of what we *should* know, especially given the above-mentioned considerably strong bonds between humans and mollusks on various grounds.

Potential Molluscan Models in EvoDevo

At present, hardly any molluscan species is available that can be reliably reared in high numbers and over many generations year round under controlled laboratory conditions from spawning through the various life cycle stages to sexual maturity. Sole exceptions are a few direct developing and phylogenetically derived gastropods with internal fertilization such as the great pond snail, *Lymnaea stagnalis*. As a consequence, research techniques that involve transgenesis or reverse genetics (RNAi, morpholinos) can still not be routinely performed on any molluscan model. Given their cryptic lifestyle and their often long period until sexual maturity, there is little hope that this will change anytime soon for most molluscan clades.

Luckily enough, however, some bivalves and gastropods can readily be spawned, yielding thousands of embryos at a time. Thus, although adult specimens still need to be collected from the wild, a few representatives have the potential to develop into models for which advanced molecular methods may be established in the future. However, long planktonic phases including larval planktotrophy usually require additional facilities to culture the respective algal diet. Here, a few examples are mentioned for which significant EvoDevo data have been published, with the focus on indirect developing species in the gastropods and bivalves, although other candidates exist and are explored in various labs (e.g., the freshwater pond snail *Lymnaea* or the Pacific oyster, *Crassostrea gigas*, for which the genome has recently been sequenced; see http://gigadb.org/dataset/100030).

A. Bivalvia: *Saccostrea kegaki*

The Japanese spiny oyster *Saccostrea kegaki* occurs along Japanese, Korean, and Taiwanese shorelines. It is an encrusting oyster which may easily be collected in rocky intertidal zones. Due to its small size, it is of no commercial interest although it is abundant and fulfills numerous criteria that render it suitable as an organism for aquaculture. In Korea, oogenesis is initiated in April when the water temperature reaches 16.5 °C. First spawning occurs in July at around 27 °C, and mature adults go into resting stage in November. Eggs are relatively small and measure 45–50 µm in diameter. Zygotes undergo first cleavage after 1.2 h postfertilization (hpf), and the blastula stage is reached at approximately 5 hpf at 24 °C. Gastrulation occurs at 7 hpf, the trochophore stage is reached by 9.5 hpf, and D-shaped larvae develop after 20–25 hpf. Larvae measure approximately 40 µm in length and are thus in principal well-suited for whole-mount in situ hybridization or immunochemical experiments, but their small size renders data interpretation (e.g., localization of gene transcripts in relation to developing organs) often difficult, even if the specimens are sectioned. Due to their largely asymmetric shape as adults, oysters are well-suited for studies concerned with the ontogenetic and molecular bases of left-right asymmetry.

B. Gastropoda: *Ilyanassa obsoleta*

The Eastern mudsnail belongs to the Caenogastropoda and occurs on mud and sand flats as well as in estuaries along the Eastern and Western coastline of the USA. Its whorled conical shell is dark brownish to black and 1.5–3 cm in length. *I. obsoleta* may also host a range of trematode species including those that cause "swimmer's itch." The Eastern mudsnail is a facultative scavenger and feeds on detritus. Animals can also be purchased from the Marine Resources Center of the Marine Biological Laboratory in Woods Hole, MA, USA (http://www.mbl.edu). Adults are easily kept in the laboratory, fertilization is internal, and large numbers of fertilized eggs are laid throughout the year, although their natural spawning season is spring to early summer. Development takes 7–10 days from egg laying to hatching of the planktotrophic veliger larvae and another 2–2.5 weeks to metamorphic competence at 23–25 °C. Embryos have a size of 200–300 μm. They are right sized for whole-mount in situ hybridization or immunochemical experiments. The adult and embryonic morphology of *I. obsoleta* is well-studied and serves as an excellent reference for future developmental analyses. *I. obsoleta* adheres to the typical spiralian developmental program, and its relatively large and transparent blastomeres facilitate manipulation of embryos (electroporation, cell ablation, dye injection). Accordingly, *I. obsoleta* has served as a model for developmental studies concerned with early cleavage, the asymmetry of cell division, and gene expression, and still continues to do so.

C. Gastropoda: *Crepidula fornicata*

Also a caenogastropod, the common slipper shell lives intertidally along the East Coast of the USA and has been introduced to several other parts of the world. Adults are suspension feeders, long-lived with 6–11 years, are easy to keep in seatables with running seawater, and tolerate fluctuations in salinity. As a protandric hermaphrodite, *C. fornicata* changes its sex from male to female during adulthood. Adult individuals are usually attached to each other in stacks with the oldest and largest animal located at the base of the stack. Fertilization is internal, and gravid adults produce numerous eggs. Egg laying may occur spontaneously or may be induced by chilling and subsequent warming of the seawater in which the adults are kept (e.g., from 14 to 25 °C). Until hatching, brood care takes place, and the 1,000–50,000 embryos per female hatch as planktotrophic veliger larvae.

Embryos may also be separated from the mother animal, allowing for clutches to be reared individually (preferably with the addition of antibiotics to avoid fungal and bacterial growth). Fertilized eggs measure about 180 μm, the first cleavage occurs after 7.3 h, and the 25-cell stage is reached at 20 h after oviposition at 21–22 °C. *C. fornicata* produces nurse eggs in the capsules. The species does not exhibit a distinct trochophore stage, and the first veliger larvae may develop after 7 days. Hatching occurs after 11–12 days, and the veliger larvae metamorphose after several weeks. Cell lineage studies have been carried out on *C. fornicata*, and its embryogenesis is well documented. Manipulation of embryos may be performed, and several experimental protocols for, e.g., microinjection of blastomeres with fluorescent dyes, are available.

D. Cephalopoda: *Sepia officinalis*

The European cuttlefish *Sepia officinalis* occurs in the Mediterranean, the Atlantic Ocean, the North Sea, and parts of the Baltic Sea. Due to its edibility, its adequate size, and its mass spawning events in spring, it is of importance for local fishery industries and a promising candidate for future aquaculture enterprises. Its diet is variable and includes crustaceans, fish, and other cuttlefish. The dorsal mantle length may reach more than 40 cm; however, in captivity, depending on aquaria size and diet, adults may only grow up to 10 cm. *S. officinalis* was one of the first cephalopods for which the life cycle was closed in captivity. In their natural environment egg clutches are attached to submersed hard substrates during spring and early summer which can readily be collected. In captivity, eggs can also be obtained from adults which may lay eggs for a period of up to 7 months per year. The gelatinous envelopes of freshly laid eggs of *S. officinalis* are approximately

1 cm in size and black in color due to the addition of ink during egg laying. They become transparent with time. The opaque envelopes are easily removed, and early developmental stages can be observed through the transparent chorion. From fertilization to hatching, it takes less than 50 days at 20 °C or less than 100 days at 15 °C. After hatching, individuals are easy to rear to sexual maturity, but tend to be cannibalistic. Research areas concerned with camouflage, communication, and related behavioral patterns including their molecular underpinnings greatly profit from *S. officinalis* as a model. Moreover, various studies on brain development have been carried out. The large-sized embryos are, however, not ideal for whole-mount immunochemical or in situ hybridization experiments. First estimates consider the genome of *S. officinalis* with 4.5 Gb as rather large among cephalopods.

E. Cephalopoda: *Idiosepius notoides*

The southern pygmy squid *Idiosepius notoides* lives in seagrass beds off Eastern Australia. Adults reach no more than 2.5 cm in size and endure rapidly changing environmental conditions such as fluctuations in temperature, oxygen, and salinity levels. Due to their minute size, pygmy squids are of no commercial value. In captivity, adults feed on small crustaceans and fish and appear not to be cannibalistic. They attach egg clutches to glass surfaces or other items provided. At 23–25 °C water temperature, development takes approximately 10 days from egg laying to the hatching squid. The first days after hatching are crucial for rearing *Idiosepius notoides*, and hatchlings have not been carried over this stage and eventually die of unknown reasons. With approximately 0.5–1 mm in size, early developmental stages are small compared to those of other cephalopods. Being rather transparent, they are ideally suited for immunochemical and in situ hybridization experiments. First estimates suggest that *I. notoides* has a relatively small genome (2.1 Gb), which renders it an ideal candidate for sequencing efforts.

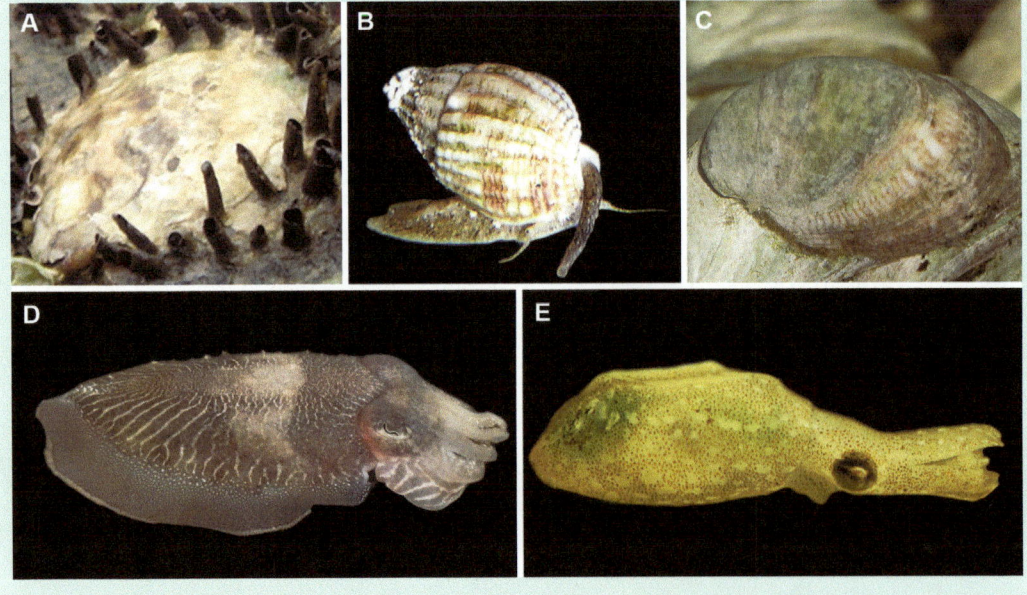

Molluscan Origin and Phylogeny

The undisputed monophyly of Mollusca implies that the wide morphological diversity exhibited by their recent (and extinct) representatives originated from a common body plan of an early ancestor, often conceptually, but nevertheless oddly, referred to as "hypothetical ancestral mollusk (HAM)." Obviously, assessing the morphological radiation that emerged from that very real last common molluscan ancestor, together with its evolutionary and developmental driving forces, firstly calls for a reconstruction of the ancestor itself. However, with a solid agreement on molluscan inter- as well as intrarelationships still largely lacking, little progress has hitherto been made to this end. This may, however, be about to change. Several independent phylogenetic analyses employing large sequence datasets have confirmed a classical hypothesis that had been largely dormant for the last 15 or 20 years, namely, that the shell-less, wormy aplacophoran taxa Neomeniomorpha and Chaetodermomorpha, together with the Polyplacophora, form a monophyletic clade, the Aculifera (Fig. 7.2). This assemblage is currently believed to form the sister taxon to all remaining mollusks – the primarily single-shelled Conchifera (Kocot et al. 2011; Smith et al. 2011; Vinther et al. 2012). Such a dichotomy at the base of the molluscan tree, however, still leaves the question open as to which of the aculiferan or conchiferan taxa bears most resemblance to their last common ancestor. Accordingly, questions such as *Did the mollusks derive from an aplacophoran-like forefather? Or, rather, from a chiton-like creature? Was it a uni-shelled, conchiferan-like species, maybe resembling a monoplacophoran?* remain unanswered.

Usually, outgroup comparisons are an important means for assessing ancestral character states of a given taxon. Potential molluscan sister groups among the Lophotrochozoa abound, but the nature of the closest molluscan ally remains controversial. Various scenarios have been proposed, including the reoccurring annelid-mollusk hypothesis, often favoring a segmented ancestral mollusk, whereby the repetitive muscular, neural, and skel-etal elements in the polyplacophoran body plan are interpreted as remnants of ancestral segmentation. However, an annelid-like posterior growth zone is absent in mollusks, and detailed comparative analyses of polychaete and aculiferan neuromuscular development and skeletogenesis have uniformly rejected the idea of a segmented molluscan ancestor (Friedrich et al. 2002; Wanninger and Haszprunar 2002a; Nielsen et al. 2007; Wanninger 2009; Haszprunar and Wanninger 2012; Scherholz et al. 2013). Morphologically, by far the strongest support is provided for a scenario that suggests a mollusk-entoproct clade based on numerous synapomorphies shared by larval entoprocts and larval and adult aplacophorans and polyplacophorans. This includes a creeping foot, anterior sensory cirri, a highly complex larval apical organ, and a tetraneurous nervous system (hence Tetraneuralia, formerly Lacunifera or Sinusoida concept; see Bartolomaeus 1993; Ax 1999; Wanninger et al. 2007; Haszprunar and Wanninger 2008; Wanninger 2009).

With (molecular) phylogeny still painting a rather blurred picture of molluscan affinities and early evolution, developmental studies may provide the key to clarify important issues. Recent work on the hitherto largely neglected aplacophorans has shown that muscle development in the neomeniomorph *Wirenia argentea* undergoes a series of polyplacophoran-like stages with distinct muscular systems being confined to both lineages alone (Scherholz et al. 2013; see below). This strongly argues for a polyplacophoran-like ancestor of neomeniomorphs and thus probably the entire Aculifera – which provides us with the basal condition for at least one of the two early molluscan branches.

In any case, and despite the contentious phylogenetic issues, the unmatched morphological plasticity of their body plan renders mollusks prime organisms for the study of the evolution of morphological novelties and their underlying molecular, cellular, and morphogenetic mechanisms. As a consequence of this wide phenotypic variation, clear-cut autapomorphies for the phylum are scarce, since all potential characters have undergone significant remodeling from their

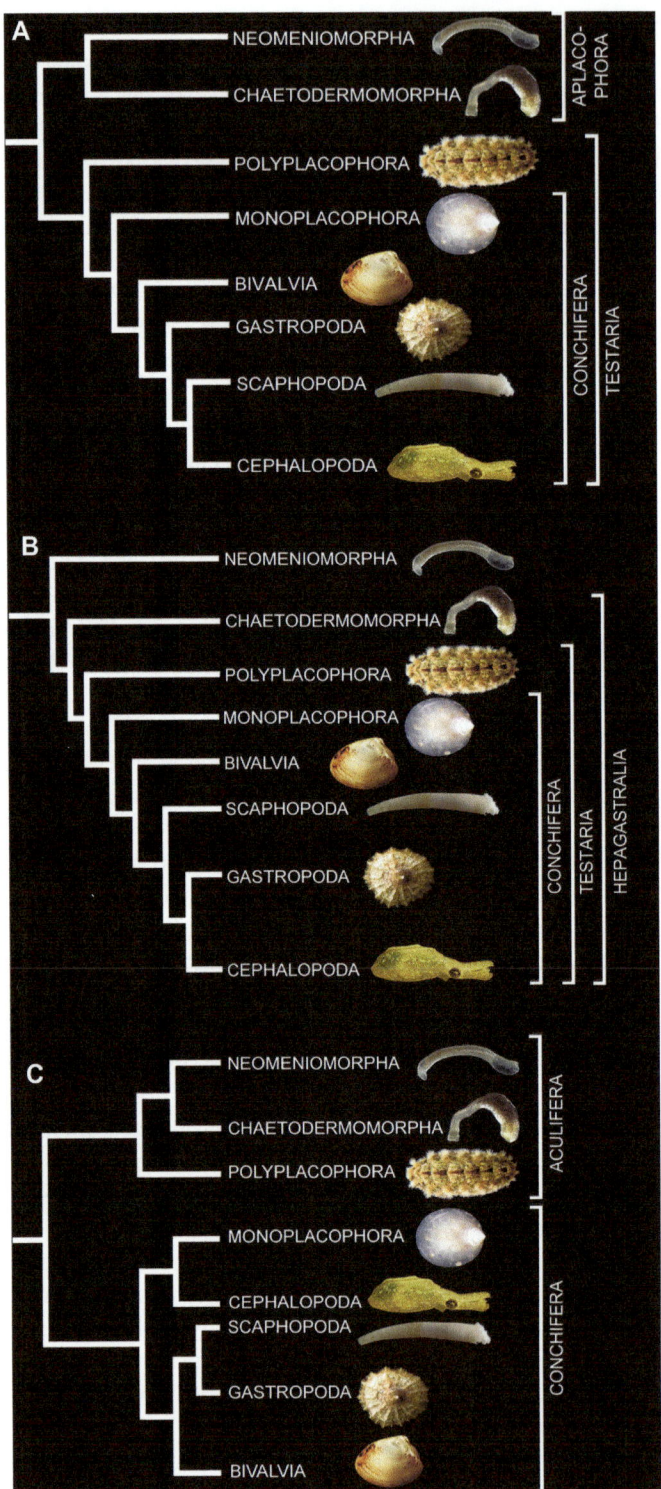

Fig. 7.2 Competing hypotheses on the interrelationships of molluscan class-level taxa. (**A**) After Waller (1998), based on morphological data. (**B**) After Haszprunar (2000), based on morphological data. (**C**) After Smith et al. (2011), based on phylogenomic data. Monoplacophoran image © Michael Schrödl, 2015. All Rights Reserved; neomeniomorph image courtesy of Maik Scherholz

ancestral state. Even the most commonly recognized molluscan feature, the radula (rasping tongue) has been lost in an entire class-level lineage, the Bivalvia, and complete reorganization of the body architecture has taken place within the heterobranch gastropods and the bivalves, often resulting in secondary flatworm-like (e.g., marine nudibranchs) or cylindrical vermiform organisms (e.g., interstitial slugs such as *Rhodope* and *Helminthope* or shipworms such as *Teredo*). Again, the molecular and developmental base that underlies these drastic phenotypic deviations from a common ground plan remains unexplored.

On the cellular level, the free-floating rhogocytes (pore cells), hemocoelic cells that have been linked to hemocyanin synthesis, excretion, and various metabolic processes, are thought to be confined to mollusks alone and may constitute the only definite feature exhibited by all recent mollusks (Haszprunar 1996; Stewart et al. 2014). On organ system level, a closer look reveals that almost all taxa exhibit a foot of some sort, which may serve as creeping, burrowing, swimming, sensory, or grooming device. A dorsoventral musculature that often attaches to the shell(s) (therefore often referred to as "shell musculature" in conchiferans) and intercrosses above the foot sole is present in all class-level taxa (Haszprunar and Wanninger 2000; Wanninger 2009). In addition, a tetraneurous nervous system is prevalent, with one pair of longitudinal nerve cords usually running ventrally and a second pair of nerve cords situated more dorsolaterally to it.

Although we are only beginning to understand the molecular, cellular, and morphogenetic mechanisms that underlie the establishment of molluscan key features and evolutionary novelties, important recent progress has been achieved. Thus, in the wake of modern-day EvoDevo, mollusks re-enter the limelight of comparative developmental research. It is thus about time to summarize the landmarks of our current knowledge on molluscan evolutionary developmental biology and point towards the numerous unclarified issues associated with this fascinating phylum that deserves our special attention.

EARLY DEVELOPMENT

Reproduction and Cleavage

Many mollusks, including the chaetodermomorphs, polyplacophorans, bivalves, scaphopods, and basal gastropods, are dioecious broadcast spawners with external fertilization and indirect development via trochophore and, in most gastropods and bivalves, veliger-type larvae (Figs. 7.3 and 7.4). Neomeniomorphs are hermaphroditic

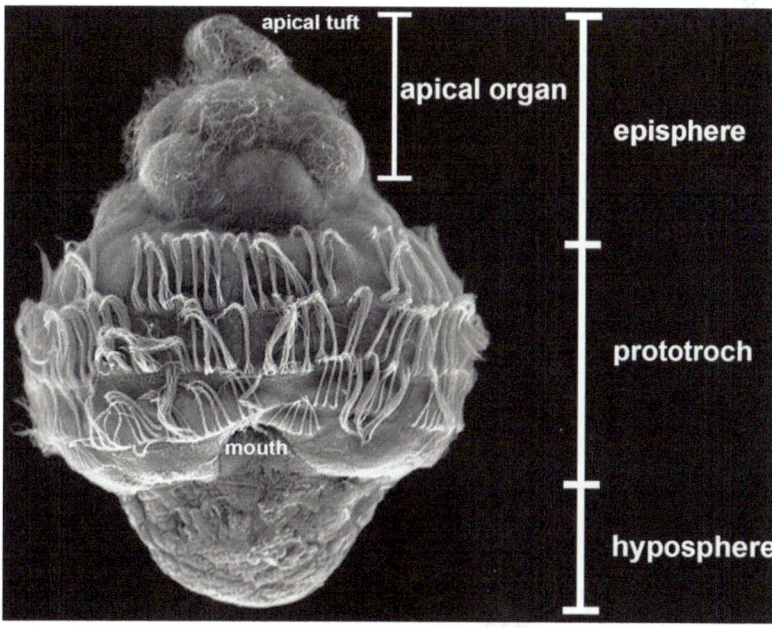

Fig. 7.3 Gross morphology of a molluscan trochophore-like larva. Here, an early scaphopod larva is depicted by scanning electron microscopy, with typical partitioning of the body into pretrochal episphere (including the apical organ), large, ciliated trochoblasts forming the prototroch, and the posttrochal hyposphere. Anterior is up and total size of the larva is 300 μm

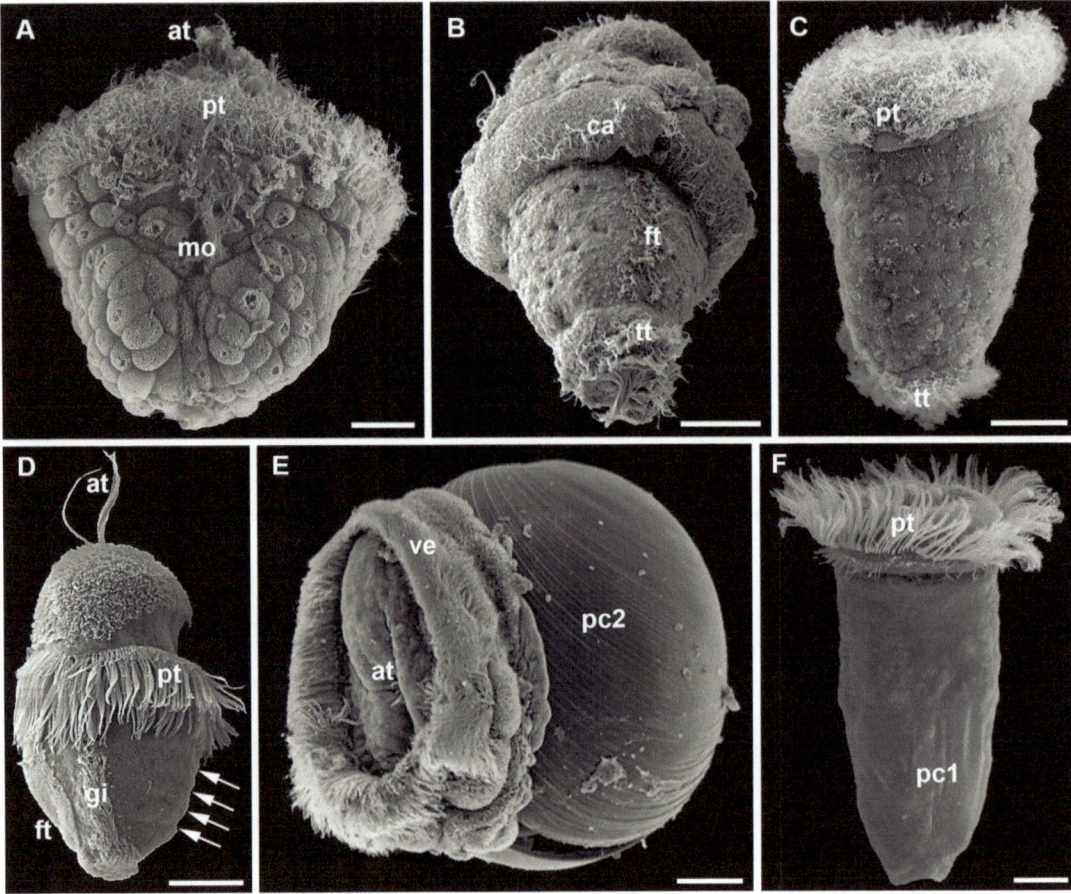

Fig. 7.4 Diversity of molluscan larvae. All images are scanning electron micrographs with anterior facing up except for (**E**) where anterior is to the left. Scale bars are 25 μm in (**A–C**) and 50 μm in (**D–F**). (**A**) Ventral view of an early trochophore larva of a basal gastropod, the patellogastropod *Lottia*, showing apical tuft (*at*), prototroch (*pt*), and mouth (*mo*). Note the individual cells in the larval hyposphere. (**B**) Ventroposterior view of a later stage larva of the neomeniomorph *Wirenia argentea* showing cells of the apical cap (*ca*), ciliary band of the developing foot (*ft*), and telotroch (*tt*) (Modified from Todt and Wanninger (2010)). (**C**) Larva of the chaetodermomorph *Chaetoderma nitidulum* with prominent prototroch (*pt*) and telotroch (*tt*).

(**D**) Late-stage polyplacophoran larva (*Mopalia muscosa*) in lateral view with long cilia of the apical tuft (*at*), prototroch (*pt*), and developing shell plates (*arrows*). The anlagen of the foot (*ft*) and mantle fold (girdle, *gi*) are visible as distinct, ciliated regions. (**E**) Lateral left view of a bivalve pediveliger larva of the shipworm *Lyrodus pedicellatus* with larval shell (protoconch II, *pc2*), velum (*ve*), and reduced apical tuft (*at*) (image courtesy of Reuben Shipway). (**F**) Late-stage larva close to metamorphic competence of the scaphopod *Antalis entalis* in ventral view with elongated embryonic shell (protoconch I, *pc1*). While the apical tuft has been lost, the cilia of the prototroch (*pt*) are still retained

and lay clutches of uncleaved zygotes which likewise develop into planktonic larvae. Both dioecy and hermaphroditism occur in the monoplacophorans, and indirect development via brooded larvae is assumed. There is a strong tendency towards hermaphroditism and direct development in marine and terrestrial heterobranch gastropods, sometimes involving brood care. Larval stages are entirely lacking in the dioecious cephalopods, and brood care is particularly pronounced in the octopods, which may carry the developing embryos either in the mantle cavity (as, e.g., in the pelagic *Argonauta*) or guard and ventilate them in rock crevices (e.g., *Octopus*).

Some free-spawners may be successfully inseminated artificially following dissection of the gonads (e.g., several patellogastropods and numerous bivalves), but often the spawning event as such is required to yield fertilizable oocytes. Follicle cells are common and oocytes are brightly colored in orange, red, yellow, or green in some species. Polyplacophoran eggs are often covered by a characteristic, richly sculptured, species-specific hull (Eernisse and Reynolds 1994). In the polyplacophoran *Mopalia muscosa*, sperm mitochondria and centrioles do not penetrate the egg surface, a phenomenon that so far appears to be unique among metazoans (Buckland-Nicks 2013). In most mollusks, meiosis is only completed after fertilization, resulting in shedding of the polar bodies at the animal pole of the zygote (Fig. 7.5; Longo 1983). Accordingly, the animal-vegetal axis of the embryo is discernible from this stage onwards. The point of fusion of the sperm cell with the oocyte determines the orientation of the first cleavage furrow. Cleavage is usually total (holoblastic) and may be equal or unequal. Cytoplasmic segregations may result in the formation of so-called polar lobes as early as prior to first cleavage (Fig. 7.5), and several gastropods (e.g., *Ilyanassa*) and the scaphopod *Antalis* exhibit a so-called trefoil stage which results from a polar lobe being formed following first cleavage (Fig. 7.5; van Dongen and Geilenkirchen 1974, 1975, 1976). After second cleavage the macromeres A, B, C, and D are established and as a consequence also the anterior-posterior axis of the embryo, which runs through the B–D cells. At this stage the typical spiral cleavage pattern is initiated, whereby the four daughter cells (micromeres) come to lie above each cleavage furrow of their mother cells at the animal pole. Each subsequent cleavage cycle results in a set of additional micromeres that involves such a 45° twist of their mitotic spindle axes relative to that of the mother cells, however, with alternating clockwise and counterclockwise chirality between generations of developing micromeres. As a result, the cells in the cleaving embryo appear spirally arranged if viewed from the animal pole (Fig. 7.5). In the exceptionally large and yolky cephalopod embryos, the spiral pattern has given way to dis-

coidal cleavage, whereby the early embryo forms as a monolayered disk of blastomeres (blastodisc) on the animal pole of the egg (Fig. 7.5). Here, the polar bodies are ejected some hours after fertilization, and the first cleavage furrow emerges close by and determines the prospective longitudinal axis (Naef 1928; Boletzky 1989; Boletzky et al. 2006). In this case, cleavage solely involves the blastodisc, and the second cleavage furrow runs perpendicular to the first cleavage plane, while the third cleavage plane is located again perpendicular to the second one (Fig. 7.5).

Cell Lineage and the 4d Cell

While a number of classical accounts on molluscan (mostly gastropod) early embryology were generated in the late nineteenth and early twentieth century (Fig. 7.6; Lacaze-Duthiers 1858; Kowalevsky 1883a, b; Lillie 1895; Conklin 1897; Heath 1898; Wilson 1898; Meisenheimer 1901; Robert 1902; Casteel 1904; Wierzejski 1905; Clement 1986; Van Dam 1986), most of our detailed knowledge concerning the subsequent fates of individual cells produced during early cleavage cycles stems from a very limited number of polyplacophoran (*Chaetopleura apiculata*) and gastropod representatives (e.g., the patellogastropod limpet *Patella vulgata*, the slipper shell *Crepidula fornicata*, and the mudsnail *Ilyanassa obsoleta*) (Render 1991, 1997; Henry et al. 2004; Hejnol et al. 2007; Goulding 2009; see also Nielsen 2004, Hejnol 2010 for review). Some data on early cell fates are available for a scaphopod (*Antalis entalis*, formerly *Dentalium dentale*; van Dongen and Geilenkirchen 1974, 1975, 1976). Rudimentary, century-old reports are available for bivalves, both on freshwater species (the painter's mussel *Unio* and the zebra mussel *Dreissena polymorpha*; Lillie 1895; Meisenheimer 1901), supplemented by few modern accounts (on the Japanese spiny oyster *Saccostrea kegaki* and the Japanese purple mussel *Septifer virgatus*), but these largely focus on the cell lineage of the bivalved shell (Kin et al. 2009; Kurita et al. 2009). No cell genealogies have hitherto been reconstructed for any neomeniomorph,

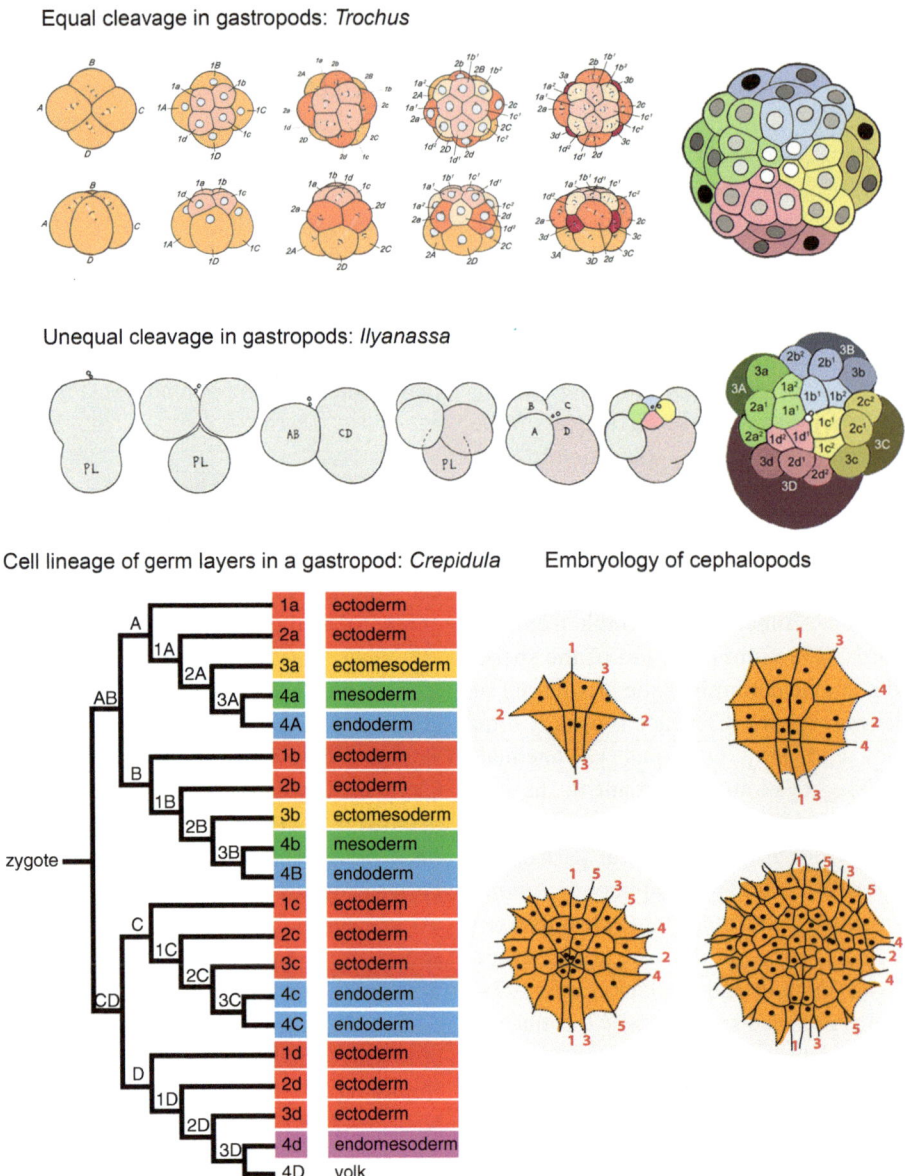

Fig. 7.5 Cleavage patterns in mollusks. *Top*: Typical spiral cleavage pattern of the equal cleaving gastropod *Trochus*, first forming four macromeres that define the four quadrants of the embryo (A–D), then the subsequent sets of micromeres emerge (A–D). Daughter cells always come to lie in the cleavage furrows of their mother cells. Lineages of the four quadrants are color-coded in the large embryo to the right. A-quadrant, *green*; B-quadrant, *blue*; C-quadrant, *yellow*; D-quadrant, *magenta*. *Upper row* of cleaving embryos: animal view; *lower row*: lateral view (slightly modified from Goulding (2009)). *Middle*: Spiral cleavage pattern of the unequal cleaving gastropod *Ilyanassa*. The embryo forms a first polar lobe (*PL*) prior to the establishment of the four quadrants (macromeres A–D), resulting in an intermediate "trefoil stage" and a

second polar lobe at second cleavage. *Small dots* indicate polar bodies. Micromeres are color-coded by quadrant in both embryos to the right. A-quadrant, *green*; B-quadrant, *blue*; C-quadrant, *yellow*; D-quadrant, *magenta*. From Goulding (2009). *Bottom left*: Micromere and macromere contributions to the three germ layers in the gastropod *Crepidula*. *Bottom right*: Discoidal cleavage of cephalopods results in a monolayer of blastomeres (blastodisc). Animal view of an embryo after the third (*upper left*), fourth (*upper right*), fifth (*lower left*), and sixth (*lower right*) cleavage. *Red numbers* indicate the first to fifth cleavage furrow. Blastomeres are located in the center of the cytoplasmic cap, while blastocones surround the latter and are continuous with the yolk syncytium (*stippled lines*) (Redrawn after Boletzky (1989))

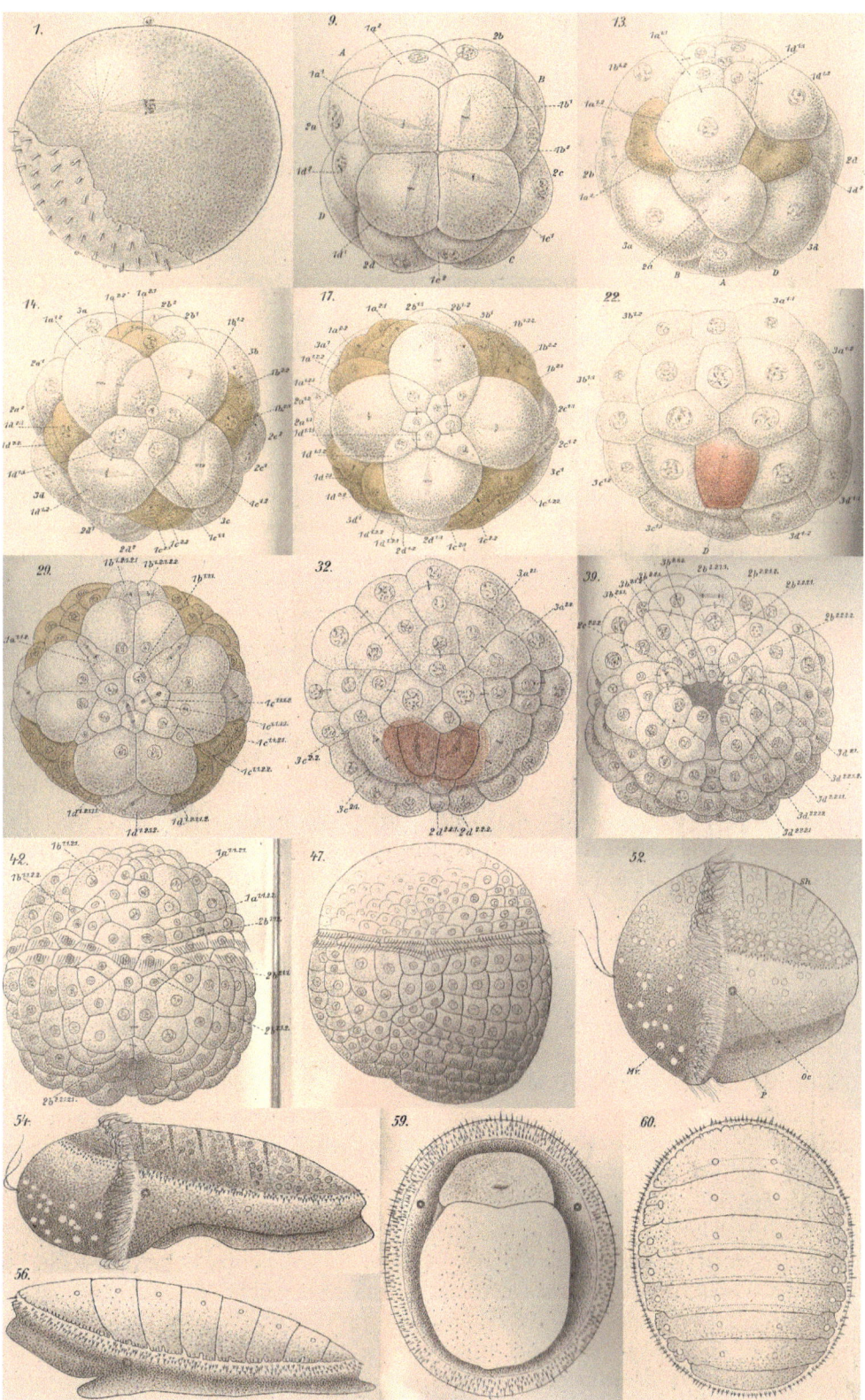

Fig. 7.6 Polyplacophoran cell lineage and postembryonic development. Camera lucida drawings of selected developmental stages from the classical work by Heath (1898). Primary trochoblasts are in *yellow*, mesoderm precursor cells in *red*

chaetodermomorph, monoplacophoran, or cephalopod species. Accordingly, comparative analyses of molluscan cell fates are largely restricted to the few taxa summarized in Table 7.1. As expected by the shared conserved spiral cleavage program, these lineages show a relatively high degree of conservation with respect to the contribution of individual blastomeres to certain germ layers (Fig. 7.5). However, differences do occur if morphogenesis of entire organ systems is traced back to the cellular level, but such detailed investigations remain restricted to even fewer species (Table 7.1).

In general, the macromeres (3A–D, 4A–D) and the micromeres 4a–d give rise to endoderm-derived structures in mollusks (see Nielsen 2004). These may include (parts of the) larval excretory systems, the heart, parts of the digestive tract including the midgut gland, stomach, and style sac, as well as various sets of muscles (Table 7.1). The mesentoblast 4d, a derivative of the mesentoblast mother cell 3D, is the spiralian-specific mesoderm founder cell from which a pair of mesodermal bands arises (see Lambert 2010 for brief review). However, detailed studies on the caenogastropod *Crepidula* have shown that 3D may also contribute to endoderm-derived features such as the gut as well as to ectodermal subportions of the larval kidney complex (Hejnol et al. 2007; Lyons et al. 2012). There is indication that 4d may also give rise to cells of the early germ line, both in *Crepidula* (Lyons et al. 2012) and in the freshwater clam *Sphaerium* (Ziegler 1885; Woods 1931, 1932), but these assumptions need detailed studies on gonadogenesis for definite proof.

A specific role in molluscan development is assigned to the 3D macromere, which is induced by the overlying micromeres to become an important signaling ("inductive") center and developmental "organizer." This happens sometime during the 24–63 cell stages and has been described for the gastropods *Patella* and *Haliotis* as well as the polyplacophoran *Acanthochiton* (van den Biggelaar and Guerrier 1979; Arnolds et al. 1983; Boring 1989; van den Biggelaar 1996; see also Wanninger et al. 2008; Nielsen 2012). In experimental studies, inhibition of 3D induction has resulted in non-

bilateral and largely disorganized embryos similar to ones raised after polar lobe deletion (van den Biggelaar 1977; Martindale et al. 1985; Martindale 1986; Kühtreiber et al. 1988; Damen and Dictus 1996; Lambert and Nagy 2003). Chemical inhibition of the 3D cell produced radialized gastropod, polyplacophoran, and, to some degree, scaphopod embryos (Gonzales et al. 2007). There is recent evidence from the caenogastropods *Ilyanassa* and *Crepidula* that the mesentoblast 4d likewise may function as an organizing signaling center (Henry and Perry 2008; Rabinowitz et al. 2008).

In the five mollusks studied in detail, the first quartet micromeres (1a–d) are destined to contribute to the formation of ectodermal structures, predominantly in pretrochal domains (Table 7.1). This includes the primary and secondary trochoblasts that form the prototroch and velum, the major larval swimming devices. The main larval sensory structure, the apical organ, that is typical for most lophotrochozoan larvae (Figs. 7.7 and 7.8A, F; see Chapters 6 and 9), also derives from the 1a–d lineages. In *Crepidula*, where cell lineage has been followed until late-stage veliger larvae, it could be shown that a type of sensory cell characteristic for lophotrochozoan apical organs, the flask-shaped receptor cells, as well as the anlagen of the adult cerebral ganglion, likewise derive from these micromeres (Hejnol et al. 2007). The larval eyes form from descendants of 1a and 1c in *Crepidula* and from 1c progenies alone in *Ilyanassa*, which also give rise to the right tentacle in this species (Render 1997). In *Crepidula*, parts of the mantle likewise stem from the 1a–c lineage, and nerves innervating cerebral structures arise from the 1d cell. Polyplacophoran larvae typically have ectodermal spicules, and the anterior ones are derived from the 1a–d lineage. In these larvae alone, the 1b micromere contributes to the mouth.

As we have seen, the first quartet micromeres are exclusively confined to ectodermal fates. The situation is somewhat different for the second quartet micromeres (2a–2d), which at least in scaphopods and gastropods appear to be involved in the formation of a specific portion of the molluscan mesoderm – the ectomesoderm. In polyplacophorans, however, these micromeres

Table 7.1 Molluscan cell fates

Class	Polyplacophora	Scaphopoda	Gastropoda		
Genus	Chaetopleura	Antalis	Patella	Crepidula	Ilyanassa
Blasto-mere			Contribution		
First quartet micromeres					
1a	Pretrochal ectoderm, spicules, prototroch	Apical organ, prototroch	Pretrochal ectoderm, prototroch	Pretrochal ectoderm, velum, apical organ (flask cells), cerebral ganglion, mantle, eye	Velum, apical organ
1b	Pretrochal ectoderm, mouth	Apical organ, prototroch	Pretrochal ectoderm, prototroch	Velum, apical organ (flask cell)	Velum, apical organ
1c	Apical tuft, pretrochal ectoderm, prototroch	Apical organ, prototroch	Pretrochal ectoderm, prototroch	Pretrochal ectoderm, velum, apical organ (flask cells), eye, cerebral ganglion, mantle	Velum, apical organ, eye, tentacle (only right one, left one unknown)
1d	Apical tuft, pretrochal ectoderm, prototroch, spicules	Apical organ, prototroch	Pretrochal ectoderm, prototroch	Pretrochal ectoderm, apical organ (incl. flask cells), head nerves	Velum, apical organ
Second quartet micromeres					
2a	Ectoderm, mouth, prototroch, spicules, eye	Prototroch, ectomesoderm	Posttrochal ectoderm (foot, mantle fold, shell gland), prototroch	Velum, pedal ganglion, statocyst, mantle	Velum, mouth, mantle, foot, statocyst
2b	Ectoderm, prototroch, mouth	Prototroch	Posttrochal ectoderm (foot, mantle fold, shell gland), prototroch, mouth	Velum, mouth, visceral nerve cords, postvelar ectoderm, mantle fold, eye nerves	Velum, mouth, mantle, foot retractor (adult?)
2c	Posttrochal ectoderm, prototroch, spicules, eye	Prototroch, ectomesoderm	Posttrochal ectoderm (foot, mantle fold, shell gland), prototroch	Velum, mantle, statocyst, heart, osphradium	Velum, mantle, statocyst, heart, mouth, foot
2d	Posttrochal ectoderm, shell gland	Posttrochal ectoderm (foot, mantle, shell gland)	Posttrochal ectoderm (foot, mantle fold, shell gland), telotroch	Mantle (incl. nerves), foot (incl. sensory cells)	Mantle, foot
Third quartet micromeres					
3a	Ectoderm, mouth	Ectoderm	Ectomesoderm	Esophagus, muscles (incl. main larval retractor, velum muscle ring)	Velum, esophagus
3b	Mouth	Ectoderm	Ectomesoderm	Esophagus, muscles (incl. velum muscle ring)	Velum, esophagus,
3c	Posttrochal ectoderm, spicules, shell gland	Ectoderm	Posttrochal ectoderm (mantle cavity), mouth, telotroch	Velum, foot, statocyst, larval kidney	Velum, foot, statocyst, mantle
3d	Posttrochal ectoderm, prototroch, spicules, shell gland	Posttrochal ectoderm, foot	Posttrochal ectoderm (mantle cavity), mouth, telotroch	Velum, foot, statocyst, larval kidney	Velum, foot, statocyst, heart

(continued)

Table 7.1 (continued)

Class			Polyplacophora	Scaphopoda		Gastropoda	
Genus			Chaetopleura	Antalis	Patella	Crepidula	Ilyanassa
	Blastomere				Contribution		
Fourth quartet micromeres	4a		Endoderm	Endoderm	Endoderm	Larval kidney	Endoderm
	4b		Endoderm	Endoderm	Endoderm	Larval kidney	Endoderm
	4c		Endoderm	Endoderm	Endoderm	Stomach, style sac	Endoderm
	4d		Mesentoblast	Mesentoblast	Mesentoblast, 2 mesodermal bands (endomesoderm)	Mesentoblast, gut, larval kidney, muscles incl. larval retractors, mesenchyme, heart, foot tissue, germ cells (?)	Mesentoblast, main larval retractors, gut (parts), heart, larval kidney
Macromeres	3A		Gut	Endoderm	Endoderm	Larval kidney, midgut gland	Midgut gland, main larval retractor
	3B		Gut	Endoderm	Endoderm	Larval kidney, midgut gland	Midgut gland, main larval retractor
	3C		Gut	Endoderm	Endoderm	Stomach, style sac, salivary gland (?)	Style sac, midgut gland, main larval retractor
	3D		Mesentoblast mother cell, gut, larval muscles	Mesentoblast mother cell	Mesentoblast mother cell	Mesentoblast mother cell, mesentoblast, gut, larval kidney, muscles incl. larval retractors, mesenchyme, heart, foot tissue, germ cells (?), yolk	Mesentoblast mother cell
	4A		Endoderm	Endoderm	Endoderm	Midgut gland	Endoderm
	4B		Endoderm	Endoderm	Endoderm	Midgut gland	Endoderm
	4C		Endoderm	Endoderm	Endoderm	Stomach, style sac, salivary gland (?)	Endoderm
	4D		Endoderm	Endoderm	Endoderm	Yolk	Yolk, midgut gland

Data from van Dongen and Geilenkirchen (1974, 1975, 1976), Render (1991, 1997), Dictus and Damen (1997), Henry et al. (2004), Nielsen (2004), Hejnol et al. (2007), Goulding (2009), Lyons et al. (2012)

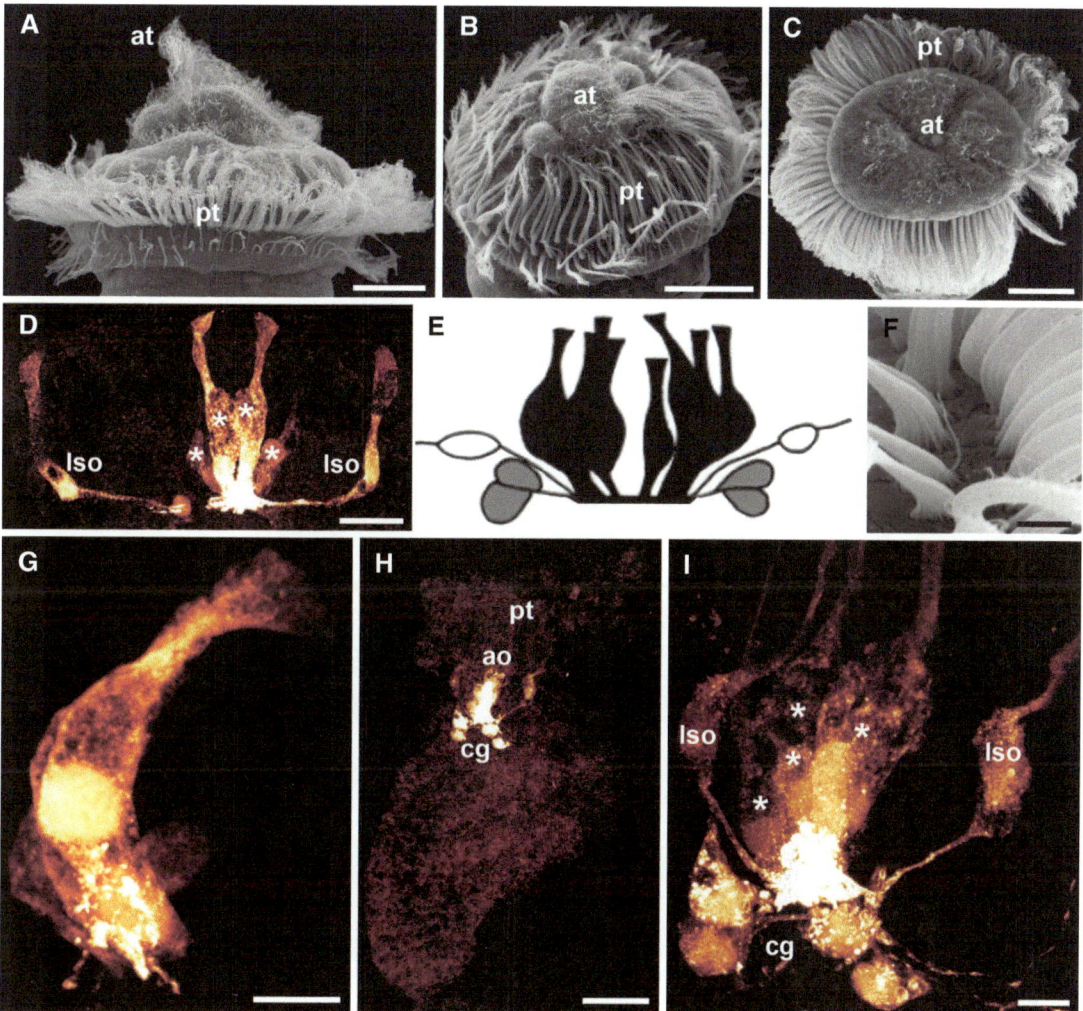

Fig. 7.7 Developmental fate of the molluscan apical organ. (**A–C**) and (**F**) are scanning electron micrographs, (**D**) and (**G–I**) are confocal projections of neural subsets stained with an anti-serotonin antibody. All images are from the scaphopod *Antalis entalis* except for (**E**), which is from the polyplacophoran *Ischnochiton hakodadensis* (= *Stenoplax heathiana*). Scale bars are 50 μm in (**A–C**) and (**H**), 25 μm in (**D**), and 10 μm in (**F, G, I**). (**A**) Lateral view of the anterior region of a trochophore larva with prototroch (*pt*) and well-developed apical tuft (*at*) that is part of the apical organ. (**B**) Anterior view of a late larva at the onset of degeneration of the apical tuft. (**C**) Anterior view of a larva close to metamorphic competence with almost fully degenerated apical tuft. (**D**) Four flask-shaped serotonin-positive cells (*asterisks*) of the apical organ, as typical for many molluscan larvae. In *Antalis* larvae, two additional cells of a putative lateral sense organ (*lso*) are interconnected with the flask cells at their base; dorsal view. (**E**) Schematic representation of the apical organ of the polyplacophoran *Ischnochiton hakodadensis* in dorsal view showing eight flask cells (*black*), four peripheral nonsensory cells, and two additional bipolar neurons (white cell bodies). (**F**) Detail of the prototroch of *Antalis* showing bundles of compound cilia. (**G**) Detail of a flask cell of the apical organ. (**H**) Degenerating apical organ. The flask cells have migrated to the area of the future cerebral ganglia, i.e., behind the prototroch; the first cells of the cerebral ganglia have already formed (*cg*). (**I**) Detail of (**H**) showing the degenerating flask cells (*asterisks*) of the apical organ and the first immunoreactive cells of the cerebral ganglia. In *Antalis entalis*, the apical organ is lost prior to metamorphic competence

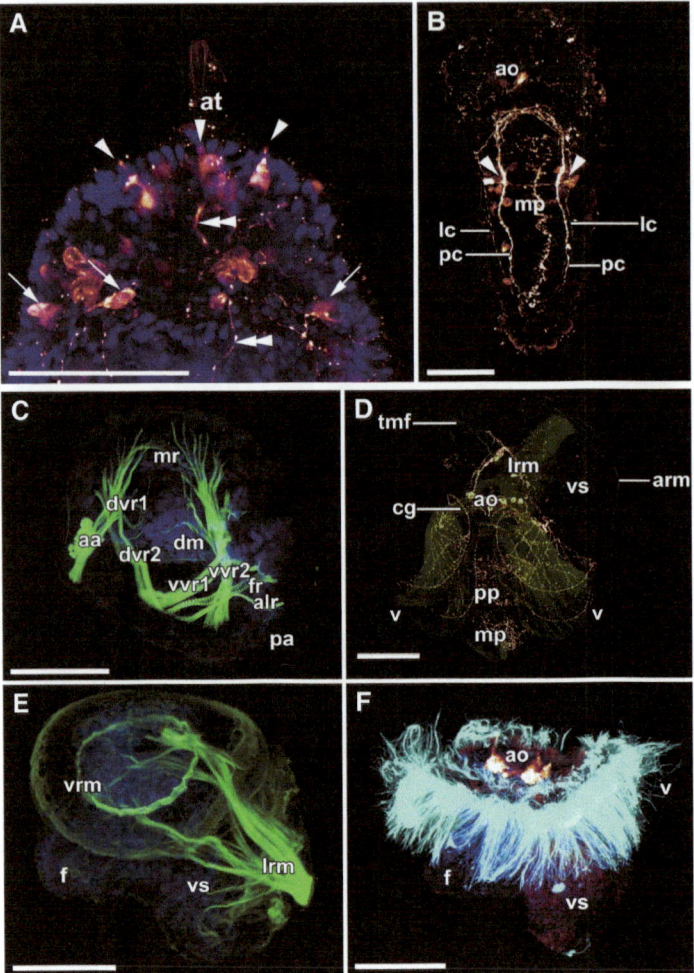

Fig. 7.8 Molluscan larval neuromuscular anatomy. Cell nuclei are stained in blue (HOECHST; lacking in **B**, **D**) and musculature is labeled green (phalloidin, F-actin). Serotonin-like immunoreactive (serotonin-LIR) nervous elements are stained in *red* and cilia (acetylated α-tubulin) are labeled in turquoise in (**F**). All images are confocal projections with anterior facing up (except **D** where it faces down). All scale bars are 50 μm. (**A**) The serotonin-LIR nervous system of the late trochophore of the chiton *Acanthochitona crinita* is composed of flask-shaped cells (*arrowheads*) in the apical organ, additional sensory cells (*arrows*), and various other cell somata which are connected via neurites (*double arrowheads*) to the apical organ or the cerebral commissure (dorsal view). (**B**) Dorsal view of the larval serotonin-LIR nervous system of the neomeniomorph *Gymnomenia pellucida* showing weak staining in the apical organ (*ao*) and lateral (*lc*) and pedal nerve cords (*pc*), with a serotonin-LIR cluster of cells connected to the latter (*arrowheads*). A dense median neural plexus (*mp*) is visible along the longitudinal axis (Image courtesy of Emanuel Redl). (**C**) The complex musculature of the veliger larva of the protobranch bivalve *Kurtiella bidentata* comprises larval retractor systems including dorsal (*dvr1*, *dvr2*) and ventral velum retractors

(*vvr1*, *vvr2*) and an accessory larval retractor (*alr*), as well as the anlagen of future adult muscles such as the anterior (*aa*) and posterior adductor muscles (*pa*), the foot retractor (*fr*), the digestive system (*dm*), and the mantle retractors (Image courtesy of Marlene Karelly). (**D**) The serotonin-LIR nervous system and the musculature of the veliger larva of the gastropod *Aplysia californica* (dorsal view, anterior faces down). The muscle system comprises musculature associated with the propodium (*pp*), metapodium (*mp*), and velum (*v*). Further muscles are the massive larval retractor muscle (*lrm*), transversal muscle fibers (*tmf*), and an accessory retractor muscle (*arm*). The larva exhibits serotonin-LIR innervation of the velum, propodium, metapodium, and visceral sac (*vs*). Serotonin-LIR cells are present in the apical organ and in the anlagen of the cerebral ganglia (*cg*). (**E**) The complex larval musculature of the patellogastropod *Lottia* is, among others, composed of a larval retractor muscle (*lrm*) that connects to the velar ring musculature (*vrm*). Lateral left view (Image courtesy of Alen Kristof). (**F**) The apical organ (*ao*) of this larva of the patellogastropod *Lottia* houses three serotonin-LIR flask-shaped cells. Lateral left view (Image courtesy of Alen Kristof). Further abbreviations: apical tuft (*at*), foot (*f*)

contribute to the (posttrochal) ectoderm. In addition, 2a, 2b, and 2c form trochoblasts. Spicule-forming cells derive from the 2a–c lineage but not from 2d, which instead contributes to parts of the seven shell fields of the larva (Henry et al. 2004). Interestingly, the larval ocelli, which, in contrast to the pretrochally positioned eyes of other spiralian larvae, are situated ventrolaterally behind the prototroch in polyplacophorans, derive from the 2a–c cells. They thus have a different developmental history than other spiralian larval eyes, which emerge from the 1a–c lineage (Henry et al. 2004). This different ontogenetic history may argue against homology of polyplacophoran and other spiralian larval eyes and should be investigated further by comparative lineage-tracing and gene expression studies.

For scaphopods, information on the fate of the second quartet micromeres is scarce. The 2a–c lineages contribute to the three-rowed prototroch and 2a and 2c in addition to the ectomesoderm, while 2d forms posttrochal ectodermal structures of the foot, mantle, and shell gland. In *Patella*, the 2a–d lineages are rather similar, with a 2d contribution to the telotroch (Dictus and Damen 1997). The situation in the two other gastropods investigated, *Crepidula* and *Ilyanassa*, mostly corresponds to *Patella*, with a few additional contribution domains identified in the two caeno-gastropods, such as nervous and sensory structures including the statocyst (2a and 2c in both species), visceral nerve cords and eye nerves (2b in *Crepidula*), sensory cells of the foot (2d in *Crepidula*), heart (2c in both), osphradium (2c in *Crepidula*), and the foot retractor (i.e., shell muscle) (2b in *Ilyanassa*) (Table 7.1). An early work on bivalve cell fates suggests that cells from the 2d lineage also contribute to the shell fields in *Dreissena polymorpha* (Meisenheimer 1901), a finding that was corroborated by a recent study on the Japanese spiny oyster *Saccostrea kegaki* (Kin et al. 2009). In the polyplacophoran *Chaetopleura* and the scaphopod *Antalis*, the third quartet micromeres give rise to various (posttrochal) ectodermal domains. In *Chaetopleura*, 3c and 3d successors are found in spicule- and shell gland-forming cells. In the gastropods, 3a and 3d contribute to not precisely characterized ectome-sodermal regions in *Patella* and to esophageal and velar cells (including muscles) in *Crepidula* and *Ilyanassa*. 3c and 3d contribute to posttrochal ectodermal domains including mouth and telotroch in *Patella* and to the velum, foot, and statocyst in *Crepidula* and *Ilyanassa*. In *Crepidula*, both micromeres also contribute to various cell types associated with the larval kidney.

Taken together, the overall cell lineage patterns appear to be rather conserved among the various polyplacophoran, scaphopod, and gastropod mollusks investigated, at least with respect to their gross morphological formation domains. Exceptions do occur, however, as manifested in the complex genealogy of the prototroch in *Patella* or the diverging lineage history of the polyplacophoran larval eyes. Unfortunately, detailed cell lineage studies in particular of neural and other internal features such as the various larval and adult muscle systems are still scarce. Even for key gross morphological features, such as the various prototrochal, metatrochal, and velar structures as well as the apical organ, more comparative data are needed and – in particular in combination with gene expression studies – should provide a powerful tool for higher-level taxon homology assessments across the Mollusca.

Gastrulation and Mesoderm Formation

In mollusks with small eggs, gastrulation occurs by invagination on the vegetal pole and subsequent formation of a fluid-filled coeloblastula. Species with large, yolk-rich eggs usually exhibit epiboly and a massive sterroblastula. The macromeres are the first cells to invaginate, followed by the fourth quartet micromeres. During subsequent development, the blastopore often narrows considerably and may even close in some species. In most cases, however, it moves in anterior direction and finally comes to lie behind the prototroch, where it develops into the definite mouth (Figs. 7.3 and 7.4A; see Nielsen 2004, 2012). In the large majority of species, the anal opening forms de novo. However, earlier reports on a "deuterostomous" condition in the river snail *Viviparus*, where the blastopore was found to remain in the posterior region where it forms the anus, while the mouth arises secondarily (Dautert 1929; Fernando 1931; Fioroni 1979), indicate that mouth/anus formation in mollusks may be rather plastic (see Nielsen 2004).

As mentioned above, the molluscan mesoderm is derived from two embryonic sources, the ectomesoderm (in gastropods mostly formed by the third quartet micromeres with minor contributions from second quartet derivatives; see Table 7.1) and the endomesoderm (derived from progenies of the 4d mesentoblast, which divides and forms one left and one right mesodermal band). However, if at all, these bands appear only vaguely visible in most representatives (Wierzejski 1905; Okada 1939; Hinman and Degnan 2002), and since both ectodermal and endomesodermal anlagen soon coalesce, it remains difficult to unequivocally assess the degree of contribution of these individual mesodermal sources to respective organs. Rather, it appears as if all mesoderm-derived organs have both ectomesodermal and endomesodermal contributors (Table 7.1).

In cephalopods, blastomeres and blastocones form the cytoplasmic cap of the early discoblastula (Boletzky et al. 2006). While all blastomeres are located in the center of the cytoplasmic cap, blastocones surround the latter (Fig. 7.5). Blastocones are ray-shaped and part of a syncytium with the uncleaved portion of the zygote (Boletzky 1989). During gastrulation, shearing movements lead to the migration of the last two outermost blastomere rows below the inner blastomeres. These last two outermost blastomere rows form the mesendoderm, and the remaining innermost blastomeres of the cytoplasmic cap remain as ectoderm. In line with this process, the equivalent to the blastopore lip is extra-embryonic ectoderm which grows over the yolk syncytium in direction of the vegetal pole (Boletzky 1989). Subsequently, various organ systems such as the anlagen of the ganglia or the arms become discrete.

LATE DEVELOPMENT

Diversity of Molluscan Larval Development

Mollusks exhibit a variety of postembryonic ontogenetic pathways. Indirect development via a trochophore-like larva is found in all recent classes except the cephalopods (potential monoplacophoran larvae remain to be described) and most likely constitutes the basal condition. Even some terrestrial gastropods with intracapsular development show rudimentary larvae that undergo metamorphosis (e.g., the mouse ear snail *Myosotella*) and several so-called direct developers show larval rudiments such as vestigial prototrochal or velar cells (e.g., the great pond snail *Lymnaea stagnalis*). In the cephalopods, early blastomere formation is followed by the embryo overgrowing the large yolk mass, eventually forming a digestive tube that is connected to the maternal yolk supply for nutrient uptake. This leads to continued growth and differentiation of the embryo which, although often only millimeter-sized, is equipped with all major organs to commence its life as active predator already at the hatchling stage.

In indirect developing mollusks, larval life typically starts with a lecithotrophic (nonfeeding) trochophore-like stage, but some species enter the pelagic realm already as ciliated gastrulae. The trochophore larva is characterized by its name-giving prototroch and an apical sensory organ that often exhibits a characteristic tuft of long cilia (Figs. 7.3 and 7.4). The prototroch (formed by rows of trochoblasts, see above and Table 7.1) constitutes the major swimming device that propels the larva through the water column, with the animal rotating clockwise around its longitudinal axis. It may comprise one to three rows of compound cilia as well as cells with accessory, smaller cilia. The prototroch subdivides the molluscan larva into the pretrochal episphere and the posttrochal hyposphere (Fig. 7.3). From the episphere, the anlagen of major adult anterior structures such as the cephalic nervous system and various sensory organs usually form already prior to metamorphosis. The mouth usually comes to lie in the anterior-most region of the hyposphere, immediately adjacent to the prototroch. The hyposphere often elongates during larval development and gives rise to the visceral body region (Wanninger et al. 1999a; Okusu 2002; Wanninger and Haszprunar 2002a, b; Todt and Wanninger 2010). It may develop spicules (in neomeniomorphs, chaetodermomorphs, and polyplacophorans), shell plates (polyplacophorans), bivalved shells (Bivalvia), or univalved shells (monoplacophorans, gastropods, scaphopods). The larval hyposphere may bear additional ciliary rings such as a terminal telotroch (Fig. 7.4B, C). The larvae of some gastropods (e.g., *Patella*) show a transitory tuft of "anal cilia" in the region of the anus, and these are sometimes homologized with the aplacophoran telotroch. Protobranch bivalves (e.g., *Nucula*, *Acila*) exhibit a so-called test cell larva, whereby cells with multiple rows of cilia cover the developing juvenile underneath (Zardus and Morse 1998), and neomeniomorph larvae produce an "apical cap" or "calymma" that covers the anterior larval region (Fig. 7.4B; Okusu 2002; Todt and Wanninger 2010).

Protonephridia have been found in the larvae of all molluscan classes investigated. Accordingly, these excretory organs together with the prototroch and apical organ are considered as part of the molluscan larval ground plan. In larvae of polyplacophorans as well as some gastropods and a few bivalves, eyes are common.

Following the trochophore stage, many gastropods and bivalves develop a secondary larva, the veliger. The name-giving velum is considered an elaborated prototroch and may form multiple lobes used both for swimming and feeding, enabling month-long planktonic life for numerous species (Fig. 7.4E). In gastropod veligers, an operculum for closure of the embryonic shell is often produced. At this stage, a strong tendency towards heterochrony is usually observed, where anlagen of a number of adult body plan features are already formed in the veliger larva, including the heart, shell muscles, neuromuscular and buccal features, osphradia, statocysts, tentacles, and other sensory organs. Torsion, the developmental process defining the Gastropoda, occurs at the transition of the late trochophore to early veliger stage (Wanninger et al. 2000).

Larval settlement is often preceded by active probing of the substrate with the apical tuft or the foot. Subsequently, metamorphosis is initiated and involves shedding of the apical cap (neomeniomorphs), test cells (protobranch bivalves), and prototroch/velum (polyplacophorans, gastropods, non-protobranch bivalves, scaphopods). The apical organ and the larval operculum (in gastropods) are lost, and the adult shell – the teleoconch – starts to form in many conchiferan taxa (see below). At this stage, the transition from planktonic larval life to the juvenile, often benthic, lifestyle takes place in most mollusks. During this period larval as well as adult excretory systems may coexist alongside each other, and cases where the larval protonephridia are transformed into adult metanephridia have been described.

Despite their direct mode of development, some octopods drift passively in the water column after hatching and are therefore often referred to as "paralarvae." These paralarvae often have specific, transitory adaptations linked to their temporary planktonic life, such as a transparent musculature and a less complex chromatophore system (Villanueva and Norman 2008). In many periodically planktonic mollusks, settlement goes hand in hand with significant behavioral changes such as loss of positive phototaxis or gain of positive geotaxis.

Neurogenesis

The development of the gastropod and cephalopod nervous system has traditionally received considerable attention. Thus, a significant number of studies, mainly concerned with the analysis of gangliogenesis based on histological sections, exist (e.g., Kölliker 1844; Faussek 1901; Martin 1965; Raven 1966; Meister 1972; Demian and Yousif 1975; Kriegstein 1977; Marquis 1989; Page 1992a, b; Lin and Leise 1996; Page and Parries 2001; Shigeno et al. 2001). In the gastropods, the ganglia form already in the larva from invaginating or delaminating ectodermal cells

(Raven 1966; Kandel et al. 1981; Mescheryakov 1990; see Croll 2000 for brief review). Thereby, the cerebral and pedal ganglia form prior to the visceral ganglia (Page 1992a, b; Lin and Leise 1996), and the buccal ganglia appear to be the last set of major ganglia to develop (Cumin 1972). After establishment of the ganglia, neuronal precursor cells proliferate from the region of gangliogenesis (Jacob 1984). Despite the often detailed descriptions of the process of ganglion formation, histology-based investigations largely failed to discover details of early molluscan neurodevelopmental processes, for example, the formation of individual neurons, particularly in early developmental stages. Accordingly, comparative molluscan (and generally invertebrate) neurodevelopmental studies employing modern optical, computational, and imaging tools (immunofluorescence labeling and confocal microscopy in combination with powerful 3D reconstruction software) have significantly altered our understanding concerning the timing and morphogenetic details underlying neurogenesis. By the use of these techniques, a solid body of data on the distribution of neuroactive substances during molluscan neurogenesis has become available. However, a pan-neural marker for molluscan (and lophotrochozoan) developing nervous systems still awaits discovery; the widely used tubulin markers or antibodies directed against horseradish peroxidase, which render comprehensive results for the study of, e.g., arthropods (see Vol. 3, Chapters 3 and 4; Vols. 4 and 5), do not work well on early developmental stages in mollusks, because they often fail to recognize existing neural subsets and/or because significant fractions of the developing nervous system are obscured by the massive presence of cilia (Fig. 7.8F). Nevertheless, neurodevelopmental work during the past two decades has shown that some commercially available antibodies, especially against the widely distributed neurotransmitter serotonin (5-hydroxytryptamine, 5-HT) as well as FMRFamide-related peptides, do label significant portions of the developing molluscan nervous system and thus provide important insights into comparative molluscan neurogenesis. As such, data on neurogenesis are now available for

representatives of all molluscan classes except for Chaetodermomorpha and Monoplacophora.

As a common trait, the adult molluscan nervous system starts to form in early developmental stages. The tetraneurous condition can be recognized in larvae of all clades, together with developing pedal commissures and the anlage of the cerebral commissure or ganglia (where present). These neural subsets do not form a strictly linear anterior-posterior formation gradient as in many annelids, thus reflecting the non-segmental ancestry of Mollusca (Friedrich et al. 2002; Voronezhskaya et al. 2002; Wanninger and Haszprunar 2003). In conchiferans, the anlagen of the pedal, visceral, pleural, and other major ganglia are usually recognized by immunocytochemical staining for serotonin and FMRFamide (Marois and Carew 1990, 1997a, b; Marois and Croll 1992; Voronezhskaya and Elekes 1993, 2003; Croll and Voronezhskaya 1996; Diefenbach et al. 1998; Croll 2000; Hinman et al. 2003; Wanninger and Haszprunar 2003; Croll and Dickinson 2004; Voronezhskaya et al. 2008; Wollesen et al. 2009, 2010, 2012; Kristof and Klussmann-Kolb 2010).

The molecular mechanisms that govern molluscan neurogenesis are still poorly understood, but the few detailed data on two vetigastropods (*Haliotis* and *Gibbula*) and one cephalopod (the decapod squid *Euprymna*) demonstrate involvement of a number of Hox and potentially also ParaHox genes in this process, although loss of function of individual genes has also been demonstrated (O'Brien and Degnan 2002a, b, 2003; Hinman et al. 2003; Lee et al. 2003; Samadi and Steiner 2009, 2010a, b). Cell lineage studies based on blastomere injection have shown that the cerebral ganglia are derived from micromeres 1a–c, while the pedal ganglia form from the 2a and the visceral nerve cords from the 2b lineage in the caenogastropod *Crepidula* (Table 7.1; Hejnol et al. 2007). Due to the lack of comparative data, it is yet impossible to assess the degree of conservation of nervous system cell lineage across the Mollusca.

The most striking and characteristic neural component in molluscan larvae is the apical organ, a common feature shared with the vast majority of lophotrochozoan species (Figs. 7.7

and 7.8; see also Chapters 6, 7, 8, 9, 10, 11, and 12; Bonar 1978; Wanninger 2009). This organ is usually the first neural structure that develops in molluscan larvae. Some terminological confusion exists because this system has also been referred to as "apical ganglion" (although it does not form a ganglionic structure). To make matters worse, "apical ganglion" has also been applied to the entire anterior complex that comprises the larval apical organ and the developing adult cerebral commissure (Nielsen 2012). In order to strictly separate these ontogenetically and structurally different neural features, "apical organ" should only be applied to the larval anterior sense organ (cf. Richter et al. 2010).

From an immunocytochemical perspective, the molluscan apical organ is relatively simple organized, although detailed three-dimensional reconstructions employing ultrathin serial sectioning and transmission electron microscopy are lacking. Thus, comparisons largely rely on immunolabeling data alone. These show that the apical organ is composed of a given number of serotonin-like immunoreactive (serotonin-LIR) and sometimes also FMRFamide-like immunoreactive (FMRFa-LIR) flask-shaped cells (sometimes called ampullary cells), from which at least some of the compound cilia that form the apical tuft emerge (Figs. 7.7 and 7.8A; Wanninger 2009; Richter et al. 2010). The number of these flask-shaped serotonin-LIR cells typically ranges between two and four, and these are often the only immunoreactive compounds that have been identified in the apical organ of basal gastropods as well as bivalves and scaphopods (Page 2002, 2006; Hinman et al. 2003; Wanninger and Haszprunar 2003; Voronezhskaya et al. 2008). An apical neuropil is situated immediately posterior to the flask cells, and at its base the adult cerebral commissure forms (Croll and Dickinson 2004). In some caeno- and heterobranch gastropods, additional sets of (ciliated or non-ciliated) non-flask-like parampullary (or peripheral) cells are found. These surround the flask cells at their base. In taxa that show such peripheral cells, the number of flask cells is usually increased, resulting in a complex cellular architecture of the apical organ

that drastically deviates from the simple condition found in prosobranch gastropods, bivalves, and scaphopods (Fig. 7.7; Croll and Dickinson 2004). Apart from the above-mentioned gastropods, such a complex apical organ is also present in polyplacophoran larvae (Friedrich et al. 2002; Voronezhskaya et al. 2002) and in the creeping larva of entoprocts (Chapter 6; Wanninger et al. 2007). This and other shared features strongly argue in favor of a mollusk-entoproct clade (Tetraneuralia), implying that a complex apical organ is plesiomorphic for Mollusca.

Recent gene expression studies on the vetigastropod *Gibbula varia* have shown expression of the Hox genes *Lox2*, *Lox4*, and *Lox5* as well as the ParaHox gene *Gsx* in cells of the apical organ (Samadi and Steiner 2010a, b). The cell lineage of conchiferan and polyplacophoran apical organs appears to be conserved to a certain degree, since data on the two caenogastropods *Ilyanassa* and *Crepidula* as well as the scaphopod *Antalis* have consistently shown its origin from all four first quartet micromeres (Table 7.1; van Dongen and Geilenkirchen 1974, 1975, 1976; Hejnol et al. 2007; Goulding 2009).

While immunoreactivity of the cells of the apical organ is lost at metamorphosis, their definite fate remains largely unknown. Most authorities probably agree that its cells are resorbed. This has been shown for the gastropod *Ilyanassa obsoleta* by applying markers for apoptotic cells (Gifondorwa and Leise 2006). However, incorporation of at least some individual cells into the adult nervous or sensory system seems possible. Postmetamorphic individuals of the nudibranch *Phestilla sibogae* appear to maintain cells of the apical organ; their eventual fate, however, is unknown (Bonar 1978).

In most indirect developing mollusks, the cerebral commissure forms at the base of the apical organ. This has fueled speculations that the apical organ plays an inductive role in the formation of the adult central nervous system. Apart from that, detecting (chemical) settlement cues has been attributed as a prime role to the apical organ, but experimental data to this end are few (Hadfield et al. 2000). The loss of immunochemical signal in the flask cells of larvae of the scaphopod *Antalis* prior to metamorphic competence indicates that the apical organ may not be necessary for successful settlement and metamorphosis in all mollusks (Wanninger and Haszprunar 2003).

Additional larval neural subsets in mollusks include a serotonin-LIR nerve underlying the prototroch and velum, again a feature shared with other lophotrochozoans (Wanninger 2009). The velum is usually heavily innervated by serotonin- and FMRFa-LIR neurites that emerge from the apical organ and/or the cerebral commissure and disappear at metamorphosis (Dickinson et al. 1999; Dickinson and Croll 2003; Wollesen et al. 2007; see Croll 2009 for review). An additional pretrochal sense organ comprising two pairs of dorsolateral and two pairs of ventrolateral ampullary cells with a ciliated lumen that stain positive for FMRFamide and, less so, for serotonin has been found in polyplacophoran trochophores. These cells are connected to the cerebral commissure. The entire ampullary system is lost at metamorphosis (Haszprunar et al. 2002).

The increased interest in gastropods as neurobiological model species, which peaked in Nobel prize-winning studies on the giant neurons of the sea hare *Aplysia* (see Kandel 2001), has led to screenings of gastropods and cephalopods for neuroactive substances. Thereby, a variety of compounds were identified in larval stages of the heterobranch gastropod *Phestilla sibogae*, including dopamine, choline acetyltransferase, and norepinephrine (Kempf et al. 1992), and an even more impressive suite in cephalopods (Messenger 1996). A so-called VD1/RPD2 alpha-neuropeptide, isolated from the pulmonate gastropod *Lymnaea stagnalis*, was found to be present in specific nerve cells in the central nervous system of this snail as well as in other conchiferans (Kerkhoven et al. 1992, 1993; Wollesen et al. 2012) and may be conserved among the entire Mollusca. Small cardioactive peptides (ScPs) are also expressed in neural subsets of gastropod and bivalve larvae and in the juvenile and adult nervous system of gastropods and cephalopods (Barlow and Truman 1992; Kempf et al. 1992; Kanda and Minakata 2006; Ellis and Kempf 2011). With antibodies against these substances available, and several other neuroactive compounds such as nitric oxide and vasopressin identified (Baratte and Bonnaud 2009; Filla et al. 2009; Bardou et al. 2010; Mattiello et al. 2012), potentially useful tools are at hand for future comparative evolutionary neurodevelopmental studies on mollusks.

Myogenesis

Mollusks are highly muscular animals with several distinct muscular subsets, including dorsoventral muscles (often called "shell muscles" or "pedal retractors"), buccal muscles to support the radula apparatus, mantle retractors, as well as more taxon-restricted systems such as enrolling muscles (in aculiferans), head retractors (in scaphopods, gastropods, and cephalopods), a dorsal rectus muscle (in polyplacophorans and larval neomeniomorph aplacophorans), and various adductors in bivalved representatives (Fig. 7.9; reviewed in Haszprunar and Wanninger 2000; see also Scherholz et al. 2013). The dorsoventral muscles are arranged as multiple, serially arranged fine bundles along the entire anterior-posterior body axis in neomeniomorphs. In the chaetodermomorphs, this arrangement is restricted to the anterior region. Polyplacophorans have eight sets of highly complicated shell muscles, correlating with the eight shell plates. Such a condition is also present in the uni-shelled monoplacophorans, which has led to heavy speculations concerning a segmented ancestry of mollusks (Lemche and Wingstrand 1959). Everything from one to eight dorsoventral muscle pairs ("foot retractors") is found among bivalves, one or two pairs are present in scaphopods, and a single pair is typical for gastropods (often referred to as "spindle muscle" in snails with a coiled shell) and cephalopods.

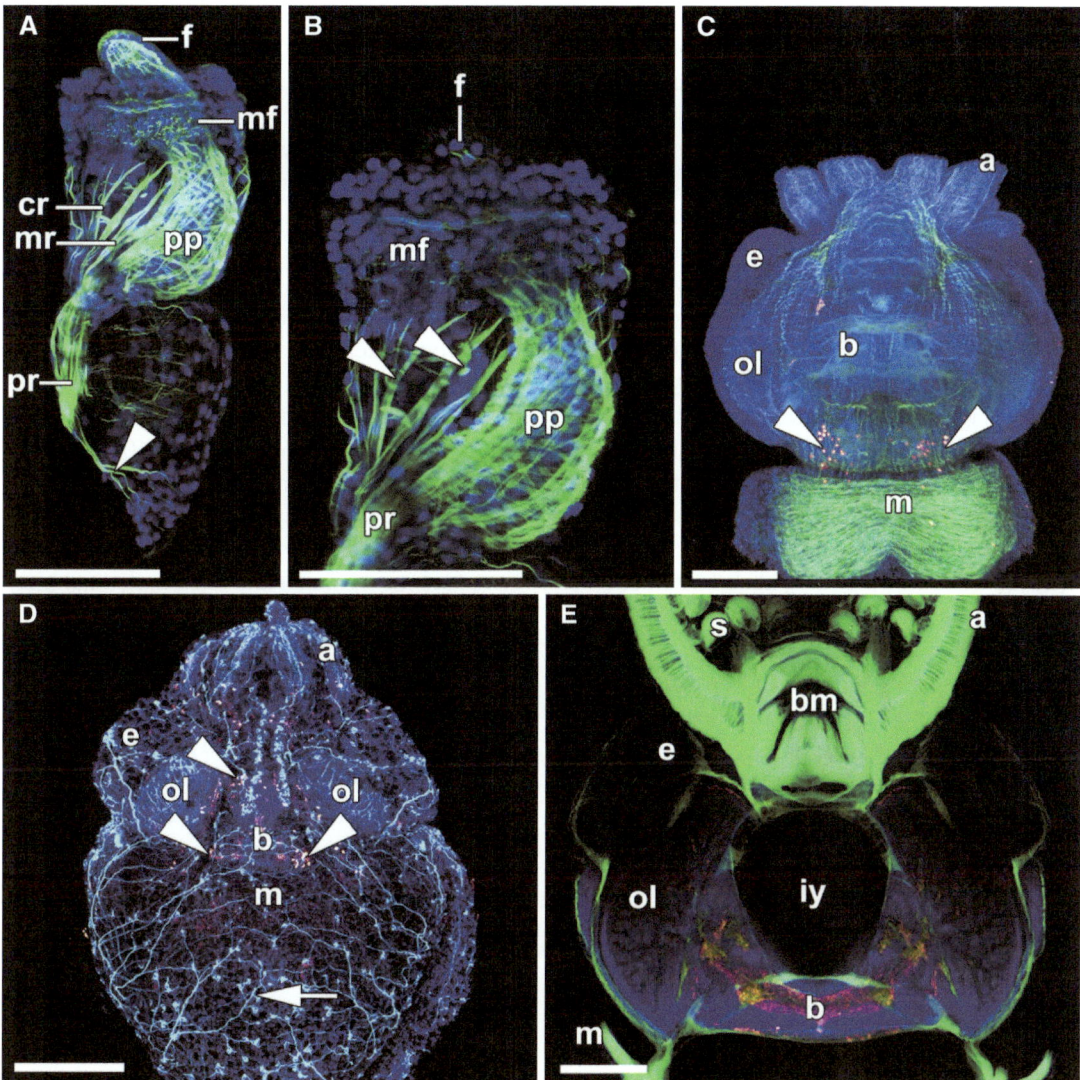

Fig. 7.9 Neuromuscular anatomy of juvenile scaphopods and embryonic and hatched cephalopods. Cell nuclei are stained in *blue* (HOECHST) and musculature and neuropil are labeled *green* (phalloidin, F-actin). Specific neural elements are stained *red* (**C–E**) or *red* and *turquoise* (**D**), respectively. All images are confocal projections with anterior (i.e., adult ventral) facing up. Scale bars are 100 μm in (**A–C**) and 200 μm in (**D, E**). (**A**) Lateral right view of an early juvenile scaphopod (*Antalis entalis*). Prominent muscles are the cephalic retractor (*cr*), mantle retractor (*mr*), and pedal retractor (*pr*). Further muscles are muscles of the pedal plexus (*pp*), those associated with the remaining foot (*f*), the mantle fold (*mf*), and a hitherto undescribed muscle inserting close to the pedal retractor which runs in posterior direction (*arrowhead*). (**B**) Magnification of the ventral region of the specimen shown in (**A**) which highlights the captacula (*arrowheads*). (**C**) This prehatching cephalopod *Idiosepius notoides* specimen exhibits muscles of the developing arms (*a*), the mantle (*m*), and the cephalic region, as well as the neuropil of the brain (*b*) and optic lobes (*ol*). Note the FMRFamide-LIR neurons (*arrowheads*) of the posterior subesophageal mass. Anterior view. (**D**) Epidermal nerve net (*arrow*) in the cephalic and mantle region of the hatching epipelagic octopod *Argonauta hians*. Anterior view. Note the serotonin-LIR cell somata in the brain (*arrowheads*). (**E**) Horizontal vibratome section through the head of a prehatching bobtail squid (*Euprymna scolopes*) showing musculature and neuropil (*green*) as well as FMRFamide-LIR elements (*red*). Note the prominent musculature of the arms (*a*), buccal mass (*bm*), mantle (*m*), and suckers (*s*). Further abbreviations: eye (*e*), internal yolk duct (*iy*)

Myogenesis in Aculiferans

The aplacophorans exhibit a typical muscular tube involving prominent longitudinal muscle bundles which are overlain by circular muscles. Additional interspersed oblique or helical muscles are present in some taxa. Such a muscular body wall is absent in non-vermiform mollusks, probably due to the evolution of shells as major skeletal elements which, together with a pronounced dorsoventral musculature, provide body stability. Interestingly, such a meshwork was also found in the anterior region of the polyplacophoran-like larva and was interpreted as an evolutionary rudiment of an ancestral body wall musculature (Fig. 7.10; Wanninger and Haszprunar 2002a). Given the similarities with other worm-like spiralians, the presence of such a muscular tube was thus considered as potentially ancestral (plesiomorphic) for Mollusca.

However, recent comparative developmental studies on polyplacophoran and neomeniomorph myogenesis have rejected this view. The data obtained demonstrated striking similarities in both clades, revealing muscle systems confined to the larvae of neomeniomorphs and polyplacophorans alone, including a single ventromedian and a pair of ventrolateral muscles (Fig. 7.10; Scherholz et al. 2013). Furthermore, it was shown that the neomeniomorph *Wirenia argentea* develops seven pairs of dorsoventral muscles at first. These form simultaneously and precede the subsequent addition of multiple muscles along the anterior-posterior axis (Scherholz et al. 2013). This is particularly interesting because polyplacophorans likewise pass through a stage with seven pairs of shell muscles which form from multiple sets of serial myocytes by secondary concentration after metamorphosis (Fig. 7.10A, C; Wanninger and Haszprunar 2002a; Scherholz et al. 2013). Both the additional multiple pairs of dorsoventral muscles in the neomeniomorph and the eighth set in polyplacophorans develop considerably later. This, together with data on aculiferan skeletogenesis (see below) and the finding of a fossil seven-shelled aplacophoran (Sutton et al. 2012), argues for a last common ancestor of neomeniomorphs and polyplacophorans with a seven-fold seriality of the dorsoventral musculature and, probably, shell plates. The synchronous appearance of the first (seven) pairs of dorsoventral myocytes in neomeniomorphs and polyplacophorans is in stark contrast to the anterior-posterior formation gradient in annelids, which is due to a posterior growth zone (Chapter 9). These fundamental differences during myogenesis corroborate all other developmental data and illustrate the non-segmented ancestry of Mollusca.

Neomeniomorph evolution from a polyplacophoran-like ancestor is further supported by additional features of neomeniomorph larval myoanatomy, such as the presence of a rectus muscle as well as a pair of lateral enrolling muscles, typical for adult chitons (Fig. 7.10; Scherholz et al. 2013). However, and in contrast to the polyplacophorans, both systems are considerably remodeled during neomeniomorph metamorphosis and are incorporated into the longitudinal body wall musculature of the juvenile. Accordingly, neomeniomorph gross anatomy as a simple, cylindrical, shell-less worm is now considered a secondary simplification that evolved from a more complex, polyplacophoran-like ancestor. The transversal musculature underlying the polyplacophoran shell plates can be interpreted as concentrated and modified units of ring muscles that were present in the larva of the last common aculiferan ancestor, similar to the apical muscular grid in the chiton larva (Wanninger and Haszprunar 2002a).

Myogenesis in Conchiferans

Myogenesis in the Conchifera follows a much simpler pattern than observed in the neomeniomorphs and polyplacophorans. Most adult muscle systems develop more or less directly and without major remodeling from precursors already formed in the larva. One of the few muscles strictly confined to the larval stage is the prototroch/velum muscle ring (Figs. 7.8E and 7.10). This muscle underlies the prototroch/velum and is present in all molluscan classes (including the aculiferans), except for the scapho-pods and cephalopods (monoplacophoran condition unknown), and thus most likely belong to the molluscan larval ground plan (Wanninger and Haszprunar 2002b; Nielsen et al. 2007; Wanninger et al. 2008; Dyachuk and Odintsova 2009; Scherholz et al. 2013).

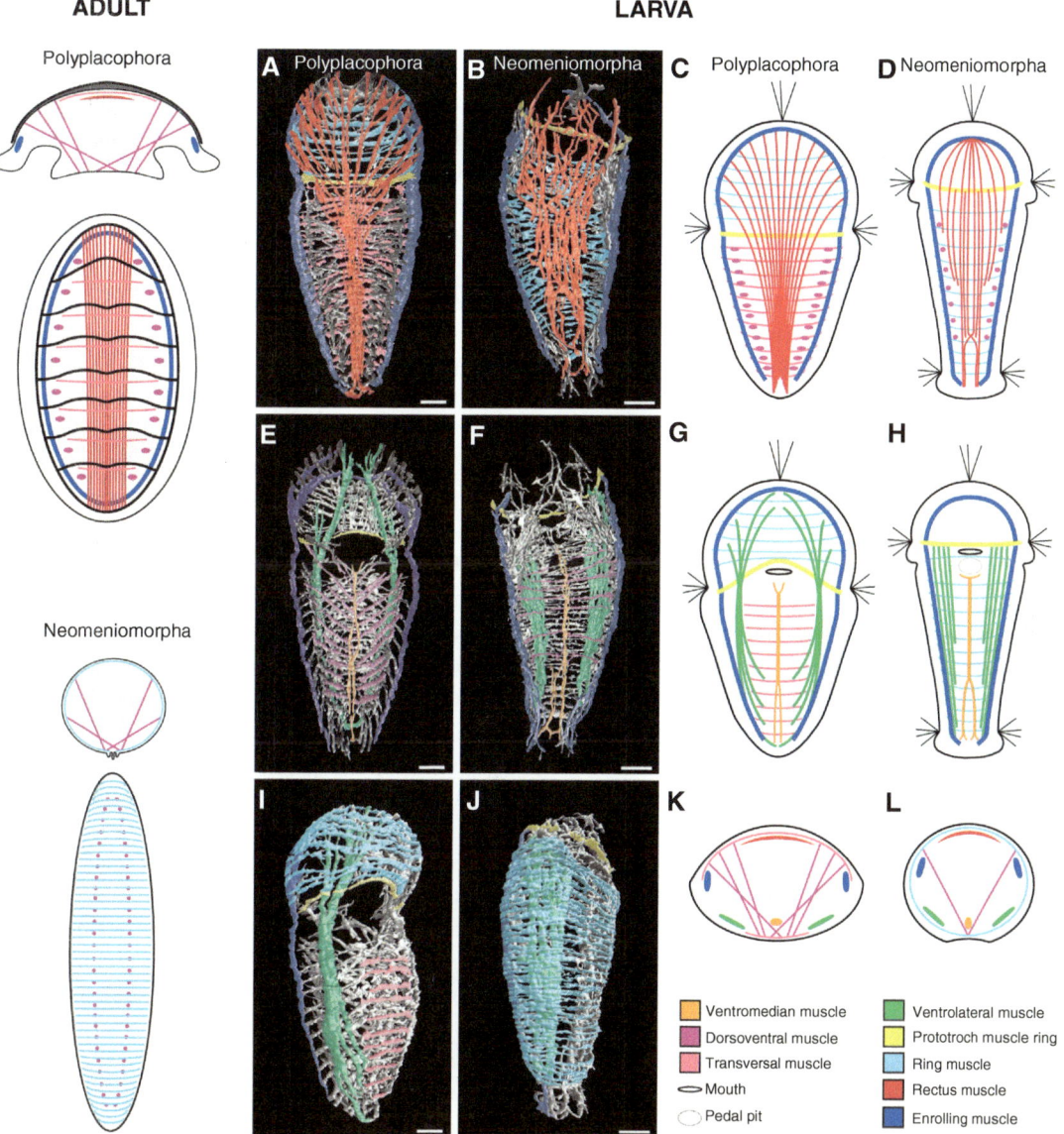

ADULT

Polyplacophora

LARVA

A Polyplacophora B Neomeniomorpha C Polyplacophora D Neomeniomorpha

Neomeniomorpha

E F G H

I J K L

☐ Ventromedian muscle   ☐ Ventrolateral muscle
☐ Dorsoventral muscle   ☐ Prototroch muscle ring
☐ Transversal muscle   ☐ Ring muscle
⬯ Mouth   ☐ Rectus muscle
⬭ Pedal pit   ☐ Enrolling muscle

Fig. 7.10 Comparative myogenesis in polyplacophorans and neomeniomorph aplacophorans. **A, B, E, F, I,** and **J** are 3D reconstructions based on confocal projections of specimens stained with phalloidin to label the musculature. (**A–D**) Dorsal muscles seen from ventral. (**E–H**) Ventral muscles seen from dorsal. (**I, J**). Ventrolateral right views. (**K, L**) Cross sections. Scale bars are 20 μm. Far left column shows adult myoanatomy of polyplacophorans and the neomeniomorph *Wirenia argentea* in cross section (*top*) and dorsal view (*bottom*). Note simple arrangement in the neomenimorph with body wall muscles (here only represented by ring muscles) and serial dorsoventral musculature, while polyplacophorans have a much more complex musculature involving a rectus mus-cle, advanced sets of dorsoventral muscles, transverse muscles, and a lateral enrolling muscle. Note also the striking similarity in the larval myoarchitecture of both taxa as depicted in (**A–L**), including numerous subsets that are lost during neomeniomorph metamorphosis in the process of secondary simplification, including the rectus, ventromedian, ventrolateral, and enrolling muscle. The latter two are postmetamorphically remodeled and contribute to the longitudinal body wall musculature of the adult. Note also the rudimentary body wall ring muscle in the anterior region of the polyplacophoran larva (**I**). See color code for identification of individual muscle sets and text for details (From Scherholz et al. (2013))

Distinct sets of larval retractor systems occur in gastropods and bivalves. These are resorbed during metamorphosis and do not contribute to the adult dorsoventral musculature. In gastropods, usually two asymmetrically positioned larval retractors are present (Wanninger et al. 1999a, b; Wollesen et al. 2008; Kristof and Klussmann-Kolb 2010). The main (or velum) retractor projects into the anterior region, where it often inserts at the velum muscle ring or other velar tissues. The second larval retractor muscle is considerably weaker and projects into the mantle. It is therefore usually termed "mantle retractor" or "accessory larval retractor." Due to their asymmetrically positioned insertion sites on the embryonic shell and their different direction of projection (into the mantle versus the prototroch or velum), it has long been speculated that their activity may play a role in ontogenetic torsion, especially since this gastropod key invention is completed very quickly in basal representatives (for in-depth discussion see Wanninger et al. 2000). Detailed investigations of muscular activity during torsion in larvae of the limpet *Patella* strongly suggest that both larval retractors, aided by hydraulic movements of the foot, are the main driving forces of this process in patellogastropods (Wanninger et al. 1999b, 2000). This was, however, not confirmed for two vetigastropods (Page 2002), illustrating that the dynamics of ontogenetic torsion may be rather plastic.

Formation of the adult dorsoventral (i.e., shell or foot) retractor muscles already starts in the veliger stage prior to the loss of the larval retractors in both gastropods and bivalves. In gastropods, one pair of adult shell muscle precursors is formed, even in patellogastropods, which have a multi-bundled horseshoe-shaped shell muscle system as adults. This indicates that one pair of adult shell muscles represents the basal gastropod condition. Since late gastropod larvae undergo a stage where both the larval and the preformed adult retractor systems exist alongside each other, and because the larval retractors are usually striated while the adult shell muscles are smooth, both systems most likely evolved independently and are thus not ontogenetically homologous (Wanninger et al. 1999a).

Myogenesis in bivalves has so far been investigated to a surprisingly little extent, but the few data available suggest the presence of highly complex retractor systems in their veliger larvae, also in certain semi-direct developing (brooding) species (Meisenheimer 1901; Altnöder and Haszprunar 2008; Wanninger et al. 2008; Dyachuk and Odintsova 2009). An additional ventral larval retractor system is present in *Dreissena*, *Mytilus*, *Pecten*, and *Lyrodus* (shipworm) larvae (Meisenheimer 1901; Cragg 1985; Dyachuk and Odintsova 2009; Wurzinger-Mayer et al. 2014), but their almost opposite projection relative to the gastropod accessory (mantle) retractor argues against homology of these muscles. Similar to the gastropods, all larval retractors are lost prior to or during metamorphosis. In bivalves, the anlagen of the various adult shell muscles (the dorsoventral or foot retractors, adductors, and mantle retractors that later form the pallial line) already appear functional in the veliger larva and form independently from the larval retractors. The high complexity of the bivalve larval muscular body plan combined with the next-to-nonexistent detailed morphological and developmental analyses calls for future studies to fully assess the plasticity of myogenesis in this molluscan class.

Scaphopod larvae have individual muscle fibers that emerge from the anlagen of the adult foot and head retractors and serve as prototroch retractors. These also disappear at metamorphosis, but since they lack distinct shell insertion sites, they are probably not homologous to any of the gastropod and bivalve larval retractor systems (Wanninger and Haszprunar 2002b).

Muscles form the major fraction of cephalopod soft tissue, and although its adult functional morphology and physiology has been subject to a great body of research (e.g., Kier 1988, 1991, 1996; Kier and Thompson 2003; Kier and Stella 2007; Kier and Schachat 2008), no recent account on cephalopod myogenesis employing fluorescence labeling, confocal microscopy, and 3D reconstruction techniques exists. The direct mode of cephalopod development suggests that myogenesis proceeds rapidly without "larval" components, but the morphogenetic and cellular dynamics of this process are unknown.

Skeletogenesis

Mollusks have evolved various ectodermal hard parts including spicules, shell plates, and external as well as internal shells. In aculiferans, the spicules are secreted by individual, invaginated cells. These are distributed across the entire dorsolateral mantle region in aplacophorans. In the eight shell plates-bearing polyplacophorans, spicules, if present, are restricted to the perinotum (girdle) (Haas 1981; Okusu 2002; Nielsen et al. 2007; Todt and Wanninger 2010). In many indirect developing conchiferans, the first-formed, unsculptured (smooth) embryonic shell (protoconch I; sometimes termed prodissoconch I in bivalves) is often followed by the adult shell (teleoconch), which may exhibit a variety of growth and/or color patterns and grows until the death of the animal (Fig. 7.11). The protoconch I often breaks off in later stages and is therefore usually only identifiable in larvae and early juveniles. Fossil remains of belemnite internal shells confirm that such an embryonic shell also belongs to the evolutionary history of cephalopods (Müller 1994).

In addition to these two common shell types, many planktotrophic caenogastropods and bivalves have a third, intermediate, often richly ornamented shell, the so-called larval shell or protoconch II (sometimes termed prodissoconch II in bivalves) (Bandel 1975). Both the protoconch II and the teleoconch, together with the often multilayered periostracum, are formed successively from cells located in folds along the mantle margin (Kniprath 1977; Checa 2000) and not from a single shell field as the protoconch I. The various layers of molluscan adult shells may include nacre (mother of pearl) in some gastropods (e.g., the tropical abalone *Haliotis*) and bivalves (e.g., the pearl oyster *Pinctada maxima*), which most likely evolved independently multiple times (Jackson et al. 2009). With a high potential interest for industrial applications (e.g., pearl production), the genetics of molluscan biomineralization and nacre formation are subject of ongoing research, and transcripts of numerous genes involved in molluscan skeleto-

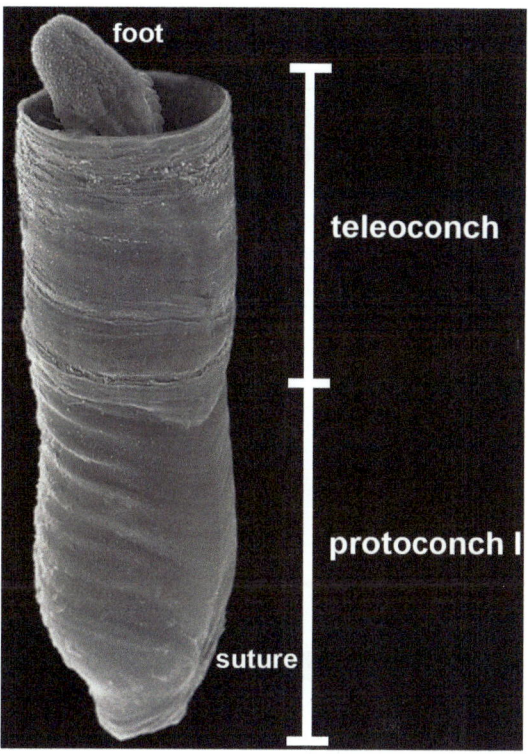

Fig. 7.11 Ontogenetic continuity of molluscan shells. Many indirect developing mollusks, such as the scaphopod depicted here, form a smooth (unsculptured) embryonic shell, the protoconch I, from a distinct shell field of the mantle. In some gastropods and bivalves (but not in scaphopods), a second shell is formed during larval life, the often richly ornamented larval shell (protoconch II). After settlement and metamorphosis, the juvenile/adult shell (teleoconch) starts to form from the mantle edge, usually showing characteristic growth patterns. In scaphopods, the protoconch I forms from a single dorsal shell field that extends laterally on both sides and subsequently encloses the entire visceral region, leaving a characteristic ventral fusion line, the suture. Size of this early juvenile *Antalis entalis* specimen is 750 μm

genesis have been identified. These suggest that the molecular pathways that underlie molluscan shell formation evolve considerably fast (Jackson et al. 2007, 2009). In a number of gastropods (e.g., marine nudibranchs and other heterobranchs), the adult shell was secondarily lost, but the protoconch I is still prevalent in the larvae of many indirect developing species. Octopods and pelagic decapod squids either have no or noncalcified remnants of adult shells, as, e.g., exemplified in the chitinous gladius of *Loligo*.

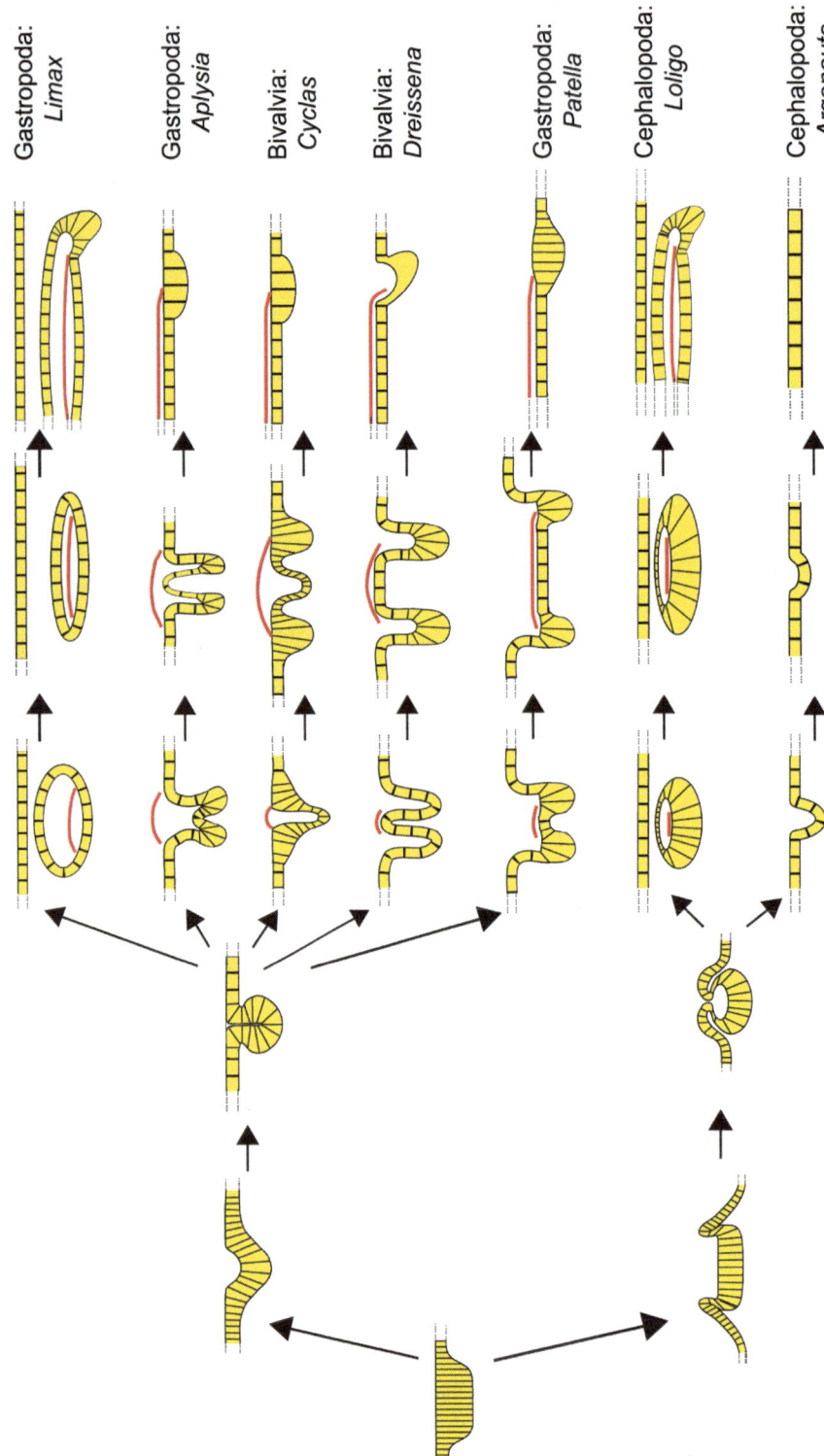

Fig. 7.12 Comparative conchiferan shell formation. Conchiferan shell fields derive from a posttrochal monolayered ectodermal cell layer (*yellow, far left*). The shell-secreting ectodermal regions form by initial invagination, followed by evagination in most taxa. Evagination may be prevented by budding off and internalization of the shell secreting tissue, which subsequently forms a shell sac in groups with an internal shell, e.g., terrestrial slugs such as *Limax*. In cephalopods, the invagination process may be (partly) prevented by the yolk-rich eggs. Here, the shell field anlage is instead overgrown by surrounding ectodermal cells. Interestingly, even in the pelagic octopod *Argonauta*, which lacks a conchiferan shell, a rudimentary shell field formation process has been described. *Red lines* indicate sites of shell secretion (Figure redrawn from Kniprath (1981) with data compiled from various sources cited therein)

Ontogeny of the various conchiferan shell types follows a considerably complex but relatively conserved pattern (Fig. 7.12). The process is highly dynamic and includes invagination and subsequent evagination of defined ectodermal domains. For a detailed comparative summary of data on a number of mollusks, see Fig. 7.12 herein as well as Kniprath (1981). Accordingly, a so-called embryonic shell field is formed from a posttrochal, thickened, monolayered ectodermal epithelium some time after gastrulation (Lankester 1873; Fol 1875; Bütschli 1877; Kowalevsky 1883b; Patten 1886; Lillie 1895; Schmidt 1895; Meisenheimer 1901; Herbers 1913). This embryonic shell field invaginates, and the organic outer layer of the shell, the periostracum, starts to form at the margins of the resulting pore, thus acting as a scaffold for shell secretion (Kniprath 1977, 1981). If an external shell is formed (i.e., in scaphopods, bivalves, and many gastropods), the shell field subsequently re-flattens (evaginates), again forming a defined region in the posterior portion of the mantle from which secretion of the (embryonic) shell starts (Fig. 7.12; Kniprath 1981). This ectodermal region is the one that is commonly referred to as the molluscan "shell field" (as opposed to the embryonic shell field, see above). During polyplacophoran shell plate formation, the invagination-evagination process seems to be lacking.

In the cephalopods, shell field invagination appears to be at least partly prevented by the massive, yolky egg. Contrary to nautiluses, which retain their external shell, the surrounding epithelium overgrows the shell field anlage in coleoids. This epithelium is subsequently budded off and internalized (Fig. 7.12; Lankester 1873, 1875; Appellöf 1898; Spiess 1972). In terrestrial pulmonate slugs an internal shell sac is formed after invagination of the embryonic shell field. The shell field is then closed, followed by secretion of the internal shell (Fig. 7.12; Gegenbaur 1852; Schmidt 1895; Meisenheimer 1896, 1898; see Kniprath 1981). In the shell-less pelagic octopus *Argonauta* (paper nautilus), an invagination and subsequent evagination can still be observed, but shell secretion does not occur (Fig. 7.12; Ussow 1874; Appellöf 1898; cf. Kniprath 1981). The characteristic, fragile brood chamber carried by the reproductive female is a secretory product of the arms and is not related to the conchiferan shells.

Early researchers have proposed that shell secretion already starts at the stage of invagination and therefore termed this morphological feature "shell gland" (Ganin 1873; Lankester 1873). However, the onset of shell secretion seems to be rather plastic and for numerous mollusks secretion has been claimed to occur only after evagination (see Kniprath 1981 for summary and discussion). In order to be less definite about the exact site and timing of primary shell secretion, the more neutral term "shell field" was subsequently used for the entire epithelial domain concerned with shell secretion, irrespective of whether in a pre-invaginated, invaginated, or evaginated state (Blochmann 1883; Schmidt 1895; Kniprath 1981). This is still the preferred terminology today, although the terms "shell field" and "shell gland" are often used interchangeably.

Comparative cell lineage data show that the micromeres 2d, 3c, and 3d contribute to the polyplacophoran shell fields, while spicules are formed from descendants of the 1a, 2a, 2c, 3c, and 3d cell. In the scaphopod *Antalis*, only 2d was identified as shell field precursor, and this was also proposed for the bivalve *Anodonta* (Herbers 1913) and various gastropods (Conklin 1897; Robert 1902; D'Asaro 1966), with an additional contribution of 2c in *Ilyanassa* (Cather 1967). In the basal gastropod *Patella*, all second quartet micromeres appear to be involved in the establishment of the shell field (Table 7.1). Accordingly, the cell lineage of the conchiferan protoconch I appears to be rather conserved, while cell lineage of the polyplacophoran (adult!) shell plates deviates from this pattern. The genealogy of the cells at the mantle edge that secrete the conchiferan adult shell (teleoconch) is unknown.

Contrary to the remaining conchiferans, bivalves have a two-partite shell whose valves are separated and interconnected by an organic ligament. However, the early shell field anlagen arise from a single ectodermal domain that is subdivided only after evagination, during which the interconnecting hinge and later the ligament form (Fig. 7.12; Ziegler 1885; Lillie 1895). Cell lineage studies have shown that the micromere 2d constitutes the shell-founding cell (often denoted "X") in bivalves. This is the first cell in bivalve embryology that divides bilaterally, and its progenitors give rise to the two-partite, symmetrical early shell. 2d (X) gives rise to its daughter micromeres X1 and X2 prior to gastrulation. Accordingly, determination of the lineage pattern of the bivalve (embryonic) shell is specified already at this early developmental stage (Kin et al. 2009; Kurita et al. 2009).

In coleoid cephalopods, the shell field develops in a roundish groove (Fig. 7.12). After establishment of the internal shell sac, a secondary subdivision into two halves takes place in octopods (Spiess 1972). In decapods (squids and cuttlefish), portions of the posterior shell sac detach bilaterally and form the fin pockets (Bandel and Boletzky 1979). In shelled non-cephalopod conchiferans, the embryonic shell is formed from a single, centrally located secretion center of the shell field (one center per valve in bivalves).

In the polyplacophoran larva the first seven shell plates are formed synchronously and appear to be of posttrochal origin, although a pretrochal contribution to the first plate is sometimes proposed (Heath 1898; Wanninger and Haszprunar 2002a; Henry et al. 2004). The eighth shell plate develops considerably after metamorphosis. Since seven-shelled fossils assigned to either the Polyplacophora or Aplacophora are also known from the fossil record, this feature is nowadays often attributed to the last common aculiferan ancestor – a hypothesis corroborated by developmental data on neomeniomorph myogenesis (see above).

The protoconch I of scaphopods arises from a single anlage as in the other conchiferans and,

together with the developing mantle, grows ventrolaterally, before eventually fusing on the ventral side. There, the suture marks the fusion zone of the embryonic shell. The tubular teleoconch is subsequently secreted from cells of the mantle edge (Fig. 7.11; Wanninger and Haszprunar 2001).

GENE EXPRESSION

Since experimental protocols involving reverse genetics (e.g., RNA interference) are only beginning to emerge for mollusks (e.g., Rabinowitz et al. 2008; Hashimoto et al. 2012), assessments of gene functions still largely rely on expression studies obtained by in situ hybridization (Figs. 7.13 and 7.14). In gastrulating molluscan embryos, *brachyury* is expressed adjacent to the blastopore and in cells of the putative ectoderm (Lartillot et al. 2002b; Kin et al. 2009). In the Japanese spiny oyster, *Saccostrea kegaki*, it was also found to be expressed in cells around the developing anus (Kin et al. 2009). Together with the mitogen-activated protein kinase (MAPK) signaling cascade, *brachyury* is likely to be involved in establishing correct cell orientation during early ontogeny, similar to its role in other bilaterians. The transcription factor *snail* and the morphogen *hedgehog* were found to be involved in (neuro)ectodermal patterning in the basal gastropod *Patella* (Nederbragt et al. 2002c). Genes expressed during molluscan mesoderm formation include *twist*, *forkhead*, *goosecoid*, *caudal*, and *vasa* (Table 7.2; Lartillot et al. 2002a; Le Gouar et al. 2003; Kakoi et al. 2008) as well as the homeobox gene *Mox*, which was found in the paired mesodermal band and cells destined to become adult (but not larval) muscles (Hinman and Degnan 2002). In the highly muscular cuttlefish *Sepia officinalis*, several muscle-specific transcription factors, among these *Nk4*, *MyoD*, and *Myf5*, are expressed during early phases of myogenesis (Grimaldi et al. 2004; Navet et al. 2008).

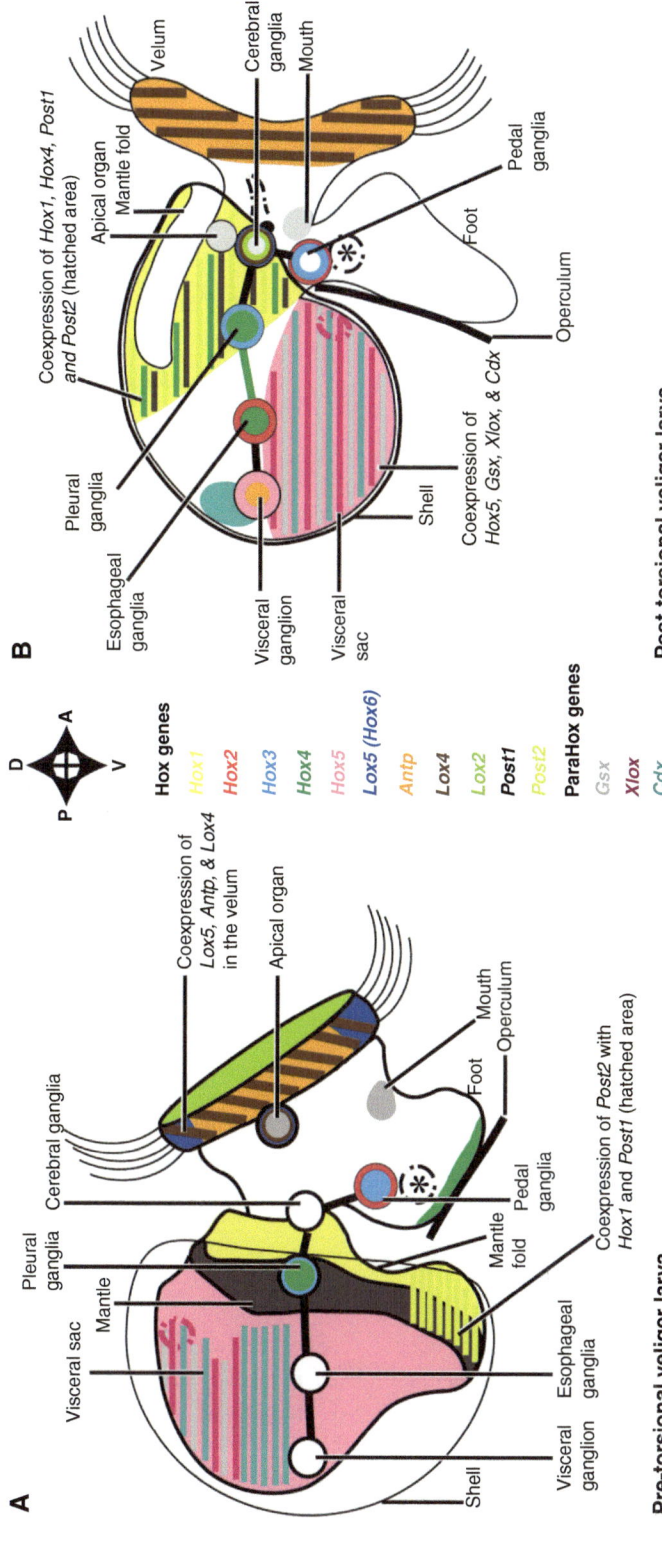

Fig. 7.13 Sketch drawings summarizing Hox and ParaHox gene expression in the pre- and posttorsional larva of the vetigastropod *Gibbula varia*. *Asterisks* mark the statocysts (Based on data from Samadi and Steiner (2009, 2010a, b))

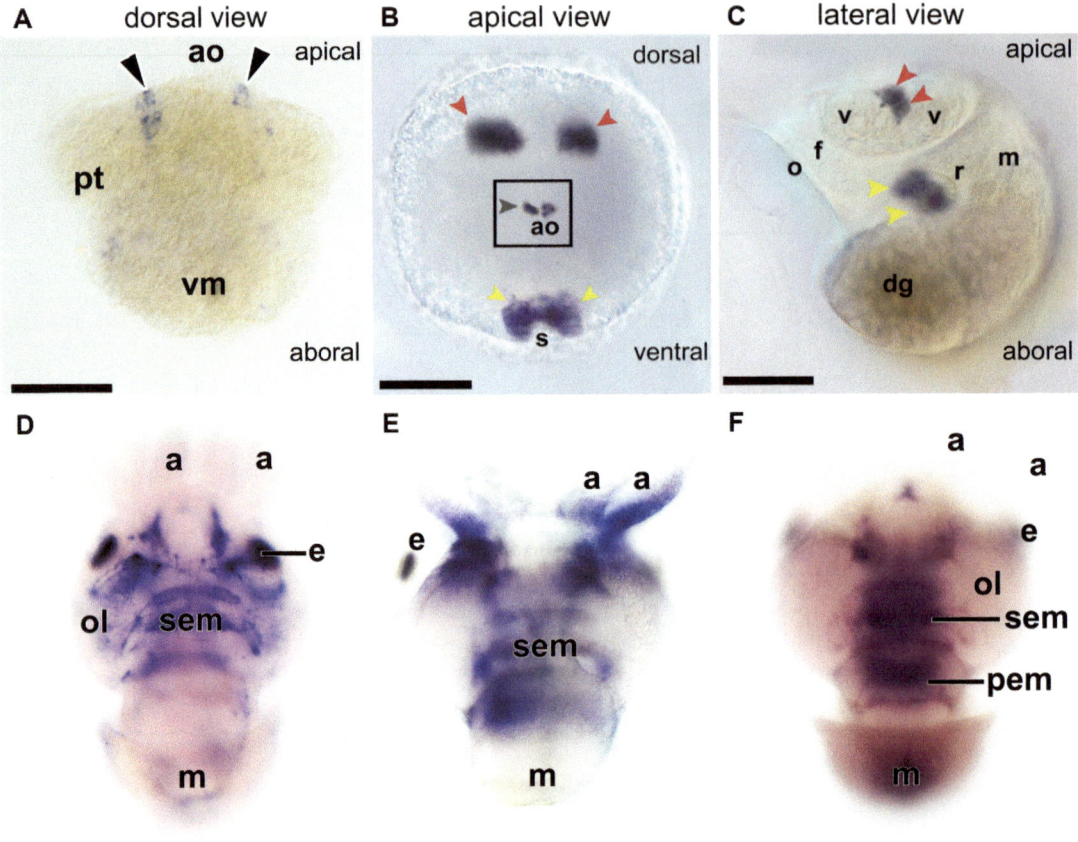

Fig. 7.14 Gene expression in developing scaphopod, gastropod, and cephalopod mollusks. (**A**) *Gsx* expression (*arrowheads*) in the apical region of an early scaphopod trochophore (*Antalis entalis*; Wollesen, unpublished data; anterior faces up). Scale bar: 100 μm. (**B**) The ParaHox gene *Gsx* is expressed in the apical organ (*gray arrowhead*), in the dorsal median episphere (*red arrowheads*), and around the stomodeum (s) (*yellow arrowheads*) in the trochophore larva of the vetigastropod *Gibbula varia*. Anterior view. Scale bar: 20 μm. (**C**) *Gsx* expression in the anlagen of the cerebral ganglia (*red arrowheads*) and the forming radula sac (*r*) (*yellow arrowheads*) of the veliger larva of the vetigastropod *Gibbula varia*. Anterior faces up. Scale bar: 25 μm (Images **B** and **C** taken and modified from Samadi and Steiner (2010b)). (**D–F**) Gene expression in prehatching embryos of the cephalopod *Idiosepius notoides*. Ventral faces up. Scale bars: 150 μm. (**D**) The ParaHox gene *Gsx* is expressed in the supraesophageal mass (*sem*) and optic lobes (*ol*). (**E**) Expression of *Pax258* in the supraesophageal mass of the central nervous system and in the arms. (**F**) *POU3* is expressed in the supraesophageal and posterior subesophageal mass (*pem*). Further abbreviations: apical organ (*ao*), arm (*a*), digestive gland (*dg*), eye (*e*), foot (*f*), mantle (*m*), operculum (*o*), prototroch (*pt*), velum (*v*), visceral mass (*vm*)

Only few early endodermal markers have been characterized in mollusks. These include *hedgehog*, a foregut patterning gene in *Sepia*, and *forkhead*, which, after expression in the 3A, 3B, and 3C macromeres, is found in endodermal cells in *Patella* (Lartillot et al. 2002a).

Eleven Hox and three ParaHox genes have been identified in mollusks. However, comparative analyses of expression patterns of these and other key developmental regulators are only beginning to emerge. Thereby, the works by Samadi and Steiner (2009, 2010a, b) on the caeno-

Table 7.2 Expression patterns of key developmental genes in Mollusca. Note the overall paucity of available expression data of Hox, ParaHox, and other homeobox genes throughout the phylum

Class/Species	Gene	Expression domain during development	Reference
Polyplacophora			
	Hox genes		
Acanthochitona crinita	Hox1–5, Hox7 (=Antp), Post2	Collinear expression in ectoderm, endoderm, and mesoderm along the anterior-posterior axis of the larval hyposphere	Fritsch et al. (2015)
	Homeobox genes		
Lepidochitona caverna	Engrailed	Shell field	Jacobs et al. (2000)
Bivalvia			
	Homeobox genes		
Transennella tantilla, Saccostrea kegaki	Engrailed	Shell field	Kin et al. (2009)
	Non-homeobox genes		
Transennella tantilla, Saccostrea kegaki	Dpp	Shell field	Kin et al. (2009)
Saccostrea kegaki	Brachyury	3-D blastomere during early cleavage, ventral region from blastopore through vegetal-most part, around prospective anus	Kin et al. (2009)
	Tektin	Primary trochoblasts, ciliary band, ciliary cells in stomach of D-shaped larva	Kakoi et al. (2008)
	Beta-tubulin	Primary trochoblasts, ciliary bands, telotroch	
	Arp2/3	Larval mesoderm	
	Vasa	Pair of 2d descendant cells, cell posterior to blastopore, endodermal cells	
	Frizzled	Two-cell stage, four 2d descendant cells	
Gastropoda			
	Hox genes		
Nipponacmea fuscoviridis	Hox1	Shell field	Hashimoto et al. (2012)

(continued)

Table 7.2 (continued)

Class/Species	Gene	Expression domain during development	Reference
Haliotis asinina	*Hox1*	Trochophore and veliger: shell field	Hinman et al. (2003)
	Hox2	Trochophore: ectodermal cells close to mouth, "anal marker", and anterior foot Veliger: pedal and esophageal ganglia, ectodermal and mesodermal (?) expression in head and foot	
	Hox3	Trochophore: semicircle of ectodermal cells close to foot Veliger: pleural and pedal ganglia, pallioviseral nerve cord	
	Hox4	Trochophore: ectodermal cells close to foot anlage Veliger: anlagen of pleural and esophageal ganglia and pedal region	
	Hox5	Trochophore: ectodermal cells in antero-ventral mantle, in ectodermal and mesodermal (?) regions of larval neck and foot Veliger: branchial, and esophageal ganglia, in vicinity of pleuropedal ganglia	
Gibbula varia	*Hox1*	Shell field	Samadi and Steiner (2009, 2010a)
	Hox2	Trochophore: ventral posttrochal neuroectodermal cell close to "anal marker" Veliger: ectodermal cell in posterior foot rudiment, pedal and esophageal ganglia	
	Hox3	Trochophore: semicircle of ectodermal cells around foot anlage Veliger: pedal and pleural ganglia	
	Hox4	Trochophore: ectodermal cells on both sides of foot rudiment (anlagen of kidneys?) Veliger: mantle, cells close to operculum, anlagen of pleural and esophageal ganglia	
	Hox5	Trochophore: ventral posttrochal area on both sides of foot anlage Veliger: mantle covering visceral mass and digestive gland	
	Lox5	Trochophore: 4 ectodermal cells close to apical organ Veliger: cells anterior and posterior of velar area, apical organ, cerebral ganglia and commissure	
	Antp	Trochophore: ciliated and non-ciliated prototrochal cells Veliger: velum, visceral ganglion	
	Lox4	Trochophore: ciliary cells of prototroch, apical organ Veliger: velum, cerebral ganglion, apical ganglion	
	Lox2	Trochophore: 2 rows of ectodermal cells in apical region Veliger: prevelar region, cerebral ganglia and commissure	
	Post1	Trochophore: shell field Veliger: mantle, shell field	
	Post2	Trochophore: shell field Veliger: mantle, shell field	

Table 7.2 (continued)

Class/Species	Gene	Expression domain during development	Reference
Haliotis rufescens	Hox5	Buccal, branchial, and esophageal ganglia	Giusti et al. (2000)
	ParaHox genes		
Patella vulgata	Cdx	Gastrulation: ectodermal cells of blastopore, paired mesentoblasts giving rise to posterior mesoderm of trochophore	Le Gouar et al. (2003)
		Trochophore: posterior neuroectoderm and parts of mesoderm	
Gibbula varia	Gsx	Apical organ, cerebral ganglia, mouth, radula, odontoblasts	Samadi and Steiner (2010b)
	Xlox, Cdx	Gut, early neuroectoderm	
	POU, Pax, and Mox genes		
Haliotis asinina	POU3	Nervous system, foot mucus cells, radula sac, statocyst	O'Brien and Degnan (2002a, b, 2003)
	POU4	Single posterior cell in mantle, ectodermal cells in developing foot, anlagen of eyes and esophageal ganglia?, anlagen of statocysts and cephalic tentacles, ctenidial and osphradial rudiments	
	Pax258	Early cleavage, dorsolateral pedal region, statocysts, ectoderm, pallial chamber, anlagen of epipodial tentacles	O'Brien and Degnan (2003)
	Mox	Paraxial mesodermal bands in trochophore, muscle system of veliger	Hinman and Degnan (2002)
	Other homeobox genes		
Patella vulgata	Otp	Nervous system (e.g., apical organ)	Nederbragt et al. (2002b)
	Otx	Nervous system, around stomodeum, precursors of primary and accessory trochoblasts	
Haliotis rufescens	Gsc	Anterior mesoderm, anterior expression in all three germ layers	Lartillot et al. (2002a)
Nipponacmea fuscoviridis	Nk2.1	Apical organ, around mouth and foregut	Dunn et al. (2007)
	Engrailed	Shell field, apical organ	Hashimoto et al. (2012)
Lymnaea stagnalis	Engrailed	Cells of shell field and in posterior region, base of foot	Iijima et al. (2008)
Patella vulgata	Engrailed	Anlage of apical organ, cells in posterior ectoderm, shell field	Nederbragt et al. (2002a)
Lottia gigantea, Biomphalaria glabrata	Pitx	Right/left ectoderm of trochophore	Grande and Patel (2009)
	Non-homeobox genes		
Nipponacmea fuscoviridis	Dpp (bmp2/4), Grainy-head, ferritin, CS1	Shell field	Hashimoto et al. (2012)
Patella vulgata	Dpp	Shell field, precursor cells of eyes, pretrochal ectoderm (apical organ?), and around blastopore	Nederbragt et al. (2002a)

(continued)

Table 7.2 (continued)

Class/Species	Gene	Expression domain during development	Reference
Haliotis asinina	Dpp	Shell field, micro- and macromeres during early cleavage	Koop et al. (2007)
Lymnaea stagnalis	Dpp	Shell field	Iijima et al. (2008)
Lottia gigantea, Biomphalaria glabrata	nodal	Right/left ectoderm of trochophore	Grande and Patel (2009)
Patella vulgata	Soxb, Wnt2/13	Nervous system	Le Gouar et al. (2004)
	Fox	Anterior mesoderm, endoderm	Lartillot et al. (2002a)
Haliotis rufescens	FoxA	Around anus	Dunn et al. (2007)
Patella vulgata	MAP kinase	3d cell during early cleavage	Koop et al. (2007)
	Hh	Nervous system (ventral midline of trochophore)	Nederbragt et al. (2002c)
	Snail	Ectoderm	Lespinet et al. (2002)
	Brachyury	3d cell during early cleavage, ectoderm, blastopore	Lartillot et al. (2002b)
Haliotis asinina	Brachyury	Blastopore, ventral midline, ventral edge of shell field of trochophore	Koop et al. (2007), Jackson et al. (2010)
	Coe	Apical organ, posterior and ventral ectodermal cells, paraxial mesodermal bands	
	Elav	Apical ectodermal cells, posttrochal lateral and posterior cells	
Haliotis rufescens	Tectin3	Apical organ	Dunn et al. (2007)
Haliotis asinina	Vasa, nanos	Micromeres during early cleavage (maternally expressed), putative mesodermal bands of trochophore	Kranz et al. (2010)
	β-catenin	Ectoderm surrounding mouth and ventrolateral ectoderm, pretrochal region	Koop et al. (2007)
	HSP90A	Ventrolateral ectoderm in trochophore	
	Serotonin-encoding gene	4 pretrochal cells in trochophore	
	Def, kak, pim-3, MLF1, Tpmt, Ther, DAu322c, DAu172c	Various domains in competent veliger	Williams and Degnan (2009)
Ilyanassa obsoleta	Onecut	Ectodermal and subectodermal cells in head and visceral mass, pedal region, mantle cavity, close to mouth	Lambert et al. (2010)

Table 7.2 (continued)

Class/Species	Gene	Expression domain during development	Reference
Cephalopoda			
	Hox genes		
Euprymna scolopes	Hox1	Pallioviseral and stellate ganglia, arms	Lee et al. (2003)
	Hox3	Pallioviseral and stellate ganglia, arms, funnel	
	Hox5	Pallioviseral ganglia, arms, funnel	
	Lox5	Arms, funnel	
	Antp	Pedal ganglia, arms	
	Lox4	Pedal and stellate ganglia, funnel, arms	
	Post1	Arms, buccal lappets, light organ	
	Post2	Pedal ganglia, arms	
	POU and Pax genes		
Idiosepius notoides	POU2, POU3, POU4, POU6	Nervous system	Fig. 7.12 herein; Wollesen et al. (2014)
Euprymna scolopes, Sepia officinalis	Pax6	Eyes, nervous system, sensory organs	Hartmann et al. (2003), Navet et al. (2009)
	Other homeobox genes		
Sepia officinalis	Nk4	Muscle system	Navet et al. (2008)
	Apt (llx2/9)	Nervous system, sensory organs, eyes	Farfán et al. (2009)
	Otx	Nervous system, eyes	Buresi et al. (2012)
Idiosepius paradoxus	Six3/6	Nervous system, lentigenic cells of eyes	Ogura et al. (2013)
	Non-homeobox genes		
Sepia officinalis	Elav1	Nervous system	Buresi et al. (2013)
	Hh	Endoderm (esophagus), nervous system, muscle system	Grimaldi et al. (2008), Navet et al. (2009)
	Reflectin genes	Developing iridosomes throughout the body	Andouche et al. (2013)
	Myf5-like and myoD-like (Muscle-specific transcription factors)	myf5: myoblasts giving rise to helical smooth-like fibers myoD: myocytes giving rise to circomyarian helical and cross-striated fibers	Grimaldi et al. (2004)

gastropod *Gibbula varia*, by Lee and coworkers (2003) on the decapod squid *Euprymna scolopes*, and by Fritsch et al. (2015) on the polyplacophoran *Acanthochitona crinita* provide the most comprehensive descriptions of Hox (and, in *Gibbula* only, ParaHox) gene expression profiles for mollusks available to date (Table 7.2, Figs. 7.13 and 7.14B, C). This is about to change soon, however, with such data being currently generated in the lab of the authors on various molluscan clades, including scaphopods and bivalves (Fig. 7.14A; Wollesen et al. unpublished). Additional data on the expression of selected genes of the Hox, POU, and Pax families are available for another gastropod, *Haliotis asinina* (O'Brien and Degnan 2002a, b, 2003; Hinman et al. 2003). The studies on the two gastropods showed that Hox and ParaHox genes play an important role in the formation of the larval apical organ, prototroch, foot, shell, statocyst, radula, and anlagen of the adult central nervous system (Table 7.2, Figs. 7.13 and 7.14B, C). In the cephalopod *Euprymna*, all but one Hox gene investigated were found to be involved in arm formation. *Hox3*, *Hox5*, *Lox4*, and *Lox5* are additionally expressed in the developing funnel (Table 7.2; Lee et al. 2003). This demonstrates that these genes have acquired novel functions in conchiferans, since Hox genes mainly act in neurogenesis and patterning of the anterior-posterior axis in the majority of bilaterians (see Jarvis et al. 2012 for review; note, however, that some Hox genes such as *Hox1*, *Post1*, and *Post2* in *Gibbula* as well as *Lox5* and *Post1* in *Euprymna* have lost their role in central nervous system formation; see Table 7.2 and Fig. 7.13), and ParaHox genes are involved in digestive tract formation (Brooke et al. 1998; Holland 2001). Novel findings on gene expression in polyplacophorans corroborate these findings, since there the Hox genes likewise show a strict anterior-posterior collinear expression pattern and are not confined to any specific morphological features (Fritsch et al. 2015). Accordingly, it appears that polyplacophorans (and maybe the aculiferans in general) have retained the original function of Hox genes in anterior-posterior body patterning,

while they were recruited secondarily into the formation of specific morphological features in the conchiferans only (Table 7.2).

As noted above, some Hox and ParaHox genes are expressed in specific larval features in scaphopods and gastropods, notably in *Gibbula varia*. This includes *Lox4* and *Antp* in the prototroch and *Lox2*, *Lox4*, *Lox5*, and *Gsx* in the apical organ (Table 7.2, Figs. 7.13 and 7.14A–C; Samadi and Steiner 2010a). Several other genes have been identified to act in gastropod apical organ formation, including *collier*, *nk2.1*, *tectin3*, *engrailed*, and *otp* (Table 7.2; Nederbragt et al. 2002a, b; Dunn et al. 2007; Jackson et al. 2010). Although the direct developing cephalopods lack an apical organ, preliminary data suggest that they express at least some of these genes in the developing (adult!) central nervous system (Fig. 7.14D–F; Wollesen, unpublished), raising the question concerning conserved versus independent recruitment of these genes in larval and/or adult neurogenesis in mollusks. In both gastropods (*Haliotis* and *Gibbula*) and the cephalopod *Euprymna*, the anterior Hox genes *Hox1–5* are not expressed in the cerebral ganglia (Table 7.2). This resembles the situation in classical bilaterian models such as *Drosophila* (Vol. 5, Chapter 1) or mouse, which do not express Hox genes in the proto- and deutocerebrum or the fore- and midbrain, respectively (Reichert 2005). Similar to these model species, the anterior-most region of the molluscan central nervous system, i.e., the anlagen of the cerebral ganglia, expresses genes such as *Otx* and *Pax258* in cephalopods (Fig. 7.14E; Reichert 2005; Buresi et al. 2012). Neuroanatomical studies on the basal cephalopod *Nautilus* suggest that the optic ganglia of coleoids constitute lateral extensions of the cerebral ganglia (Young 1971; Shigeno et al. 2008), a finding that is also corroborated by co-expression of *Otx* and *Pax6* in both ganglia (Hartmann et al. 2003; Navet et al. 2009; Buresi et al. 2012). In the trochophore larva of the basal gastropod *Patella*, however, *Otx* is neither expressed in the apical organ nor in the developing cerebral ganglia (Nederbragt et al. 2002b). Instead, ring-like arranged cells expressing *Otx* were found in the pretrochal region, but it is unclear whether or not they contribute to neural structures (Nederbragt et al. 2002b).

Pax258 is expressed in the supraesophageal mass of the pygmy squid *Idiosepius notoides* (Fig. 7.14E; Wollesen, unpublished) and in the cerebral ganglia of adult *Haliotis*, but not in its apical organ (O'Brien and Degnan 2000, 2003). Several POU genes are expressed in the central nervous system of *Drosophila* and mouse (Treacy and Rosenfeld 1992), and this was also found in *Idiosepius* and *Haliotis*, indicating an ancestral role of these genes in bilaterian neurogenesis (Fig. 7.14F). In any case, some POU genes (e.g., *POU3* and *POU4*) appear to have been recruited into the formation of novel molluscan traits including the statocysts, tentacles, ctenidia, and osphradia (Table 7.2).

The transcription factor *engrailed* is consistently expressed in cells demarcating the early shell field(s) from the remaining mantle epithelium in polyplacophorans, gastropods, scaphopods, bivalves, and cephalopods. Several other genes are known to be expressed in shell formation domains including *decapentaplegic* (*Dpp*) in gastropods and bivalves, as well as *Hox1*, *Post1*, and *Post2* in the gastropod *Gibbula* (Table 7.2, Fig. 7.13). Research into the molecular regulatory processes of molluscan shell formation has recently received considerable attention, especially with respect to nacre formation in gastropods and bivalves, whereby an impressive genetic toolkit has been identified. This suggests independent evolution of the nacreous layer in various gastropod and bivalve lineages (Jackson et al. 2010).

Although still rather patchy, the data on molluscan gene expression currently available provide some distinct molecular markers for conchiferan (larval) structures that are of evolutionary relevance for Mollusca and the entire Lophotrochozoa. While these are only first steps into reconstructing the genetic regulatory network that underlies the development of these characters, they provide important reference points for homology hypotheses of various molluscan and lophotrochozoan features including

apical organs, trochi, and shells, which now can be assessed by comparative gene expression studies. With respect to some key players in bilaterian development, the Hox and ParaHox genes, it seems that these have played an important role in the evolution of conchiferan-specific morphological novelties and that they may have been an important evolutionary driving force of conchiferan or maybe even molluscan diversification and evolutionary success (see Wanninger et al. 2008). It is therefore of particular interest to reveal the full potential of Hox and ParaHox gene functions by future comparative studies on representatives of the various molluscan classes.

BREAKING SYMMETRY: EVODEVO OF GASTROPOD HANDEDNESS

Asymmetries are widely distributed and obvious in many bilaterian clades, notably in deuterostomes including the echinoderms and vertebrates. Within mollusks, the gastropods are a particularly drastic example of a handed body plan, which is manifested by a dextrally (right-handed or clockwise) versus sinistrally (left-handed or anticlockwise) coiled shell and gut, together with a respective right or left position of the anal and genital pores. Interestingly, in some pulmonates such as the great pond snail *Lymnaea* and the ram's horn snail *Biomphalaria*, both dextrally and sinistrally coiled individuals occur in the same population in the wild. Early investigations into the phenomenon as to how handedness is determined in these species have shown that this break in symmetry is maternally inherited and that the dextral phenotype is dominant over the sinistral one (Boycott and Diver 1923; Sturtevant 1923; Boycott et al. 1930; Freeman and Lundelius 1992). Although mapping of the responsible locus, termed *sinistral*, has been carried out, its identity remains unknown (Asami et al. 2008; Liu et al. 2013; Namigai et al. 2014). In the quest to reveal the cytokinetic dynamics and the genetic foundations of this phenomenon, important progress has recently been achieved

(Shibazaki et al. 2004; Grande and Patel 2009; Kuroda et al. 2009; see Patel 2009, Lambert 2010 for brief summary). Accordingly, the nodal signaling pathway, known to act in the correct establishment of left-right asymmetries in deuterostomes, is involved in defining handedness in gastropods (Grande and Patel 2009). In accordance with the deuterostome models, *nodal* and the downstream acting transcription factor *Pitx* are asymmetrically expressed in gastropods, depending on their right or left chirality: In embryos of the limpet *Lottia gigantea* with a dextral body plan, both genes are expressed on the right side of the embryo, while their expression is to the left in sinistral embryos of *Biomphalaria glabrata* (Grande and Patel 2009). By chemically inhibiting nodal signaling, it was found that in those *Biomphalaria* embryos that had survived the treatment, *Pitx* expression was lost. The juveniles developed tubular rather than the left-handed coiled shells of the usual, non-treated specimens. Interestingly, non-coiled shells were only obtained if nodal signaling was blocked before the blastula stage, while trochophore larvae treated with the drug developed normal shells (Grande and Patel 2009). Since asymmetrical expression of *nodal* is already a result of broken symmetry in the gastropod embryo and not its cause, the key symmetry-breaking events must take place prior to the first signaling events, i.e., before the blastula stage.

In two *Lymnaea* species, dextral versus sinistral coiling coincides with a clockwise (dextral) versus anticlockwise (sinistral) direction at third cleavage (Freeman and Lundelius 1992; Shibazaki et al. 2004). In order to experimentally test whether this corresponds to the first symmetry-breaking event responsible for handedness in these gastropods, Kuroda et al. (2009) manipulated the (genetically determined) direction of third cleavage in both sinistral and dextral *Lymnaea* embryos. Thereby, the arrangement of the first quartet micromeres was altered during their formation at third cleavage by imposing a 90° shift, such that they came to lie in opposite direction with respect to their genetically predefined position. As a consequence, artificial

"dextralized" and "sinistralized" embryos were produced (Fig. 7.15). In the surviving embryos that had maintained the artificially induced cleavage pattern, the entire body organization including shell coiling was reversed. The manipulated (female) snails that made it to fertile adults gave rise to offspring with "correct" body organization, i.e., the genetically inherited handedness and not the one artificially imposed on the mother animal.

Embryos with altered cleavage directions at the first or second round of cleavage reverted to their inherited cleavage direction at third cleavage, thus providing evidence that the first symmetry-breaking signaling events take place at the eight-cell stage and not earlier (Kuroda et al. 2009). The two genes of the nodal pathway involved in molecular left-right patterning, *nodal* and *Pitx*, were expressed according to the expected chirality as predicted by the blastomere arrangement in the eight-cell embryo, thus demonstrating that the blastomere configuration at this stage defines the localization of embryonic nodal signaling (Fig. 7.15; Kuroda et al. 2009). Whether these cytogenetic and molecular determinants of left-right asymmetry are conserved among Mollusca, Spiralia, or even the entire protostomes and deuterostomes, and thus were present in the last common ancestor of all Bilateria, remains to be seen once more comparative studies become available.

Our current knowledge on molluscan developmental biology as summarized here is testimony of the numerous pathways evolution has taken in shaping the wide phenotypic diversity of Mollusca. Deciphering the developmental underpinnings of these mechanisms provides an important window into the evolutionary ancestry of this fascinating phylum. We have only begun to scratch the surface of reconstructing molluscan evolutionary history, but with the recently established and ever advancing molecular, morphological, and in silico methods at hand, and more and more genomes and transcriptomes being generated, the stage is set for present and future generations of biologists to engage in studies on the evolutionary developmental biology of this fascinating phylum.

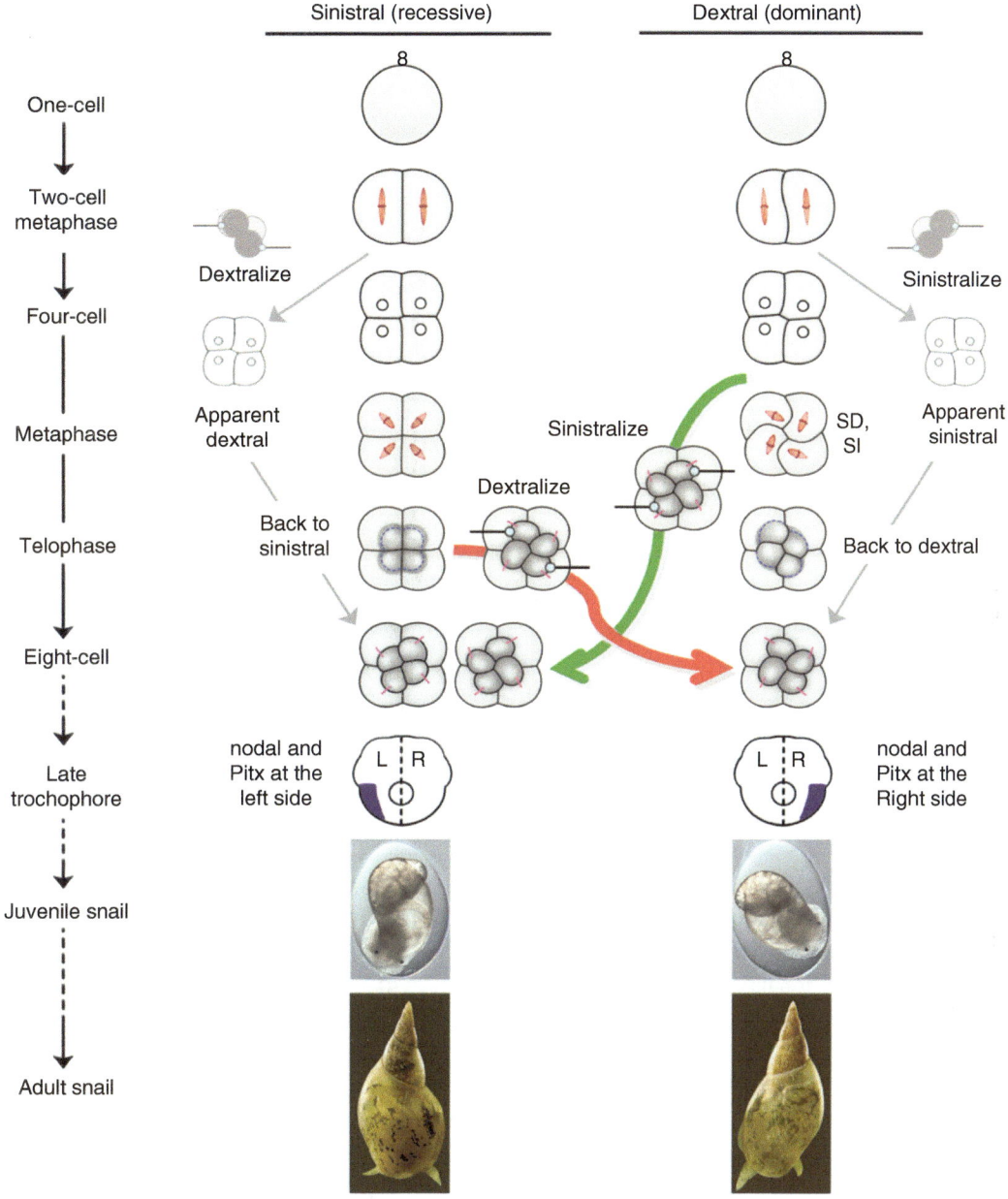

Fig. 7.15 Alteration of handedness in the gastropod *Lymnaea stagnalis* by micromanipulation. Experimental reversal of chirality during the first two cleavage rounds does not result in organisms with reversed chirality. This is because embryos whose cleavages have been manipulated at such early stages revert at the eight-cell stage to their original handedness (*gray arrows*). However, reversing chirality during third cleavage results in "inverted juveniles" and eventually in fertile adults (*red* and *green arrows*). Note oppositely coiled shells (photographs) in these individuals (the entire digestive tract likewise shows a mirror image in these healthy snails with respect to non-manipulated ones; not depicted). *Nodal* and *Pitx*, two major genes involved in left (*L*)-right (*R*) patterning in many bilaterian animals, are also reversibly expressed in these micromanipulated specimens. Interestingly, the dominant dextral snails exhibit spiral deformation (*SD*, a helical deformation of the blastomeres at the third cleavage metaphase-anaphase directly linked to the handedness-determining genes) and spindle inclination (*SI*, a spiral orientation of the four spindles, as a consequence of spiral deformation, before cleavage furrow ingression), while the recessive sinistral snails do not show them. However, dextralized snails can be produced from sinistral ones without spiral deformation (Figure reproduced from Kuroda et al. (2009) with permission from the publisher)

OPEN QUESTIONS

- Neuromuscular development in Chaetodermomorpha: How does it conform to the "aculiferan pattern" as exemplified by Polyplacophora and Neomeniomorpha?
- Neuromuscular development in protobranch bivalves: How do developmental patterns of test cell larvae compare to direct versus indirect development of other mollusks?
- Development in Monoplacophora: How is the seriality of neuromuscular and excretory structures achieved?
- Comparative tempo-spatial expression patterns of Hox, ParaHox, and other key developmental genes in Neomeniomorpha, Chaetodermomorpha, Monoplacophora, Polyplacophora, Bivalvia, and Scaphopoda: What are the functions of individual genes, and how do they govern development of specific morphological features?
- Cell lineages and gene expression profiles of test cells and velar and trochal structures: How conserved or plastic are they among molluscan (and lophotrochozoan) larvae?
- Cell lineages, gene expression profiles, and submicroscopic 3D architecture of apical organs: How (dis)similar are they among Mollusca, other Lophotrochozoa, and more distant taxa, e.g., Cnidaria?
- What are the functional roles of apical organs?
- Cephalopod neuro- and myogenesis: Are there "larval" remnants?
- Genetic signatures of cephalic appendages of scaphopods, gastropods, and cephalopods: Are they homologous?
- Gene expression profiles of embryonic and adult conchiferan shells and aculiferan spicules: What are the key players that govern molluscan biomineralization processes? How does this compare to other animals that form mineralized skeletons?

Acknowledgments AW expresses his sincere thanks to the numerous colleagues that have offered their time to discuss various issues on molluscan and metazoan morphology, development, and evolution over the past many years, often at most peculiar times and in most inspiring locations, in particular Gerhard Haszprunar (Munich), Bernie Degnan (Brisbane), Jens Hoeg (Copenhagen), Claus Nielsen (Copenhagen), Pedro Martinez (Barcelona), Christiane Todt (Bergen), his coauthor of this paper and longtime colleague Tim Wollesen (Vienna), and many others, including the present and past students and postdocs in his labs in Copenhagen and Vienna. AW is also grateful for the generous support of the Faculty of Life Sciences, University of Vienna, during the past four years as well as the Danish Science Foundation (FNU) and the Carlsberg Foundation for previous support during his Copenhagen years. He also warmly acknowledges funding of our Early Stage Research Training Network MOLMORPH during the years 2005–2009 by the European Commission. The Austrian Science Fund (FWF) is thanked for current support of a project on aplacophoran EvoDevo (grant number P24276-B22). TW thanks Sonia Victoria Rodríguez Monje (Vienna) and all members of the Wanninger lab for help and discussions as well as the crews of the Néomysis (Roscoff) and the RV Hans Brattström (Bergen) for assistance with the collection of animals. TW kindly thanks Andreas Wanninger, coauthor of this book chapter, and Bernie Degnan (Brisbane) for their great support during the last years. The authors thank Jonathan Henry (Urbana), Reuben Shipway (Nahant), Hiroshi Wada (Tsukuba), Michael Schrödl (Munich), Emanuel Redl, Maik Scherholz, Alen Kristof, Marlene Karelly, and Marion Hüffel (all Vienna) for providing images used in this chapter.

References

Altnöder A, Haszprunar G (2008) Larval morphology of the brooding clam *Lasaea adansonii* (Gmelin, 1791) (Bivalvia, Heterodonta, Galeommatoidea). J Morphol 269:762–774

Anderson PD, Bokor G (2012) Conotoxins: potential weapons from the sea. J Bioterr Biodef 3:120

Andouche A, Bassaglia Y, Baratte S, Bonnaud L (2013) Reflectin genes and development of iridophore patterns in *Sepia officinalis* embryos (Mollusca, Cephalopoda). Dev Dyn 242:560–571

Appellöf A (1898) Über das Vorkommen innerer Schalen bei den achtarmigen Cephalopoden (Octopoda). Bergens Mus Arb 12:1–15

Arnolds WJA, van den Biggelaar JAM, Verdonk NH (1983) Spatial aspects of cell interactions involved in the determination of dorsoventral polarity in equally cleaving gastropods and regulative abilities of their embryos, as studied by micromere deletions in *Lymnaea* and *Patella*. Roux's Arch Dev Biol 192:75–85

Asami T, Gittenberger E, Falkner G (2008) Whole-body enantiomorphy and maternal inheritance of chiral reversal in the pond snail *Lymnaea stagnalis*. J Hered 99:552–557

Ax P (1999) Multicellular animals. Springer, Berlin

Bandel K (1975) Embryonalgehäuse karibischer Meso- und Neogastropoden (Mollusca). Abh Math Naturw Kgl Akad Wiss Mainz 1:1–133

Bandel K, Boletzky SV (1979) A comparative study of the structure, development and morphological relationships of chambered cephalopod shells. The Veliger 21:313–354

Baratte S, Bonnaud L (2009) Evidence of early nervous differentiation and early catecholaminergic sensory system during *Sepia officinalis* embryogenesis. J Comp Neurol 517:539–549

Bardou I, Maubert E, Leprince J, Chichery R, Dallérac G, Vaudry H, Agin V (2010) Ontogeny of oxytocin-like immunoreactivity in the cuttlefish, *Sepia officinalis*, central nervous system. Dev Neurosci 32:19–32

Barlow LA, Truman JW (1992) Patterns of serotonin and SCP immunoreactivity during metamorphosis of the nervous system of the red abalone, *Haliotis rufescens*. J Neurobiol 23:829–844

Bartolomaeus T (1993) Die Leibeshöhlenverhältnisse und Nephridialorgane der Bilateria – Ultrastruktur, Entwicklung und Evolution. University of Göttingen, Göttingen

Blochmann F (1883) Beiträge zur Kenntnis der Entwicklung der Gastropoden. I. Zur Entwicklung von *Aplysia limacina* L. Z Wiss Zool 38:392–410

Boletzky S (1989) Recent studies on spawning, embryonic development, and hatching in the Cephalopoda. Adv Mar Biol 25:85–115

Boletzky S, Erlwein B, Hofmann DK (2006) The *Sepia* egg: a showcase of cephalopod embryology. Vie Et Milieu – Life Environ 56:191–201

Bonar DB (1978) Ultrastructure of a cephalic sensory organ in larvae of the gastropod *Phestilla sibogae* (Aeolidacea Nudibranchia). Tissue Cell 10:153–165

Boring L (1989) Cell-cell interactions determine the dorsoventral axis in embryos of an equally cleaving opisthobranch mollusc. Dev Biol 136:239–253

Boycott AE, Diver C (1923) On the inheritance of sinistrality in *Lymnaea peregra*. Proc R Soc Lond B Biol Sci 95:207–213

Boycott AE, Diver C, Garstang SL, Hardy MAC, Turner FM (1930) The inheritance of sinistrality in *Lymnaea peregra*. Philos Trans R Soc Lond B Biol Sci 219:51–130

Brooke NM, Garcia-Fernandez J, Holland PWH (1998) The ParaHox gene cluster is an evolutionary sister of the Hox gene cluster. Nature 392:920–922

Buckland-Nicks (2013) Acorena. Revista de Estudos Acoreanos. Suplemento 8. Book of abstracts. World Congress of Malacology

Buresi A, Baratte S, Da Silva C, Bonnaud L (2012) *Orthodenticle/otx* ortholog expression in the anterior brain and eyes of *Sepia officinalis* (Mollusca, Cephalopoda). Gene Expr Patterns 12:109–116

Buresi A, Canali E, Bonnaud L, Baratte S (2013) Delayed and asynchronous ganglionic maturation during cephalopod neurogenesis as evidenced by *Sofelav1* expression in embryos of *Sepia officinalis* (Mollusca, Cephalopoda). J Comp Neurol 521:1482–1496

Bütschli O (1877) Entwicklungsgeschichtliche Beiträge. I. Zur Entwicklungsgeschichte von *Paludina vivipara*. Z Wiss Zool 29:216–231

Casteel DB (1904) The cell-linage and early larval development of *Fiona marina*, a nudibranch mollusk. Proc Acad Nat Sci Phila 56:325–405

Cather JN (1967) Cellular interactions in the development of the shell gland of the gastropod, *Ilyanassa*. J Exp Zool 166:205–223

Checa A (2000) A new model for periostracum and shell formation in Unionidae (Bivalvia, Mollusca). Tissue Cell 32:405–416

Clement AC (1986) The embryonic value of the micromeres in *Ilyanassa obsoleta*, as determined by deletion experiments. III. The third quartet cells and the mesentoblast cell. Int J Invertebr Reprod Dev 9:155–168

Conklin EG (1897) The embryology of *Crepidula*, a contribution to the cell lineage and early development of some marine gastropods. J Morphol 13:1–226

Cragg SM (1985) The adductor and retractor muscles of the veliger of *Pecten maximus* (L.) (Bivalvia). J Molluscan Stud 51:276–283

Croll RP (2000) Insights into early molluscan neuronal development through studies of transmitter phenotypes in embryonic pond snails. Microsc Res Tech 49:570–578

Croll RP (2009) Developing nervous systems in molluscs: navigating the twists and turns of a complex life cycle. Brain Behav Evol 74:164–176

Croll RP, Dickinson AJG (2004) Form and function of the larval nervous system in molluscs. Invertebr Reprod Dev 46:2–3

Croll RP, Voronezhskaya EE (1996) Early elements in gastropod neurogenesis. Dev Biol 173:344–347

Cumin R (1972) Normentafel zur Organogenese von *Lymnaea stagnalis* L. mit besonderer Berücksichtigung der Mitteldarmdrüse. Rev Suisse Zool 79:709–774

D'Asaro CN (1966) The egg capsules, embryogenesis, and early organogenesis of a common oyster predator, *Thais haemastoma floridana* (Gastropoda: Prosobranchia). Bull Mar Sci 16:884–914

Damen P, Dictus WJAG (1996) Organiser role of the stem cell of the mesoderm in prototroch patterning in *Patella vulgata* (Mollusca, Gastropoda). Mech Dev 56:41–60

Dautert E (1929) Die Bildung der Keimblätter bei *Paludina vivipara*. Zool Jb Anat 50:433–496

Demian ES, Yousif F (1975) Embryonic development and organogenesis in the snail *Marisa cornuarietis* (Mesogastropoda: Ampullariidae). V. Development of the nervous system. Malacologia 15:29–42

Dickinson AJG, Croll RP (2003) Development of the larval nervous system of the gastropod *Ilyanassa obsoleta*. J Comp Neurol 466:197–218

Dickinson AJG, Nason J, Croll RP (1999) Histochemical localization of FMRFamide, serotonin and catecholamines in embryonic *Crepidula fornicata* (Gastropoda, Prosobranchia). Zoomorphology 119:49–62

Dictus WJAG, Damen P (1997) Cell-lineage and clonal-contribution map of the trochophore larva of *Patella vulgata* (Mollusca). Mech Dev 62:213–226

Diefenbach TJ, Koss R, Goldberg JI (1998) Early development of an identified serotonergic neuron in *Helisoma trivolvis* embryos: serotonin expression, de-expression, and uptake. Dev Neurobiol 34:361–376

Dunn EF, Moy VN, Angerer LM, Angerer RC, Morris RL, Peterson KJ (2007) Molecular paleoecology: using gene regulatory analysis to address the origins of complex life cycles in the late Precambrian. Evol Dev 9:10–24

Dyachuk V, Odintsova N (2009) Development of the larval muscle system in the mussel *Mytilus trossulus* (Mollusca, Bivalvia). Dev Growth Differ 51:69–79

Eernisse DJ, Reynolds PD (1994) Chapter 3: Polyplacophora. In: Harrison FW (ed) Microscopic anatomy of invertebrates. Wiley-Liss, New York, pp 56–110

Ellis I, Kempf SC (2011) Characterization of the central nervous system and various peripheral innervations during larval development of the oyster *Crassostrea virginica*. Invertebr Biol 130:236–250

Farfán C, Shigeno S, Nödl MT, de Couet HG (2009) Developmental expression of *apterous/Lhx2/9* in the sepiolid squid *Euprymna scolopes* supports an ancestral role in neural development. Evol Dev 11:354–362

Faussek V (1901) Untersuchungen über die Entwicklung der Cephalopoden. Mitt Zool Stat Neapel 14:83–237

Fernando W (1931) The origin of the mesoderm in the gastropod *Viviparus* (=*Paludina*). Proc R Soc Lond B Containing Pap Biol Character 107:381–390

Filla A, Hiripi L, Elekes K (2009) Role of aminergic (serotonin and dopamine) systems in the embryogenesis and different embryonic behaviors of the pond snail, *Lymnaea stagnalis*. Comp Biochem Physiol C 149:73–82

Fioroni VP (1979) Phylogenetische Abänderungen der Gastrula bei Mollusken. Z Zool Syst Evol Forsch 1:82–100

Fol H (1875) Études sur le développement des mollusques. Sur le développement des ptéropodes. Archs Zool Exp Gén 4:1–214

Freeman G, Lundelius JW (1992) Evolutionary implications of the mode of D quadrant specification in coelomates with spiral cleavage. J Evol Biol 5:205–247

Friedrich S, Wanninger A, Brückner M, Haszprunar G (2002) Neurogenesis in the mossy chiton, *Mopalia muscosa* (Gould) (Polyplacophora): evidence against molluscan metamerism. J Morphol 253:109–117

Fritsch M, Wollesen T, Oliveira ALd, Wanninger A (2015) Unexpected co-linearity of Hox gene expression in an aculiferan mollusk. BMC Evol Biol. doi:10.1186/s12862-015-0414-1

Ganin M (1873) Zur Lehre von den Keimblättern bei den Weichtieren. Warschauer Ber 1:115–140

Gegenbaur C (1852) Beiträge zur Entwicklungsgeschichte der Land-Pulmonaten. Z Wiss Zool 3:371–411

Gifondorwa DJ, Leise EM (2006) Programmed cell death in the apical ganglion during larval metamorphosis of the marine mollusc *Ilyanassa obsoleta*. Biol Bull 210:109–120

Giusti AF, Hinman VF, Degnan SM, Degnan BM, Morse DE (2000) Expression of a *Scr/Hox5* gene in the larval central nervous system of the gastropod *Haliotis*, a nonsegmented spiralian lophotrochozoan. Evol Dev 2:294–302

Gonzales EE, van der Zee M, Dictus WJ, van den Biggelaar JAM (2007) Brefeldin A and monensin inhibit the D quadrant organizer in the polychaete annelids *Arctonoe vittata* and *Serpula columbiana*. Evol Dev 9:416–431

Goulding MQ (2009) Cell lineage of the *Ilyanassa* embryo: evolutionary acceleration of regional differentiation during early development. PloS One 4:e5506

Grande C, Patel NH (2009) Lophotrochozoa get into the game: the nodal pathway and left/right asymmetry in Bilateria. Cold Spring Harb Symp Quant Biol 74:281–287

Grimaldi A, Tettamanti G, Rinaldi L, Brivio MF, Castellani D, de Eguileor M (2004) Muscle differentiation in tentacles of *Sepia officinalis* (Mollusca) is regulated by muscle regulatory factors (MRF) related proteins. Dev Growth Differ 46:83–95

Grimaldi A, Tettamanti G, Acquati F, Bossi E, Guidali ML, Banfi S, Monti L, Valvassori R, de Eguileor M (2008) A *hedgehog* homolog is involved in muscle formation and organization of *Sepia officinalis* (Mollusca) mantle. Dev Dyn 237:659–671

Haas W (1981) Evolution of calcareous hardparts in primitive molluscs. Malacologia 21:403–418

Hadfield MG, Meleshkevitch EA, Boudko DY (2000) The apical sensory organ of a gastropod veliger is a receptor for settlement cues. Biol Bull 198:67–76

Hartmann B, Lee PN, Kang YY, Tomarev S, de Couet HG, Callaerts P (2003) *Pax6* in the sepiolid squid *Euprymna scolopes*: evidence for a role in eye, sensory organ and brain development. Mech Dev 120:177–183

Hashimoto N, Kurita Y, Wada H (2012) Developmental role of *dpp* in the gastropod shell plate and co-option of the *dpp* signaling pathway in the evolution of the operculum. Dev Biol 366:367–373

Haszprunar G (1996) The Mollusca: coelomate turbellarians or mesenchymate annelids? In: Taylor JD (ed) Origin and evolutionary radiation of the Mollusca. Oxford University Press, Oxford, pp 3–28

Haszprunar G (2000) Is the Aplacophora monophyletic? A cladistic point of view. Amer Malac Bull 15:115–130

Haszprunar G, Wanninger A (2000) Molluscan muscle systems in development and evolution. J Zool Syst Evol Res 38:157–163

Haszprunar G, Wanninger A (2008) On the fine structure of the creeping larva of *Loxosomella murmanica*: additional evidence for a clade of Kamptozoa (Entoprocta) and Mollusca. Acta Zool (Stockholm) 89:137–148

Haszprunar G, Wanninger A (2012) Molluscs. Curr Biol 22:R510–R514

Haszprunar G, Friedrich S, Wanninger A, Ruthensteiner B (2002) Fine structure and immunocytochemistry of a new chemosensory system in the chiton larva (Mollusca: Polyplacophora). J Morphol 251:210–218

Heath H (1898) The development of *Ischnochiton*. Zool Jb Anat Ontog Tiere 12:567–656

Hejnol A (2010) A twist in time: the evolution of spiral cleavage in the light of animal phylogeny. Integr Comp Biol 50:695–706

Hejnol A, Martindale MQ, Henry JQ (2007) High-resolution fate map of the snail *Crepidula fornicata*: the origins of ciliary bands, nervous system, and muscular elements. Dev Biol 305:63–76

Henry JJ, Perry KJ (2008) MAPK activation and the specification of the D quadrant in the gastropod mollusc, *Crepidula fornicata*. Dev Biol 313:181–195

Henry JQ, Okusu A, Martindale MQ (2004) The cell lineage of the polyplacophoran, *Chaetopleura apiculata*: variation in the spiralian program and implications for molluscan evolution. Dev Biol 272:145–160

Herbers K (1913) Entwicklungsgeschichte von *Anodonta cellensis* Schröt. Z Wiss Zool 108:1–174

Hinman VF, Degnan BM (2002) *Mox* homeobox expression in muscle lineage of the gastropod *Haliotis asinina*: evidence for a conserved role in bilaterian myogenesis. Dev Genes Evol 212:141–144

Hinman V, O'Brien EK, Richards GS, Degnan BM (2003) Expression of anterior Hox genes during larval development of the gastropod *Haliotis asinina*. Evol Dev 5:508–521

Holland PWH (2001) Beyond the Hox: how widespread is homeobox gene clustering? J Anat 199:13–23

Iijima M, Takeuchi T, Sarashina I, Endo K (2008) Expression patterns of *engrailed* and *dpp* in the gastropod *Lymnaea stagnalis*. Genes Evol 218:237–251

Jackson DJ, Wörheide G, Degnan BM (2007) Dynamic expression of ancient and novel molluscan shell genes during ecological transitions. BMC Evol Biol 7:160

Jackson DJ, McDougall C, Woodcroft B, Moase P, Rose RA, Kube M, Reinhardt R, Rokhsar DS, Montagnani C, Joubert C, Piquemal D, Degnan BM (2009) Parallel evolution of nacre building gene sets in molluscs. Mol Biol Evol 27:591–608

Jackson DJ, Meyer MP, Seaver E, Pang K, McDougall C, Moy VN, Gordon K, Degnan BM, Martindale MQ, Burke RD, Peterson KJ (2010) Developmental expression of *COE* across the Metazoa supports a conserved role in neuronal cell-type specification and mesodermal development. Dev Genes Evol 220:221–234

Jacob MH (1984) Neurogenesis in *Aplysia californica* resembles nervous system formation in vertebrates. J Neurosci 4:1225–1239

Jacobs DK, Wray CG, Wedeen CJ, Kostriken R, DeSalle R, Staton JL, Gates RD, Lindberg DR (2000) Molluscan *engrailed* expression, serial organization, and shell evolution. Evol Dev 2:340–347

Jarvis E, Bruce HS, Patel NH (2012) Evolving specialization of the arthropod nervous system. Proc Natl Acad Sci U S A 109:10634–10639

Kakoi S, Kin K, Miyazaki K, Wada H (2008) Early development of the Japanese spiny oyster (*Saccostrea kegaki*): characterization of some genetic markers. Zool Sci 25:455–464

Kanda A, Minakata H (2006) Isolation and characterization of a novel small cardioactive peptide-related peptide from the brain of *Octopus vulgaris*. Peptides 27:1755–1761

Kandel ER (2001) The molecular biology of memory storage: a dialogue between genes and synapses. Science 294:1030–1038

Kandel ER, Kriegstein A, Schacher S (1981) Development of the central nervous system of *Aplysia* in terms of the differentiation of its specific identifiable cells. Neuroscience 5:2033–2063

Kempf SC, Chun GV, Hadfield MG (1992) An immunocytochemical search for potential neurotransmitters in larvae of *Phestilla sibogae* (Gastropoda, Ophisthobranchia). Comp Biochem Physiol 101C:299–305

Kerkhoven RM, Croll RP, Ramkema MD, Van Minnen J, Bogerd J, Boer HH (1992) The VD1/RPD2 neuronal system in the central nervous system of the pond snail *Lymnaea stagnalis* studied by in situ hybridization and immunocytochemistry. Cell Tissue Res 267:551–559

Kerkhoven RM, Ramkema MD, Van Minnen J, Croll RP, Pin T, Boer HH (1993) Neurons in a variety of molluscs react to antibodies raised against the VD1/RPD2 α-neuropeptide of the pond snail *Lymnaea stagnalis*. Cell Tissue Res 273:371–379

Kier WM (1988) The arrangement and function of molluscan muscle. In: Trueman ER, Clarke MR (eds) The Mollusca: form and function. Academic, New York, 11:211-252

Kier WM (1991) Squid cross-striated muscle: the evolution of a specialized muscle fiber type. Bull Mar Sci 49:389–403

Kier WM (1996) Muscle development in squid: ultrastructural differentiation of a specialized muscle fiber type. J Morphol 229:271–288

Kier WM, Schachat FH (2008) Muscle specialization in the squid motor system. J Exp Biol 211:164–169

Kier WM, Stella MP (2007) The arrangement and function of *Octopus* arm musculature and connective tissue. J Morphol 268:831–843

Kier WM, Thompson JT (2003) Muscle arrangement, function and specialization in recent coleoids. Berl Paläobiol Abhandl 3:141–162

Kin K, Kakoi S, Wada H (2009) A novel role for *dpp* in the shaping of bivalve shells revealed in a conserved molluscan developmental program. Dev Biol 329:152–166

Kniprath E (1977) Zur Ontogenese des Schalenfeldes von *Lymnaea stagnalis*. Wilhelm Roux Arch Entw Mech 181:11–30

Kniprath E (1981) Ontogeny of the molluscan shell field: a review. Zool Scr 10:61–79

Kocot KM, Cannon JT, Todt C, Citarella MR, Kohn AB, Meyer A, Santos SR, Schander C, Moroz L, Lieb B, Halanych KM (2011) Phylogenomics reveals deep molluscan relationships. Nature 477:452–456

Kölliker A (1844) Entwickelungsgeschichte der Cephalopoden. Verlag von Meyer und Zeller, Zürich

Koop D, Richards GS, Wanninger A, Gunter HM, Degnan BM (2007) The role of MAPK signaling in patterning and establishing axial symmetry in the gastropod *Haliotis asinina*. Dev Biol 311:200–212

Kowalevsky MA (1883a) Embryogénie du *Chiton polii* (Philippi) avec quelques remarques sur le développement des autres chitons. Ann Mus Hist Nat Marseille Zool 1:1–46

Kowalevsky MA (1883b) Étude sur l'embryogénie du Dentale. Ann Mus Hist Nat Marseille Zool 1:1–54

Kranz AM, Tollenaere A, Norris BJ, Degnan BM, Degnan SM (2010) Identifying the germline in an equally cleaving mollusc: *Vasa* and *Nanos* expression during embryonic and larval development of the vetigastropod *Haliotis asinina*. J Exp Zool B Mol Dev Evol 314:267–279

Kriegstein AR (1977) Development of the nervous system of *Aplysia californica*. Proc Natl Acad Sci U S A 74:375–378

Kristof A, Klussmann-Kolb A (2010) Neuromuscular development of *Aeolidiella stephanieae* Valdéz, 2005 (Mollusca, Gastropoda, Nudibranchia). Front Zool 7:5

Kühtreiber WM, Vantil EH, Van Dongen CAM (1988) Monensin interferes with the determination of the mesodermal cell-line in embryos of *Patella vulgata*. Roux's Arch Dev Biol 197:10–18

Kurita Y, Deguchi R, Wada H (2009) Early development and cleavage pattern of the Japanese purple mussel, *Septifer virgatus*. Zool Sci 26:814–820

Kuroda R, Endo B, Masanri A, Shimuzu M (2009) Chiral blastomere arrangement dictates zygotic left-right asymmetry pathway in snails. Nature 462: 790–794

Lacaze-Duthiers FJH (1858) Histoire de l'organisation, du développement, des mœurs et des rapports zoologiques du Dentale. Librairie de Victor. Masson, Paris

Lambert JD (2010) Developmental patterns in spiralian embryos. Curr Biol 20:R72–R77

Lambert JD, Nagy LM (2003) The MAPK cascade in equally cleaving spiralian embryos. Dev Biol 263:231–241

Lankester ER (1873) Summary of the zoological observations made in Naples in the winter of 1871–72. Ann Mag Nat Hist 11:81–97

Lankester ER (1875) Observations on the development of the Cephalopoda. Q J Microsc Sci 15:37–47

Lartillot N, Le Gouar M, Adoutte A (2002a) Expression patterns of *fork head* and *goosecoid* homologues in the mollusc *Patella vulgata* supports the ancestry of the anterior mesendoderm across Bilateria. Dev Genes Evol 212:551–561

Lartillot N, Lespinet O, Vervoort M, Adoutte A (2002b) Expression pattern of *Brachyury* in the mollusc *Patella vulgata* suggests a conserved role in the establishment of the AP axis in Bilateria. Development 129:1411–1421

Le Gouar M, Lartillot N, Adoutte A, Vervoort M (2003) The expression of a *caudal* homologue in a mollusc, *Patella vulgata*. Gene Expr Patterns 3:35–37

Le Gouar M, Guillou A, Vervoort M (2004) Expression of a *SoxB* and a *Wnt2/13* gene during the development of the mollusc *Patella vulgata*. Dev Genes Evol 214: 250–256

Lee PN, Callaerts P, de Couet HG, Martindale MQ (2003) Cephalopod Hox genes and the origin of morphological novelties. Nature 424:1061–1065

Lemche H, Wingstrand KG (1959) The anatomy of *Neopilina galatheae* Lemche, 1957. Galathea Rep 3:9–71

Lespinet O, Nederbragt AJ, Cassan M, Dictus WJ, van Loon AE, Adoutte A (2002) Characterization of two *snail* genes in the gastropod mollusc *Patella vulgata*. Implications for understanding the ancestral function of the *snail*-related genes in Bilateria. Dev Genes Evol 212:186–195

Lillie FR (1895) The embryology of the Unionidae: a study in cell-linage. J Morphol 10:1–100

Lin M-F, Leise EM (1996) Gangliogenesis in the prosobranch gastropod *Ilyanassa obsoleta*. J Comp Neurol 374:180–193

Liu MM, Davey JW, Banerjee R, Han J, Yang F, Aboobaker A, Blaxter ML, Davison A (2013) Fine mapping of the pond snail left-right asymmetry (chirality) locus using rad-seq and fibre-fish. PloS One 8:e71067

Longo FJ (1983) Meiotic maturation and fertilization. In: Verdonk NH, Van den Biggelaar JAM, Tompa AS (eds) The Mollusca, vol III. Academic, New York, pp 49–89

Lyons DC, Perry KJ, Lesoway MP, Henry JQ (2012) Cleavage pattern and fate map of the mesentoblast, 4d, in the gastropod *Crepidula*: a hallmark of spiralian development. EvoDevo 3:21

Marois R, Carew TJ (1990) The gastropod nervous system in metamorphosis. J Neurobiol 21: 1053–1071

Marois R, Carew TJ (1997a) Projection patterns and target tissues of the serotonergic cells in larval *Aplysia californica*. J Comp Neurol 386:491–506

Marois R, Carew TJ (1997b) Ontogeny of serotonergic neurons in *Aplysia californica*. J Comp Neurol 386:477–490

Marois R, Croll RP (1992) Development of serotonin-like immunoreactivity in the embryonic nervous system of the snail *Lymnaea stagnalis*. J Comp Neurol 322:255–265

Marquis VF (1989) Die Embryonalentwicklung des Nervensystems von *Octopus vulgaris* Lam. (Cephalopoda, Octopoda), eine histologische Analyse. Verh Naturforsch Ges Basel 99:23–76

Martin R (1965) On the structure and embryonic development of the giant fiber system of the squid *Loligo vulgaris*. Z Zellforsch 67:77–85

Martindale MQ (1986) The organizing role of the D quadrant in an equal cleaving spiralian, *Lymnaea stagnalis* as studied by UV laser deletion of macromeres at intervals between 3rd and 4th quartet formation. Int J Invertebr Reprod Dev 9:229–242

Martindale MQ, Doe CQ, Morrill JB (1985) The role of animal-vegetal interaction with respect to the determination of dorsoventral polarity in the equal-cleaving spiralian *Lymnaea palustris*. Roux's Arch Dev Biol 194:281–295

Mattiello T, Costantini M, Di Matteo B, Livigni S, Andouche A, Bonnaud L, Palumbo A (2012) The dynamic nitric oxide pattern in developing cuttlefish *Sepia officinalis*. Dev Dyn 241:390–402

Meisenheimer J (1896) Entwicklungsgeschichte von *Limax maximus* L. I. Teil: furchung und Keimblätterbildung. Z Wiss Zool 62:415–468

Meisenheimer J (1898) Entwicklungsgeschichte von *Limax maximus* L. II. Teil: Die Larvenperiode. Z Wiss Zool 63:573–664

Meisenheimer J (1901) Entwicklungsgeschichte von *Dreissensia polymorpha* Pall. Z Wiss Zool 69:1–137

Meister G (1972) Organogenese von *Loligo vulgaris* LAM. Mollusca, Cephalopoda, Teuthoidea, Myopsida, Loliginidae. Zool Jb Anat 89:247–300

Mescheryakov VN (1990) The common pond snail *Lymnaea stagnalis* L. In: Dettlaff DA, Vassetzky SG (eds) Animal species for developmental studies. Plenum Press, New York, pp 69–132

Messenger JB (1996) Neurotransmitters of cephalopods. Invert Neurosci 2:95–114

Müller AH (1994) Lehrbuch der Paläozoologie. Band II. Invertebraten. II Teil: Mollusca 2 – Arthropoda 1, 4. Auflage. Fischer Verlag, Jena

Naef A (1928) Die Cephalopoden. Fauna Flora Golfo Napoli 35:149–863

Namigai EKO, Kenny NJ, Shimeld SM (2014) Right across the tree of life: the evolution of left-right asymmetry in the Bilateria. Genesis 52:458–470

Navet S, Bassaglia Y, Baratte S, Martin M, Bonnaud L (2008) Somatic muscle development in *Sepia officinalis* (Cephalopoda – Mollusca): a new role for NK4. Dev Dyn 237:1944–1951

Navet S, Andouche A, Baratte S, Bonnaud L (2009) *Shh* and *Pax6* have unconventional expression patterns in embryonic morphogenesis in *Sepia officinalis* (Cephalopoda). Gene Expr Patterns 9:461–467

Nederbragt AJ, te Welscher P, van den Driesche S, van Loon AE, Dictus WJAG (2002a) Novel and conserved roles for *orthodenticle/otx* and *orthopedia/otp* orthologs in the gastropod mollusc *Patella vulgata*. Dev Genes Evol 212:330–337

Nederbragt AJ, van Loon AE, Dictus WJAG (2002b) Expression of *Patella vulgata* orthologs of *engrailed* and *dpp-bmp2/4* in adjacent domains during molluscan shell development suggests a conserved compartment boundary mechanism. Dev Biol 246:341–355

Nederbragt AJ, van Loon AE, Dictus WJAG (2002c) *Hedgehog* crosses the snail's midline. Nature 417:811–812

Nielsen C (2004) Trochophore larvae: cell-lineages, ciliary bands, and body regions. 1. Annelida and Mollusca. J Exp Zool B Mol Dev Evol 302:35–68

Nielsen C (2012) Animal evolution: interrelationships of the living phyla. Oxford University Press, Oxford

Nielsen C, Haszprunar G, Ruthensteiner B, Wanninger A (2007) Early development of the aplacophoran mollusc *Chaetoderma*. Acta Zool (Stockholm) 88:231–247

Nixon M, Young JZ (2003) The brains and lives of cephalopods. Oxford University Press, New York

O'Brien EK, Degnan BM (2000) Pax, POU and Sox genes are expressed in the ganglia of the tropical abalone *Haliotis asinina*. Mar Biotech 2:545–557

O'Brien EK, Degnan BM (2002a) Pleiotropic developmental expression of *HasPOU-III*, a class III *POU* gene, in the gastropod *Haliotis asinina*. Mech Dev 114:129–132

O'Brien EK, Degnan BM (2002b) Developmental expression of a class IV *POU* gene in the gastropod *Haliotis asinina* supports a conserved role in sensory cell development in bilaterians. Dev Genes Evol 212:394–398

O'Brien EK, Degnan BM (2003) Expression of *Pax258* in the gastropod statocyst: insights into the antiquity of metazoan geosensory organs. Evol Dev 5:572–578

Ogura A, Yoshida MA, Moritaki T, Okuda Y, Sese J, Shimizu KK, Sousounis K, Tsonis PA (2013) Loss of the *six3/6* controlling pathways might have resulted in pinhole-eye evolution in *Nautilus*. Sci Rep 3:1432

Okada K (1939) The development of the primary mesoderm in *Sphaerium japonicum biwaense* Mori. Sci Rep Tohoku Univ Biol 14:25–48

Okusu A (2002) Embryogenesis and development of *Epimenia babai* (Mollusca Neomeniomorpha). Biol Bull 203:87–103

Page LR (1992a) New interpretation of a nudibranch central nervous system based on ultrastructural analysis of neurodevelopment in *Melibe leonina*. I. Cerebral and visceral loop ganglia. Biol Bull 182:348–365

Page LR (1992b) New interpretation of a nudibranch central nervous system based on ultrastructural analysis of neurodevelopment in *Melibe leonina*. II. Pedal, pleural, and labial ganglia. Biol Bull 182:366–381

Page LR (2002) Apical sensory organ in larvae of the patellogastropod *Tectura scutum*. Biol Bull 202:6–22

Page LR (2006) Early differentiating neuron in larval abalone (*Haliotis kamtschatkana*) reveals the relationship between ontogenetic torsion and crossing of the pleurovisceral nerve cords. Evol Dev 8:458–467

Page LR, Parries SC (2001) Comparative study of the apical ganglion in planktotrophic caenogastropod larvae: ultrastructure and immunoreactivity to serotonin. J Comp Neurol 418:383–401

Parkhaev PY (2008) The early Cambrian radiation of Mollusca. In: Ponder WF, Lindberg DR (eds) Phylogeny and evolution of the Mollusca. University of California Press, Berkely

Patel NH (2009) Developmental biology: asymmetry with a twist. Nature 462:727–728

Patten W (1886) The embryology of *Patella*. Arb Zool Inst Univ Wien 6:149–174

Ponder WF, Lindberg DR (eds) (2008) Phylogeny and evolution of the Mollusca. University of California Press, Berkeley

Rabinowitz JS, Chan XJ, Kingsley EP, Duan Y, Lambert JD (2008) Nanos is required in somatic blast cell lineages in the posterior of a mollusk embryo. Curr Biol 18:331–336

Raven CP (1966) Morphogenesis: the analysis of molluscan development. Pergamon Press, Oxford

Reichert H (2005) A tripartite organization of the urbilaterian brain: developmental genetic evidence from *Drosophila*. Brain Res Bull 66:491–494

Render J (1991) Fate maps of the first quartet micromeres in the gastropod *Ilyanassa obsoleta*. Development 113:495–501

Render J (1997) Cell fate maps in the *Ilyanassa obsoleta* embryo beyond the third division. Dev Biol 189: 301–310

Richter S, Loesel R, Purschke G, Schmidt-Rhaesa A, Scholtz G, Stach T, Vogt L, Wanninger A, Brenneis G, Döring C, Faller S, Fritsch M, Grobe P, Heuer CM, Kaul S, Møller OS, Müller CHG, Rieger V, Rothe BH, Stegner MEJ, Harzsch S (2010) Invertebrate neurophylogeny: suggested terms and definitions for a neuroanatomical glossary. Front Zool 7:29

Robert A (1902) Recherches sur le développement des troques. Arch Zool Exp Gén 10:269–538

Samadi L, Steiner G (2009) Involvement of Hox genes in shell morphogenesis in the encapsulated development of a top shell gastropod (*Gibbula varia* L.). Dev Genes Evol 219:523–530

Samadi L, Steiner G (2010a) Expression of Hox genes during the larval development of the snail, *Gibbula varia* (L.)—further evidence of non-colinearity in molluscs. Dev Genes Evol 220:161–172

Samadi L, Steiner G (2010b) Conservation of ParaHox genes' function in patterning of the digestive tract of the marine gastropod *Gibbula varia*. BMC Dev Biol 10:74

Scherholz M, Redl E, Wollesen W, Todt C, Wanninger A (2013) Aplacophoran mollusks evolved from ancestors with polyplacophoran-like features. Curr Biol 23:2130–2134

Schmidt F (1895) Beiträge zur Kenntnis der Entwicklungsgeschichte der Stylommatophoren. Zool Jb Anat 8:318–341

Shibazaki Y, Shimizu M, Kuroda R (2004) Body handedness is directed by genetically determined cytoskeletal dynamics in the early embryo. Curr Biol 14:1462–1467

Shigeno S, Tsuchiya K, Segawa S (2001) Embryonic and paralarval development of the central nervous system of the loliginid squid *Sepioteuthis lessoniana*. J Comp Neurol 437:449–475

Shigeno S, Sasaki T, Moritaki T, Kasugai T, Vecchione M, Agata K (2008) Evolution of the cephalopod head complex by assembly of multiple molluscan body parts: evidence from *Nautilus* embryonic development. J Morphol 269:1–17

Smith SA, Wilson NG, Goetz FE, Feehery C, Andrade SCS, Rouse GW, Giribet G, Dunn CW (2011) Resolving the evolutionary relationships of molluscs with phylogenomic tools. Nature 480:364–367

Spiess PE (1972) Organogenese des Schalendrüsen-komplexes bei einigen coleoiden Cephalopoden des Mittelmeeres. Rev Suisse Zool 79:167–226

Stewart H, Westlake HE, Page LR (2014) Rhogocytes in gastropod larvae: developmental transformation from protonephridial terminal cells. Invertebr Biol 133:47–63

Sturtevant AH (1923) Inheritance of direction of coiling in *Lymnaea*. Science 58:269–270

Sutton MD, Briggs DEG, Siveter DJ, Siveter DJ, Sigwart JD (2012) A silurian armoured aplacophoran and implications for molluscan phylogeny. Nature 490:94–97

Todt C, Wanninger A (2010) Of tests, trochs, shells, and spicules: development of the basal mollusk *Wirenia argentea* (Solenogastres) and its bearing on the evolution of trochozoan larval key features. Front Zool 7:6

Treacy MN, Rosenfeld MG (1992) Expression of a family of POU-domain protein regulatory genes during development of the central nervous system. Annu Rev Neurosci 15:139–165

Ussow M (1874) Zoologisch-embryologische Untersuchungen. Die Kopffüßler. Arch Naturgesch 40:329–372

Van Dam WI (1986) Embryonic development of *Bithynia tentaculata* L. (Prosobranchia, Gastropoda). I. Cleavage. J Morphol 188:289–302

Van den Biggelaar JAM (1977) Development of dorso-ventral polarity and mesentoblast determination in *Patella vulgata*. J Morphol 154:157–186

Van den Biggelaar JAM (1996) Cleavage pattern and mesentoblast formation in *Acanthochiton crinitus* (Polyplacophora, Mollusca). Dev Biol 174:423–430

Van den Biggelaar JAM, Guerrier P (1979) Dorsoventral polarity and mesentoblast determination as concomitant results of cellular interactions in the mollusk *Patella vulgata*. Dev Biol 68:462–471

Van Dongen CAM, Geilenkirchen WLM (1974) The development of *Dentalium* with special reference to the significance of the polar lobe. I, II, III. Division chronology and development of the cell pattern in *Dentalium dentale* (Scaphopoda). Proc Kongl Ned Akad Wet C 77:57–100

Van Dongen CAM, Geilenkirchen WLM (1975) The development of *Dentalium* with special reference to the significance of the polar lobe. IV. Division chronology and development of the cell pattern in *Dentalium dentale* after removal of the polar lobe at first cleavage. Proc Kongl Ned Akad Wet C 78:358–375

Van Dongen CAM, Geilenkirchen WLM (1976) The development of *Dentalium* with special reference to the significance of the polar lobe. V and VI. Differentiation of the cell pattern in lobeless embryos of *Dentalium vulgare* (da Costa) during late larval development. Proc Kongl Ned Akad Wet C 79:245–266

Villanueva R, Norman MD (2008) Biology of the planktonic stages of benthic octopuses. Oceanogr Mar Biol 46:105–202

Vinther J, Jell P, Kampouris G, Carney R, Racicot RA, Briggs DEG (2012) The origin of multiplacophorans – convergent evolution in aculiferan molluscs. Palaeontology 55:1007–1019

Voronezhskaya EE, Elekes K (1993) Distribution of serotonin-like immunoreactive neurons in the embryonic nervous system of lymnaeid and planorbid snails. Neurobiology 1:371–383

Voronezhskaya EE, Elekes K (2003) Expression of peptides encoded by the FMRFamide gene in the developing nervous system of *Lymnaea stagnalis*. Cell Tissue Res 314:297–313

Voronezhskaya EE, Tyurin SA, Nezlin LP (2002) Neuronal development in larval chiton *Ischnochiton hakodadensis* (Mollusca: Polyplacophora). J Comp Neurol 444:25–38

Voronezhskaya EE, Nezlin LP, Odintsova NA, Plummer JT, Croll RP (2008) Neuronal development in larval mussel *Mytilus trossulus* (Mollusca: Bivalvia). Zoomorphology 127:97–110

Waller TR (1998) Origin of the molluscan class Bivalvia and a phylogeny of major groups. In: Johnston PA, Haggart JW (eds) Bivalves: an eon of evolution. University of Calgary Press, Calgary, pp 1–45

Wanninger A (2009) Shaping the things to come: ontogeny of lophotrochozoan neuromuscular systems and the Tetraneuralia concept. Biol Bull 216:293–306

Wanninger A, Haszprunar G (2001) The expression of an engrailed protein during embryonic shell formation of the tusk-shell, *Antalis entalis* (Mollusca, Scaphopoda). Evol Dev 3:312–321

Wanninger A, Haszprunar G (2002a) Chiton myogenesis: perspectives for the development and evolution of larval and adult muscle systems in molluscs. J Morphol 251:103–113

Wanninger A, Haszprunar G (2002b) Muscle development in *Antalis entalis* (Mollusca, Scaphopoda) and its significance for scaphopod relationships. J Morphol 254:53–64

Wanninger A, Haszprunar G (2003) The development of the serotonergic and FMRF-amidergic nervous system in *Antalis entalis* (Mollusca, Scaphopoda). Zoomorphology 122:77–85

Wanninger A, Ruthensteiner B, Dictus WJAG, Haszprunar G (1999a) The development of the musculature in the limpet *Patella* with implications on its role in the process of ontogenetic torsion. Invertebr Reprod Dev 36:211–215

Wanninger A, Ruthensteiner B, Lobenwein S, Salvenmoser W, Dictus WJAG, Haszprunar G (1999b) Development of the musculature in the limpet *Patella* (Mollusca, Patellogastropoda). Dev Genes Evol 209:226–238

Wanninger A, Ruthensteiner B, Haszprunar G (2000) Torsion in *Patella caerulea* (Mollusca, Patellogastropoda): ontogenetic process, timing, and mechanisms. Invertebr Biol 119:177–187

Wanninger A, Fuchs J, Haszprunar G (2007) The anatomy of the serotonergic nervous system of an entoproct creeping-type larva and its phylogenetic implications. Invertebr Biol 126:268–278

Wanninger A, Koop D, Moshel-Lynch S, Degnan BM (2008) Molluscan evolutionary development. In: Ponder WF, Lindberg DR (eds) Phylogeny and evolution of the Mollusca. University of California Press, Berkeley, pp 425–443

Wierzejski A (1905) Embryologie von *Physa fontinalis* L. Z Wiss Zool 83:502–706

Williams EA, Degnan SM (2009) Carry-over effect of larval settlement cue on postlarval gene expression in the marine gastropod *Haliotis asinina*. Mol Ecol 18:4434–4449

Wilson EB (1898) Considerations of cell-lineage and ancestral reminiscence. Ann N Y Acad Sci 11:1–27

Wollesen T, Wanninger A, Klussmann-Kolb A (2007) Neurogenesis of the cephalic sensory organs of *Aplysia californica*. Cell Tissue Res 330:361–379

Wollesen T, Wanninger A, Klussmann-Kolb A (2008) Myogenesis in *Aplysia californica* (Cooper, 1863) (Mollusca, Gastropoda, Opisthobranchia) with special focus on muscular remodeling during metamorphosis. J Morphol 269:776–789

Wollesen T, Loesel R, Wanninger A (2009) Pygmy squids and giant brains: mapping the complex cephalopod CNS by phalloidin staining of vibratome sections and whole-mount preparations. J Neurosci Methods 179:63–67

Wollesen T, Cummins SF, Degnan BM, Wanninger A (2010) FMRFamide gene and peptide expression during CNS development of the cephalopod mollusk, *Idiosepius notoides*. Evol Dev 12:113–130

Wollesen T, Sukhsangchan C, Seixas P, Nabhitabhata J, Wanninger A (2012) Analysis of neurotransmitter distribution in brain development of benthic and pelagic octopod cephalopods. J Morphol 273:776–790

Wollesen T, McDougall C, Degnan BM, Wanninger A (2014) POU genes are expressed during the formation of individual ganglia of the cephalopod central nervous system. EvoDevo 5:41

Woods FH (1931) History of the germ cells in *Sphaerium striatinum* (Lam.). J Morphol 51:545–595

Woods FH (1932) Keimbahn determinants and continuity of the germ cells in *Sphaerium striatinum* (Lam). J Morphol 53:345–365

Wurzinger-Mayer A, Shipway JR, Kristof A, Schwaha T, Cragg SM, Wanninger A (2014) Developmental dynamics of myogenesis in the shipworm *Lyrodus pedicellatus* (Mollusca: Bivalvia). Front Zool 11:90

Young JZ (1971) The anatomy of the nervous system of *Octopus vulgaris*. Clarendon Press, Oxford

Zardus JD, Morse PD (1998) Embryogenesis, morphology and ultrastructure of the pericalymma larva of *Acila castrensis* (Bivalvia: Protobranchia: Nuculoida). Invertebr Biol 117:221–244

Ziegler HE (1885) Die Entwicklung von *Cyclas cornea* Lam. (*Sphaerium corneum* L.). Z Wiss Zool 41:525–569

Nemertea

Jörn von Döhren

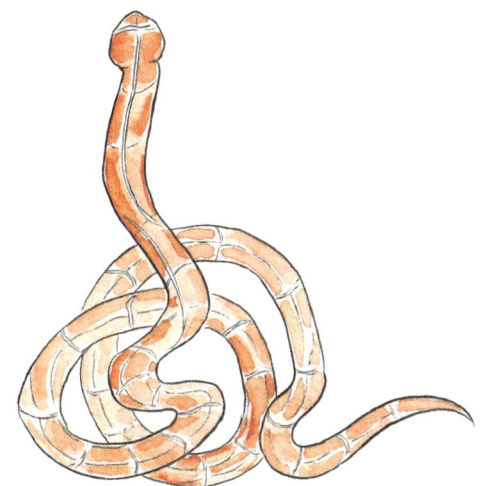

Chapter vignette artwork by Brigitte Baldrian.
© Brigitte Baldrian and Andreas Wanninger.

J. von Döhren
Rheinische-Friedrich-Wilhlems Univerität Bonn,
Institut für Evolutionsbiologie und Ökologie,
An der Immenburg 1, Bonn 53121, Germany
e-mail: jdoehren@evolution.uni-bonn.de

A. Wanninger (ed.), *Evolutionary Developmental Biology of Invertebrates 2: Lophotrochozoa (Spiralia)*
DOI 10.1007/978-3-7091-1871-9_8, © Springer-Verlag Wien 2015

INTRODUCTION

Anatomy and Systematics

Nemertea is a clade of unsegmented, worm-shaped Spiralia comprising about 1.300 described species (Fig. 8.1A–F; Kajihara et al. 2008). The vast majority inhabits marine benthic habitats, but several species are limnic, terrestrial, or marine pelagic. Most species have been described as predators although a number of parasitic, commensalic, and probably even scavengers are known (Gibson 1972). Prey is captured by means of an eversible proboscis that may be armed with one to numerous calcareous stylets in some clades. The proboscis apparatus comprises the proboscis and the rhynchocoel. It represents the apomorphic character that has led to the alternative name Rhynchocoela. The rhynchocoel is a dorsally located, fluid-filled secondary body cavity surrounded by muscle layers housing the proboscis. It opens to the tip of the head via a tube-shaped rhynchodeum (Fig. 8.2A, B). Additional characters that unequivocally qualify Nemertea as monophyletic are the ring-shaped brain surrounding the proboscis insertion instead of the esophagus, a pair of laterally located longitudinal medullary cords, and an endothelialized blood-vascular system. Apart from that nemertean anatomy is marked by characters that are arguably plesiomorphic for Spiralia (Turbeville 2002). These include a largely compact arrangement of the tissue, a medullary cord type organization of the nervous system; a body wall muscle tube comprising minimally two, an outer circular and an inner longitudinal, muscle layers; and one to several paired lateral protonephridia that are not arranged in a segmental fashion. Characters that place Nemertea closer to Trochozoa are a regionalized through-gut with mouth, foregut, midgut, and anus and the presence of glial type cells in the nervous system (Turbeville and Ruppert 1985; Turbeville 1991).

Nemertean ingroup systematics is presently in the consolidation phase with many subclades still being unstable. Traditionally, four higher-ranking taxa have been distinguished: Paleonemertea (Fig. 8.1A, B), Heteronemertea (Fig. 8.1C, D), Hoplonemertea (Fig. 8.1E, F), and Bdellonemertea, the latter comprising only one genus of commensal representatives (Malacobdella) (Coe 1943; Gibson 1972). Paleonemertea and Heteronemertea have been classified as Anopla due to their proboscis being uniformly organized and lacking a stylet armature. In Bdellonemertea a stylet armature of the proboscis is also absent, but this has been interpreted as secondary reduction due to the commensalic lifestyle of this group. Hence, Bdellonemertea and Hoplonemertea have been classified as Enopla, characterized by a primarily armed proboscis, the brain being positioned behind the mouth opening and a more intimate connection of the proboscis insertion and the foregut. In Hoplonemertea, two clades have been identified due to their proboscis armature: Monostilifera and Polystilifera. In Monostilifera the proboscis is armed with a single, comparably large stylet in its middle section. In this clade the mouth opens to the ventral face of the rhynchodeum making its distal part a bifunctional rhynchostomodeum. Polystilifera are characterized by a proboscis that is armed in its middle portion with a cushion equipped with multiple, relatively small stylets. The connection of the mouth opening with the rhynchodeum varies between species. Both rhynchodeum and mouth open independently of each other but in nearby positions in many pelagic forms (Pelagica). In most benthic polystiliferan species (Reptantia), a gradual fusion of both openings by sharing a common atrial chamber at the tip of the head is present. Recent molecular analyses and reassessment of morphological data, however, give a different picture putting the traditional classification in jeopardy (von Döhren et al. 2010; Bartolomaeus and von Döhren 2010; Andrade et al. 2012, 2014; Kvist et al. 2014). Of the traditional higher-ranking taxa, only Hoplonemertea and Heteronemertea are recovered (Fig. 8.3). Paleonemertea has been recognized as nonmonophyletic with Hubrechtiidae being more closely related to Heteronemertea than to the remaining paleonemertean clades (Fig. 8.3). The presence of a specialized larva, the pilidium, in both Hubrechtella dubia (Hubrechtiidae) and most heteronemertean species leads to them being combined in the clade Pilidiophora (Fig. 8.3).

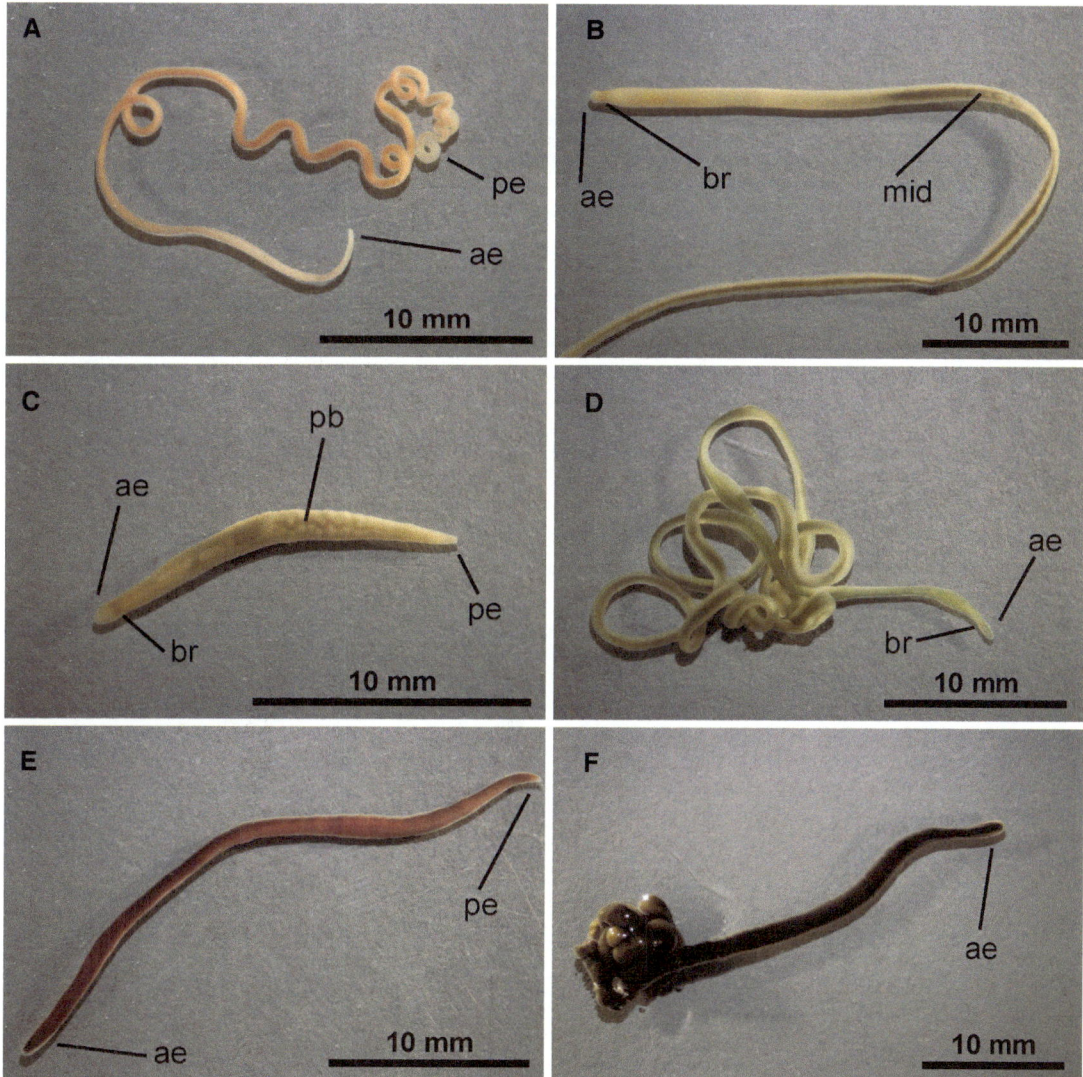

Fig. 8.1 Diversity of nemertean species, living specimens. (**A**) *Procephalothrix oestymnicus* (Cephalothricidae, Paleonemertea), adult. (**B**) *Carinina ochracea* (Carininidae, Paleonemertea), adult. (**C**) *Amphiporus lactifloreus* (Monostilifera, Hoplonemertea), juvenile. (**D**) *Emplectonema gracile* (Monostilifera, Hoplonemertea), adult. (**E**) *Lineus ruber* (Heteronemertea, Pilidiophora), adult. (**F**) *Riseriellus occultus* (Heteronemertea, Pilidiophora), adult. Note: in lightly colored or unpigmented species the brain ring (*br*) is visible through the body wall. *ae* anterior end, *br* brain ring, *mid* midgut region, *pb* proboscis apparatus in rhynchocoel, *pe* posterior end (© Dr. J. von Döhren, All Rights Reserved)

The remaining paleonemertean species have weak support as monophylum based on molecular data (Andrade et al. 2012, 2014; Kvist et al. 2014). Anatomically, Paleonemertea represent the most inhomogeneous taxon within Nemertea with some species having an additional inner circular muscle layer, while others vary with respect to the position of the brain and lateral medullary cords relative to the body wall muscles. In fact, there is no unequivocally apomorphic morphological character that supports paleonemertean monophyly (Fig. 8.3). Within the armed clades (Enopla), traditional taxonomy was not confirmed either. On the one hand, Bdellonemertea proved to be an apparently secondarily reduced member of Monostilifera according to molecular phylogenetic analyses,

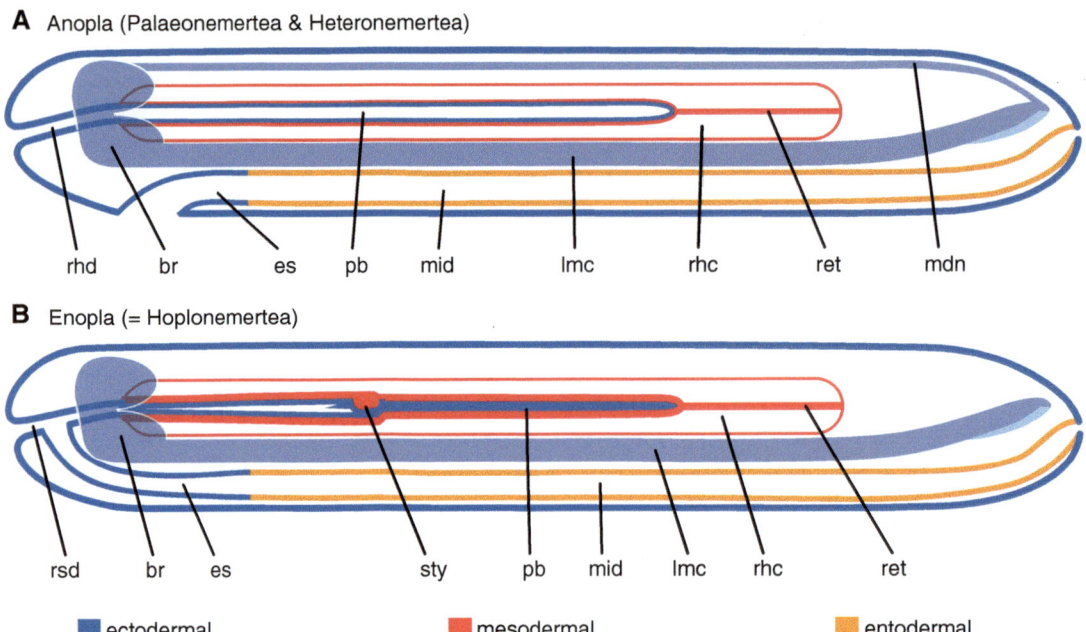

Fig. 8.2 Schematic representation (side view) of the nemertean body plan (**A**) Anoplan organization (Paleo- and Heteronemertea). (**B**) Enoplan organization (Hoplonemertea). Color coding indicates the germ layer that the structure originates from. Note: body wall musculature, nephridia, blood-vascular system, sensory organs, and gland cells have been omitted for clarity. *br* brain, *lmc* lateral medullary nerve cords, *mdn* middorsal nerve, *mid* midgut, *es* esophagus, *pb* proboscis, *ret* retractor muscle of proboscis, *rhc* rhynchocoel, *rhd* rhynchodeum, *rsd* rhynchostomodeum, *sty* stylet apparatus (© Dr. J. von Döhren, All Rights Reserved)

rendering Hoplonemertea and Enopla synonymous, while on the other hand with the same data the monophyly of Polystilifera remains a matter of debate (Thollesson and Norenburg 2003; Sundberg and Strand 2007; Andrade et al. 2012, 2014; Kvist et al. 2014). Currently, there is strong support for a clade Neonemertea comprising Hoplonemertea and Pilidiophora (Figs. 8.1C–F and 8.3; Thollesson and Norenburg 2003; Andrade et al. 2012, 2014; Kvist et al. 2014). All members of this clade share the presence of a median dorsal blood vessel between alimentary canal and rhynchocoel (Gibson 1972).

A number of organs, characteristic of nemerteans, cannot be placed robustly in an evolutionary scenario due to their disparate distribution within Nemertea. Cerebral organs connected to the brain are a feature of many species, although they are quite different in shape and position. While being located behind the brain in tubulanid Paleonemertea, Pilidiophora, and reptant

Polystilifera, they are situated far in front of the brain in Monostilifera (Gibson 1972). Lateral sensory organs have a morphology that is reminiscent of cerebral sense organs in paleonemerteans but are located more posteriorly on the sides of the animal in the vicinity of the nephropores. Lateral organs are typical of tubulanid paleonemerteans but have also been recorded from some heteronemertean species (Gibson 1972). Frontal organs that are connected to the almost ubiquitously present cephalic glands and represented by a single epidermal pit in most neonemertean species or by three epidermal pits arranged in a triangular pattern in lineid Heteronemertea might be apomorphic for Neonemertea, although a row of median epidermal pits located in the head region have also been described in the paleonemertean genus *Carinoma* (Gibson 1972). Pigmented eyes in adults are characteristic of neonemertean species as well but show very high variation in size,

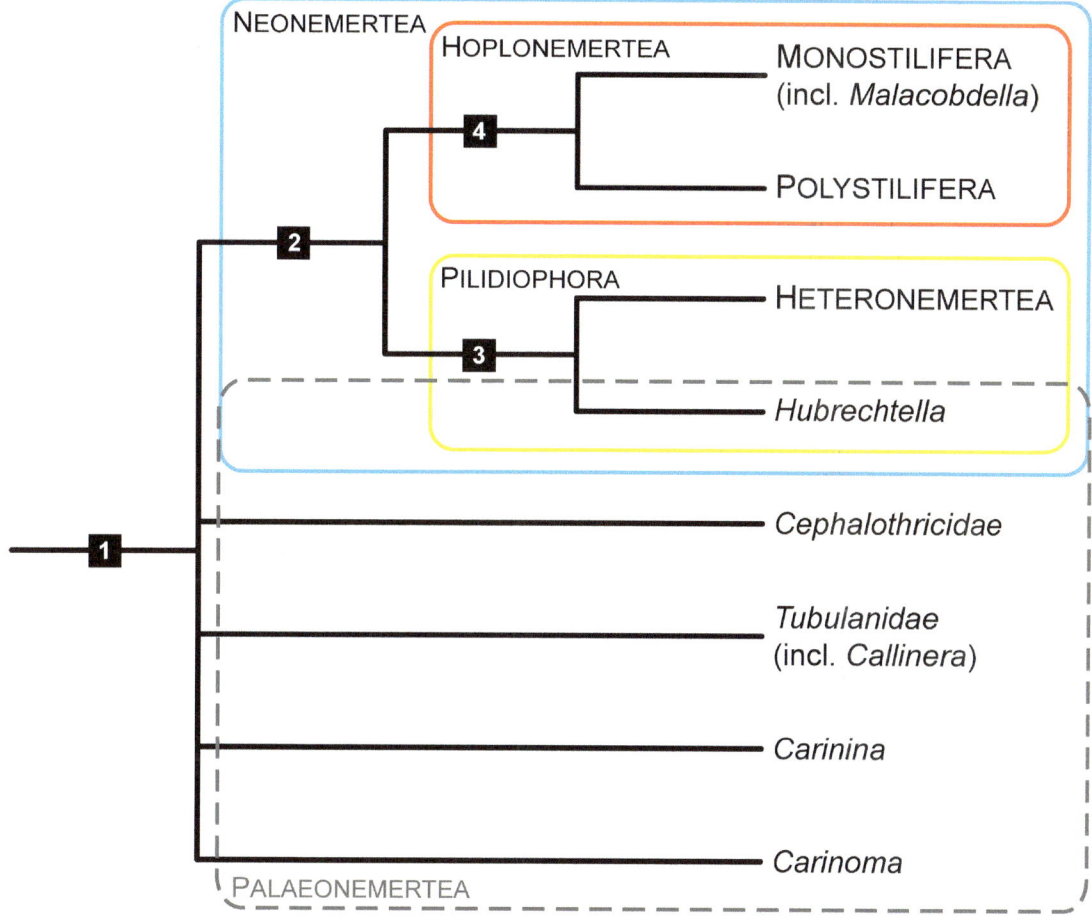

Fig. 8.3 Consensus phylogeny of Nemertea (modified after Andrade et al. 2012, 2014). Numbers indicate apomorphic characters for respective clades. (*1*) Nemertea: dorsal, eversible proboscis housed in fluid-filled rhynchocoel, ring-shaped brain located around the proboscis opening, blood-vascular system lined with endothelium; (*2*) Neonemertea: middorsal blood vessel; (*3*) Pilidiophora: pilidium larva; (*4*) Hoplonemertea: proboscis equipped with stylet apparatus. Note: for the paleonemertean taxa, no apomorphic character exists (© Dr. J. von Döhren, All Rights Reserved)

complexity, and number even between arguably closely related species. Moreover, the eyes in adults have also been described in a few isolated paleonemertean species (Gibson 1972).

Reproductive Biology

The majority of nemertean species are gonochoristic. Only comparably few (mostly non-marine) hermaphrodites have been described (Friedrich 1979). Free-living nemerteans have the tendency to aggregate during their reproductive season, and there have been anecdotic reports of nuptial dances reminiscent of those of epitokous polychaetes in *Nipponnemertes pulcher* (Berg 1972). In most species, females spawn their eggs freely into the water. Egg masses deposited in mucus sheaths are reported from species living in marine (e.g., *Tetrastemma candidum*, *Antarctonemertes phyllospadicola*), intertidal (e.g., *Lineus ruber*, *Lineus viridis*), freshwater and terrestrial habitats (e.g., *Prostoma jenningsi*, *Apatronemertes albimaculosa*), as well as from species that are living as parasites on or in other animals (e.g., *Carcinonemertes* species on several crustaceans) (Thiel and Junoy 2006). In the monostiliferous hoplonemertean *Amphiporus incubator*, the female remains in the secreted mucus sheath with the developing offspring apparently providing them with nourishment by means of complete histolysis of its intestinal tract (Joubin 1914).

In general, spermatozoa are released freely into the water as well. In those species that aggregate during reproduction and in mucus spawning species males typically come into close contact with either the eggs or the female to ensure successful fertilization (Bartolomaeus 1984; Thiel and Junoy 2006). In many species viscid mucous secretions around the worms putatively aid in scaling down the space into which eggs and sperm are shed. This behavior has occasionally been termed "pseudocopulation" (e.g., *Carcinonemertes epialti, Lineus ruber, Lineus viridis*) (Gontcharoff 1951; Bartolomaeus 1984; Roe 1984; Stricker 1986). In *Lineus viridis*, one to several males enter a mucus sheath that is secreted by the female. Into this gelatinous mass, additional mucus layers and the eggs are shed (Gontcharoff 1951; Bartolomaeus 1984). During pseudocopulation in some species (e.g., *Carcinonemertes epialti*), sperm may enter the female gonads through the gonopores, and internal fertilization occurs. Internal fertilization with direct sperm transfer (i.e., true copulation) has been assumed to occur in some pelagic polystiliferans. Several structures have been interpreted as accessory sperm transfer structures such as muscular penes in *Phallonemertes murrayi*, sucker-like attachment organs in *Plotonemertes adhaerens*, and specialized glandular epithelia around the male gonopores in *Balaenanemertes chuni* and some other pelagic polystiliferans (Thiel and Junoy 2006). In species of the monostiliferous hoplonemertean genus *Carcinonemertes*, the efferent ducts of the testes open into a common duct, termed Takakura's duct, that widens into a seminal vesicle which opens into the intestine near the anal opening (Gibson 1972). In some *Carcinonemertes* species, the anal opening is surrounded by a flattened or concave muscular area that is used to transfer sperm from the anus to the female gonopores (Roe 1984). Some accounts of viviparous species from all phylogenetic lineages demonstrate that vivipary is not uncommon in this phylum. In these species internal fertilization has to be expected (Thiel and Junoy 2006).

Egg sizes in nemerteans range from 50 μm in *Carinina (Procarinina) remanei* (Nawitzki 1931) to 2.5 mm in *Dinonemertes investigatoris* (Coe 1926), with most being between 100 and 300 μm in diameter (Friedrich 1979). With the relatively scarce data at hand, there seems to be no strict correlation of egg size to the various phylogenetic entities, but there is a tendency of paleonemertean eggs being on the smaller side of the spectrum, while larger egg sizes are encountered in hoplonemertean species (e.g., *Pantinonemertes (Geonemertes) agricola*: 350–450 μm (Coe 1904); *Nipponnemertes pulcher*: 280–340 μm (Berg 1972)). While there is usually only a very thin and delicate chorion surrounding the eggs of anoplan species, the eggs of hoplonemertean species are invested with a chorion (also termed "vitelline envelope") that is thicker and usually set off from the egg membrane by a fluid-filled space of different dimensions (Stricker et al. 2001). In many free-spawning species, a glutinous mucus layer surrounds the egg chorion to attach the eggs to the substrate. In some species this mucus is reported to dissolve in the water soon after the eggs have been shed. In this case the mucus possibly enhances attraction and movement of the sperm to the egg.

Spermatozoa in nemertean species generally comprise a sperm head consisting of an apical acrosomal vesicle, a condensed nucleus, a mitochondrial mass, and diplosomal centrioles. From the distal centriole, a single flagellum with a regular $9 \times 2 + 2$ axoneme emanates. Only in the pelagic polystiliferan *Nectonemertes mirabilis* aflagellate sperm cells together with separate flagella have been observed in the testes (Stricker and Folsom 1997). Morphology of the sperm head is very variable with all components varying in both length and width (Stricker and Folsom 1997; von Döhren and Bartolomaeus 2006; von Döhren et al. 2010). Acrosomal vesicles and mitochondria may be dislocated from the terminal poles of the sperm cell in some species. In most paleonemertean and hoplonemertean species, the single mitochondrion represents the product of the fusion of numerous mitochondria during spermiogenesis. Elongated headed sperm has been hypothesized to be correlated with either internal fertilization (e.g., *Antarctonemertes phyllospadicola, Carcinonemertes epialti, Cephalothrix rufifrons*) or the investment of the egg with a tough, resistant vitelline membrane in free-spawning species (e.g., *Cerebratulus lacteus*) (Stricker and Folsom 1997). In some hoplonemertean species, much of the elongation of the sperm head is due to a conspicuously elongated acrosomal vesicle (von Döhren et al. 2010). Interestingly, in these species eggs are invested with both a vitelline envelope and a fairly thick glutinous mucus layer that is resisting degradation well beyond the gastrulation of the embryo inside (e.g., *Paranemertes peregrina*, *Emplectonema gracile*).

In externally fertilizing species, eggs are commonly shed arrested in the prophase of the first meiotic division, i.e., the germinal vesicle is clearly visible. In oviparous species with internal fertilization, this is also the case only in *Carcinonemertes epialti* where the eggs are reported to be shed in early cleavage stages (Stricker et al. 2001). Freely spawned eggs typically show a somewhat compressed morphology due to them having been tightly packed in the ovaries. In some eggs, especially when they are artificially extracted from the female, there is a cytoplasmic protrusion representing the spot where the egg was contacting the ovarian lining during oogenesis (Iwata 1960; Stricker et al. 2001). In contact with seawater, they round up and usually undergo meiotic maturation indicated by germinal vesicle breakdown (GVBD) (Stricker et al. 2001). In some species the cytoplasmic protrusion separates from the egg but remains in its vicinity; in other species it disappears soon after fertilization. Abolishment of prophase I arrest during GVBD in the heteronemertean pilidiophoran *Cerebratulus lacteus* is mediated by intracellular signaling of nitric oxide (NO), cyclic guanosine monophosphate (cGMP), and an atypical protein kinase C (aPKC). Adenosine monophosphate-activated protein kinase (AMPK) blocks GVBD, but it can be resumed by the action of cyclic adenosine monophosphate (cAMP) and protein kinase A (PKA) signaling. By these alternative pathways, an inactive form of the maturation promoting factor (pre-MPF) is activated. The active maturation promoting factor (MPF) triggers GVBD which is accompanied by a drastic reorganization of the endoplasmic reticulum (ER) of the egg into numerous ER microdomains about 5 μm in diameter, distributed evenly within the cytoplasm (Stricker et al.

2013 and references therein). The egg, arrested in the metaphase of the first meiotic division, has then an eccentrically located nucleus and is ready for fertilization. Fusion of the sperm with the egg induces a sudden cortical calcium flash followed by repetitive calcium oscillations in the egg due to the action of a soluble sperm factor delivered from the sperm to the egg. The first two calcium waves start at the point where the sperm has fused with the egg, while the following waves are elicited from a pacemaker region in the vegetal cortex of the egg opposite of the site of prospective polar body extrusion. Along with the calcium waves, there is a global disassembly of the ER microdomains. The calcium oscillations continue through first polar body formation and cease before the second polar body is extruded (Stricker et al. 2013 and references therein). Polar bodies are situated at the opposite pole of the cytoplasmic protrusion if one is present. This hints at the animal-vegetal axis of the egg being already established in the gonad during oogenesis (Fig. 8.4A; Henry and Martindale 1997). Polar body formation is followed by decondensation of the male pronucleus and subsequent karyogamy (Fig. 8.4B).

EARLY DEVELOPMENT

Diversity of Cleavage in Nemertea

Although exhibiting quite diverse larval types, embryonic development of nemerteans is comparably uniform. In general, cleavage is of the holoblastic, equal (homoquadrant), spiral type (Fig. 8.4C–F). The first cleavage division is meridional, passing through the anterior-posterior plane as indicated by the location of the polar bodies. It results in a pair of equally sized blastomeres. They are initially rounded, with little contact to each other. The embryo soon becomes compact prior to second cleavage division (Friedrich 1979; Henry and Martindale 1997, 1998a; Maslakova et al. 2004a; Maslakova 2010a). In the heteronemertean species *Lineus ruber*, the first cleavage division has been reported to be unequal resulting in a pair of slightly or significantly differently sized blastomeres (Fig. 8.4D; Nusbaum and Oxner 1913). This phenomenon, however, is not

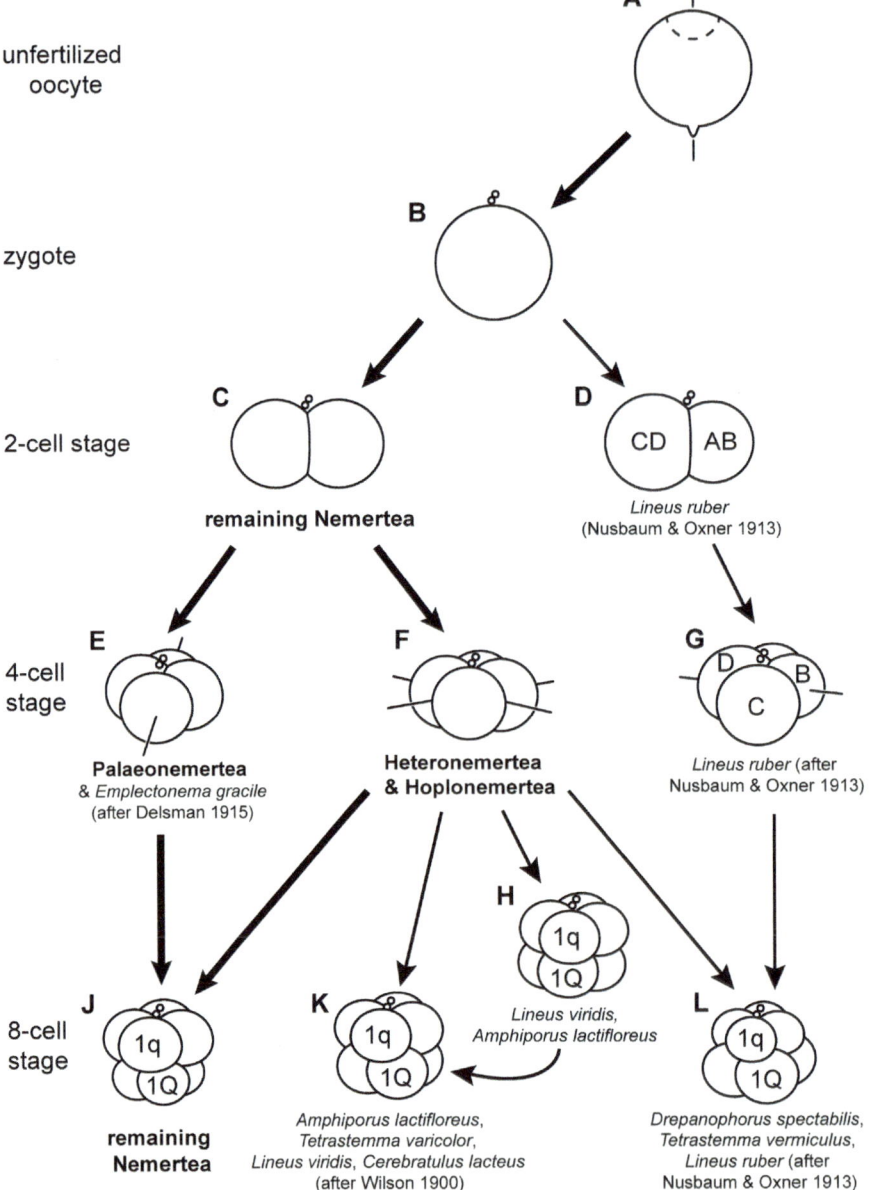

Fig. 8.4 Diversity of the cleavage in Nemertea. *Bold arrows* indicate the most common sequences. (**A**) Unfertilized egg. The animal-vegetal axis is running through the germinal vesicle (*dashed line*), and the cytoplasmic stalk on the opposite vegetal pole of the oocyte. (**B**) Zygote. The animal pole is marked by the polar bodies; the vegetally located cytoplasmic protrusion has been cast off. (**C**) Two-cell stage, equal first cleavage division. (**D**) Two-cell stage, unequal first cleavage division resulting in a smaller *AB* and a larger *CD* blastomere. (**E**) Four-cell stage after second cleavage division with equal sized blastomeres and cross-furrows (see also Fig. 8.5A). One dorsoventral axis is present running through the vegetal cross-furrow blastomeres. (**F**) Four-cell stage after second cleavage division with equal sized blastomeres lacking a cross-furrow. Two alternative dorsoventral axes are present. (**G**) Four-cell stage after second cleavage division with unequal-sized blastomeres (*C* and *D*) and cross-furrows. The dorsoventral axis runs through the larger *D* and the smaller *B* quadrant. (**H**) Eight-cell stage after third cleavage division with equally sized animal (*1q*) and vegetal (*1Q*) blastomeres, showing an initially radial-like arrangement of blastomeres (see also Fig. 8.5B). (**J**) Eight-cell stage after third cleavage division with animal blastomeres ("micromeres," *1q*) being larger than vegetal blastomeres ("macromeres," *1Q*) most common in nemertean species studied to date. (**K**) Eight-cell stage after third cleavage division with animal blastomeres ("micromeres," *1q*) being of the same size as vegetal blastomeres ("macromeres," *1Q*). (**L**) Eight-cell stage after third cleavage division with animal blastomeres ("micromeres," *1q*) being smaller than vegetal blastomeres ("macromeres," *1Q*). Citations in brackets indicate conflicting reports that have been given for the respective species. For further explanations, see text (© Dr. J. von Döhren, All Rights Reserved)

seen in all embryos of a given clutch and has therefore to be attributed to a certain variability of cleavage in this species. It is unclear to what extend the differing blastomere size influences further development. The plane of the second cleavage division is meridional and perpendicular to the first, dividing the blastomeres into four daughter cells of equal size (Fig. 8.4E, F). In the case of the unequally cleaving embryos of *Lineus ruber*, the second cleavage division results in a four-cell stage in which one pair of blastomeres, the progeny of the smaller two-cell stage blastomere, is smaller than the progeny of the larger two-cell stage blastomere (Fig. 8.4G). This is in contrast to what is observed in, e.g., unequally cleaving annelids or mollusks (see Chapters 7 and 9) in which typically one of the four-cell stage blastomeres, the precursor of the dorsal (D) quadrant, is larger than the other three (Henry and Martindale 1998b). This also speaks for the phenomenon of unequal cleavage in Nemertea not being homologous to the unequal cleavage of Annelida or Mollusca. The four-cell stage blastomeres are initially rounded with little connecting surface to each other but soon move toward each other just like in the two-cell stage. The occurrence of cross-furrows at the four-cell stage has so far been reported from only a few species. The paleonemertean species *Carinoma armandi tremaphoros* and *Procephalothrix oestrymnicus* show distinct cross-furrows (Figs. 8.4E and 8.5A; Maslakova et al. 2004a). In the former species, the cross-furrows are a result of a slightly leotropic (sinistral) second cleavage division with one of the vegetal cross-furrow cells representing the precursor of the future dorsal (D) quadrant. However, it is not clear whether the specification of the D quadrant in this species is mediated by the segregation of cytoplasmic components or by inductive cellular interactions (Maslakova et al. 2004a). In neonemertean species, cross-furrows are reported from *Lineus ruber* (Heteronemertea) and *Emplectonema gracile* (Hoplonemertea), although in the latter species a more recent study does not confirm the existence of a cross-furrow (Fig. 8.4E, G; Delsman 1915; Iwata 1960). In *Lineus ruber* the cross-furrow has been recorded only in four-cell stage embryos with unequal-

sized blastomere pairs. It is situated between one of the larger and the opposing smaller blastomere in the majority of examined cases (Fig. 8.4G; Nusbaum and Oxner 1913; Schmidt 1964). It is not completely clear, however, how the cross-furrow is related to the further course of development. The closely related heteronemertean *Cerebratulus lacteus* does not show any sign of cross-furrows at this stage nor do any of the other examined heteronemertean species (Fig. 8.4F; Friedrich 1979; Henry and Martindale 1997, 1998a; Maslakova 2010a). Due to its variability in occurrence, it can be assumed that in *Lineus ruber* the cross-furrow merely represents an intraspecific variation of development that might not have any influence on the determination of the future embryonic quadrants.

The third cleavage division is synchronous and equatorial. It results in an eight-celled embryo (Fig. 8.4H–L). In most species the third cleavage division is described as clearly dexiotrophic (dextral), but there are some accounts of animal blastomere quartets being positioned initially exactly opposite to their respective vegetal counterparts (Figs. 8.4H and 8.5B), resembling a radial cleavage pattern as found in, e.g., cnidarians (Vol. 1, Chapter 6), ectoprocts (Chapter 11), phoronids (Chapter 10), brachiopods (Chapter 12), or invertebrate deuterostomes (Vol. 6). In *Amphiporus lactifloreus* and *Lineus viridis* (as *Lineus obscurus*), the aligned animal and vegetal blastomeres shift position after segregation to come to lie as if generated by a regular dexiotropic cleavage division (Fig. 8.4K; Barrois 1877). This peculiar behavior might be due to the high yolk content of the large eggs in both of the mentioned species. In *Cephalothrix rufifrons* there is certain variability regarding the angle at which the eight-cell blastomeres are positioned, but subsequent cleavage divisions restore the spiral pattern by being regularly alternating leotropic (sinistral) and dexiotropic (Smith 1935). Regarding the relative sizes of the blastomeres at the eight-cell stage, there is some noteworthy variation among species investigated. In most species the animal eight-cell stage blastomeres (micromeres) are slightly or even considerably larger than cells of the vegetal blastomere quartet (macromeres) (Fig. 8.4J; Friedrich 1979; Henry and

Martindale 1997; Maslakova et al. 2004a), while in *Tetrastemma vermiculus*, *Drepanophorus specta-bilis*, and *Lineus ruber*, the relative size differences accord to the general nomenclature, i.e., the animal micromeres are smaller than the vegetal macro-meres (Fig. 8.4L; Lebedinsky 1897; Nusbaum and Oxner 1913; Schmidt 1964). Both animal and vegetal blastomere quartets have been reported to be of roughly equal size in *Tetrastemma vari-color* (Hoplonemertea), *Cerebratulus lacteus*, and *Lineus viridis* (Figs. 8.4K and 8.5B; Barrois 1877 for *Lineus viridis* as *Lineus obscurus*; Hoffman 1877; Wilson 1900). In *Cephalothrix rufifrons* there is no externally visible size difference of animal and vegetal blastomeres in the eight-cell stage, but sections reveal that the macromeres are slightly larger, their additional volume projecting interiorly into the blastocoel (Smith 1935). The fourth cleavage division leading to the 16-cell stage embryo is generally reported to be synchro-nous. The transition to the 16-cell stage in *Lineus ruber* passes through a series of stages in which first the two larger of the animal blastomeres divide in a leotropic manner, followed by the smaller pair. Finally, the macromere quartet (2A-D) divides to accomplish the 16-cell stage. The 32-cell stage in

this species is again reached by a transitory stage with the macromeres and the vegetal-most micro-mere quartet dividing first, followed by the divi-sion of the two animal-most micromere quartets (Nusbaum and Oxner 1913). Interestingly, there seems to be a tendency in this species to arrange the micromere quartet daughter cells in a planar rather than a spiral pattern, a phenomenon that has also been described for the equatorial-most blas-tomeres in *Amphiporus lactifloreus* (Barrois 1877; Nusbaum and Oxner 1913). Relative asynchronies regarding the timing of division of blastomere quartets have been described in a number of other species, e.g., *Carinoma tremaphoros*, *Cerebratulus lacteus*, *Cerebratulus marginatus*, *Emplectonema gracile*, *Malacobdella grossa*, and *Tubulanus nothus* (Wilson 1900; Zeleny 1904; Delsman 1915; Hammarsten 1918; Dawydoff 1928; Maslakova et al. 2004a). However, while in *Lineus ruber* the cleavage divisions of the vegetal pole precede those of the animal pole, the reverse is true in the other exemplified species. Nevertheless, cleavage gener-ally generates four quartets of animal blastomeres along with their respective progeny as well as one quartet of comparably small vegetal blastomeres. The paleonemertean *Tubulanus nothus* represents

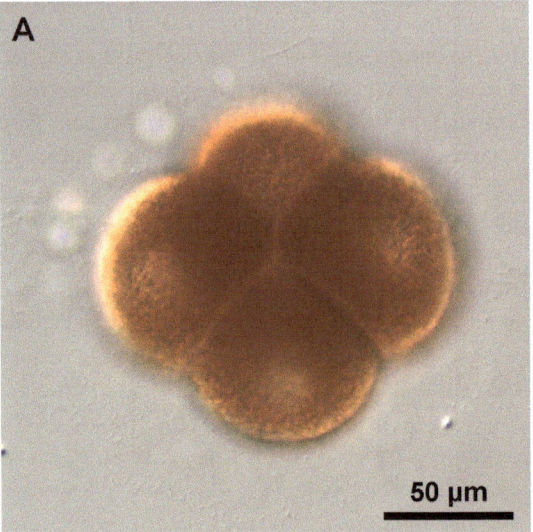
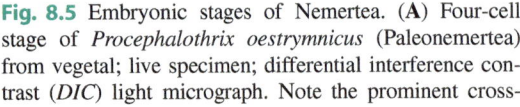
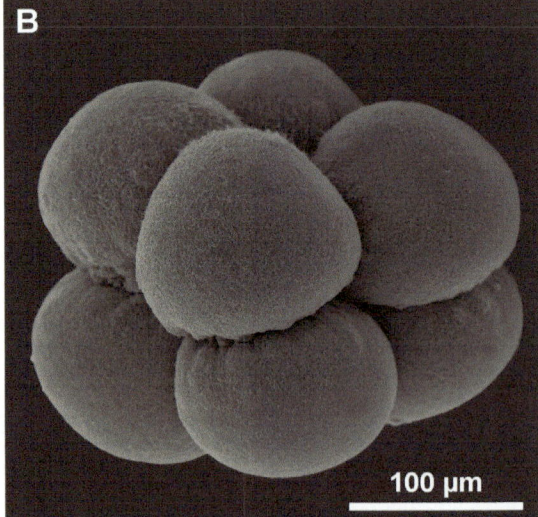

Fig. 8.5 Embryonic stages of Nemertea. (**A**) Four-cell stage of *Procephalothrix oestrymnicus* (Paleonemertea) from vegetal; live specimen; differential interference con-trast (*DIC*) light micrograph. Note the prominent cross-furrow. (**B**) Eight-cell stage of *Lineus viridis* (Pilidiophora), from vegetal; scanning electron micrograph. Note the almost radial arrangement of equally sized animal and veg-etal blastomeres (© Dr. J. von Döhren, All Rights Reserved)

an exception in that the regular spiral cleavage pattern is largely given up after the fifth cleavage division, resulting in a chaotic cleavage pattern in which neither fourth micromere nor macromere quartets are discernible (Dawydoff 1928).

Determination of the Future Body Axes

Cleavage of nemerteans does not only show the characteristic pattern, but it also complies with the general characteristics exhibited in Spiralia, such as stereotypy and determination. Contributions and capabilities of the embryo to form the future body parts have been most thoroughly studied in the heteronemertean pilidiophoran *Cerebratulus lacteus* (Table 8.1; Henry and Martindale 1997, 1998a). Compared to the spiral cleavage pattern of annelids and mollusks, however, there are some remarkable differences in terms of regulation and contribution of blastomeres to future organ systems in this species (Henry and Martindale 1997 and references therein). The larval anterior-posterior axis is already set up within the ovary and is outlined by the stalk-like process on the vegetal pole and the germinal vesicle situated on the opposite, animal pole of the egg (Fig. 8.4A). During meiotic maturation the polar bodies form near the animal pole. After fertilization the first two cleavage divisions pass through the animal-vegetal axis which thus becomes the anterior-posterior axis of the larva. Experimental alteration of the first cleavage plane by compressing fertilized eggs during first cleavage division results in its decoupling from the larval anterior-posterior and dorsoventral axis, indicating that the cleavage planes in normal development are not the cause of the future larval axes but follow a scaffold of the embryo that is precociously set up in the egg (Henry and Martindale 1995, 1996a). It has been shown that morphogenetic factors are evenly distributed within the egg prior to fertilization and that a progressive restriction due to segregation of these morphogenetic factors along the preformed animal-vegetal axis is executed after fertilization. The determinants seem to be fully segregated at the third cleavage division although vegetal determinants (as exemplified by formation of a larval gut) seem to be already confined to the vegetal pole of the embryo after the second cleavage division, i.e., at the four-cell stage (Zeleny 1904; Yatsu 1909; Hörstadius 1937; Freeman 1978). The acceleration

Table 8.1 Clonal contributions of blastomeres during early cleavage (up to the 64-cell stage) in Nemertea

Taxon	Paleonemertea	Heteronemertea	Hoplonemertea
Blastomere	Contribution		
First quartet	Apical tuft ($1q^1$), ectoderm, 28 "trochoblasts" ($1q^1$, $1q^2$), eye ($1c^1$)	Apical tuft, larval ectoderm, ciliated band, cephalic disks (1a, b), larval nervous system (1c, d)	Apical tuft, ectoderm
Second quartet	12 "trochoblasts," esophagus, primary somatoblast (2d), mesoderm[a]	Larval ectoderm, ciliated band, esophagus, larval nervous system (2a, c, d)	Ectoderm, mesoderm ($2a^{111}$–d^{111})
Third quartet	?	Esophagus, larval muscles (3a, b), larval ectoderm (3c, d), larval nervous system (3c, d)	Ectoderm
Fourth quartet	Mesoderm (3D)[a]	Gut (4a–c), adult mesoderm (4d)	Gut
Macromeres	Gut	Gut	Gut
Species	*Carinoma tremaphoros*	*Cerebratulus lacteus*	*Malacobdella grossa*
References	Maslakova et al. (2004a); [a]*Tubulanus nothus*: Dawydoff (1928)	Henry and Martindale (1998a)	Hammarsten (1918)

[a]Marks data by Dawydoff (1928) for *Tubulanus nothus*

or retardation of this segregation of determinants by stimulation or inhibition of the formation of mitotic asters during cleavage hints at a role the cytoskeleton, especially the microtubules, plays in the segregation process (Freeman 1978; Goldstein and Freeman 1997).

Embryonic Regulative Capacities of Nemertea

Additional remarkable modifications of the stereotypic spiral cleavage are seen in the capacity of the embryo to regulate as well as in the contribution of the blastomeres to the larval body. There is, however, considerable difference regarding regulative capabilities in Pilidiophora as opposed to other embryos. Embryos of *Cerebratulus lacteus* that have been halved at the two-cell stage are capable of forming completely normal but miniature larvae, while embryos dissected at the four-cell stage do not regulate to complete larvae. Cleavage after isolation of blastomeres is resumed in the stereotypic spiral fashion, instead of being reinitiated from a point that is analogous to earlier stages of cleavage (e.g., zygote in the case of isolation at the two-cell stage) (Wilson 1900, 1903; Zeleny 1904; Yatsu 1910; Hörstadius 1937). Embryos of the hoplonemertean *Nemertopsis bivittata* in which blastomeres have been deleted at either the two-cell or the four-cell stage develop into characteristically deficient larvae, indicating that regulation does not take place to such an extent as in *Cerebratulus lacteus* (Martindale and Henry 1995). Compared to other spiralian taxa, the inability to regulate as seen in *Nemertopsis bivittata* seems to be the ancestral state and to be attributed to an early determination of the clonal contributions of blastomeres to the juvenile tissues.

The regulative capacity of Pilidiophora represents a modification of the stereotypic determination in spiral cleavage and can be attributed to the mode of development in which cells remain undifferentiated until forming the juvenile rudiments much later in development. As a consequence of later determination of cell fates, the cells already during cleavage retain a certain ability to compensate for loss of blastomeres. The molecular mechanism underlying this delayed determination is still unknown.

Cell Lineage

Contributions of blastomeres to the larval/adult body plan have most thoroughly been studied in *Cerebratulus lacteus* for which the complete cell lineage is known up to the pilidium state (Table 8.1; Henry and Martindale 1998a). Partial cell lineages are available for the paleonemertean *Carinoma tremaphoros* and the hoplonemerteans *Nemertopsis bivittata* and *Malacobdella grossa* (Hammarsten 1918; Martindale and Henry 1992, 1995; Henry and Martindale 1994; Maslakova et al. 2004a). In all species examined, all blastomeres of the first quartet contribute equally to forming the apical pit with a tuft of long cilia as well as the majority of the larval epidermis. Due to the large size of the first quartet micromeres, the ectodermal domains are positioned in a dorsolateral and ventrolateral orientation, respectively, instead of being clearly dorsal, ventral, and lateral as in other spiralian animals (see Chapters 7 and 9; Henry and Martindale 1997, 1998a, 1999). In four-cell stages of *Cerebratulus lacteus* and *Nemertopsis bivittata* that do not possess cross-furrows, quadrant identities cannot be predicted, while in *Carinoma tremaphoros* the A and C quadrants are formed by animal cross-furrow blastomeres, and the B and D quadrants by vegetal cross-furrow blastomeres (Henry and Martindale 1997, 1998a; Maslakova et al. 2004a). In *Cerebratulus lacteus*, the D quadrant along with the dorsoventral axis is induced by the first quartet micromeres after the third cleavage division. Deletion of all first quartet micromeres results in radialized larvae, while deletion of a single or two adjacent first quartet micromeres leads to the D quadrant being determined in a position where the first quartet macromere had contact with most of the remaining first quartet micromeres during the eight-cell stage (Henry 2002). In *Cerebratulus lacteus* the first quartet micromeres contribute to the apical epidermis down to the circumoral ring of elongated cilia including the outer side of the lateral larval lappets, while in *Carinoma tremaphoros* the first quartet micromeres form the apical epidermis down to the apical row of trochoblast cells (Table 8.1; Henry and Martindale 1998a; Maslakova et al. 2004a, b). The second micromere quartet cells contribute equally to most of the remainder of the epidermis, represented in *Cerebratulus lacteus* by parts of the ciliated band, the inner side of the lappets, and parts of the esophagus (Table 8.1; Henry and Martindale 1998a). In *Carinoma tremaphoros*, contributions of the second micromere quartet are unequal. In this species each of the second micromere quartet cells forms three of the posterior trochoblast cells each and parts of the esophagus. The entire epidermis posterior to the trochoblast ring is formed by the progeny of the dorsal second quartet micromere (2d), the so-called primary somatoblast (Table 8.1; Maslakova et al. 2004a, b). The role of this blastomere is in accord to what is reported from other spiralian taxa. In the hoplonemertean *Nemertopsis bivittata*, there is no somatoblast; all first and second quartet micromeres contribute equally to the ectodermal domains (Martindale and Henry 1995). The band of long cilia in the pilidium has to be regarded as not being homologous to the prototroch of the typical trochophore larva. In the latter, the prototroch is composed by a limited number of cells which originate from the first and second quartet micromeres, while there are a large number of cells in the ciliated band of the pilidium that are additionally contributed by the C and D quadrants of the third quartet micromeres (Maslakova 2010b). Moreover, the ciliated marginal band cells lack the expression of a trochoblast-specific β-tubulin found in other trochophores (van den Biggelaar et al. 1997). Due to the position of Pilidiophora within Nemertea and the absence of prototroch-type long cilia in either hoplonemertean or paleonemertean species, the long cilia that adorn the marginal band of the pilidium thus have to be regarded as newly evolved (Maslakova 2010b).

Contributions of blastomeres after the fifth cleavage division, i.e., the 32-cell stage, are known only from the pilidium larva of *Cerebratulus lacteus* (Henry and Martindale

1998a). As a characteristic of spiralian development, mesoderm is derived from two sources. The so-called ectomesoderm is derived from micromeres of either the second or the third quartet, while the so-called endomesoderm originates from the fourth quartet micromere of the D quadrant, the so-called 4d mesendoblast (Table 8.1; Boyer et al. 1996; Boyer and Henry 1998; Henry and Martindale 1999). In other nemertean species, there have been different accounts on the origin of the mesoderm. While some researchers derive the mesoderm exclusively from ectomesodermal sources such as the second micromere quartet (e.g., Hammarsten 1918 for *Malacobdella grossa*), others claim both ectomesodermal and endomesodermal sources to be present (e.g., second micromere quartet and 3D in *Tubulanus nothus*; see Table 8.1 and Dawydoff 1928). The mesoderm from both a mesendoblast (4d) and a multipolar delamination from the archenteron after gastrulation but without ectodermal contributions has also been described (e.g., Nusbaum and Oxner 1913 for *Lineus ruber*). More recent studies on the cell lineage of the pilidium of *Cerebratulus lacteus* have identified the third quartet micromeres of the A and B quadrant as source of the ectomesoderm forming the array of larval muscles as well as some undifferentiated mesenchymal cells scattered in the blastocoel (Table 8.1; Henry and Martindale 1996b, 1998a). It is known from isolation experiments that blastomeres isolated at the four-cell stage are capable of regulating for mesenchymal tissues in further development. This indicates that inductive interactions from the A and B quadrants inhibit the formation of mesodermal cell types in the progeny of their dorsally located counterparts (i.e., 3c and 3d; Martindale and Henry 1995).

Endomesoderm is formed by the fourth quartet micromere of the D quadrant (Table 8.1). In the pilidium, endomesoderm is represented by a pair of loosely organized mesodermal bandlets situated underneath the epidermis at the junction of the esophagus and stomach and some scattered, undifferentiated mesenchymal cells (Henry and Martindale 1996b, 1998a). The differentiation of adult mesoderm could not be followed but is assumed to originate from a population of undifferentiated, dormant mesenchyme cells (Henry and Martindale 1997). In paleonemertean and hoplonemertean larvae, assumed cell-cell interactions as operating during the development of the pilidium to specify the dorsoventral axis clearly do not inhibit that all quadrants contribute equally to the muscle layers displayed in the larvae (Iwata 1957; Martindale and Henry 1995). To what extent the musculature is derived from both ectomesodermal and endomesodermal sources in nonpilidial types of development has been a matter of debate and remains to be clarified. The ectodermally derived nervous system of the pilidium of *Cerebratulus lacteus* is formed by progeny of the blastomeres 1c, 1d, 2a, 2c, 2d, 3c, and 3d with all of them contributing to the marginal ciliary neuropil (Table 8.1). The oral and the suboral neuropils are formed without contribution of the first quartet micromeres (Henry and Martindale 1998a). Due to the nervous system generating blastomeres being situated mostly in dorsal positions in the embryo, the authors suggested inductive interactions of other blastomeres specifying neuronal fates of blastomeres in the pilidium as has been shown in D quadrant specification.

There are no data available on the cell lineage of the nervous system in paleonemerteans and hoplonemerteans. The only sensory structures that provenance from blastomeres is known of are the eyes of the monostiliferan hoplonemertean *Nemertopsis bivittata* and the carinomid paleonemertean *Carinoma tremaphoros* (Martindale and Henry 1995; Maslakova et al. 2004b). Although apparently different with respect to both number and ultrastructure (one pair of rhabdomeric eyes in the monostiliferan hoplonemertean *Paranemertes peregrina* versus a single ciliary eye in the carinomid paleonemerteans *Carinoma tremaphoros* and *Carinoma mutabilis*), there seems to be a common pattern in deriving at least one eye from the C quadrant ($1c^1$ in *Carinoma tremaphoros*, right eye in *Nemertopsis bivittata* from so-called RD, i.e., C quadrant; see Martindale and Henry 1995; Henry and Martindale 1997, 1999; Maslakova et al. 2004a). The second eye in *Nemertopsis bivittata* is derived from the oppositely located (i.e. A) quadrant. In the latter species, there seem to be induc-

tive potential located in two adjacent quadrants necessary to induce eye formation (Martindale and Henry 1995; Henry and Martindale 1997). Although there is considerable variation with respect to some blastomere lineages in nemerteans, their mode of cleavage clearly operates within the framework of a spiral cleavage as seen in many other spiralians. In contrast to most other spiralian species, relative blastomere size, inductive interactions, and, to a certain degree, regulation seem to play a prominent role in nemertean development (Henry and Martindale 1999; Henry 2002).

Gastrulation

Accelerated divisions of the progeny of the animal quartets and subsequent flattening of the cells lead to the widening of the blastocoel into which the more columnar vegetal progeny projects to different degrees. This process results in a coeloblastula in nearly all species studied, whereas the shape of the blastula and the dimensions of the blastocoel may vary. The blastula may be domed, but in the majority of species, it is flattened, resulting in an either rather spacious or highly compressed blastocoel (Friedrich 1979). While the blastula is radially symmetric in most species, it attains a somewhat rectangular appearance (having been termed blastosquare) due to some of the progeny of animal blastomere quartets jutting out to the sides in *Cephalothrix ruffrons*, *Malacobdella grossa*, and *Micrura alaskensis* (Hammarsten 1918; Smith 1935; Maslakova 2010a). At this time of development, some species show signs of beginning ciliation, while in other species ciliation only starts when gastrulation sets in. Some species with intracapsular development develop cilia even later after gastrulation (e.g., *Pantinonemertes* (*Geonemertes*) *australiensis*, *Antarctonemertes phyllospadicola*) (Hickman 1963; Maslakova and von Döhren 2009). The apical pole in some species with pelagic stages already shows first rudiments of organs, namely, the apical pit characterized by a group of columnar cells that are housed in a shallow depression. In species with intracapsular development, an apical organ rudiment is usually absent (Friedrich 1979; Senz and Tröstl 1999). A pair of large, sometimes slightly invaginated cells situated on both sides of the apical pit has been reported in *Procephalothrix simulus*, *Emplectonema gracile*, *Prosorhochmus viviparus*, and *Malacobdella grossa*, already marking the bilateral symmetry of the blastula (Friedrich 1979). The large cells have been interpreted as the first rudiments of the nervous system (Iwata 1960). On the vegetal pole of the embryo, one (*Cerebratulus lacteus*: Coe 1899; Wilson 1903; *Lineus ruber*: Nusbaum and Oxner 1913, but see Schmidt 1964; *Tubulanus nothus*: Dawydoff 1928) or two pairs (*Procephalothrix filiformis*: Iwata 1960; *Tetrastemma vermiculus* and *Drepanophorus spectabilis*: Lebedinsky 1897) of distinct blastomeres, having been interpreted as primary mesoblast cells, are reported. In other species, however, the vegetal-most blastomeres are indistinguishable in size and form (Friedrich 1979).

Gastrulation proceeds by invagination of the blastomeres of the vegetal pole, although epibolic processes caused by the proliferation of the ectodermal components have been indicated to be involved. During gastrulation the shape of the embryo changes to become more domed. Some species, especially hoplonemerteans with very yolky eggs, gastrulate by polar ingression of the vegetal cells (*Malacobdella grossa*, *Prostoma graecense*, *Argonemertes* (*Geonemertes*) *australiensis*, *Gononemertes australiensis*) (Hammarsten 1918; Reinhardt 1941; Hickman 1963; Egan and Anderson 1979).

LATE DEVELOPMENT

Diversity of Larval Types in Nemertea

Although most nemerteans develop via planktonic stages, their development has traditionally been characterized as direct and indirect development, according to the transformation of the larva to form the juvenile body (Fig. 8.6). While in nemertean, so-called direct development morphological transformations are gradual, resulting in a smooth transition from the pelagic stage to the juvenile,

so-called indirect development is characterized by a catastrophic metamorphosis in which the larval epidermis is shed as a whole and usually eaten by the juvenile that has developed inside (Cantell 1967; Maslakova 2010a). Larval types in Pilidiophora are quite diverse, including several morphotypes of pilidia, the Desor larva of *Lineus viridis*, the Schmidt larva of *Lineus ruber*, the Iwata larva of *Micrura akkeshiensis*, and other unspecified pelagic, lecithotrophic larvae of some *Micrura* species (Norenburg and Stricker 2002; Schwartz and Norenburg 2005; Schwartz 2009; Maslakova 2010b). Larvae of hoplonemertean and paleonemertean species have traditionally been classified as so-called planuliform larvae due to their superficial resemblance to the planula larvae of cnidarians (Fig. 8.6A–C). Several hypotheses have been put forward to derive the pilidium from planuliform nemertean or even other, non-nemertean larval types, but the evolution of this aberrant larval form is presently still enigmatic (Iwata 1972; Jägersten 1972; Hiebert et al. 2010).

Recent findings, however, hint at a more complex picture. The planuliform larval type includes the decidula larva that possesses a transitory larval epidermis whose cells are replaced by cells of the definite juvenile epidermis and the hidden trochophore that is characterized by a distorted preoral belt of large cells that have been homologized with the trochoblasts of annelids and mollusks (Maslakova et al. 2004b; Maslakova 2010b). The decidula is typically present in Hoplonemertea but has also been suggested to occur in the tubulanid paleonemertean *Tubulanus punctatus* (Iwata 1960; Maslakova 2010b and references therein). The hidden trochophore has been found in *Carinoma tremaphoros* based on cell lineage studies (Maslakova et al. 2004a, b). The larval type of the remaining paleonemertean species has never been specified. Species with intracapsular development typically show characteristics of the larval type present in their respective phylogenetic lineage (Maslakova and Malakhov 1999; Maslakova and von Döhren 2009; von Döhren 2011). Larvae of hoplo- and paleonemerteans develop from the gastrula by differential growth of one side of the body which becomes the dorsal side of the animal. As a consequence, the blastoporal region moves to

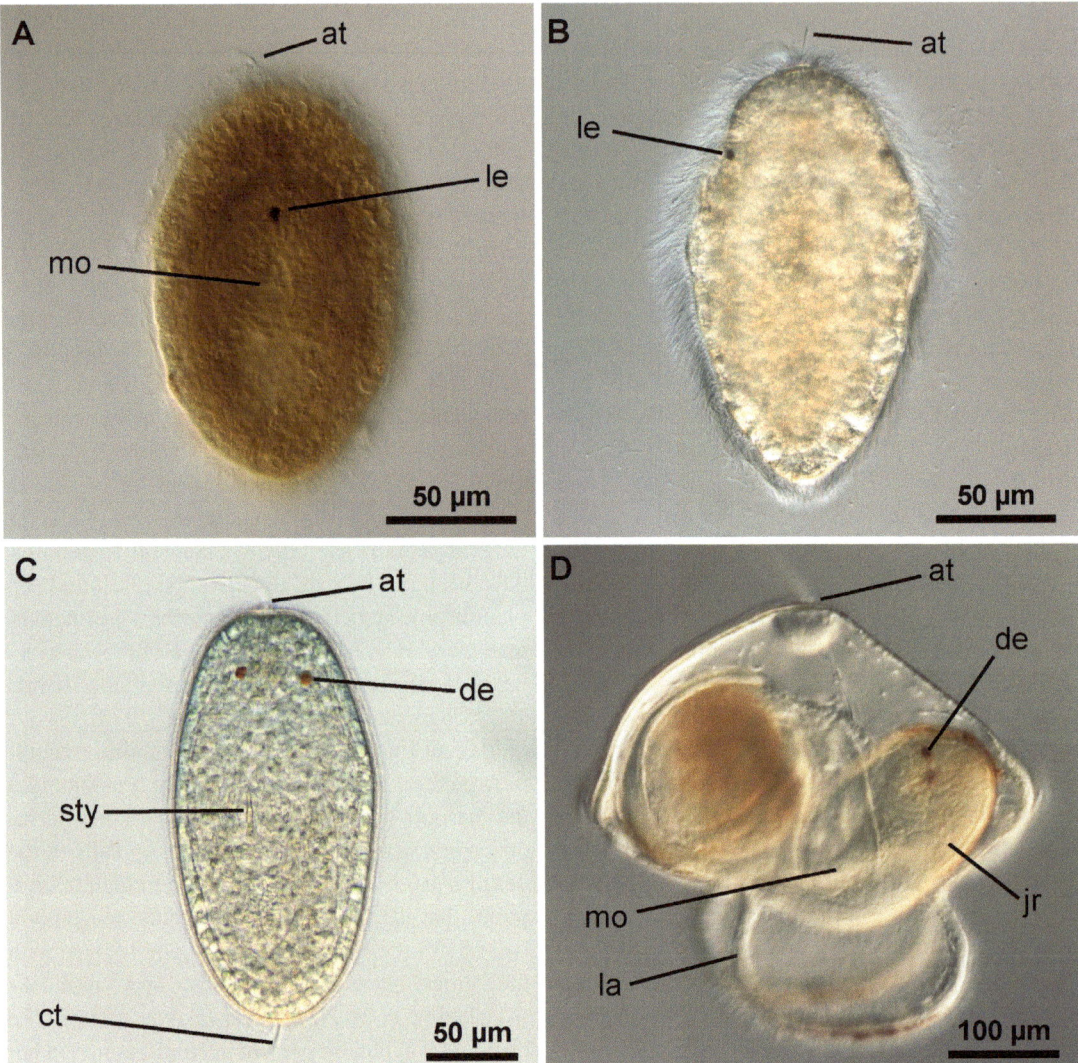

Fig. 8.6 Pelagic larvae of Nemertea. (**A**) Five-day-old larva of *Carinina ochracea* (Paleonemertea), fixed and osmicated specimen, differential interference contrast (*DIC*) light micrograph. Note the single median ventral eye. (**B**) Three-day-old larva (at 18 °C) of *Procephalothrix oestrymnicus* (Paleonemertea), live specimen, DIC. Note the pair of dorsal eyes. (**C**) Four-day old larva of *Emplectonema gracile* (Hoplonemertea), live specimen, bright field. Note the first pair of dorsal eyes. The stylet apparatus has already formed inside the larva. (**D**) Six-week-old, advanced pilidium of *Riseriellus occultus* (Pilidiophora), live specimen, DIC. Note the advanced juvenile with eyes inside the pilidium. The imaginal disks are fused; only the dorsal part of the juvenile has yet to form. *at* apical tuft, *ct* caudal tuft, *de* definite (adult) eye, *jr* juvenile rudiment, *la* lateral lappet, *le* larval eye, *mo* mouth opening, *sty* stylet (© Dr. J. von Döhren, All Rights Reserved)

a more frontal position marking the ventral side of the larva, thus changing the angle between the apical and the former vegetal pole of the embryo.

The pilidium was originally described by Müller (1847) as a previously unknown, putatively larval animal from the North Sea. It was consequently given a binomen: pilidium gyrans. Although being aware of the larval nature of the organism, the tradition of binomial nomenclature was adopted by subsequent researchers (e.g., Bürger 1894; Dawydoff 1940; Cantell 1969; Chernyshev et al. 2013). More than half a dozen morphotypes have been described of which the original pilidium gyrans represent the archetype but not necessarily the ancestral state (Figs. 8.6D and 8.7). Apart from different shapes of the lappets, the dome, and the bands of elongated cilia, the pilidial types have been distinguished by the orientation that the juvenile rudiment assumes

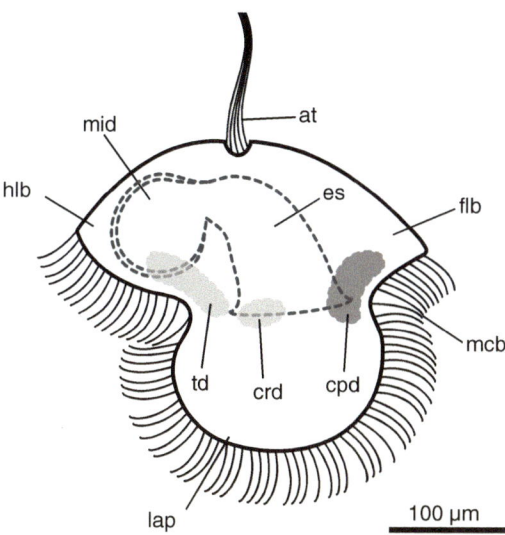

Fig. 8.7 Schematic representation of a young pilidium gyrans (lateral view). The positions of the prospective paired epidermal invaginations forming the juvenile are shown in *shaded areas* (*black*, exumbrellar invagination; *light grey*, subumbrellar invagination). *at* apical tuft, *cpd* cephalic disk rudiment, *crd* cerebral organ disk rudiment, *flb* forelobe of pilidium, *hlb* hindlobe of pilidium, *lap* lateral lappet of pilidium, *mcb* marginal band of elongated cilia, *mid* midgut, *es* esophagus, *td* trunk disk rudiment (© Dr. J. von Döhren, All Rights Reserved)

with respect to the larval apical-vegetal axis. In the archetypical pilidium (e.g., pilidium gyrans), the juvenile anterior-posterior axis assumes a roughly rectangular orientation to the larval apical-vegetal axis as indicated by the larval apical tuft and vegetal mouth opening. There is a hypothetical series to derive the archetypical pilidial forms (pilidium gyrans, pilidium pyramidale) from types in which the larval and juvenile main axes are roughly parallel, such as the pilidium recurvatum (Jägersten 1972). However, more recent findings indicate that the vast diversity of pilidial morphotypes might be a consequence of the uncoupling of morphogenetic processes between the larval stage and the formation of the juvenile (Hiebert and Maslakova 2015).

Comparative Larval Development

Paleo- and Hoplonemertean Larvae

In larvae of paleonemertean species, the blastoporal region differentiates into an ectodermal esophagus and an endodermal, sac-like midgut,

both of which are functional. In the majority of hoplonemertean species, the blastopore has been reported to close. A secondary, ectodermal mouth opening along with an esophageal tube is subsequently formed in various ways according to different authors (Friedrich 1979; Maslakova and von Döhren 2009). In some hoplonemertean species, especially those with intracapsular development, no functional mouth opening is formed until much later in development (Salensky 1914; Hammarsten 1918; Hickman 1963; Iwata 1960; Egan and Anderson 1979). The larvae are initially ovoid in shape and gradually elongate with age. While the surface of the larva is covered with cilia of equal length, the apical pole of the larva is equipped with an apical tuft of longer cilia housed in an apical depression, the apical pit. In many species the posterior end of more advanced stage larvae is endowed with another tuft of elongated cilia termed caudal tuft or posterior cirrus (Fig. 8.6C; Iwata 1960; Stricker and Reed 1981; Maslakova and von Döhren 2009). One to several pairs of lateral cirri located anterior of and at the level of the mouth opening have been described for members of the cephalothricids (Iwata 1960). In most larvae a pair of epidermal invaginations on either side of the apical pit is formed which has variously been interpreted as rudiments of the nervous system or the cerebral organs (Lebedinsky 1897; Salensky 1909; Hammarsten 1918; Smith 1935; Iwata 1960; Maslakova and von Döhren 2009; but see Hiebert et al. 2010; Maslakova 2010b).

Some larvae of paleonemerteans and most hoplonemertean larvae possess eyes (Fig. 8.6A–C). They are usually simple pigment cup ocelli composed of only a few photoreceptor cells surrounded by one to a few shading pigment cells (von Döhren 2008). While in *Carinoma* species and *Carinina ochracea* there is a single almost median ventral eye anterior to the mouth opening, cephalothricids and most hoplonemertean species possess a pair of eyes situated dorsally, in front of the mouth opening (Fig. 8.6A–C; Hammarsten 1918; Iwata 1960; Maslakova et al. 2004a, b; Maslakova and von Döhren 2009; Hiebert et al. 2010; Maslakova 2010b). In the larva of *Quasitetrastemma* (*Tetrastemma*) *stimpsoni* and *Quasitetrastemma*

(*Tetrastemma*) *nigrifrons*, three pairs of dorsally located eyes have been described (Chernyshev 2008). In paleonemertean larvae, the eyes are epidermal and possess ciliary receptor cells, while hoplonemertean species possess subepidermal eyes equipped with rhabdomeric receptors. Since the adults of those paleonemertean species studied lack eyes, the eyes present in the larvae have to be interpreted as transitory larval organs (von Döhren 2008). In hoplonemerteans, the subepidermal eyes persist to adulthood, increasing in size and number in most species. Only in *Quasitetrastemma* species the one pair of eyes has been hypothesized to be fused to the neighboring pair or completely reduced during development (Chernyshev 2008).

Pilidiophoran Larvae: The Pilidium

The archetypical pilidium develops from the swimming gastrula by growth of the lateral rims of the invaginated vegetal field to form the lateral lappets. A ring of elongated cilia, the marginal ciliary band, is present around the invaginated ciliary field, marking border between the outer (exumbrellar, also: epispheric) and the inner (subumbrellar, also: hypospheric) surfaces of the larva (Fig. 8.7; Friedrich 1979; Maslakova 2010a). In some forms of pilidia, the anterior and posterior parts of the ciliated ring may form additional anterior and posterior lobes in the larva. Denticle-like structures and chromatophores may adorn the ciliated ring. Depending on the species, they may increase in number with age of the larva (Cantell 1969; Norenburg and Stricker 2002; Lacalli 2005). The exumbrellar part, housing the apical pit with a tuft of elongated cilia, expands to form the typical dome shape of the pilidium (Figs. 8.6D and 8.7). The esophagus is formed from ectodermal cell material being dragged in during blastopore invagination (Friedrich 1979). The larval gut is sac-like and located interior of the posterior lobe. The gut is open to the esophagus via a narrow opening with a valve that develops from the blastopore. Along the posterior wall of the esophagus, there is a pair of prominent longitudinal ridges that exhibit elongated cilia. They have been reported to serve as main structures to mediate transport of food particles

through the esophagus to the gut (Maslakova 2010a). The larva is composed of two epithelial layers consisting of a single row of multiciliated, large flat cells. In the region of the apical tuft, the marginal ciliary band, and the longitudinal ridges, the cells attain a more columnar shape. Inside the larva, an extensive, comparably loose extracellular matrix with interspersed neuronal and mesodermal cell clusters fills the body.

Both the larval musculature and nervous system are formed early in development, enabling the larva to take up food particles and to locomote. For the most part, nerves and muscles are not taken over into the juvenile organization (Maslakova 2010a), but some neural components are incorporated into the postmetamorphic nervous system (Hindinger et al. 2013). The first structures of the nervous system that differentiate in the pilidium are a prominent ring-shaped neuropil underlying the marginal ciliary band (marginal band neuropil), a ring-shaped neuropil encircling the opening of the esophagus into the gut (oral ring neuropil), as well as a pair of neurite bundles connecting the oral ring neuropil on both sides of the larva with the marginal band neuropil at the posterior base of the lateral lappets. A pair of linear plexuses (circumesophageal plexus) extending from the anterior to the posterior base of the lateral lappets alongside the opening of the esophagus connecting to the marginal band neuropil has been described in some species (Lacalli and West 1985; Hay-Schmidt 1990; Henry and Martindale 1998a; Maslakova 2010a; Zaitseva and Flyachinskaya 2010; Hindinger et al. 2013). The epidermis is underlain by a loosely arranged epidermal plexus that is in contact with the marginal band neuropil and exhibits a prominent pair of monociliated, serotonergic, putatively sensory neurons, laterally next to the apical pit (Maslakova 2010a; Hindinger et al. 2013). The apical pit itself is devoid of neuronal elements (Cantell et al. 1982; Lacalli and West 1985; Chernyshev et al. 2013). From some of the mesodermal cells, the larval musculature forms underneath the epidermis, while others remain in an undifferentiated, almost amoeboid state (Henry and Martindale 1998a; Maslakova 2010a). The most prominent structure of the larval musculature is a muscle ring situated underneath the marginal ciliary band. The muscle ring branches at the bases of the lateral lappets to send a portion along the rim of the lappets, while the other encircles the esophageal opening. A ring-shaped sphincter muscle underlies the opening between the esophagus and the gut. Underneath the epidermis there is a prominent array of muscles that radiate from the dome into all parts of the larva. By branching they form an oblique muscle network being most prominent in the lateral lappets (Maslakova 2010a). A muscle extending from the apical pit downward (apical muscle or central muscle), branching around the esophagus, and serving to contract the larval apex has been described in most of the larval forms (Bürger 1897–1907; Lacalli 2005; Zaitseva and Flyachinskaya 2010).

The juvenile develops inside the pilidium from a set of epidermal invaginations of the larval envelope, termed imaginal disks. A minimum of three paired, bilaterally located disks are described: the most frontally located cephalic disks, the middle cerebral organ disks, and the trunk disks situated ventral of the gut posterior of the lateral lappets (Fig. 8.7). After invaginating from the larval epidermis, the imaginal disks are pinched off becoming flattened, fluid-filled cavities. While the cells of the proximal wall of the imaginal disk are initially more columnar and differentiate into definite, juvenile epidermis, the distal layer retains its flat, thin structure reminiscent of the larval epidermis. The distal, thin cell layer is referred to as amnion (Friedrich 1979; Maslakova 2010a). In addition to the three pairs of epidermal invaginations, two rudiments have been observed, of which the ectodermal origin has not been unambigouously demonstrated: the anterior-most proboscis rudiment and the posterior-most dorsal rudiment, both of which might not be generally present in all species (Friedrich 1979; Maslakova 2010a). The imaginal disks do not, however, form at the same time, but there is a stereotypic sequence of appearance. The first pair of disks that forms is the cephalic disks. Unlike the remaining imaginal disks, the cephalic disks are invaginated from the exumbrellar epidermis as derivatives of the A and B quadrants of the first

quartet micromeres, hence 1a and 1b (Fig. 8.7; Henry and Martindale 1998a; Maslakova 2010a). After some time, the trunk disks are formed, followed by the cerebral organ disks. Both pairs of imaginal disks are unanimously reported to invaginate from the subumbrellar larval epidermis (Fig. 8.7; Friedrich 1979; Maslakova 2010a). Shortly after, an unpaired anterior-most putative proboscis rudiment is observed sitting dorsal of the cephalic disks underneath the larval epidermis. The exact origin of the putative proboscis rudiment is unknown. It has been described as epidermal invagination similar in structure to the paired imaginal disks, as a delamination of the larval epidermis or as a cluster of undifferentiated mesenchymal cells forming underneath the larval epidermis (Bürger 1894; Schmidt 1937; Maslakova 2010a). At the time the cephalic disks fuse with each other, incorporating the putative proboscis rudiment and thus forming the head rudiment, the anterior end of each trunk disk fuses with the posterior rim of the cerebral organ disk of the respective side. During fusion of the proximal cell layer of the imaginal disk, the amnion layers are separated from the proximal layer to fuse to each other, forming a continuous amniotic layer distal to the respective rudiments. In most species an additional, unpaired, posterior-most dorsal rudiment forms (Friedrich 1979; Maslakova 2010a). In contrast to the paired imaginal disks, the dorsal rudiment arises as a delamination from the larval epidermis or a mesodermal cell layer instead of an invagination from the larval epidermis. Although initially single layered, the dorsal rudiment soon becomes double layered, the outer layer forming an amnion as is seen in the imaginal disks that develop as invaginations of the larval epidermis (Bürger 1894; Salensky 1912; Schmidt 1937; Maslakova 2010a). During further development the dorsal rudiment fuses with the dorsoposterior rims of the bilaterally still separate trunk rudiments which themselves fuse posteriorly soon afterwards. As well as during the fusion of the cephalic disks, a continuous amnion layer is formed.

The next stage is accomplished by the fusion of the lateral posterior rims of the head rudiment to the lateral anterior rims of the trunk rudiment represented by the cerebral organ disks. Thus, a continuous torus-shaped rudiment is formed around the larval esophagus (Maslakova 2010a). The anterior dorsal rim of the trunk rudiment subsequently fuses with the posterior dorsal margin of the head rudiment, completing the formation of the juvenile inside the pilidium. The juvenile is almost completely separated from the larval tissues by the amniotic cavity which is lined by the continuous amnion layer. The only contact of pilidium and juvenile is by the mouth opening that is shared by larva and juvenile. After the development of the juvenile inside the pilidial envelope is completed, the juvenile escapes from its envelope by vigorous movements that rupture the pilidial tissues (Cantell 1967; Maslakova 2010a). This process has been termed catastrophic metamorphosis and is usually a matter of a few minutes. The larval envelope that is still connected to the mouth of the escaping juvenile is usually eaten by the juvenile (Cantell 1967; Maslakova 2010a).

Aberrant Pilidiophoran Larvae

Within Pilidiophora there seems to be an evolutionary tendency toward lecithotrophic stages (Maslakova and Hiebert 2014). The larva of *Micrura akkeshiensis*, named Iwata larva after its discoverer, is a pelagic stage but does not take up food during its development (Iwata 1958). Thus, the larva does not show either lappets or bands of elongated cilia used for food capture in the pilidium. It develops from a comparably flat gastrula that elongates and attains a pyriform shape. The animal pole is marked by an apical pit with an apical tuft of elongated cilia. The opposite vegetal pole is marked by the site of the blastopore. In this type of development, the blastoporal opening is not shifted to either side of the gastrula during further development but remains in its original position. Around the blastopore four larger, mesodermal cells have been observed which proliferate to form a mesodermal mass between the epidermis and the invaginated archenteron.

Apart from an apical tuft and an evenly ciliated epidermis, no distinct larval organs have been reported. Similar to the pilidium, the epidermis covering the larva is a transitory larval envelope in which the juvenile develops. However, instead of developing at an angle to the larval body, the juvenile inside is parallel to the larva but with its head directed toward the side of the blastopore. Furthermore, in the Iwata larva, the number of imaginal disks is restricted to five, a pair of frontal head disks, a pair of posteroventral trunk disks, and an unpaired posterior dorsal disk, all of which form from invaginations of the larval epidermis. An anterior-posterior sequence of their formation is likely (Iwata 1958).

Similar nonfeeding larval types have been described in *Micrura rubramaculosa*, *Micrura verrilli*, and at least two other yet undescribed *Micrura* species; although in the larva of the former species there is an additional equatorial band of elongated cilia present and the larval animal-vegetal axis coincides with the adult anterior-posterior axis (Schwartz and Norenburg 2005; Schwartz 2009). Another nonfeeding larval form possessing an equatorial as well as a posterior band of elongated cilia that have been termed proto- and telotroch analogously to the ciliated bands of the trochophore larva has been described for an undescribed lineid heteronemertean (Maslakova and von Dassow 2012; Maslakova and Hiebert 2014). Unfortunately, there is virtually nothing known on the development in any of the abovementioned larval types.

Although being the first developmental stage described in nemerteans, the Desor larva of *Lineus viridis* might represent one of the most derived larval types. Despite of showing entirely intracapsular development within an egg capsule, it shows distinct characteristics typical of pilidiophoran development, including imaginal disks and the devouring of the larval epidermis during a catastrophic metamorphosis (Friedrich 1979; von Döhren 2011). However, since this larval type is lecithotrophic, no food is taken up until metamorphosis; hence, accessory feeding structures as seen in the pilidium are absent. Following gastrulation a spherical larva is formed that is uniformly ciliated, lacking any apical tuft or band of elongated cilia. Reminiscent of the pilidium larva, three pairs of imaginal disks are formed by invagination along with two rudiments formed by delamination from the larval epidermis (Nusbaum and Oxner 1913; Schmidt

1964). They correspond in position to the imaginal disks and rudiments seen in the pilidium. In contrast to canonical pilidial development, cephalic and trunk disks appear almost simultaneously in the Desor larva, followed by the proboscis rudiment, the cerebral organ disks, and finally the dorsal rudiment (Schmidt 1964). The sequence of fusion only differs from that seen in the pilidium in that the trunk disks first fuse with each other and with the dorsal rudiment before fusing with the cerebral organ disks (Friedrich 1979). At that time, the larva elongates posteriorly, attaining a rhomboid shape.

The larva of the closely related species *Lineus ruber* is smaller but has essentially the same larval morphology (Schmidt 1964). It has been given a different name due to its reported divergent mode of feeding. In contrast to the Desor larvae of *Lineus viridis*, the so-called Schmidt larva of *Lineus ruber* is adelphophagic, feeding on putatively deficiently developing sibling embryos inside the same egg capsule. Hence, differences in timing of the development of the proboscis and the alimentary tract with its associated musculature in later-stage larvae have been reported (Schmidt 1964).

Nemertean Larval Types – Evolutionary Considerations

Including the other nemertean larval types, there is a series of transitions from the paleonemertean larvae to the most derived developmental types, namely, the Iwata and Desor larva (Iwata 1972). This series, however, has to be taken with some reservation. On the one hand, the basally branching, non-heteronemertean pilidiophoran *Hubrechtella dubia* possesses a pilidium of the pilidium auriculatum type which is most similar to the pilidium gyrans type (Cantell 1969), while on the other hand, most pilidial morphotypes cannot be assigned to a certain species or clade, thus rendering phylogenetic inferences needed to set up an evolutionary series impossible. The aberrant pilidium recurvatum type has been assigned to members of the heteronemertean pilidiophoran Baseodiscidae but has also been identified as the larval type of the putatively most basal heteronemertean genus

Riserius (Cantell 1969; Thollesson and Norenburg 2003; Hiebert et al. 2013a, b).

An apical organ situated basal to the apical pit present in all nemertean larval types has variously been interpreted as homologous to the apical organ in several trochozoan and lophophorate species, especially in Annelida and Mollusca but also in Brachiopoda (see Chapters 7, 9, and 12 as well as Nielsen 2013 and references therein). In many spiralian larvae, the apical organ is composed of the apical tuft and a usually low number of flask-shaped, serotonergic neurons at its base. These flask-shaped neurons generally degenerate prior to metamorphosis (Nielsen 2013 and references therein). In nemertean species a single flask-shaped serotonergic neuron has been reported only for hoplonemertean species although it does not project directly into the apical pit but reaches the epidermal surface in its vicinity (Fig. 8.8A; Chernyshev and Magarlamov 2010). Moreover, this flask-shaped neuron persists through metamorphosis as part of the brain ring (von Döhren J 2015, unpublished). In paleonemertean and pilidiophoran species, no flask-shaped serotonergic neurons have hitherto been found. In the pilidium, there are two spherical, putatively sensoric serotonergic neurons in the vicinity of the apical pit, while the single apical serotonergic neuron of paleonemertean species has an ovoid shape and does not make contact with the epidermis (Fig. 8.8B). While the ovoid apical serotonergic neuron in Paleonemertea disappears when the brain rudiment starts to form, the apical serotonergic neurons of Pilidiophora are shed at metamorphosis along with the larval envelope. Similar to the situation seen in the larvae of polyclad Platyhelminthes, homology of the serotonergic apical neurons is unclear (Rawlinson 2010). Two evolutionary scenarios seem possible: either the apical serotonergic neurons of Nemertea are homologous to those in Trochozoa, but there have been various substantial transformations in both Paleonemertea and Pilidiophora, or apical serotonergic neurons have evolved multiple times independently in Nemertea. Current hypotheses regarding the phylogeny of Nemertea favor the second

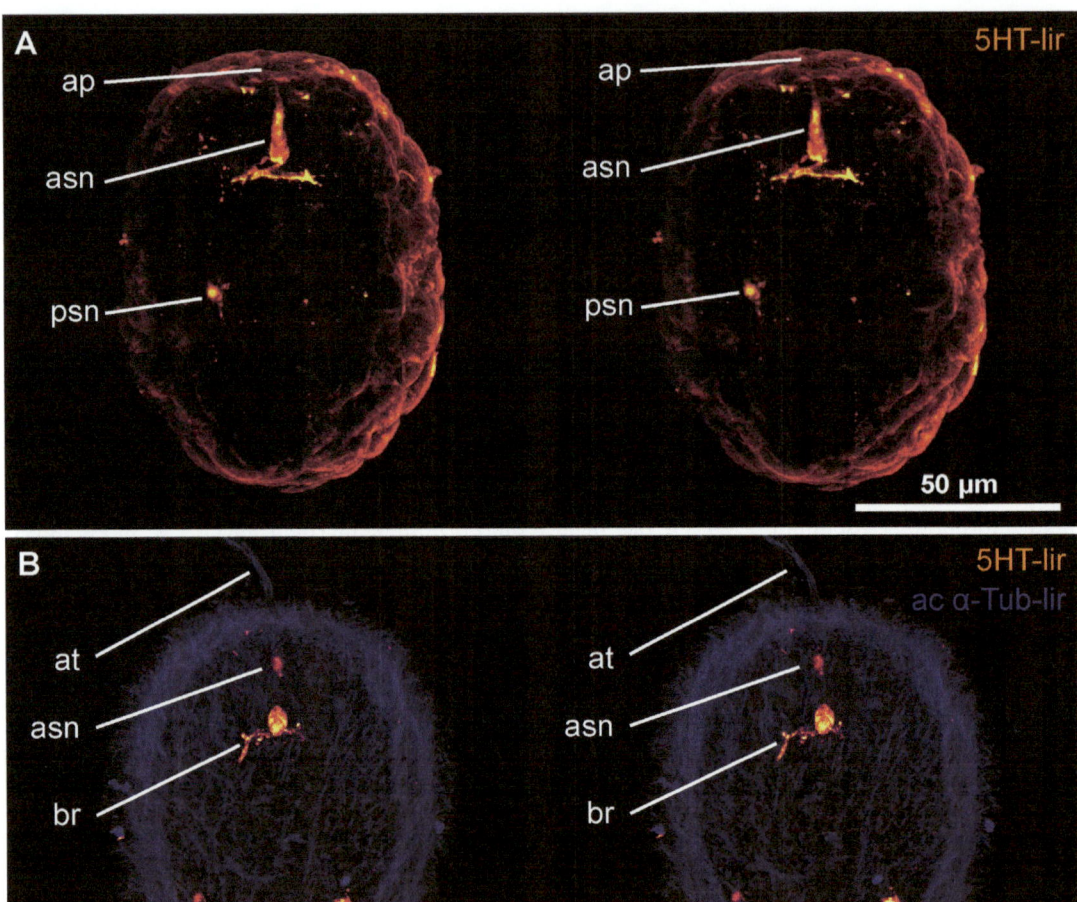

Fig. 8.8 Stereo pair images of confocal microscopy stacks of neural staining of nemertean larvae. (**A**) *Emplectonema gracile (Hoplonemertea)*, one-day-old larva stained with antibodies against serotonin (5HT-lir). Note: the flask-shaped apical serotonergic neuron (*asn*) projects in the vicinity of the apical pit (*ap*). An additional, more posterior serotonergic neuron (*psn*) is visible at this stage. (**B**) *Procephalothrix oestrymnicus (Paleonemertea)*, three-day-old larva (at 12 °C) stained with antibodies against serotonin (5HT-lir) and acetylated α-Tubulin (ac α-Tub-lir). Note: there is no connection of the apical serotonergic neuron (*asn*) to the apical pit that the apical tuft (*at*) emanates from. The rudiment of the brain (*br*) is visible as well as a few more posteriorly located neurons (*psn*) showing 5HT-lir. *ap* apical pit, *asn* apical serotonergic neuron, *at* apical tuft, *br* brain rudiment, *psn* posterior serotonergic neuron (© Dr. J. von Döhren, All Rights Reserved)

alternative scenario as being more parsimonious (Andrade et al. 2012).

Formation of the Adult Body Plan

Data on the formation of the definite adult organization in Nemertea are rather heterogeneous so that generalizations are at best preliminary at this point in time. Comprehensive information is restricted to classical accounts on about a dozen species, most of which are neonemerteans with arguably aberrant, i.e., intracapsular, development (e.g., *Lineus viridis/ruber*: Barrois 1877; Arnold 1898; Hubrecht 1885, 1886; Nusbaum and Oxner 1913; Schmidt 1964; *Prosorhochmus viviparus*: Salensky 1882–1883, 1909, 1914; *Pantinonemertes* (*Geonemertes*) *agricola*: Coe 1904; *Prostoma graecense*: Reisinger 1926; Reinhardt 1941; *Argonemertes* (*Geonemertes*) *australiensis*: Hickman 1963; *Prosorhochmus adriaticus*: Senz and Tröstl 1999). Organs that are undisputedly of ectodermal origin are the epidermis with the associated cephalic glands, the nervous system and its associated sense organs, the outer (when seen extruded) epithelium of the proboscis, and the rhynchodeum. The esophagus and anus have been reported to be ectodermal derivatives by the majority of authors although there has been disagreement regarding the origin of the esophagus according to older accounts (Barrois 1877; Salensky 1886, 1909, 1912, 1914; Arnold 1898). Of mesodermal origin are the muscular systems, comprising the various body wall muscle layers, the muscle layers and the inner epithelium of the proboscis facing the rhynchocoel, the rhynchocoel epithelium and its underlying muscle layers, as well as the endothelialized blood-vascular system. The intestinal tract exclusive of the esophagus and anus is derived from the endoderm (Fig. 8.2A, B). On the origin and formation of the protonephridia in Nemertea, there has been considerable disagreement as comprehensive data are presently missing.

Epidermis

In Neonemertea there are two generations of epidermal cells that appear sequentially and of which only the second is kept as the definite adult epidermis. The first generation of cells covers the entire body as a larval epidermis in all Pilidiophora. It has been shown to be present in Hoplonemertea according to more recent findings (Maslakova 2010b). While the larval epidermis is kept as intact larval envelope in which the juvenile develops until metamorphosis in Pilidiophora, in Hoplonemertea cells of the larval epidermis are intercalated by definite epidermal cells and later disappear (Maslakova and Malakhov 1999; Maslakova and von Döhren 2009; Maslakova 2010b). In *Pantinonemertes californiensis* larval epidermal cells have been shown to be shed (Hiebert et al. 2010), a mechanism that has also been suspected to occur in the development of *Pantinonemertes* (*Geonemertes*) *agricola* (Coe 1904). Although it has been argued to be homologous to the pilidial larval envelope, the larval epidermis of Hoplonemertea has been given the name "decidula" due to its differing fate (Maslakova 2010b).

A first generation of epidermal cells in paleonemertean species has been hinted at in *Tubulanus punctatus* (Iwata 1960), while in *Carinoma tremaphoros*, there is a population of transitory trochoblast cells forming a distorted preoral belt that has been termed "vestigial prototroch" (Maslakova et al. 2004b). The definite, adult epidermis consists of multiciliated cells, several types of gland cells, basal granular cells, and basal neurites originating from subepidermal neurons constituting a pseudostratified epithelium in paleonemertean and heteronemertean species (Turbeville 1991). In Hoplonemertea, due to the presence of a basal cup cell layer, the pseudostratified appearance of the epidermis is even more pronounced (Norenburg 1985; Turbeville 1991). The heteronemertean dermis is formed late in development, i.e., after metamorphosis, by some gland cells sinking underneath the level of the ciliated epidermis cells. Thus, proximal of the epidermal basal extracellular matrix, a distinct layer of differing extent containing gland cells intermingled with muscle cells and neurons appears to be present at light microscopic resolution (Norenburg 1985).

Nervous System

The nervous system in paleo- and hoplonemertean species seems to have two areas of origin. While the first nervous elements represented by a peripheral plexus are most likely derived from cells that sink in from the apical and the posterior ectoderm of late gastrulae of paleonemertean and hoplonemertean species, the brain and lateral medullary cords have been reported to originate from paired large ectodermal blastomeres bilateral to the apical pit (Salensky 1909, 1914; Hammarsten 1918; Smith 1935; Iwata 1960). The brain primordia sink underneath the epidermis, first forming a pair of narrow epidermal invaginations that soon lose contact to the epidermis and proliferate (Coe 1904; Salensky 1909, 1914; Hammarsten 1918; Smith 1935; Iwata 1960).

The different positions of the brain and lateral medullary cords in different clades have been reported to originate from the depth to which the nervous system rudiments sink in. Remaining in a distal position to mesodermal tissues in *Tubulanus punctatus*, the nervous system rudiments sink in underneath the mesodermal tissues in *Procephalothrix* species and Hoplonemertea (Iwata 1960). From the invaginated cell clusters, the ventral and dorsal brain lobes on either side are formed that are later connected horizontally by the ventral and dorsal commissural tracts (Coe 1904; Salensky 1909, 1914; Hammarsten 1918; Smith 1935; Iwata 1960). Formation of the brain lobes, lateral medullary cords, and ventral commissural tract prior to the dorsal commissural tract has been reported for the monostiliferous hoplonemertean *Prosorhochmus adriaticus* (Senz and Tröstl 1999). In Pilidiophora, the brain and lateral medullary cords are formed independent of the extensive larval nervous system (Salensky 1886, 1912; Maslakova 2010a; Hindinger et al. 2013). The first nervous system rudiment is observable in pilidia as a paired epidermal invagination or delamination on the proximal side of either of the cephalic disks prior to their fusion (Salensky 1886, 1912). With the fusion of the disks, the brain lobes on either side are secondarily connected by formation of both the ventral and later the dorsal commissural tract (Salensky 1912).

In the Desor larva, the brain is formed later in development when the cephalic disks have already fused with each other and with the trunk rudiment. The brain is formed by a uniform, ring-shaped rudiment that later differentiates into the four brain lobes and their connecting commissural tracts. A temporal sequence of development can only be seen by the preceding separation of the ventral commissural tract from the surrounding ectodermal layer (Nusbaum and Oxner 1913). The lateral medullary cords are reported to be outgrowths of the ventral brain lobes in most species. Immunohistochemistry data reveal that the serotonergic neurons of the brain ring start to show immunoreactivity in the ventral lobes and the ventral commissural tract first, while peptidergic (FMRF-like) immunoreactivity is first observed in the dorsal brain lobes and commissural tract. Nervous structures innervating the mouth opening and the esophagus are observable already in larval stages of paleonemertean species. In *Cephalothrix rufifrons* the esophageal nerves have been derived from neuron precursors in the vicinity of the mouth opening, while in *Carinina ochracea* esophageal nerves are putatively outgrowths of the lateral medullary cords (Smith 1935; von Döhren J 2015, unpublished). While paired ectodermal invaginations have been interpreted as the main rudiment of the brain and lateral medullary cords in most species, there have been differing accounts in a few species from all clades.

In the intracapsularly developing species *Argonemertes* (*Geonemertes*) *australiensis* and *Prostoma graecense*, the first nervous system rudiments are represented by a cluster of cells that is not formed by an invagination (Reinhardt 1941; Hickman 1963), while in *Cephalothrix rufifrons* paired groups of cells from the ectoderm in the vicinity of the stomodeum have been suspected to contribute to the ventral brain lobes (Smith 1935). In *Tetrastemma vermiculus* and *Drepanophorus spectabilis*, the brain is reported to originate from a ventrolateral and a dorsolateral pair of epidermal thickenings that disconnect from the epidermis to each form a brain lobe. Connection of the isolated brain compartments is later accomplished by secondarily forming neuroectodermal ridges (Lebedinsky 1897). The same mode of formation has been stated for the lateral medullary cords in the abovementioned species and *Argonemertes* (*Geonemertes*) *australiensis* as well as for the median dorsal nerve of *Drepanophorus spectabilis* (Lebedinsky 1897; Hickman 1963). Likewise, in Pilidiophora, independent rudiments of the dorsal brain lobes from the cephalic disks and the ventral brain lobes as well as the lateral medullary cords from the trunk rudiments have been reported in pilidia by Bürger (1894). The relevance of these rather particular findings, however, has been doubted by several other contemporary authors (Salensky 1912, 1914; Nusbaum and Oxner 1913).

Development of the proboscis nerves or the nerves innervating the sense organs has never been concisely described, but they have been reported to form comparably late during development in *Prosorhochmus adriaticus* (Senz and Tröstl 1999).

Body Wall Musculature

The body wall muscle layers develop from mesenchymal cells situated between the epidermis and the gut rudiment. In both paleonemertean and hoplonemertean larvae, the sequence of formation of musculature is uniform, the inner longitudinal muscles being differentiated prior to the outer circular layer (Salensky 1914; Iwata 1960). Additional dorsoventral muscles and the musculature of the anterior head region are reported to develop much later in the hoplonemertean species *Prosorhochmus adriaticus* (Senz and Tröstl 1999). There is, however, considerable disagreement over the nature of the mesoderm. In *Tetrastemma vermiculus* and *Drepanophorus spectabilis*, four longitudinal, mesodermal bands originating from four mesendoblasts form all of the body wall and the splanchnic musculature (Lebedinsky 1897). In pilidia, however, there are four somatic rudiments underneath the two cephalic and the two trunk disks, but only one splanchnic mesodermal rudiment surrounding the larval gut formed from initially scattered mesenchymal cells (Salensky 1912).

Classical accounts on the formation of muscle layers in heteronemertean Pilidiophora differ both concerning the origin of muscle layers as well as regarding the sequence of their formation. Hubrecht (1885, 1886) derived the muscle layers in the Desor larva from mesoderm. The first muscle layer to form is the outer longitudinal layer, that is typical of Heteronemertea, followed by the inner circular and longitudinal layers. According to other authors the dermal muscles and the outer longitudinal layer are derived from the proximal layer of the imaginal disks, while the inner circular and longitudinal muscles are derived from the underlying mesenchymal cells (Bürger 1894; Salensky 1912). A similar origin of muscle layers has been reported in the Iwata larva of *Micrura akkeshiensis*, although the sequence of formation of muscle layers is reversed: first, the inner longitudinal and circular layers are formed, while outer longitudinal and dermal muscles have not been seen until 39 days after metamorphosis (Iwata 1958). Recent fluorescent

labeling studies have revealed that the sequence of formation of muscle layers is uniform in all species studied. The inner longitudinal body wall muscle layer is formed prior to the outer circular layer. Furthermore, in the Desor larva, the outer longitudinal muscles as well as the dermal muscles develop later during post-larval development. Moreover, these findings argue against an ectodermal origin of the dermal and outer longitudinal muscle layers, since muscle cells of both the dermal and the longitudinal muscle layers are observed prior to the formation of the dermis (von Döhren 2008).

Proboscis Apparatus

Data on the development of the proboscis are only available for neonemertean species (Friedrich 1979; Senz and Tröstl 1999; Maslakova 2010a, b). In paleonemerteans the formation of the proboscis has never been witnessed during development. Therefore, the onset of the formation of the proboscis has been assumed to occur quite late, close to or even after the shift to the benthic lifestyle. While in heteronemertean pilidiophorans the proboscis epithelium is formed from a cone-shaped, proximal process of the fused cephalic disks, it is generally reported to be formed as an epidermal invagination close to the apical organ in hoplonemerteans.

According to classical accounts on pilidia, the cephalic disks fuse and the proboscis is either formed by an invagination of the proximal layer of the head rudiment or by a separate proboscis disk that is incorporated into the head rudiment during fusion of the cephalic disks (Bürger 1894; Salensky 1912; Schmidt 1937). A more recent account shows a different picture (Maslakova 2010a). While the cephalic disks fuse to form the head rudiment by proliferation of their proximal cell layer, the putative proboscis rudiment is represented by a separate cluster of cells underneath the larval epidermis. The arguably mesodermal proboscis rudiment makes its way ventrally to come to lie adjacent to the inner side of the proximal cell layer of the head rudiment. A portion of the proximal layer of the head rudiment invaginates to form a cone-shaped structure that protrudes into the larva so that the putative proboscis rudiment sits on it like a shallow cap, forming a compound proboscis bud. It has been reported that in the proboscis bud, the cone-shaped structure merely forms the ectodermal components, i.e., the outer epithelium and the gland cells of the proboscis. Its muscular layers, the inner epithelium, as well as the rhynchocoel epithelium and the rhynchocoel muscles originate from the cap-like putative proboscis rudiment by means of schizocoely. The rudiment of the proboscis apparatus elongates on the dorsal side of the esophagus until it reaches the level of the larval gut. It consists of two components. An outer tube-shaped structure, the forming rhynchocoel wall, is clearly separated from an inner, invaginated structure representing the developing proboscis (Maslakova 2010a). After elongation of the proboscis, the dorsal margin of the trunk rudiment extends anteriorly over the larval gut and the proboscis rudiment.

The hypothetical homology of the nemertean rhynchocoel with secondary body cavities of other coelomate spiralians is said to be supported by the mode of its development (Maslakova 2010b). The hypothesis of a schizocoelous development of the rhynchocoel, however, demands a complete fusion of the ectodermal cone with the mesodermal cap of the proboscis bud. The components of the proboscis rudiment, however, seem to be separated in all stages investigated; schizocoelous processes within the mesodermal component have not been reported (Maslakova 2010a). In the Desor larva, the proboscis epithelium develops from an independent proboscis rudiment, while its musculature, the rhynchocoel and the rhynchocoel muscles, are derived from the mesoderm. It forms a continuous layer around the proboscis to separate into a proximal layer forming the proboscis muscles and the inner proboscis epithelium and an outer layer developing into the rhynchocoel epithelium and the associated muscle layers. In the part where no separation of the layers occurs, the retractor muscle develops (Arnold 1898; Nusbaum and Oxner 1913). Whether this separation is by delamination or schizocoely is not reported. Fluorescent labeling of F-actin reveals that the formation of the proboscis musculature precedes that of the rhynchocoel in the Desor larva of *Lineus viridis* (von Döhren J 2015, unpublished). In the Iwata larva of *Micrura akkeshiensis*, the proboscis rudiment is derived from an invagination of the secondary epidermis, and its formation corresponds to the classical description on the development of the proboscis in the pilidium larva (Iwata 1958).

In hoplonemerteans the proboscis is formed from an epidermal invagination. In *Malacobdella grossa*, *Emplectonema gracile*, *Oerstedia dorsalis*, *Gononemertes australiensis*, *Carcinonemertes epialti*, and *Paranemertes peregrina*, the proboscis rudiment detaches from the epidermis, while in *Drepanophorus spectabilis*, *Tetrastemma vermiculus*, *Prosorhochmus viviparus*, and *Pantinonemertes* (*Geonemertes*) *agricola*, the connection to the epidermis persists (Lebedinsky 1897; Coe 1904; Salensky 1914; Hammarsten 1918; Iwata 1960; Egan and Anderson 1979; Stricker and Reed 1981). In the former group of species (except *Paranemertes peregrina*), the rhynchodeum is formed by an independent invagination located more frontally, in some species underneath the apical pit, while in the latter group the rhynchodeum is differentiated from the anterior part of the epidermal proboscis rudiment invagination. The tripartite organization of the armed hoplonemertean proboscis becomes apparent by a strong stylet bulb with a narrow canal separating an anterior tube shaped from a posterior sacculate portion. In the area of the stylet bulb, the stylet armature is later formed (Bürger 1895; Coe 1904; Stricker and Reed 1981; Stricker and Cloney 1982; Stricker 1985; Senz and Tröstl 1999). While the mesodermal components of the proboscis apparatus are derived from the surrounding mesodermal cells in most accounts by delamination of a mass of mesodermal cells, Lebedinsky (1897) identified a dorsal and a ventral mesodermal strip independent of the somatic mesoderm that form from paired mesoblast cells at the junction of the proboscis invagination in *Drepanophorus spectabilis* and *Tetrastemma vermiculus*. The rudiments become hollow, associate with the proboscis invagination, grow around it, and fuse. The inner wall of the fused mesodermal rudiment forms the proboscis musculature and its inner epithelium, the outer rhynchocoel wall, and the associated muscle layers. The hollow space between the two-layered rudiment represents the rhynchocoel.

Alimentary Canal

The intestinal tract in nemerteans is a one-way through-gut. It comprises the midgut and its derivatives as well as a histologically different foregut. An extensive hindgut connecting the midgut with the anus is absent (Bürger 1895; Gibson 1972). The midgut and its derivatives develop from the embryonic endoderm. In species with gastrulation by invagination, the gut persists in the larva as a hollow cavity, while in species that gastrulate by polar ingression, the midgut rudiment is regularly a solid mass of cells that establish the gastric cavity later during development (Friedrich 1979). In paleonemerteans and hoplonemerteans, the future mouth is moved to the future ventral side by accelerated growth of the dorsal side of the body, while in canonic pilidiophoran development, it remains in its original position (Iwata 1957, 1960, 1985; Friedrich 1979; Maslakova 2010a, b). During and following gastrulation, parts of the ectoderm have regularly been reported to be dragged inside the gastrula with the blastopore to form the ectodermally derived esophagus (Friedrich 1979). In paleonemertean species and Pilidiophora that develop via a pilidium, the blastopore remains open, marking the connection between the esophagus and the blindly ending, sac-like midgut. All or most of the larval intestinal tract is taken over from the larval to the juvenile organization. According to Salensky (1912), the esophagus of the pilidium is divided into a distal and a proximal part by a constrictor muscle in later-stage larvae. During metamorphosis only the proximal part is taken over into the juvenile; the distal part is shed along with the pilidial envelope. In the Iwata larva of *Micrura akkeshiensis*, the blastopore remains open although no food is taken up. The definite mouth opening and esophagus develop from a rudiment composed of multilayered cells located in the dorsal wall of the stomodeum. By elongating in a slight curve, the rudiment becomes the esophagus of the juvenile. The stomodeal part distal of the esophageal rudiment is shed along with the larval epidermis during metamorphosis (Iwata 1958).

During intracapsular development of *Lineus ruber* and *Lineus viridis*, the esophagus development differs markedly. While in the former species the esophagus develops early and retains its functionality, the esophagus in the latter is reported to be separated from the midgut. A definite connection is formed late, i.e., after metamorphosis, by a secondary esophagus from two groups of cells located laterally near the mouth opening (Arnold 1898; Nusbaum and Oxner 1913; Schmidt 1964).

The development of the mouth opening and pharynx in hoplonemerteans is much more diverse and complicated. In general, the blastopore is said to be closed early in development. In some species (e.g., *Emplectonema gracile*, *Oerstedia dorsalis*), the blastopore reopens and a functioning larval gut is formed (Iwata 1960). In other species the larval gut is formed by a secondary invagination of the epidermis, forming the esophagus in a more anterior position than the blastopore (*Drepanophorus spectabilis*, *Tetrastemma vermiculus*, *Malacobdella grossa*, *Paranemertes peregrina*) (Lebedinsky 1897; Hammarsten 1918; Maslakova and von Döhren 2009). In *Tetrastemma vermiculus* and *Drepanophorus spectabilis*, the blastopore remains open even after the secondary esophagus has formed. It has been reported to close later so that the remaining pouch develops into the intestinal cecum present in most hoplonemerteans (Lebedinsky 1897). While the mouth and proboscis pore are separate in most polystiliferan hoplonemerteans, in monostiliferan hoplonemerteans the larval mouth subsequently closes, and the esophagus gains connection with the rhynchodeum forming the rhynchostomodeum that is typical of this clade. In *Prosorhochmus viviparus*, *Prostoma graecence*, *Argonemertes* (*Geonemertes*) *australiensis*, and *Gononemertes australiensis*, neither a reopening nor a secondary esophagus forms, resulting in a midgut that is closed for most of the time of the development (Salensky 1909, 1914; Hickman 1963; Egan and Anderson 1979). A functioning intestinal opening is accomplished by fusion of the midgut with the ventral part of the rhynchodeum.

The fusion of the esophagus with the rhynchodeum is accomplished in various ways. While in *Emplectonema gracile*, *Oerstedia dorsalis*, and *Gononemertes australiensis* the esophagus and the proboscis rudiments gain access to the independently invaginated rhynchodeum, the esophagus fuses to the distal part of the proboscis rudiment that later becomes the rhynchodeum in *Prosorhochmus viviparus* (Salensky 1909, 1914; Iwata 1960; Egan and Anderson 1979). In *Tetrastemma vermiculus*, *Drepanophorus spectabilis*, and *Pantinonemertes* (*Geonemertes*) *agricola*, the anterior-most part of the proboscis rudiment forms a ventral invagination that fuses with the intestine developing into the adult esophagus (Lebedinsky 1897; Coe 1904). In *Paranemertes peregrina*, the proboscis rudiment gains its connection prior to the closure of the larval mouth opening (Maslakova and von Döhren 2009). In the aberrant commensal *Malacobdella grossa*, there is no true rhynchodeum. The proboscis rudiment connects to the esophagus and the larval mouth is closed. A secondary mouth opening is formed anterior of the larval mouth (Hammarsten 1918). Apart from the disparate development in *Malacobdella grossa*, there is no conceivable reason for the diversity shown in esophagus development within the otherwise relatively uniform clade of Hoplonemertea. It is therefore very likely that the diversity shown in the development of the esophagus and rhynchostomodeum depends rather on the different researchers than on diverging lines of development (Friedrich 1979).

In some hoplonemerteans, an anus is formed early in development by means of a caudal epidermal invagination or ingression (Lebedinsky 1897; Hammarsten 1918; Iwata 1960; Friedrich 1979; Maslakova 2010b). Apparently, in the majority of hoplonemerteans, as well as in Pilidiophora and paleonemertean species studied, the anus is formed much later as the formation of an anal opening has not been reported in these. Initially, the intestinal tract is an undifferentiated tube. The morphological differentiation of the alimentary canal in hoplonemerteans comprising the ectodermal esophagus, stomach, and pyloric tube as well as the mesodermal intestinal ceca and lateral diverticula forms comparatively late (Friedrich 1979; Senz and Tröstl 1999).

Nephridia

Excretory organs have been observed to develop early in paleonemertean larvae of *Carinoma mutabilis* and *Procephalothrix oestrymnicus* as simple protonephridia anterior of the mouth opening (Bartolomaeus et al. 2014). They consist of few (two to three) multiciliated terminal cells constituting the site of ultrafiltration and two to three multiciliated cells that form the nephroduct which modifies the ultrafiltrate. Distally, the nephroduct opens through the epidermis at the level of the mouth opening via a nephropore cell in both species (Bartolomaeus et al. 2014). The protonephridia develop from a subepidermal rudiment underneath the trochoblast cells and are fully formed prior to the degeneration of the so-called vestigial prototroch (Bartolomaeus et al. 2014). In hoplonemertean species protonephridia have been observed in larvae but in a more posterior position behind the mouth opening. Nothing is known about the ultrastructure of the protonephridia in hoplonemertean larvae. Structure, position, and time of development of the protonephridia in paleonemerteans are reminiscent of the "head kidneys" of the trochophore larva, although it is not clear whether the nemertean protonephridia share the same fate as transitory organs that are restricted to the larval organization.

The protonephridia of hoplonemertean larvae correspond in position to the respective organs in the adults. Therefore, it is very probable that the excretory organs in hoplonemerteans represent early developmental stages of the adult organs being elaborated during development to attain the adult morphology. In Pilidiophora, branched protonephridia located slightly anterior of the midgut have been observed as early as 2 weeks after fertilization (von Dassow and Maslakova 2013). In later stages the protonephridia become sandwiched between the intestinal tract and the distally developing juvenile rudiment (Maslakova 2010a). In *Lineus viridis* the first protonephridia were observed in the juvenile after metamorphosis. At this time no trace of the blood-vascular system can be observed. The protonephridia are branched structures with a single terminal cell on the proximal end of each branch. Initially monociliated, the terminal cell becomes soon multiciliated. The nephroduct is intercellular, and the nephropore is formed by four specialized epidermal cells (Bartolomaeus 1985). Since the first protonephridia formed correspond in structure and position to those seen in the adults, it can be assumed that protonephridia in Pilidiophora are definite persisting adult organs (Bartolomaeus et al. 2014). Although it has been shown that contrary to classical accounts the protonephridia in the pilidium are not derived from bilateral invaginations of the esophagus, their origin as invaginations of the subumbrellar epidermis could not be substantiated either (Hubrecht 1885, 1886; Arnold 1898; Salensky 1912; Maslakova 2010a).

In the Iwata larva of *Micrura akkeshiensis*, the origin of the protonephridia has been identified as a pair of cell groups located in the ventral wall of the stomodeum on the level of the juvenile mouth opening (Iwata 1958). Their further development, however, has not been followed. While the abovementioned data hint at an ectodermal origin of the protonephridia, a mesodermal origin has been hypothesized for the protonephridia in the Desor larva. The rudiment is represented by a cluster of cells which later forms narrow tubular structures situated between the esophagus, the cerebral organs, and the ventral rhynchocoel wall (Nusbaum and Oxner 1913). Further development and opening of the nephridial rudiments to the exterior have not been observed.

Sensory Organs

Sensory organs comprise cerebral sense organs present in many paleonemertean, nearly all pilidiophoran and most hoplonemertean species; frontal organs and eyes, both of which are present in the majority of Neonemertea; and lateral organs that are confined to some tubulanid species (Gibson 1972). Cerebral organs differ in their complexity and position relative to the brain in different taxa. Cerebral organs comprise a ciliated, blind ending canal, running from the epidermis proximally to end in a mass of neuronal tissue that is connected via a nerve to the brain. While they are comparably small, simple canals situated in front of the brain in Monostilifera or behind the brain in *Tubulanus* and *Carinina* species, cerebral organs are large, compound organs comprising ciliated cells,

gland cells, and neurons located behind the brain in Pilidiophora and reptant Polystilifera. In the heteronemertean pilidium and the Desor larva, the cerebral organs develop from the cerebral organ disks, a paired rudiment invaginated from the subumbrellar larval ectoderm. They are situated bilaterally between the frontal cephalic disks and the posterior trunk disks and are the last of the invaginated rudiments to form in the respective larvae. The cerebral organ disks first fuse posteriorly with the trunk disks and subsequently with the already fused head rudiment. The invagination of the original disks is retained to later form the cerebral organ canal, while gland cells and neurons form in the proximal portion of the rudiment. The neurons connect to the developing dorsal brain lobes already before the juvenile body has completely formed (Salensky 1912; Nusbaum and Oxner 1913; Maslakova 2010a; Hindinger et al. 2013). Although nothing is known about the formation of cerebral organs in non-heteronemertean Pilidiophora such as *Hubrechtella* species, it can be assumed that the process is similar as described in the heteronemertean pilidium judging from the corresponding larval type and the morphological similarity of the cerebral organs in the respective taxa. In the Iwata larva, the cerebral organs are not derived from paired epidermal larval invaginations but from paired lateral invaginations of the stomodeum. The connection to the stomodeum is soon obliterated, and the cerebral organs open secondarily into the cephalic furrow formed in the cephalic rudiment. A connection to the dorsal brain lobes is accomplished prior to metamorphosis (Iwata 1958). Cerebral organs in larvae of hoplonemertean species have been reported to develop from a pair of narrow invaginations of thickened epidermal areas that can be observed early in development during the larval phase (Lebedinsky 1897; Iwata 1960; Maslakova and von Döhren 2009). In *Paranemertes peregrina* and *Emplectonema gracile*, they are situated bilaterally on both sides of the larva on the level of the mouth opening, while in *Tetrastemma vermiculus* and *Drepanophorus spectabilis*, the epidermal invaginations have been reported to be located between the dorsal and the ventral brain lobes (Lebedinsky 1897; Iwata 1960; Maslakova and von Döhren 2009). Epidermal invaginations that have initially been interpreted as cerebral organ rudiments in some hoplonemertean species have later been interpreted as invaginations that were hypothesized to be homologous to the imaginal disks of Pilidiophora (Maslakova 2010b). In the direct developing hoplonemertean *Prosorhochmus adriaticus*, the internal nervous and glandular portion of the cerebral organ forms first, while its connection to the exterior via the cerebral organ canal and pore opening in the cephalic furrows is established later in development (Senz and Tröstl 1999). In *Malacobdella grossa* larval cerebral organ rudiments have been suspected although the adult does not possess cerebral organs. The cerebral organ rudiments comprise paired epidermal cell clusters that are situated a little posterior of the mouth opening. After being invaginated they give off some neuronal cells to contribute to the ventral brain lobes. The remainder of the rudiment disappears shortly after (Hammarsten 1918). Although cerebral organs exist in some paleonemertean species, neither their formation nor any rudiments have ever been observed (Iwata 1960). While this appears logical in species lacking cerebral organs, it is somewhat astonishing judging from the fact that cerebral organ rudiments are formed early in other nemertean species. Moreover, recent phylogenetic analyses suggest cerebral organs to be an ancestral character in Nemertea that was reduced repeatedly (Thollesson and Norenburg 2003; Andrade et al. 2012). It would therefore be conceivable to interpret the bilateral apical invaginations that had been attributed as rudiment of the brain in paleonemertean species as a joint rudiment giving rise to not only the brain and lateral medullary cords but also to the cerebral organs. In species that do not possess cerebral organs as adults, e.g., *Procephalothrix simulus*, they are reduced later in development.

In Hoplonemertea and some paleonemertean species, the eyes are formed during the larval phase (Fig. 8.6A–C; Iwata 1960; Stricker and Reed 1981; Martindale and Henry 1995; Norenburg and Stricker 2002; Maslakova et al.

2004b; Chernyshev 2008; Maslakova and von Döhren 2009). But while the eyes in the larva are transitory in most paleonemertean species, they persist and become more in number in Hoplonemertea (Fig. 8.4C). In Heteronemertea pigmented eyes are confined to the juvenile rudiment; no larval type of Pilidiophora is reported to have eyes (Fig. 8.6D; Cantell 1969; Norenburg and Stricker 2002; von Döhren and Bartolomaeus 2007). The eyes in neonemertean species are subepidermal, rhabdomeric eyes (Jespersen and Lützen 1988; von Döhren and Bartolomaeus 2007; von Döhren 2008). In *Lineus viridis* the eyes develop from an unpigmented, subepidermal rudiment comprising a small number of cells of two types. A bundle of rhabdomeric receptor cells is surrounded by undifferentiated, unpigmented corneal progenitor cells. From the latter type of cells, the closed optical cavity is formed that houses the receptor cells. Pigmented cells that form a pigment cup to one side are formed as the eye begins to function. The eye spot enlarges as the number of all cell types increases (von Döhren and Bartolomaeus 2007). Contrary to the statement that the eyes in Heteronemertea form by fragmentation of existing eyes, additional eyes in *Lineus viridis* juveniles are formed de novo in the same manner as described above (Gontcharoff 1960; von Döhren and Bartolomaeus 2007).

The frontal sensory organ is typical of Neonemertea being represented by a single protrusible epidermal pit in most species, while comprising three triangularly arranged protrusible pits in lineid heteronemertean species (Gibson 1972). The frontal organ is commonly associated with the head glands that discharge their glandular products through canals opening between the epithelial cells of the frontal organ. In some species the head glands open independently of the frontal organ by numerous smaller ducts to the epidermis (Gibson 1972). According to Lebedinsky (1897), the frontal organ develops from the apical pit of the larva in *Drepanophorus spectabilis* and *Tetrastemma vermiculus*, although this has been doubted by other authors (Bürger 1897–1907; Hammarsten 1918). In *Malacobdella grossa* the apical pit gives rise only to the cephalic glands (Hammarsten 1918).

Various sensory structures are restricted to certain lineages, such as tactile cirri and statocysts in *Ototyphlonemertes* species; lateral sensory organs in some *Tubulanus*, *Callinera*, and *Micrella* species; or so-called integumentary and subcutaneous organs of pelagic Polystilifera (Coe 1927; Gibson 1972, 1982). There are no data available about the development of these organs.

Blood-Vascular System

The development of the blood-vascular system occurs comparably late in Nemertea after the majority of musculature has already formed. While, according to classical accounts, the blood vessels in Pilidiophora precede metamorphosis, no trace of blood vessels could be found in the postmetamorphic juvenile of *Lineus viridis* according to more recent data (Bürger 1897–1907; Nusbaum and Oxner 1913; Bartolomaeus 1985). There is no account on formation of the blood vessels in paleonemertean species, while in Hoplonemertea data are only available for species with intracapsular development (Salensky 1914; Reinhardt 1941; Turbeville 1986). In classical accounts, the blood vessels have either been derived from gaps remaining or been reopened in the blastocoel or from postulated embryonic coelems (e.g., Bürger 1897–1907; Salensky 1914; Nusbaum and Oxner 1913). The problem in these classical accounts is that the blastocoel was considered to be devoid of matrix or that the matrix within it was later liquefied. This led to the misinterpretation that nemerteans possess extensive primary and/or secondary body cavities. Instead, nemerteans should be considered as largely compact with the exception of the rhynchocoel and the blood vessels that represent secondary body cavities (Gibson 1972; Turbeville and Ruppert 1985; Turbeville 1991). More recent data on two hoplonemertean species suggest that the blood vessels are formed in the parenchymatous tissue underlying the body wall musculature by forming solid bands of cells (Reinhardt 1941; Turbeville 1986). According to light microscopic data in *Prostoma graecense*, the bands of radially arranged cells become hollow

by resorbing encased yolk material (Reinhardt 1941). Ultrastructural data on *Prosorhochmus americanus*, however, provide a different picture (Turbeville 1986). The solid bands of the blood vessel rudiments become hollow by a process that is reminiscent of the schizocoelous mode of formation observed in annelid coelomic cavities. In contrast to annelid schizocoely, the cells of the nemertean blood vessels acquire intercellular junctions after the onset of cavitation, whereas the respective intercellular junctions in annelids are formed at the onset of cavitation (Turbeville 1986; Koch et al. 2014). Due to these differences, the question of homology of these laterally located endothelialized, hollow compartments to the secondary body cavities found in Trochozoa can be at best preliminarily answered; more comparative data on the development of the blood vessels in other nemertean taxa are needed to consolidate this hypothesis.

In summary, the development of Nemertea is generally working on the regular spiralian scaffold; although in organs that represent interfaces of two germ layers (e.g., proboscis, foregut, nephridia), there is considerable disagreement over their respective origin and formation. Until more comparative data are collected, concise statements regarding the evolution of organ system development in Nemertea cannot be made.

GENE EXPRESSION

Gene expression studies in development of Nemertea are presently scarce and restricted to the pilidiophoran species *Cerebratulus lacteus*, *Lineus viridis*, *Micrura alaskensis*, and *Ramphogordius* (*Lineus*) *sanguineus*. The data available refer to the expression of *β-catenin*, a trochoblast-specific tubulin-4 gene from the basal gastropod *Patella vulgata*, homeobox-containing genes (Hox genes, as well as *Otx,* and *Cdx*), genes encoding photopigments (*opsins*), and genes of the so-called retinal determination gene network (RDGN) (*Pax-6, Six1/2, Six3/6, Six4/5, Dach*) (Loosli et al. 1996; van den Biggelaar et al. 1997; Klerkx 2001; Charpignon 2007; Henry et al. 2008; Döring 2012; Hiebert and Maslakova 2015).

Axis Specification Genes

A transcriptomic assessment of the pilid-
iophoran *Micrura alaskensis* revealed a single
Hox cluster comprising nine Hox genes while
only six to seven Hox genes have been identi-
fied by genomic screening of the pilidiophoran
Ramphogordius (Lineus) sanguineus (Kmita-
Cunisse et al. 1998; Hiebert and Maslakova
2015). In *Micrura alaskensis (Ramphogordius
(Lineus) sanguineus)*, the genes of the Hox
gene cluster have been identified as *MaLab
(LsHox1), MaPb, MaHox3 (LsHox3), MaDfd
(LsHox4), MaScr, MaLox5 (LsHox6), MaAntp
(LsHox7), MaLox4*, and *MaPost2 (LsHox9)*. In
Ramphogordius (Lineus) sanguineus a seventh
putative Hox gene, *LsHox8* that is possibly dislocated from the chromosomal Hox gene region
was found using an alternative set of primers
(Kmita-Cunisse et al. 1998). Data on the expression of Hox genes during development are only
available for *Micrura alaskensis* (Hiebert and
Maslakova 2015). Hox gene expression largely
complies both temporally and spatially to the
canonical bilaterian fashion, albeit with noteworthy exceptions: All pilidial tissues are completely devoid of Hox genes expression during
all stages of development. Hox gene expression
in *Micrura alaskensis* does not start before the
trunk disks are formed. It then proceeds posteriorly. Thus the anterior part of the juvenile comprising the cephalic disks and the cerebral organ
disks as well as the later developing head rudiment and the proboscis develop without showing
Hox gene expression. The absence of Hox gene
expression in the pilidial tissues has been interpreted as an indication of a functional decoupling of the early, larval from the later juvenile
development phases (Hiebert and Maslakova
2015). While giving a possible explanation of
the diversity of pilidial morphotypes the genetic
mechanism for the patterning of the pilidial
envelope remains unclear. Preliminary results of
inhibition studies hint at an involvement of the
Wnt and the *fibroblast growth factor* pathways
in patterning the early larval shape (Hiebert
and Maslakova 2015). Recent findings indicate
that *β-catenin*, a downstream component of the

Wnt pathway, is both necessary and sufficient
to promote endoderm fates in the vegetal cells
during cleavage of *Cerebratulus lacteus* (Henry
et al. 2008). *β-catenin* is expressed only in the
vegetal-most quartet at each cleavage division
and is passed on through the cleavage cycles to
finally end up in the blastomeres forming the
endoderm. Experimental overexpression leads
to vegetalized embryos that fail to form typical,
anteriorly positioned apical tufts, while block-
ing of *β-catenin* leads to animalized larvae with
supernumerary apical tufts. Animalized larvae
gastrulate only to a limited extent. It has been
hypothesized that the animal fates of blastomeres
represent the default condition which is shifted
to vegetal fates under the influence of *β-catenin*
(Henry et al. 2008). A trochoblast-specific tubulin gene (*tub4*) was assessed for its expression
in the cells that form the marginal band of elongated cilia on the lappets around the pilidium of
Cerebratulus lacteus (van den Biggelaar et al.
1997; Klerkx 2001). However, no expression
signal was recorded in cells bearing elongated
cilia, underpinning the hypothesis that the marginal band of cells bearing elongated cilia is
not homologous to the prototroch of the trochophore (van den Biggelaar et al. 1997; Maslakova
2010a). An Otx-class and a Cdx-class gene were
identified in *Ramphogordius (Lineus) sanguineus (Ls-Otx, Ls-Cdx)* (Kmita-Cunisse et al. 1998;
Charpignon 2007). A *Cdx* ortholog (*MaCdx*) has
also been found in *Micrura alaskensis* (Hiebert
and Maslakova 2015). *Ls-Otx* was found to be
expressed during the development of postmetamorphic juvenile *Lineus viridis* (Fig. 8.9A;
Charpignon 2007). First, it is expressed in the
entire brain ring and the anterior-most part of the
lateral medullary cords, the cerebral organs and
their canals, and additionally, but weaker, in the
gut. In older stages there is an *Ls-Otx* expression in the frontal organ and in the frontal organ
nerves. Expression in the brain and cerebral
organs becomes weaker, although in the cerebral
organ canals, *Ls-Otx* is still clearly expressed. At
this stage of development, no expression signal
is detectable in the lateral medullary cords or
the gut (Charpignon 2007). *Ls-Cdx* is strongly
expressed in *Lineus viridis* juveniles consistently

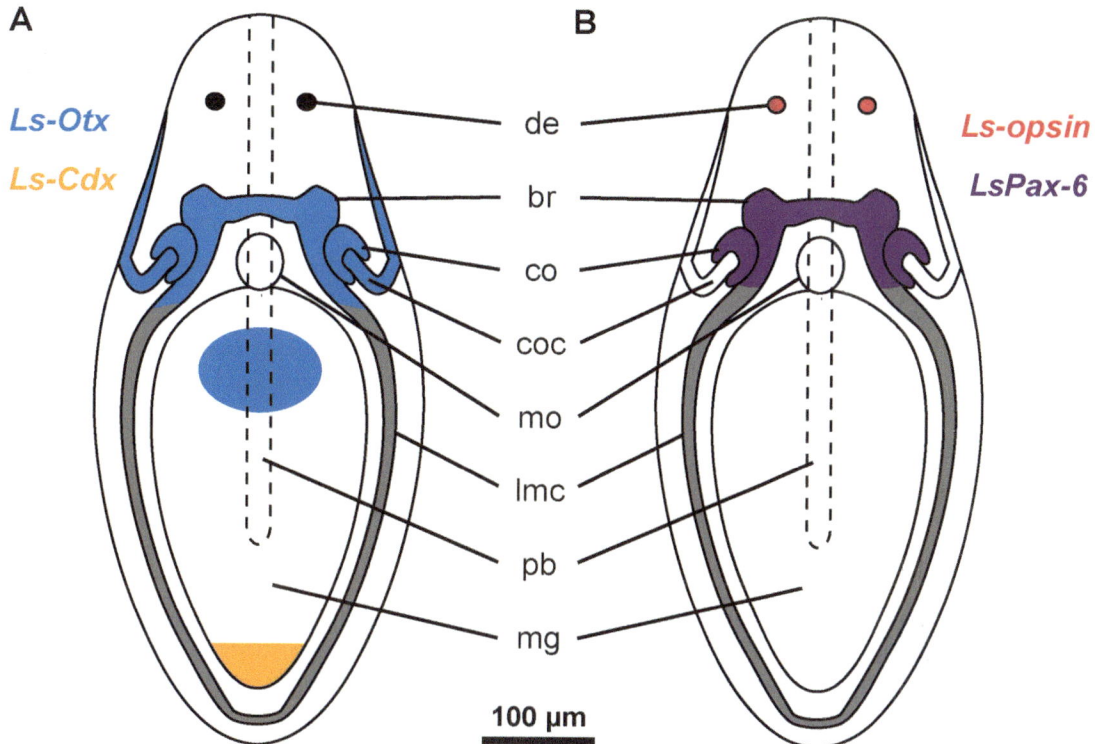

Fig. 8.9 Schematic representation of gene expression of several genes from *Ramphogordius* (*Lineus*) *sanguineus* in postmetamorphic juveniles of *Lineus viridis*. (**A**) Expression of *Ls-Otx* and *Ls-Cdx*. (**B**) Expression of *Ls-opsin* and *LsPax-6*. Note: *LsPax-6* is not expressed in the developing eyes of *Lineus viridis*. *br* brain ring, *co* cerebral organ, *coc* cerebral organ canal, *de* developing eye, *lmc* lateral medullary cord, *mg* midgut, *mo* mouth opening, *pb* proboscis rudiment (© Dr. J. von Döhren, All Rights Reserved)

throughout development in the posterior-most part of the body. The signal is located in internal tissues, arguably in the developing intestine (Fig. 8.9A; Charpignon 2007). In *Micrura alaskensis MaCdx* is first expressed in the trunk disk being shifted posteriorly during the course of development to be finally located as a ring shaped pattern around the base of the developing caudal cirrus. In this species however, expression of *MaCdx* in the intestinal tissue is absent (Hiebert and Maskakova 2015).

Retinal Determination Gene Network Genes

The most extensive expression data have been gathered on the eye development in *Lineus viridis* (Loosli et al. 1996; Charpignon 2007; Döring 2012). One opsin gene has been isolated from *Ramphogordius* (*Lineus*) *sanguineus*, while two opsin genes have been found in *Lineus viridis* (Charpignon 2007; Döring 2012). The former, *Ls-opsin*, represents an unusual G-protein-coupled-receptor (GPCR) opsin that does not cluster with canonical rhabdomeric-type opsins. Its expression in *Lineus viridis* is restricted to the developing eyes, while expression in adult eyes is absent (Fig. 8.9B; Charpignon 2007). The opsins identified in *Lineus viridis* (*LiVi-ops1*, *LiVi-ops2*) cluster with peropsins and photoisomerases (*LiVi-ops1*) or in a basal position to these together with *opsin5* (*LiVi-ops2*) (Döring 2012). Although a G-protein alpha subunit Q gene (*LiVi-Gq*) was also found in the same species, expression experiments for any of the three genes were unsuccessful (Döring 2012). Genes marking the canonical neurotransmitters employed by photoreceptor cells of *Platynereis dumerilii*, the vesicular acetylcholine receptor gene in larval eyes, and the vesicular glutamate receptor gene in adult eyes have both been extracted from juveniles of *Lineus viridis* (*LiVi-vacht* and *LiVi-vglut*, respectively). Expression data on these genes in *Lineus viridis* have not been reported (Döring 2012).

Expression of genes of the RDGN during development was studied in *Lineus viridis* and includes the genes *LsPax-6* and *Lv-Six1/2*, *Lv-Six3/6*, *Lv-Six4/5*, and *Dach*, although for the latter an antibody reaction was studied instead of in situ hybridization (Loosli et al. 1996; Charpignon 2007). In *Micrura alaskensis* a *Six3/6* ortholog (*MaSix3/6*) was found to be expressed during early larval development (Hiebert and Maslakova 2015). *LsPax-6* is expressed in *Lineus viridis* during postmetamorphic development in the brain and the cerebral organs, but not in the eye region (Fig. 8.9B; Loosli et al. 1996; Charpignon 2007). Due to differential splicing and a thus changed open reading frame, two isoforms have been hypothesized to exist. One of them is shorter, missing the PST domain. The possible role of this shorter isoform has been suspected to be differential regulation of developmental processes. However, there is no proof of the existence of the theoretical isoform (Charpignon 2007). Of the remaining genes assessed, only *Lv-Six1/2* and the monoclonal *Drosophila melanogaster* antibody against *Dach* show an expression of these genes in the developing eyes of *Lineus viridis*. The former gene is also expressed in the lateral medullary cords near the brain and in the frontal organs. *Lv-Six3/6* also shows expression in the frontal organ, but its strongest expression is observed in the brain lobes. Expression of *MaSix3/6* is restricted to early developmental stages of *Micrura alaskensis*. In the blastosquare stage several cells on one pole of the embryo are labelled while in the feeding pilidium stage some expression signals are detectable near the apical pit along with few additional signals situated in the anterior region of the developing lateral lappets (Hiebert and Maslakova 2015). In *Lineus viridis Lv-Six4/5* is expressed in the posterior region of the developing brain and the cerebral organs but also in two bilateral stripes running from the brain lobes anteriorly. A correspondence with the nerves connecting the eyes with the brain seems likely (Charpignon 2007).

Other Genes

Genes that have been identified in *Ramphogordius* (*Lineus*) *sanguineus* but have not been subjected to expression studies include *Ls-Bmp2/4*, *Ls-Engrailed*, *Ls-Msx*, *LsNK*, *LsPax-2/5/8*, *Ls-Snail*, and *Ls-Twist* (Charpignon 2007). Gene expression of the canonical developmental pathways is very limited both with respect to the gene products involved and to the diversity of developmental trajectories found in Nemertea. Currently, a sound assessment of gene expression networks in Nemertea is impossible.

OPEN QUESTIONS

- How does the high yolk content in Pelagica oocytes influence embryonic cleavage?
- How are the dorsal quadrant and the mesoderm precursor cell specified in Paleonemertea and Hoplonemertea?
- Is there an apical neuronal structure that is homologous to the apical organ of Trochozoa in larvae of Nemertea?
- How did the larval and adult eyes in Nemertea evolve?
- What is the developmental origin of nemertean protonephridia?
- What is the developmental fate of the first formed protonephridia – transitory larval or definite adult organs?
- Where does the adult mesoderm originate from – entirely endomesodermal or with ectodermal components?
- Is there a taxon-specific mode of muscle formation in Nemertea?
- How do the different components of the proboscis and rhynchocoel develop?
- Is there a generalizable mode of foregut formation in Hoplonemertea?

- How does the blood-vascular system develop in Paleonemertea and Pilidiophora?
- How are key developmental regulators such as Hox and ParaHox genes expressed in the various nemertean subclades?

References

Andrade S, Strand M, Schwartz M, Chen H, Kajihara H, von Döhren J, Sun SC, Junoy J, Thiel M, Norenburg JL, Turbeville JM, Giribet G, Sundberg P (2012) Disentangling ribbon worm relationships: multi-locus analysis supports traditional classification of the phylum Nemertea. Cladistics 28(2):141–159

Andrade SC, Montenegro H, Strand M, Schwartz ML, Kajihara H, Norenburg JL, Turbeville JM, Sundberg P, Giribet G (2014) A transcriptomic approach to ribbon worm systematics (Nemertea): resolving the Pilidiophora problem. Mol Biol Evol 31(12):3206–3215

Arnold G (1898) Zur Entwicklungsgeschichte des *Lineus gesserensis* OF Müller. Trav Soc Imp Nat St Petersbourg Sect Zool Physiol 28:1–30

Barrois JH (1877) Mémoire sur l'embryologie des némertes. Ann Sci Nat Zool 6(11):1–232

Bartolomaeus T (1984) Zur Fortpflanzungsbiologie von *Lineus viridis* (Nemertini). Helgoländer Meeresuntersuchungen 38(1):185–188

Bartolomaeus T (1985) Ultrastructure and development of the protonephridia of *Lineus viridis* (Nemertini). Microfauna Marina 2(6):261–283

Bartolomaeus T, von Döhren J (2010) Comparative morphology and evolution of the nephridia in Nemertea. J Nat Hist 44(37–40):2255–2286

Bartolomaeus T, Maslakova S, von Döhren J (2014) Protonephridia in the larvae of the paleonemertean species *Carinoma mutabilis* (Carinomidae, Nemertea) and *Cephalothrix* (*Procephalothrix*) *filiformis* (Cephalothricidae, Nemertea). Zoomorphology 133(1):43–57

Berg G (1972) Studies on *Nipponnemertes* Friedrich, 1968 (Nemertini, Hoplonemertini). Zool Scr 1(4):211–225

Boyer BC, Henry JQ (1998) Evolutionary modifications of the spiralian developmental program. Am Zool 38(4):621–633

Boyer BC, Henry JQ, Martindale MQ (1996) Dual origins of mesoderm in a basal spiralian: cell lineage analyses in the polyclad turbellarian Hoploplana inquilina. Dev Biol 179(2):329–338

Bürger O (1894) Studien zu einer Revision der Entwicklungsgeschichte der Nemertinen. Ber Nat Ges Freiburg i Br 8:111–141

Bürger O (1895) Die Nemertinen des Golfes von Neapel und der angrenzenden Meeres-Abschnitte. R. Friedländer & Sohn, Berlin

Bürger O (1897–1907) Nemertini. In: Bronn EG (ed), Bronn's Klassen und Ordnungen des Tierreichs, Bd, 4, CF Winter'sche Verlagshandlung, Leipzig, pp 1–151

Cantell CE (1967) Devouring of larval tissues during metamorphosis of pilidium larvae (Nemertini). Arkiv för Zoologi 18(5):489

Cantell CE (1969) Morphology, development and biology of the pilidium larvae (Nemertini) from the Swedish West Coast. Zool Bidr Uppsala 38:61–111

Cantell CE, Franzén Å, Sensenbaugh T (1982) Ultrastructure of multiciliated collar cells in the pilidium larva of *Lineus bilineatus* (Nemertini). Zoomorphology 101(1):1–15

Charpignon V (2007) Homeobox-containing genes in the nemertean *Lineus*: key players in the antero-posterior body patterning and in the specification of the visual structures. PhD Thesis, University of Basel

Chernyshev AV (2008) Larval development of nemerteans of the genus *Quasitetrastemma* (Nemertea: Monostilifera). Russ J Mar Biol 34(4):258–262

Chernyshev AV, Magarlamov TY (2010) The first data on the nervous system of hoplonemertean larvae (Nemertea, Hoplonemertea). Dokl Biol Sci 430(1):48–50

Chernyshev AV, Astakhova AA, Dautov SS, Yushin VV (2013) The morphology of the apical organ and adjacent epithelium of pilidium prorecurvatum, a pelagic larva of an unknown heteronemertean (Nemertea). Russ J Mar Biol 39(2):116–124

Coe WR (1899) On the development of the pilidium of certain nemerteans. Trans Connecticut Acad 10:235–262

Coe WR (1904) The anatomy and development of the terrestrial nemertean (*Geonemertes agricola*) of Bermuda. Proc Boston Soc Nat Hist 31(10):531–571

Coe WR (1926) The pelagic nemerteans. Mem Mus Comp Zool Harv Coll 49:1–244

Coe WR (1927) The nervous system of pelagic nemerteans. Biol Bull 53(2):123–138

Coe WR (1943) Biology of the nemerteans of the Atlantic coast of North America. Trans Connecticut Acad 35:129–328

Dawydoff C (1928) Sur l'embryologie des Protonémertes. CR Hebd Séances Acad Sci Paris 186:531–533

Dawydoff C (1940) Les formes larvaires de polyclades et de némertes du plancton indochinois. Bull Biol Fr Belg 74:443–496

Delsman HC (1915) Eifurchung und Gastrulation bei *Emplectonema gracile* Stimpson. Tijdschr Ned Dierk Vereen 14:68–114

Döring C (2012) Tracing metazoan phylogeny through the analysis of light sensitive organs. PhD Thesis, University of Osnabrück

Egan EA, Anderson DT (1979) The reproduction of the entozoic nemertean *Gononemertes australiensis* Gibson (Nemertea: Hoplonemertea: Monostylifera) – gonads, gametes, embryonic development and larval development. Mar Freshw Res 30(5):661–681

Freeman G (1978) The role of asters in the localization of the factors that specify the apical tuft and the gut of the nemertine *Cerebratulus lacteus*. J Exp Zool 206(1):81–107

Friedrich H (1979) Nemertini. In: Seidel F (ed) Morphogenese der Tiere, vol 3, D5-I. Gustav Fischer, Verlag, Jena

Gibson R (1972) Nemerteans. Hutchinson Univ Library, London

Gibson R (1982) British Nemerteans: keys and notes for the identification of the species. In: Kermack DM, Barnes RSK (eds) Synopses of the British Fauna (new series), vol 24. Cambridge University Press, Cambridge, UK

Goldstein B, Freeman G (1997) Axis specification in animal development. Bioessays 19(2):105–116

Gontcharoff M (1951) Biologie de la régénération et de la reproduction chez quelques Lineidae de France. Ann Sci Nat Zool 11(13):149–235

Gontcharoff M (1960) Le développement post-embryonnaire et la croissance chez Lineus ruber et Lineus viridis (Némertes Lineidae). Ann Sci Nat Zool 12(2):225–279

Hammarsten OD (1918) Beitrag zur Embryonalentwicklung der Malacobdella grossa (Müll). PhD Thesis, University of Stockholm

Hay-Schmidt A (1990) Catecholamine-containing, serotonin-like and neuropeptide FMRFamide-like immunoreactive cells and processes in the nervous system of the pilidium larva (Nemertini). Zoomorphology 109(5):231–244

Henry JJ (2002) Conserved mechanism of dorsoventral axis determination in equal-cleaving spiralians. Dev Biol 248(2):343–355

Henry JQ, Martindale MQ (1994) Establishment of the dorsoventral axis in nemertean embryos: evolutionary considerations of spiralian development. Dev Genet 15(1):64–78

Henry JQ, Martindale MQ (1995) The experimental alteration of cell lineages in the nemertean Cerebratulus lacteus: implications for the precocious establishment of embryonic axial properties. Biol Bull 189(2):192–193

Henry JQ, Martindale MQ (1996a) Establishment of embryonic axial properties in the nemertean, Cerebratulus lacteus. Dev Biol 180(2):713–721

Henry JQ, Martindale MQ (1996b) The origins of mesoderm in the equal-cleaving nemertean worm Cerebratulus lacteus. Biol Bull 191(2):286–288

Henry JJ, Martindale MQ (1997) Nemerteans, the ribbon worms. In: Gilbert SF, Raunio AM (eds) Embryology. Constructing the organism. Sinauer, New York, pp 151–166

Henry JJ, Martindale MQ (1998a) Conservation of the spiralian developmental program: cell lineage of the Nemertean, Cerebratulus lacteus. Dev Biol 201(2): 253–269

Henry JQ, Martindale MQ (1998b) Evolution of cleavage programs in relationship to axial specification and body plan evolution. Biol Bull 195:363–366

Henry JJ, Martindale MQ (1999) Conservation and innovation in spiralian development. Hydrobiologia 402: 255–265

Henry JQ, Perry KJ, Wever J, Seaver E, Martindale MQ (2008) β-Catenin is required for the establishment of vegetal embryonic fates in the nemertean Cerebratulus lacteus. Dev Biol 317(1):368–379

Hickman VV (1963) The occurrence in Tasmania of the land nemertine Geonemertes australiensis Dendy. In:

papers and proceedings of the Royal Society of Tasmania. 97:63–76

Hiebert LS, Gavelis G, von Dassow G, Maslakova SA (2010) Five invaginations and shedding of the larval epidermis during development of the hoplonemertean Pantinonemertes californiensis (Nemertea: Hoplonemertea). J Nat Hist 44(37–40):2331–2347

Hiebert LS, Maslakova SA (2015) Hox genes pattern the anterior-posterior axis of the juvenile but not the larva in a maximally indirect developing invertebrate, Micrura alaskensis (Nemertea). BMC Biol 13:23

Hiebert TC, von Dassow G, Hiebert LS, Maslakova SA (2013a) Long-standing larval mystery solved: pilidium recurvatum is the larva of Riserius sp. a basal heteronemertean (Heteronemertea; Pilidiophora; Nemertea). Integr Comp Biol 53(suppl1):e92

Hiebert TC, Dassow G, Hiebert LS, Maslakova S (2013b) The peculiar nemertean larva pilidium recurvatum belongs to Riserius sp. a basal heteronemertean that eats Carcinonemertes errans, a hoplonemertean parasite of Dungeness crab. Invertebr Biol 132(3): 207–225

Hindinger S, Schwaha T, Wanninger A (2013) Immunocytochemical studies reveal novel neural structures in nemertean pilidium larvae and provide evidence for incorporation of larval components into the juvenile nervous system. Front Zool 10(1):31

Hoffman CK (1877) Beitrage zur Kenntnis der Nemertinen. I. Zur Entwicklungsgeschichte von Tetrastemma varicolor Oerst. Niederl Arch Zool 3(3):205–215

Hörstadius S (1937) Experiments on determination in the early development of Cerebratulus lacteus. Biol Bull 73(2):317–342

Hubrecht AAW (1885) Zur Embryologie der Nemertinen. Zool Anz 8:470–472

Hubrecht AAW (1886) Contributions to the embryology of the Nemertea. Q J Microsc Sci 26:417–448

Iwata F (1957) On the early development of the nemertine, Lineus torquatus Coe. J Fac Sci Hokkaido Univ Ser 6 Zool 13(1–4):54–58

Iwata F (1958) On the development of the nemertean Micrura akkeshiensis. Embryologia 4(2):103–131

Iwata F (1960) Studies on the comparative embryology of nemerteans with special reference to their interrelationships. Publ Akkeshi Mar Biol Stat 10:1–51

Iwata F (1972) Axial changes in the nemertean egg and embryo during development and its phylogenetic significance. J Zool 168(4):521–526

Iwata F (1985) Foregut formation of the nemerteans and its role in nemertean systematics. Am Zool 25(1):23–36

Jägersten G (1972) Evolution of the metazoan life cycle. Academic, London

Jespersen Å, Lützen, J (1988) Fine structure of the eyes of three species of hoplonemerteans (Rhynchocoela: Enopla). NZ J Zool 15(2):203–210

Joubin L (1914) Némertiens. In: Charcot J (ed) Deuxième Expéd. Antarc. Française 1908–1910. Masson et Cie, Paris, pp 1–33

Kajihara H, Chernyshev AV, Sun SC, Sundberg P, Crandall FB (2008) Checklist of nemertean genera and species published between 1995 and 2007. Species Divers 13:245–274

Klerkx JHEM (2001) Molecular analysis of early specification in the mollusc *Patella vulgata*. PhD Thesis, University of Utrecht

Kmita-Cunisse M, Loosli F, Bièrne J, Gehring WJ (1998) Homeobox genes in the ribbonworm *Lineus sanguineus*: evolutionary implications. Proc Natl Acad Sci 95(6):3030–3035

Koch M, Quast B, Bartolomaeus T (2014) Coeloms and nephridia in annelids and arthrophods. In: Wägele JW, Bartolomaeus T (eds) Deep metazoan phylogeny: the backbone of the tree of life. De Gruyter, Berlin, pp 273–284

Kvist, S, Laumer CE, Junoy J, Giribet, G (2014) New insights into the phylogeny, systematics and DNA barcoding of Nemertea. Invertebrate Systematics 28(3):287–308

Lacalli TC (2005) Diversity of form and behaviour among nemertean pilidium larvae. Acta Zool 86(4):267–276

Lacalli TC, West JE (1985) The nervous system of a pilidium larva: evidence from electron microscope reconstructions. Can J Zool 63(8):1909–1916

Lebedinsky J (1897) Beobachtungen über die Entwicklungsgeschichte der Nemertinen. Archiv für mikroskopische Anatomie 49(1):503–556

Loosli F, Kmita-Cunisse M, Gehring WJ (1996) Isolation of a *Pax-6* homolog from the ribbonworm *Lineus sanguineus*. Proc Natl Acad Sci 93(7):2658–2663

Martindale MQ, Henry JJQ (1992) Evolutionary changes in the program of spiralian embryogenesis: fates of early blastomeres in a direct-developing nemertean worm. Am Zool 32:79A

Martindale MQ, Henry JQ (1995) Modifications of cell fate specification in equal-cleaving nemertean embryos: alternate patterns of spiralian development. Development 121(10):3175–3185

Maslakova SA (2010a) Development to metamorphosis of the nemertean pilidium larva. Front Zool 7(1):30

Maslakova SA (2010b) The invention of the pilidium larva in an otherwise perfectly good spiralian phylum Nemertea. Integr Comp Biol 50(5):734–743

Maslakova SA, Hiebert TC (2014) From trochophore to pilidium and back again-a larva's journey. Int J Dev Biol 58:585–591

Maslakova SA, Malakhov VV (1999) A hidden larva in nemerteans of the order Hoplonemertini. Dokl Biol Sci 366:314–317

Maslakova SA, von Dassow G (2012) A non-feeding pilidium with apparent prototroch and telotroch. J Exp Zool B Mol Dev Evol 318(7):586–590

Maslakova SA, von Döhren J (2009) Larval development with transitory epidermis in *Paranemertes peregrina* and other hoplonemerteans. Biol Bull 216(3):273–292

Maslakova SA, Martindale MQ, Norenburg JL (2004a) Fundamental properties of the spiralian developmental program are displayed by the basal nemertean *Carinoma tremaphoros* (Palaeonemertea, Nemertea). Dev Biol 267(2):342–360

Maslakova SA, Martindale MQ, Norenburg JL (2004b) Vestigial prototroch in a basal nemertean, *Carinoma tremaphoros* (Nemertea; Palaeonemertea). Evol Dev 6(4):219–226

Müller J (1847) Fortsetzung des Berichts über einige neue Thierformen der Nordsee. Arch Anat Physiol 45(2):157–179

Nawitzki W (1931) *Procarinina remanei*: eine neue Paläonemertine der Kieler Förde. Zool Jb Anat 54:159–234

Nielsen C (2013) Life cycle evolution: was the eumetazoan ancestor a holopelagic, planktotrophic gastraea. BMC Evol Biol 13(1):1–18

Norenburg JL (1985) Structure of the nemertine integument with consideration of its ecological and phylogenetic significance. Am Zool 25(1):37–51

Norenburg JL, Stricker SA (2002) Phylum Nemertea. In: Young CM, Sewell MA, Rice ME (eds) Atlas of marine invertebrate larvae. Academic, London, pp 163–177

Nusbaum J, Oxner M (1913) Die Embryonalentwicklung des *Lineus ruber* Müll. Ein Beitrag zur Entwicklungsgeschichte der Nemertinen. Z Wiss Zool 107:78–191

Rawlinson KA (2010) Research Embryonic and post-embryonic development of the polyclad flatworm *Maritigrella crozieri*; implications for the evolution of spiralian life history traits. Front Zool 7:12

Reinhardt H (1941) Beiträge zur Entwicklungsgeschichte der einheimischen Süßwassernemertine *Prostoma graecense* (Böhmig). Vierteljahrsschr der Naturf Ges in Zürich 46:184–254

Reisinger E (1926) Nemertini. In: Schulze P (ed) Biologie der Tiere Deutschlands, Liefg 17, Teil 5. Verlag von Gebrüder Borntraeger, Berlin

Roe P (1984) Laboratory studies of feeding and mating in species of *Carcinonemertes* (Nemertea: Hoplonemertea). Biol Bull 167(2):426–436

Salensky W (1882–1883) Zur Entwicklungsgeschicht der *Borlasia vivipara* Uljanin. Biol Centralbl 2:740–745

Salensky W (1886) Bau und Metamorphose des Pilidium. Z Wiss Zool 43:481–511

Salensky W (1909) Über die embryonale Entwicklung des *Prosorochmus viviparus* Uljanin (*Monopora vivipara*). Bull Acad Imp Sci, St Petersburg (6) 3(5):325–340

Salensky W (1912) Über die Morphogenese der Nemertinen. I. Entwicklungsgeschichte der Nemertine im Inneren des Pilidium. Mém Acad Imp Sci St Petersburg (8) 30(10):1–74

Salensky W (1914) Die Morphogenese der Nemertinen. II. Über die Entwicklungsgeschichte des *Prosorochmus viviparus*. Mém Acad Imp Sci St Petersburg (8) 33(2):1–36

Schmidt GA (1934) Ein zweiter Entwicklungstypus von *Lineus gesserensis ruber* OF Müller (Nemertini). Zool Jb Abt Anat 58:607–660

Schmidt GA (1937) Bau und Entwicklung der Pilidium von *Cerebratulus pantherinus* und *marginatus* und die Frage der morphologischen Merkmale der

Hauptformen der Pilidien. Zool Jb Anat Ontog 62: 423–448

Schmidt GA (1964) Embryonic development of littoral nemertines *Lineus desori* (mihi, species nova) and *Lineus ruber* (OF Mülleri, 1774, GA Schmidt, 1945) in connection with ecological relation changes of mature individuals when forming the new species *Lineus ruber*. Zool Pol 14:75–122

Schwartz ML (2009) Untying a Gordian knot of worms: systematics and taxonomy of the Pilidiophora (phylum Nemertea) from multiple data sets. PhD Thesis, George Washington University)

Schwartz ML, Norenburg JL (2005) Three new species of *Micrura* (Nemertea: Heteronemertea) and a new type of heteronemertean larva from the Caribbean Sea. Caribb J Sci 41:528–543

Senz W, Tröstl R (1999) Beiträge zur Entwicklung von *Prosorhochmus adriaticus* Senz, 1993 (Nemertini: Holponemertini: Monostilifera). Annalen des Naturhistorischen Museums in Wien. Serie B für Botanik und Zoologie. 437–443

Smith JE (1935) Memoirs: the early development of the nemertean *Cephalothrix rufifrons*. Q J Microsc Sci 2(307):335–381

Stricker SA (1985) The stylet apparatus of monostiliferous hoplonemerteans. Am Zool 25(1):87–97

Stricker SA (1986) An ultrastructural study of oogenesis, fertilization, and egg laying in a nemertean ectosymbiont of crabs, *Carcinonemertes epialti* (Nemertea, Hoplonemertea). Can J Zool 64(6): 1256–1269

Stricker SA, Cloney RA (1982) Stylet formation in nemerteans. Biol Bull 162(3):387–403

Stricker SA, Folsom MW (1997) A comparative ultrastructural analysis of spermatogenesis in nemertean worms. Hydrobiologia 365(1–3):55–72

Stricker SA, Reed CG (1981) Larval morphology of the nemertean *Carcinonemertes epialti* (Nemertea: Hoplonemertea). J Morphol 169(1):61–70

Stricker SA, Smythe TL, Miller L, Norenburg JL (2001) Comparative biology of oogenesis in nemertean worms. Acta Zool 82(3):213–230

Stricker SA, Cline C, Goodrich D (2013) Oocyte maturation and fertilization in marine nemertean worms: using similar sorts of signaling pathways as in mammals, but often with differing results. Biol Bull 224(3):137–155

Sundberg P, Strand M (2007) Annulonemertes (phylum Nemertea): when segments do not count. Biol Lett 3(5):570–573

Thiel M, Junoy J (2006) Mating behavior of nemerteans: present knowledge and future directions. J Nat Hist 40(15–16):1021–1034

Thollesson M, Norenburg JL (2003) Ribbon worm relationships: a phylogeny of the phylum Nemertea. Proc R Soc Lond B Biol Sci 270(1513):407–415

Turbeville JM (1986) An ultrastructural analysis of coelomogenesis in the hoplonemertine *Prosorhochmus americanus* and the polychaete *Magelona sp.* J Morphol 187(1):51–60

Turbeville JM (1991) Nemertinea. In: Harrisson W, Bogitsh BJ (eds) Microscopic anatomy of invertebrates, vol 3, Platyhelminthes and Nemertinea. Wiley-Liss, New York, pp 285–328

Turbeville JM (2002) Progress in nemertean biology: development and phylogeny. Integr Comp Biol 42(3):692–703

Turbeville JM, Ruppert EE (1985) Comparative ultrastructure and the evolution of nemertines. Am Zool 25(1):53–71

van den Biggelaar JA, Dictus WJ, van Loon AE (1997) Cleavage patterns, cell-lineages and cell specification are clues to phyletic lineages in Spiralia. Semin Cell Dev Biol 8(4):367–378

von Dassow G, Maslakova SA (2013) How the pilidium larva pees. Integr Comp Biol 53(suppl1):e386

von Döhren J (2008) Zur Phylogenie der Nemertea: vergleichende Untersuchungen der Reproduktion und Entwicklung. PhD Thesis, Free University, Berlin

von Döhren J (2011) The fate of the larval epidermis in the Desor-larva of *Lineus viridis* (Pilidiophora, Nemertea) displays a historically constrained functional shift from planktotrophy to lecithotrophy. Zoomorphology 130(3):189–196

von Döhren J, Bartolomaeus T (2006) Ultrastructure of sperm and male reproductive system in *Lineus viridis* (Heteronemertea, Nemertea). Zoomorphology 125(4): 175–185

von Döhren J, Bartolomaeus T (2007) Ultrastructure and development of the rhabdomeric eyes in *Lineus viridis* (Heteronemertea, Nemertea). Zoology 110(5):430–438

von Döhren J, Beckers P, Vogeler R, Bartolomaeus T (2010) Comparative sperm ultrastructure in Nemertea. J Morphol 271(7):793–813

Wilson CB (1900) The habits and early development of *Cerebratulus lacteus* (Verrill). A contribution to physiological morphology. Q J Microsc Sci 43:97–198

Wilson EB (1903) Experiments on cleavage and localization in the nemertine egg. Roux's Archiv für Entwicklungsmechanik der Organismen 16(3): 411–460

Yatsu N (1909) Observations on ookinesis in *Cerebratulus lacteus*, Verrill. J Morphol 20(3):353–401

Yatsu N (1910) Experiments on the cleavage in the egg of *Cerebratulus*. J Coll Sci Imp Univ Tokyo 27(10):1–19

Zaitseva OV, Flyachinskaya LP (2010) In vivo studies of development of the main functional systems in the heteronemertean pilidium larva. J Evol Biochem Physiol 46(4):396–406

Zeleny C (1904) Experiments on the localization of developmental factors in the nemertine egg. J Exp Zool 1(2):293–329

Annelida

9

Christoph Bleidorn, Conrad Helm, Anne Weigert,
and Maria Teresa Aguado

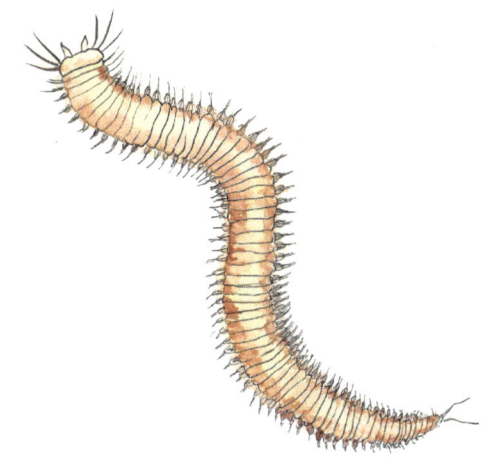

Chapter vignette artwork by Brigitte Baldrian.
© Brigitte Baldrian and Andreas Wanninger.

C. Bleidorn (✉) • C. Helm • A. Weigert
Molecular Evolution and Systematics of Animals,
Institute of Biology, University of Leipzig,
Talstraße 33, Leipzig D-04103, Germany
e-mail: bleidorn@uni-leipzig.de

M.T. Aguado
Departamento de Biología, Facultad de Ciencias,
Universidad Autónoma de Madrid,
Canto Blanco, Madrid 28049, Spain

INTRODUCTION

Annelids are a taxon of protostomes comprising more than 17,000 worldwide-distributed species, which can be found in marine, limnic, and terrestrial habitats (Zhang 2011). Their phylogeny was under discussion for a long time, but recent phylogenomic analyses resulted in a solid backbone of this group (Struck et al. 2011; Weigert et al. 2014). According to these analyses, most of the annelid diversity is part of Errantia or Sedentaria, which both form reciprocally monophyletic sister groups (Fig. 9.1) and are now known as Pleistoannelida (Struck 2011). The Sedentaria also include the Clitellata, Echiura, and Pogonophora (Siboglinidae) as derived annelid taxa. Outside Sedentaria and Errantia, several groups can be found in the basal part of the annelid tree, namely, Sipuncula, Amphinomida, Chaetopteridae, Magelonidae, and Oweniidae. The latter two taxa together represent the sister taxon of all other annelids. Given this hypothesis, it has to be assumed that the early diversification of extant annelids took place at least in the Lower Cambrian (520 Ma ago) (Weigert et al. 2014). The phylogenetic position of Myzostomida, a group of commensals or parasites of echinoderms (and, rarely, cnidarians), remains still uncertain. Whereas there is strong support for an annelid ancestry, its exact position awaits to be determined (Bleidorn et al. 2014). Likewise, the phylogenetic position of several interstitial taxa is still under debate (Westheide 1987; Worsaae and Kristensen 2005; Worsaae et al. 2005; Struck 2006). A position of Diurodrilidae outside Annelida, as suggested by Worsaae and Rouse (2008), was rejected by molecular data (Golombek et al. 2013), and the position of the enigmatic *Lobatocerebrum* and *Jennaria* remains unresolved (Rieger 1980, 1991). Likewise, the position of Annelida within Protostomia is still uncertain. However, recent phylogenomic analyses recover a clade uniting annelids with Mollusca, Nemertea, Brachiopoda, and Phoronida, but without strong support for any

sister group relationship (Edgecombe et al. 2011).

Annelids show a huge diversity of body plans, and it is difficult to describe a consistent anatomy matching most of this variety (Fig. 9.2). Most annelids are coelomate organisms, possessing multiple segments which occur repetitively along the anterior-posterior body axis (Purschke 2002). If segmentation is present, the annelid body is divided into a prostomium, an either homonomously (i.e., identical segments) or heteronomously (i.e., segments differ from each other) segmented trunk and a pygidium (Fauchald and Rouse 1997). In many annelid taxa, the prostomium contains the brain; however, in Clitellata the brain may be found in the following segments (Bullock 1965). The head of the annelids may bear appendages, as palps or antennae, but these are lacking in a number of taxa. The mouth can be found in the first segment which is termed peristomium. Several members of the Errantia as well as Amphinomida bear sclerotized mandibular structures, which may be replaced by a mechanism resembling molting (Paxton 2005). The segments of many annelids contain a pair of nephridia (usually metanephridia), coelomic cavities, ganglia, and ventral and dorsal groups of chitinous chaetae which might be organized in parapodia (Purschke 2002; Bartolomaeus et al. 2005). Segments are generated by a posterior growth zone which is located in front of the pygidium (Nielsen 2004). The pygidium contains the anus, which is usually either dorsally or terminally located and is often equipped with pairs of cirri.

Annelids show a wide variety in the organization of their nervous system (Bullock 1965; Orrhage and Müller 2005; Müller 2006). Müller (2006) proposed a nervous system with paired circumesophageal connectives, four cerebral commissures, five connectives, and numerous commissures in the ventral nerve cord as a hypothetical ground pattern. However, many variations of this pattern exist, and many taxa have not been investigated at all. Accordingly, alternative hypotheses suggest that a strict rope-ladder-like nervous system with segmental

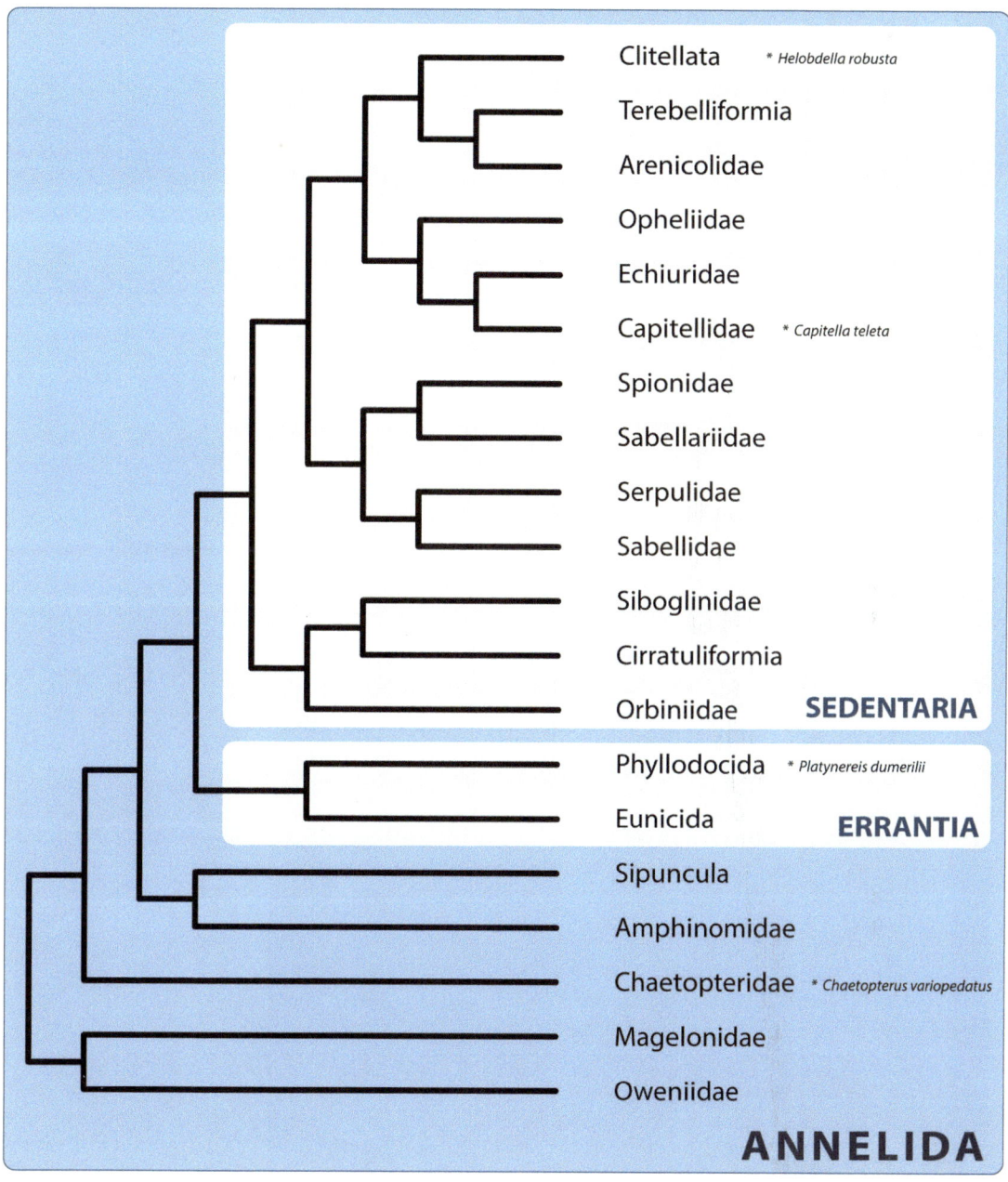

Fig. 9.1 Phylogeny of Annelida based on Weigert et al. (2014). Placement of well-investigated model annelids indicated with *asterisks*

ganglia interconnected by a pair of connectives and commissures was not present in the last common ancestor of annelids (Purschke et al. 2014). Instead, an orthogonal arrangement of the peripheral nervous system and the presence of additional longitudinal nerves might constitute the ground pattern of Annelida (Lehmacher et al. 2014).

Annelids show varying grades of brain complexity which may comprise a number of ganglia.

Fig. 9.2 The diversity of marine Annelida. (**A**) *Brada villosa*, Flabelligeridae. (**B**) *Glycera capitata*, Glyceridae. (**C**) *Lepidonotus squamatus*, Polynoidae. (**D**) *Nereis pelagia*, Nereididae. (**E**) *Chaetopterus* sp., Chaetopteridae. (**F**) Syllinae indet., Syllidae. (**G**) *Branchiomma arctica*, Sabellidae. (**H**) *Lumbrineris* sp., Lumbrineridae. (**I**) *Amblyosyllis* sp., Syllidae. (**J**) *Phyllodoce* sp., Phyllodocidae. (**K**) Polynoidae indet. (**L**) Sabellidae indet., Sabellidae. (**M**) *Spirobranchus giganteus*, Serpulidae. (**N**) *Lysidice* sp., Eunicidae. (**O**) *Serpula* sp., Serpulidae. (**P**) *Travisia* sp., *Travisia* (All images provided by Alexander Semenov (www.clione.ru). © Alexander Semenov, 2015. All Rights Reserved)

Mushroom bodies have been reported for several taxa of the Phyllodocida (Heuer et al. 2010). As for the nervous system, many different variations can be found in the muscular system. The presence of an outer layer of circular muscle fibers and an inner layer of four longitudinal bands of muscle fibers is often regarded as a possible ground pattern (Tzetlin and Filippova 2005; Lehmacher et al. 2014), but circular muscles are missing in several annelid taxa and may thus not constitute a basal annelid feature. Additionally, other muscular fiber bundles referred to as oblique, diagonal, bracing, or dorso-ventral fibers might be present and are often compensating missing circular musculature (Purschke and Müller 2006).

However, many taxa, such as myzostomids, sipunculids, or echiurids, are clearly deviating from the pattern described above in various aspects, and all of them seem to have lost segmentation convergently. Interestingly, all these examples still show some traces hinting to a secondary loss of segmentation (Purschke et al. 2000; Hessling 2003; Kristof et al. 2008; Helm et al. 2014). Other taxa such as clitellates and many other sedentarians lost their parapodia. Siboglinidae (Pogonophora + Vestimentifera) show many reductions as adaptation to their lifestyle in close association with bacterial endosymbionts (Schulze and Halanych 2003). Loss of key characters in Annelida is well-documented and is regarded as one of the problems to converge to a well-accepted phylogeny of the whole group (Purschke et al. 2000; Bleidorn 2007; Miyamoto et al. 2013).

Several systems of sensory organs are described for annelids, including a type of chemosensory organ called "nuchal organ." This type of sensory organ can be found in the posterior part of the prostomium and usually consists of ciliated supporting cells, sensory cells, and retractor muscles (Purschke 1997). Clitellates as well as several other annelids such as the basal branching Oweniidae and Magelonidae lack nuchal organs completely.

Many annelids possess some kind of light receptive photoreceptors which show great structural diversity (Purschke et al. 2006). Generally, rhabdomeric, ciliary, and phaosomous photoreceptor cell types are distinguished, and they might represent either larval or adult eyes (Purschke et al. 2006; Arendt et al. 2009). Larval eyes are simple organized, and the eye spots of the trochophore of *Platynereis dumerilii* consist of a rhabdomeric photoreceptor cell and a pigment cell which provide a direct coupling of light-sensing ciliary locomotory control (Jekely et al. 2008). Eyes of adult annelids might be present on the head, palps, segments (usually laterally), or even the pygidium (Purschke et al. 2006).

Annelids show a variety of reproductive strategies, and sexual as well as asexual reproduction is well-documented for many taxa (Wilson 1991; Bely 2006). For sexual reproduction, different types of free spawning, brooding, and encapsulation of embryos in cocoons can be distinguished, and all types involve either planktotrophic or lecithotrophic developmental stages (Thorson 1950; Wilson 1991). Multiple modes of development (poecilogeny) are reported for some annelid species, with the spionid *Streblospio benedicti* as the best-investigated example (Levin 1984; Zakas and Wares 2012). By far the most spectacular diversity of reproductive modes can be found across syllids, including swarming and external fertilization, internal fertilization, viviparity, and parthogenesis, as well as different forms of hermaphroditism (Franke 1999). Several annelid taxa show a pronounced sexual dimorphism resulting in dwarf male forms as found, for example, in some echiurids, siboglinids, or antonbruunids (Spengel 1879; Hartman and Boss 1965; Worsaae and Rouse 2010). Not surprisingly, annelids are also a prime example for the investigation of heterochrony, and several putative paedomorphic taxa have been hypothesized (Westheide 1987; Struck 2006; Bleidorn 2007; Osborn et al. 2007).

Platynereis dumerilii as a Model for Evolutionary Developmental Biology

Platynereis dumerilii is a marine annelid belonging to the errant family Nereididae, which emerged as a thoroughly investigated model species. The life cycle of this indirect developing gonochoric species with planktotrophic larvae is well established and controllable in the lab. Immature, atokous worms live in self-constructed tubes. Sexually mature, epitokous individuals, which appear morphologically different to atokous individuals, leave the tube and begin swimming to find partners for spawning during the night. The day of swarming is controlled by an endogenous lunar cycle which can be triggered artificially in the lab. Cultures were established in the lab in 1953 and are bred since then without interruption. Experimental techniques such as cell ablation, whole-mount in situ hybridization, RNA interference, and Morpholino knockdowns are routinely applicable. First transgenic lineages have been created, and a project sequencing the genome is underway for this species (http://4dx.embl.de/platy/). Comparative genomic studies suggest that the genome of *P. dumerilii* retains a more ancestral organization compared to other protostomian model organisms such as *Drosophila melanogaster* or *Caenorhabditis*

elegans. Important insights into the evolution of segmentation, vision, and the nervous system in Bilateria were provided by evolutionary developmental studies on *P. dumerilii*, and since a number of labs now use this animal as a model, important results with considerable relevance for our understanding of animal evolution and development are likely to keep emerging in the near future.

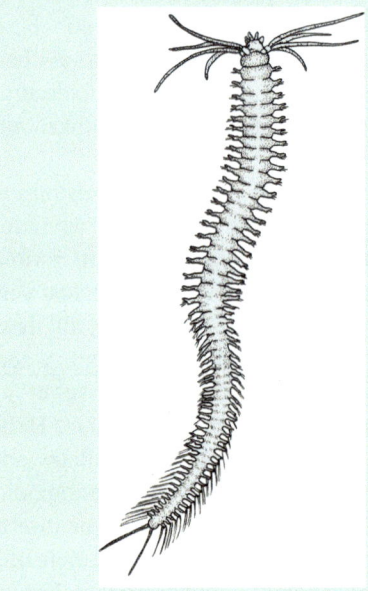

Atokous juvenile of *Platynereis dumerilii* (After Fischer and Dorresteijn (2004))

EARLY DEVELOPMENT

Egg Structure and Fertilization

Annelids with planktotrophic development usually have small, non-yolky eggs, whereas species with lecithotrophic development bear larger and yolk-rich eggs (Irvine and Seaver 2006). Ultrastructural studies of annelid eggs are scarce given the immense diversity of this group. One of the best-investigated examples is the egg of parchment worms of the genus *Chaetopterus*. Different regions are distinguished based on staining properties, divided into a cortical ectoplasm,

endoplasm, and hyaloplasm (Lillie 1906, 1909), the latter two regions forming the cytoplasm. The ectoplasm contains large membrane-bound spherules, nuage (a germ line-specific organelle containing several proteins), and intracellular membrane systems. The endoplasm contains yolk and lipid, interspersed with mitochondria, granular bodies, and endoplasmatic reticulum. In contrast, the hyaloplasm (or teloplasm) is characterized by the absence of granular bodies (Eckberg 1981; Jeffery 1985). Eggs of many annelid species show a clear polarity with an accumulation of developmental factors in the cortex of future polar regions (Dorresteijn 2005). The spatial distribution of maternal mRNA in the ectoplasm has been described for *Chaetopterus* (Jeffery and Wilson 1983; Jeffery 1985). After fertilization and before initiation of the first cleavage, a reorganization of the yolk-free hyaloplasm (teloplasm) has been observed in several annelids (Dorresteijn 1990; Weisblat and Huang 2001). It has been shown for *Platynereis dumerilii* that after attachment of the sperm to the egg surface, cortical granules released by exocytosis from the ectoplasm start forming an egg jelly on the outside, which removes supernumerary sperm from its surface (Dorresteijn 1990). Ultrastructural investigations of the ectoplasm of eggs of *P. dumerilii* and the clitellate *Theromyzon rude* reveal an extensive framework of actin filaments that are involved in remodeling the egg surface after fertilization (Fernandez et al. 1987; Kluge et al. 1995).

Following fertilization a reorganization of the endoplasm can be observed. The distribution of the two cytoplasmic domains can be categorized into different types, which seem to be restricted to certain annelid taxa (Shimizu 1999). Most investigated non-clitellate annelids (e.g., chaetopterids, nereidids, and onuphids) show a stratification of the endoplasm into two domains, and, in most cases, the clear hyaloplasm (teloplasm) is localized at the animal pole (Wilson 1892; Huebner and Anderson 1976; Jeffery and Wilson 1983). In contrast, three domains can be distinguished in the clitellate endoplasm, with teloplasm localized at both the animal and vegetal poles of the egg (Shimizu 1999; Weisblat and Huang 2001). These cytoplasmatic movements are coordinated by complex cytoskeletal mechanisms which even seem to vary among taxa. Whereas in the leech *Helobdella triserialis*, microtubules are shown to play an important role, movement of the teloplasm in the oligochaete *Tubifex* is orchestrated by an actin network (Astrow et al. 1989; Shimizu 1995).

Cleavage

Annelids develop by spiral cleavage which is characterized by cleavage furrows which are oblique to the egg axis due to an inclination of the mitotic spindle (see Chapter 7). This cleavage starts with two orthogonal cell divisions which generate four blastomeres, called A, B, C, and D (Costello and Henley 1976). Correlating with the differing developmental modes in annelids, blastomeres usually exhibit the same size (equal cleavage) in species with indirect development and planktotrophic larvae, whereas direct developers show pronounced differences in blastomere size (unequal cleavage) (Anderson 1966; Arenas-Mena 2007). However, exceptions to this trend exist. For example, *Platynereis dumerilii* and *Platynereis massiliensis* both show unequal spiral cleavage patterns, even though the former species develops indirectly and the latter directly (Schneider et al. 1992). In most cases unequal cleavage is achieved due to positioning of the mitotic spindle. However, in some annelids this cleavage pattern is facilitated due to the presence of membrane-bound polar lobes (Freeman and Lundelius 1992). Such a polar lobe is also reported for the myzostomid *Myzostoma cirriferum* (Eeckhaut and Jangoux 1993). The different modes of spiral cleavage across annelids have been thoroughly reviewed in Dorresteijn (2005). The future axis of the developing embryo is already determined, with A and C corresponding respectively to the left and right side of the embryo and blastomeres B to D defining the antero-ventral to postero-dorsal axis (Nielsen 2004). Due to uneven cleavage, starting from the third cell division, a shifting in the angle of the mitotic spindles, which alternates during subsequent divisions, becomes obvious in almost all annelids. These shifts are either dextral (clockwise) or sinistral (anti-clockwise) and lead to the name-giving spiral arrangement pattern of blastomeres. By oblique divisions animal and vegetal daughter cells are generated, referred to as micromere quartets and macromere quartets. Some annelids such as the opheliid *Armandia brevis* show equal cleavage also in the third cleavage, generating micro- and macromeres of equal size (Hermans 1964). In the oweniid *Owenia collaris* and some leeches, micromeres are larger than macromeres in the eight-cell stage, a pattern which is also known from several nemerteans

(Dohle 1999; Smart and Von Dassow 2009). Several authors introduced the idea that a specific, phylogenetically conserved pattern of blastomeres can be seen at this stage, termed the "annelid cross." This pattern is regarded as typical for most annelids (including echiurans) but cannot be found in sipunculids, which show the so-called molluscan cross. However, a continuum of different variants between these patterns is demonstrated, and consequently, these concepts have been neglected for phylogenetic purposes (see also Chapter 6; Maslakova et al. 2004a; Nielsen 2004). The cell fate of individual blastomeres is conserved across annelids, and a specific nomenclature is used to trace the fate of blastomeres throughout development, using capital letters for macromeres and small letters for micromeres (Conklin 1897; Costello and Henley 1976; Nielsen 2004). A number is used as prefix to designate the quartet of which the macromeres or micromeres originated from. The micromeres continue to divide, and daughter cells inherit the name of their mother cell, with modification, to trace its origin (animal vs. vegetal) (Fig. 9.3). An alternative cell nomenclatural system is in use for leeches (Dohle 1999). Many clitellates show a cleavage pattern that nearly obscures the original spiral mode of cleavage. In these cases where yolk content and egg size are reduced, the embryo is nourished by the surrounding fluid within the cocoon (Dohle 1999). Several siboglinids show elongated eggs with a high yolk content leading to an aberrant pattern of spiral cleavage (Southward 1999). In most annelids, the cleavage pattern shifts from spiral to bilaterally symmetric after the formation of the fourth quartet of micromeres (Meyer and Seaver 2010). In *P. dumerilii* cell fates of sister blastomeres along the animal-vegetal axis are specified by levels of beta-catenin. High levels specify vegetal sister cell fates, while lower levels specify animal sister cell fates. Interestingly, no beta-catenin asymmetry is observed after the first bilaterally symmetrical and transverse cell divisions (Schneider and Bowerman 2007).

Descendants of the first micromere quartet (1a–1d) form larval head structures including the apical organ, larval eyes, and the head ectoderm as well as the primary trochoblasts (Nielsen 2004). The trochoblasts received their name as they will give rise to the prototroch, and this has been demonstrated for several annelids,

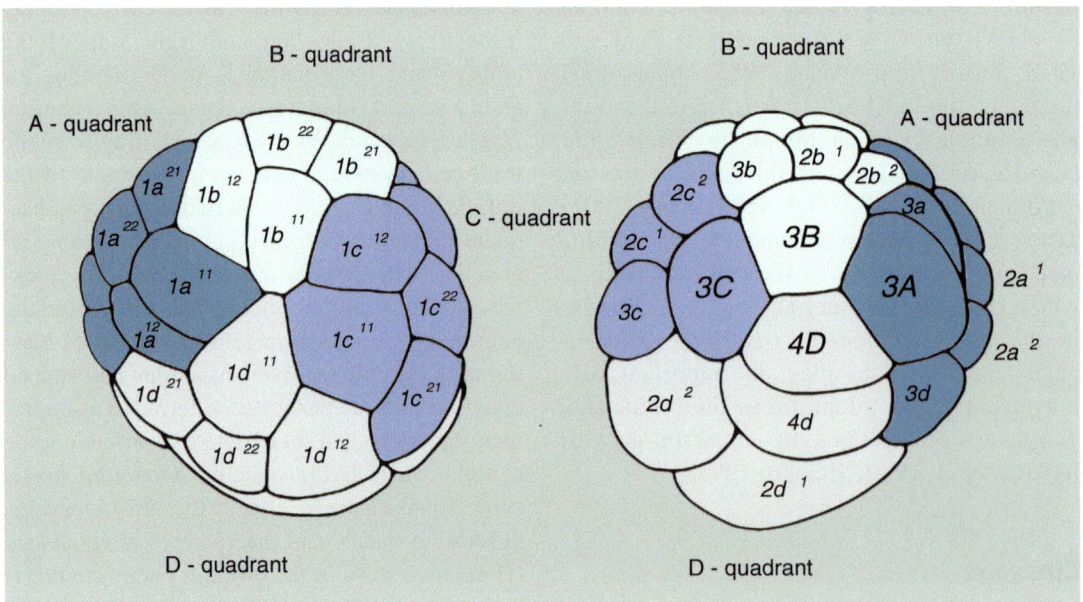

Fig. 9.3 Diagram illustrating the spiral cleavage cell nomenclature in the 33-cell stage of an unequally cleaving embryo of *Arenicola cristata* (Child 1900). The four quadrants (*A–D*) are indicated by colors

e.g., capitellids, dinophilids, and nereidids (Wilson 1892; Eisig 1898; Nelson 1904). Trochophore larvae of Myzostomida lack or show a reduced prototroch (Rouse 1999), but as it has been shown for the nemertean *Carinoma tremaphorus*, this need not be reflected in the formation of trochoblasts (Maslakova et al. 2004b). Three sets of trochoblast cells are involved in prototroch formation, a pattern which is highly conserved across Spiralia (Henry et al. 2007). Besides the primary and accessory trochoblasts, which are derived from the first micromere quartet, this includes secondary trochoblasts formed by some descendants of the second micromere quartet (2a–2c). Some annelids deviate from this pattern, e.g., the terebellid *Amphitrite ornata*, who lacks the accessory trochoblasts (Damen and Dictus 1994). Other descendants of the second micromere quartet generally develop into the foregut (stomodaeum) as well as part of the ectoderm (Nielsen 2004). The 2d cell is the somatoblast, developing into the major part of the body ectoderm posterior of the prototroch (Meyer and Seaver 2010). Using cell ablation studies, it has been shown for *Capitella teleta* that the 2d cell is responsible for organizing activity during early embryonic development, as well as bilateral symmetry and dorso-ventral axis organization of the head, and formation of neural, foregut, and mesoderm tissue (Amiel et al. 2013). In clitellates, four pairs of ectoteloblasts (called N, O, P, Q) are descendants of the 2d cell and give rise to four germbands including smaller cells (Dohle 1999; Goto et al. 1999). Cells derived from the third micromere quartet (3a–3d) form the foregut and ectomesoderm and might be the origin of protonephridia (Nielsen 2004; Ackermann et al. 2005). Interestingly, in *C. teleta*, mesodermal bands are generated by 3c and 3d (Meyer et al. 2010).

Usually the mesoderm and endoderm are formed by cells of the fourth micromere quartet (4a–4d), as characteristic for spiral cleavage in Lophotrochozoa in general (Chapter 7; Gline et al. 2011). Of special interest is the fate of the 4d cell, which has been called "mesenteloblast" or "primary mesoblast" (Wilson 1898). This cell gives rise to the adult mesoderm in most spiralians including mollusks or entoprocts (Chapters

6 and 7). In Clitellata, a fifth teloblast (M) is derived from the 4d cell, specifying a mesodermal germband (Goto et al. 1999). Progenitors of 4d form bilaterally symmetrical mesodermal anlagen in *Platynereis dumerilii* (Ackermann et al. 2005; Fischer and Arendt 2013). Prior to gastrulation, four secondary mesoblast cells bud from descendants of the 4d cell and show the morphology and gene expression signature of primary germ cells (Rebscher et al. 2012). These primary germ cells stay in mitotic arrest until individuals enter gametogenesis (Lidke et al. 2014). In *C. teleta* the 4d cell generates few muscle cells, primordial germ cells, and the anus (Meyer et al. 2010). It has been suggested that mesoteloblast-like mesodermal stem cells forming continuous mesodermal bands are part of the Pleistoannelida ground pattern (Fischer and Arendt 2013).

Gastrulation

The process of gastrulation in annelids has been reviewed in detail by several authors (Okada 1957; Anderson 1973; Weisblat and Huang 2001; Irvine and Seaver 2006), and the following descriptions provide a generalized pattern found in clitellate and non-clitellate annelids.

Gastrulation of embryos with less yolk starts with the invagination (embolic gastrulation) of putative midgut cells, and epithelia derived from the micromere cap grow toward the ventral side. Mesoteloblasts can be found in a posterior position in the blastocoel, whereas ectoteloblasts are located adjacent to them, below the larval ectoderm. The fate of the blastopore differs across annelid taxa, and protostomy, where the blastopore becomes the mouth, is found in most annelid taxa. Notably, deuterostomy, where the blastopore becomes the anus, has been demonstrated for eunicids (Åkesson 1967). The concept of amphistomy, in which both the mouth and the anus are derived from the corresponding ends of the blastopore, which was claimed to be present in *Polygordius* (Arendt and Nübler-Jung 1997), might not occur in any organism at all (Hejnol and Martindale 2009). Some differences apply for the gastrulation of yolk-rich annelid embryos.

Here, the process is rather described as epiboly, where the micromere cap grows over putative midgut cells and teloblasts (Irvine and Seaver 2006). In clitellates, the blastopore is found in the point where the germinal bands coalesce to form the germinal plate.

LATE DEVELOPMENT

Larval Ciliary Bands

Larval morphological characters vary across different annelid families (Fig. 9.4). A trochophore has distinct larval ciliary regions forming prominent bands or tufts, and the presence of a prototroch is regarded as defining (Bhaud and Cazaux 1987; Rouse 1999). However, detailed investigations concerning the homology of the respective ciliated regions within the different annelid families are lacking. At the anterior end of the episphere, the apical tuft marks the position of the larval apical organ, a feature well known for most invertebrate taxa with ciliated larvae (Marlow et al. 2014). Appearing early in development, the apical tuft forms a sensory region that is located in the direction of larval movement but often disappears in early larval stages (see Chapter 7 for details on apical tuft morphology). Although an apical tuft is widespread within annelids, larvae without an apical tuft are known for most cirratulids, histriobdellids, lopadorhynchids, orbiniids, sabellids, and tomopterids (Rouse 1999).

The prominent prototroch is represented by an equatorial ring consisting of usually compound cilia formed by a group of specific trochoblasts (Damen and Dictus 1994). Situated anterior to the mouth opening, a prototroch is known for most annelids, mollusks, and entoprocts (Nielsen 2012). Dividing the larval body in an anterior episphere and a posterior hyposphere (see Chapter 7), the prototroch is present mainly in planktotrophic annelid larvae and some lecithotrophic stages but absent in direct developing taxa such as clitellates, aelosomatids, and histriobdellids (Rouse 1999). In some annelid taxa, the cilia of the prototroch may cover almost the whole episphere in early developmental stages (e.g., in *Chaetopterus*; see Fig. 9.4L). The prototroch may be formed by equatorially arranged ciliary tufts (*Myzostoma cirriferum*; see Fig. 9.4G), or the whole larva may be covered by cilia, and a defined prototroch is hardly distinguishable, e.g., in early larvae of the eunicid *Marphysa* (Fig. 9.4E). An epispheral ciliated band is represented by the meniscotroch, which is only known for Phyllodocida (Bhaud and Cazaux 1982; Rouse 1999). Forming a tuft of short cilia, the meniscotroch is located in a ventral position within the episphere. Posterior to the latter structure, some annelids possess a ciliated band situated anterior to the prototroch – the akrotroch (Häcker 1896). Forming a complete ring separated from the apical tuft and the prototroch, an akrotroch can be found in syllids, orbiniids (e.g., in *Scoloplos armiger*, see Fig. 9.5), onuphids, cirratulids, and several Eunicida (Rouse 1999).

Situated posteriorly to the prototroch, the metatroch is represented by a ciliated ring that often beats opposed to the latter one and lies in a pre-segmental (= peristomial) position (Strathmann 1993; Nielsen 2012). Being present in most annelid families, a metatroch seems to be absent in Echiura (Fig. 9.4J) and Opheliidae. For Capitellidae, Siboglinidae, and Syllidae, the presence of a metatroch is still discussed (Rouse 2000a). Planktotrophic larvae of several polynoid scale worms possess another bundle of long cilia, the oral brush, which seems also to be involved in feeding mechanisms (Phillips and Pernet 1996).

A prominent ventral ciliary band is represented by the neurotroch, which is known at least in some annelid families including many sedentarian taxa, e.g., Orbiniidae (Fig. 9.5), Sabellidae (Fig. 9.6), and Maldanidae. In both planktotrophic and lecithotrophic developmental stages, the neurotroch forms a distinct ventral ciliated area interconnecting proto- and telotroch, which often appears later in larval development (Rouse 1999). The telotroch, defined as a posterior ring of cilia used for locomotion (Strathmann 1993), also appears later in development and is known for both planktotrophic and lecithotrophic developmental stages (Strathmann 1993). The telotroch marks the

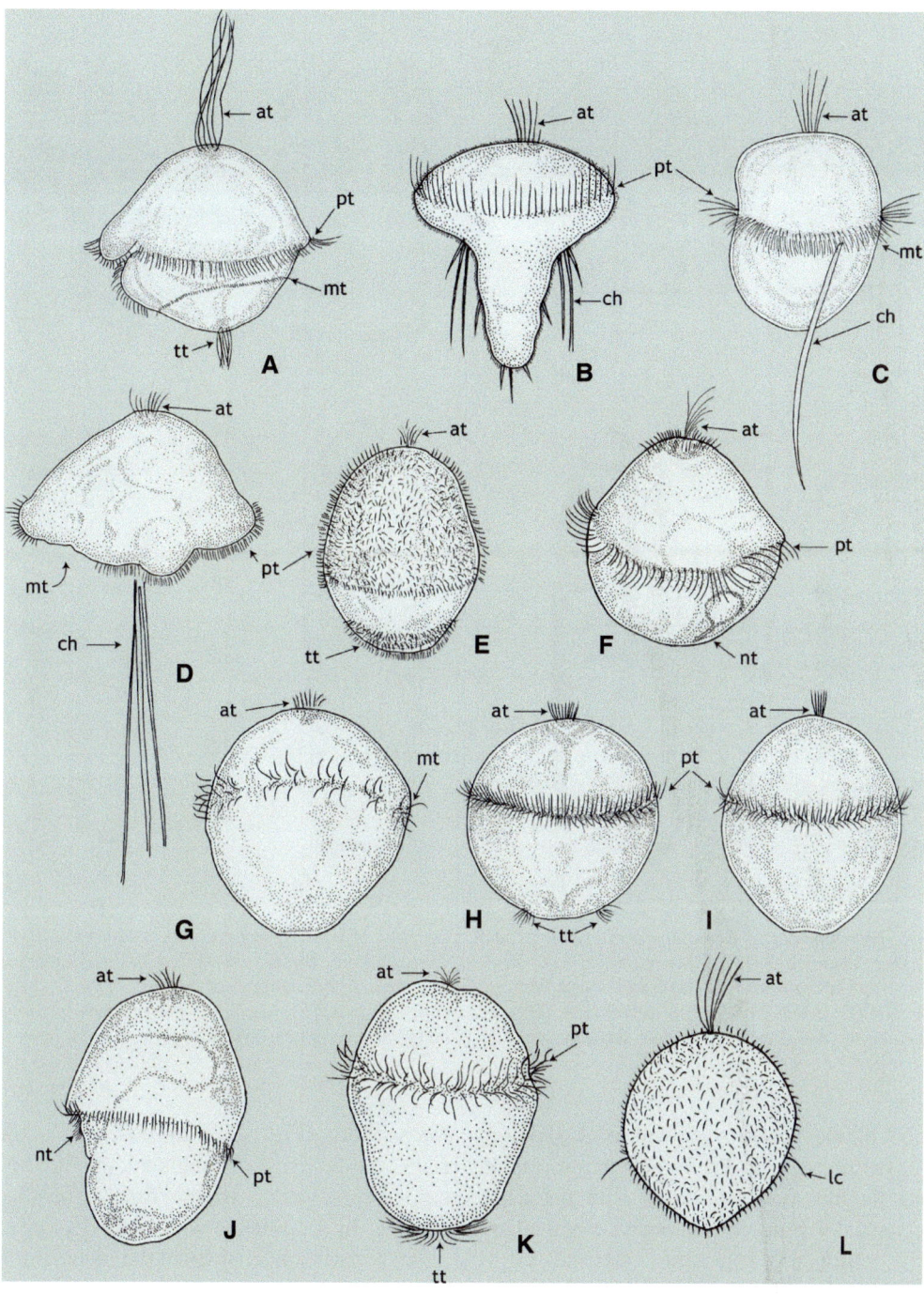

Fig. 9.4 Diversity of annelid trochophore larvae. Anterior (apical) is up in all aspects. (**A**) *Polygordius* sp. (Polygordiidae) after Woltereck (1904). (**B**) *Magelona filiformis* (Magelonidae) after Wilson (1982). (**C**) *Eurythoe complanata* (Amphinomidae) after Kudenov (1974). (**D**) *Owenia collaris* (Oweniidae) after Smart and Von Dassow (2009). (**E**) *Marphysa sanguinea* (Eunicidae) after Prevedelli et al. (2007). (**F**) *Phyllodoce maculata* (Phyllodocidae) after Voronezhskaya et al. (2003). (**G**) *Myzostoma cirriferum* (Myzostomida) after Eeckhaut and Jangoux (1993). (**H**) *Platynereis dumerilii* (Nereididae) after Fischer and Dorresteijn (2004). (**I**) *Phascolosoma perlucens* (Sipuncula) after Jaeckle and Rice (2002). (**J**) *Urechis caupo* (Echiura) after Pilger (2002). (**K**) *Osedax* sp. (Siboglinidae) after Rouse et al. (2009). (**L**) *Chaetopterus variopedatus* (Chaetopteridae) after Henry (1986). Abbreviations: *at* apical tuft, *ch* chaetae, *lc* lateral cilia, *mt* metatroch, *nt* neurotroch, *pt* prototroch, *tt* telotroch

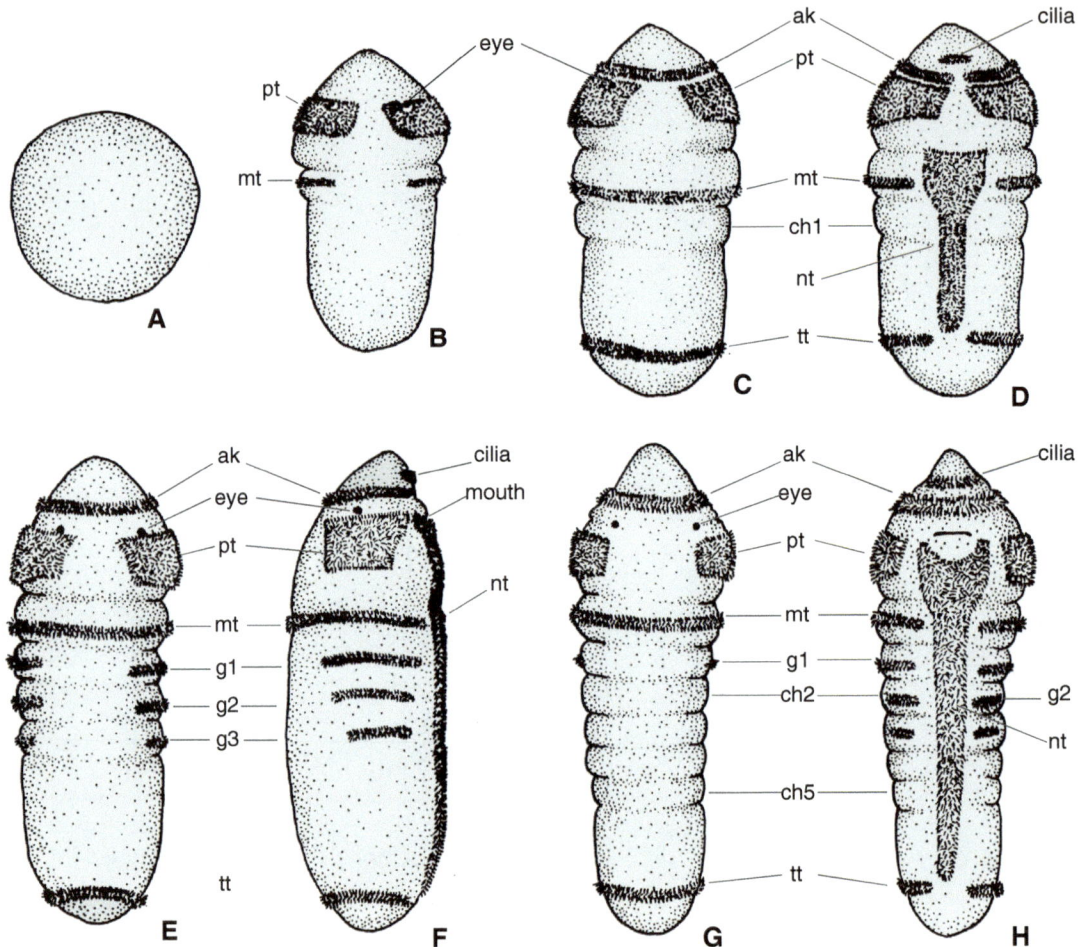

Fig. 9.5 Development of *Scoloplos armiger* (intertidal clade) after Anderson (1959). Anterior (apical) is up in all aspects. (**A**) Unfertilized egg. (**B**) Early 4-day embryo, dorsal view. (**C**) Late 5-day embryo, dorsal view. (**D**) Late 5-day embryo, ventral view. (**E**) Late 6-day embryo, dorsal view. (**F**) Late 6-day embryo, lateral view. (**G**) Early 7-day embryo, dorsal view. (**H**) early 7-day embryo, ventral view. Abbreviations: *ak* akrotroch; *ch 1, ch 2, ch 5* chaetiger 1, 2, 5; *g 1, g 2, g 3* gastrotrochs of chaetigers; *mt* metatroch; *nt* neurotroch; *pt* prototroch; *tt* telotroch

position of the posterior growth zone (Nielsen 2012). Further ciliated bands that may occur in several families are the gastro- and nototroch (= segmentally arranged ventral and dorsal ciliary bands), which are sometimes referred to as paratrochs (Bhaud and Cazaux 1982).

Post-trochophore Development and Larval Forms

Within annelid ontogeny, the metatrochophoral stage usually follows the prototrochophore/ trochophore (Fig. 9.7). In this developmental stage, the first signs of segmentation are visible, e.g., the formation of the first parapodia and chaetae. In accordance with individual development, several subdivisions of the metatrochophoral stage are possible (Fischer et al. 2010). In some terebellids and pectinariids, the metatrochophore builds a tube and is called aulophore (Bhaud and Cazaux 1982).

In Oweniidae a special type of trochophore occurs, the so-called mitraria (Fig. 9.4D). The mitraria larva exhibits prominent proto- and metatrochal bands, as well as an apical tuft.

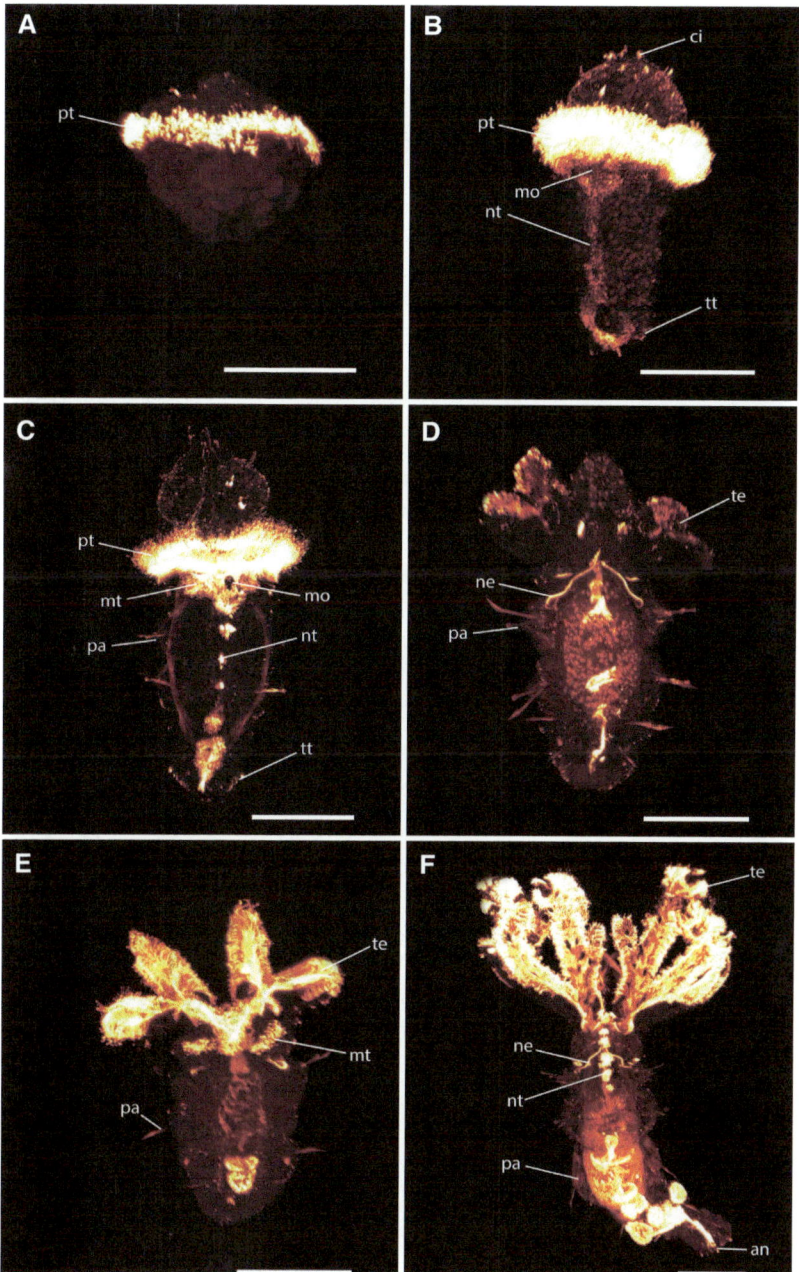

Fig. 9.6 Development of the lecithotrophic developmental stages of *Megalomma vesiculosum* revealed by anti-tubulin staining. All images are in ventral view except of (**D**) which is in dorsal view. Anterior is up. Confocal maximum projections. (**A**) The early, nonfeeding trochophore exhibits a prominent prototroch (*pt*). An apical tuft is lacking. (**B**) The later trochophore gains a well-developed prototroch (*pt*) and cilia (*ci*) at the anterior pole. Furthermore, the ventral neurotroch (*nt*) and the posterior telotroch (*tt*) develop at this stage. (**C**) The early metatrochophore exhibits three pairs of parapodia (*pa*) and a metatroch (*mt*). Neurotroch (*nt*), prototroch (*pt*), and telotroch (*tt*) are still present in this free-swimming but nonfeeding stage. (**D**) Shortly before metamorphosis, the late nectochaete has lost the main ciliary bands and starts development of the adult tentacles (*te*). (**E**) After metamorphosis the juveniles settle within a tube and start feeding. The parapodia (*pa*), the tentacles (*te*), and remnants of the metatroch (*mt*) are exhibited. (**F**) Late juvenile worms show an adult-like morphology. The tentacles (*te*) are well-developed, and the animals start to elongate by posterior segment addition. *an* anus, *ci* cilia, *mo* mouth opening, *mt* metatroch, *ne* nephridia, *nt* neurotroch, *pa* parapodia, *pt* prototroch, *tt* telotroch. Scale bars = 100 μm (© Conrad Helm, 2015. All Rights Reserved)

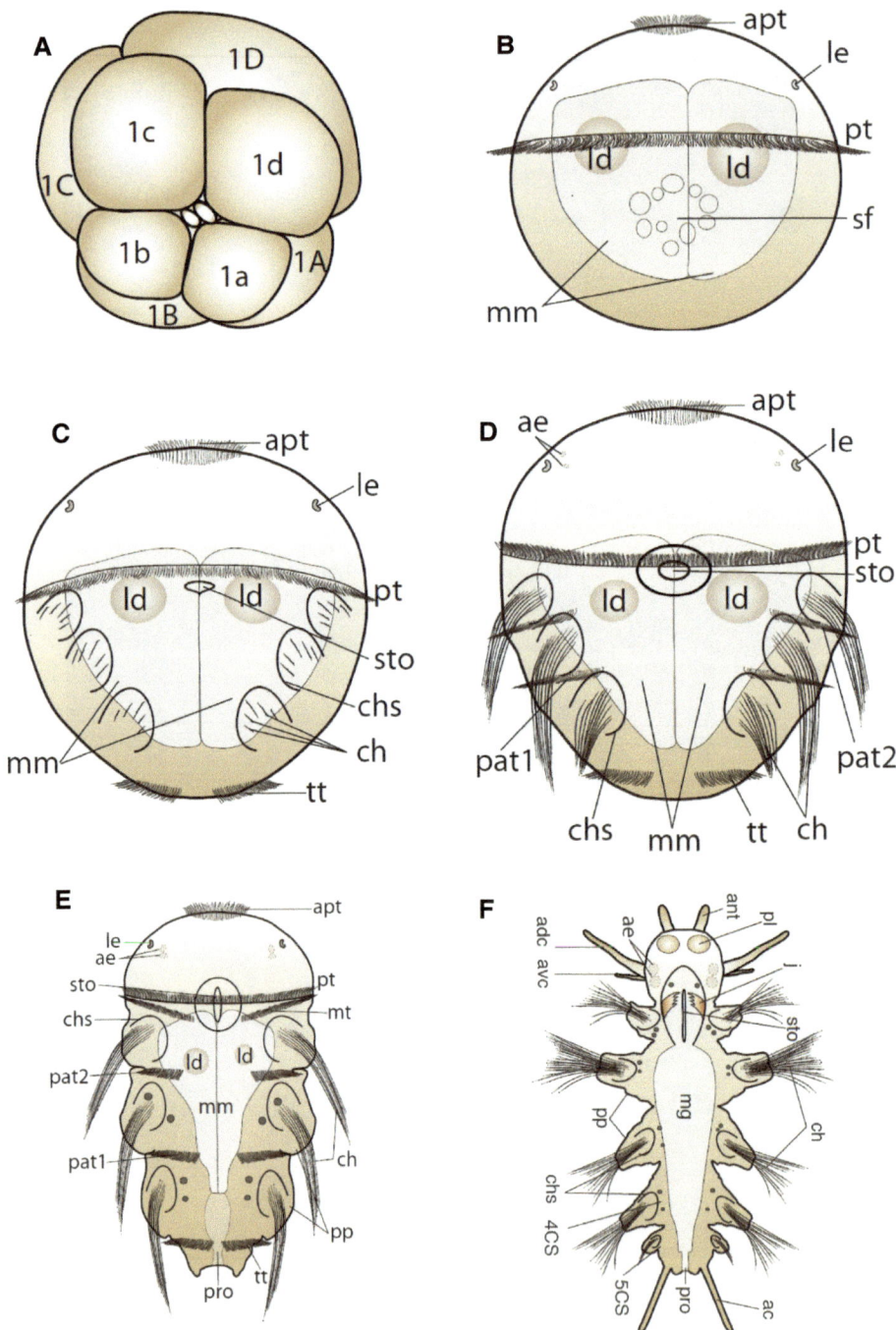

Fig. 9.7 Development of *Platynereis dumerilii*. (**A**) Cleaving embryo, where the third cleavage forms micromeres and macromeres. (**B**) Early trochophore, with prototroch and apical tuft. (**C**) Late trochophore, with simultaneous appearance of the first three larval segments. (**D**) Mid metatrochophore, with developing chaetae reaching over the body wall. (**E**) Early nectochaete, with formation of the metatroch and elongation of the trunk. (**F**) Juvenile, with rapidly growing jaws and further addition of body segments. Combined after Fischer et al. (2010). Abbreviations: *ac* anal cirrus, *ae* adult eyes, *adc* anterior dorsal cirrus, *ant* antenna, *apt* apical tuft, *avc* anterior ventral cirrus, *ch* chaetae, *chs* chaetal sac, *j* jaws, *ld* lipid droplet, *le* larval eye, *mg*, midgut, *mm* macromere, *mt* metatroch, *pat 1*, first paratroch, *pat2* second paratroch, *pl* palps, *pp* parapodia, *pro* proctodeum, *pt* prototroch, *sf* stomodeal field, *sto* stomodeum, *tt* telotroch, *4CS* 4th chaetigerous segment, *5CS* 5th chaetigerous segment

Notably, all ciliary bands are monociliated, an unusual feature for annelid larvae. The hyposphere of the mitraria is strongly reduced, and the juvenile segmental body develops within the larval body (Wilson 1932; Smart and Von Dassow 2009). Another unusual larval type is represented by the rostraria in Amphinomidae and Euphrosinidae (Mileikovsky 1960, 1961). After a trochophore stage with a proto- and metatroch and an apical tuft (Fig. 9.4C), the episphere of the metatrochophore elongates, and tentacles are formed for feeding (Jägersten 1972). Remarkably elongated metatrochophore stages can be found within siboglonids (Southward 1999). The metatrochophore of investigated vestimentiferan siboglinids is sessile and bears a prostomium, a peristomium, and two chaetigers. A prototroch, a neurotroch, and an apical organ are present as well as juvenile/adult organs such as tentacles and pyriform glands (Bright et al. 2013).

The end of the metatrochophore stage is usually marked by the point when the parapodia are fully developed. If present, the next larval stage is represented by the nectochaete (Fig. 9.7E), which is characterized by the presence of functional parapodia which are used mainly for swimming. In *Poecilochaetus* (Spionidae) this stage is sometimes called nectosoma; in other spionids, it refers to the chaetosphaera stage (Bhaud and Cazaux 1982). In this stage or the previous one, most larvae start body elongation and segment formation through the posterior growth zone (Irvine and Seaver 2006). A unique swimming larval form called pelagosphaera is known for Sipuncula (Rice 1976). The body of this larval type can be divided into three body regions: head, mid region including metatroch, and a large trunk. These larvae can be either lecithotrophic or planktotrophic forms, whereas the latter can live up to 6 months in the plankton (Jaeckle and Rice 2002).

After the nectochaetal (or pelagosphaera) stage, metamorphosis occurs. During metamorphosis, the free-swimming larvae change behavioral characteristics and start an adult-like lifestyle as juveniles. This metamorphic step can differ drastically between various species and is markedly pronounced in larvae of *Polygordius*, which rupture the larval body that is later either eaten or discarded (Rouse 2006). Such a transition of lifestyles seems to be less distinct or missing in annelids with lecithotrophic developing stages, where ciliary bands are absorbed or the chaetal morphology changes (Figs. 9.5 and 9.6).

Although late development and subsequent metamorphosis may differ between several taxa, in almost all annelid larvae, the larval episphere becomes the adult prostomium, and the posterior hyposphere becomes the pygidium and the posterior growth zone (Fig. 9.8). The remaining hyposphere forms the peristomium, which lacks chaetae in adult annelids (Nielsen 2004). The segmented body between the peristomium and the pygidium develops by segment formation from the posterior growth zone (Irvine and Seaver 2006).

Larval Feeding Modes

Several types of larval feeding behaviors and developmental modes occur in different annelid families, mostly divided into either feeding and free-swimming planktotrophic larvae with "indirect" development or nonfeeding and mostly less motile embryonic and juvenile forms with "direct" development. The latter rely on maternal sources of nutrition in the form of yolk stored in the egg during oogenesis, feeding on yolk-rich nurse eggs, or translocation of nutrition directly from the parent (Qian and Dahms 2006). An overview of larval feeding in annelids is summarized by Rouse (2006). As direct development occurs without a metamorphosis, developing stages are usually referred to as "embryonic" or "juveniles," avoiding the term "larvae" (Nielsen 2009; Winchell et al. 2010). However, debates remain about the definition of the term "larva," and alternative terminologies exist (McEdward and Janies 1993; Pechenik 1999).

Many authors regard a biphasic life cycle including a planktotrophic trochophore larva as representing the plesiomorphic condition of annelid development (Heimler 1988; Nielsen 2012). Nevertheless, this idea is doubted by some authors (Haszprunar et al. 1995; Rouse 2000a, b). Based on cladistic analyses and ancestral state

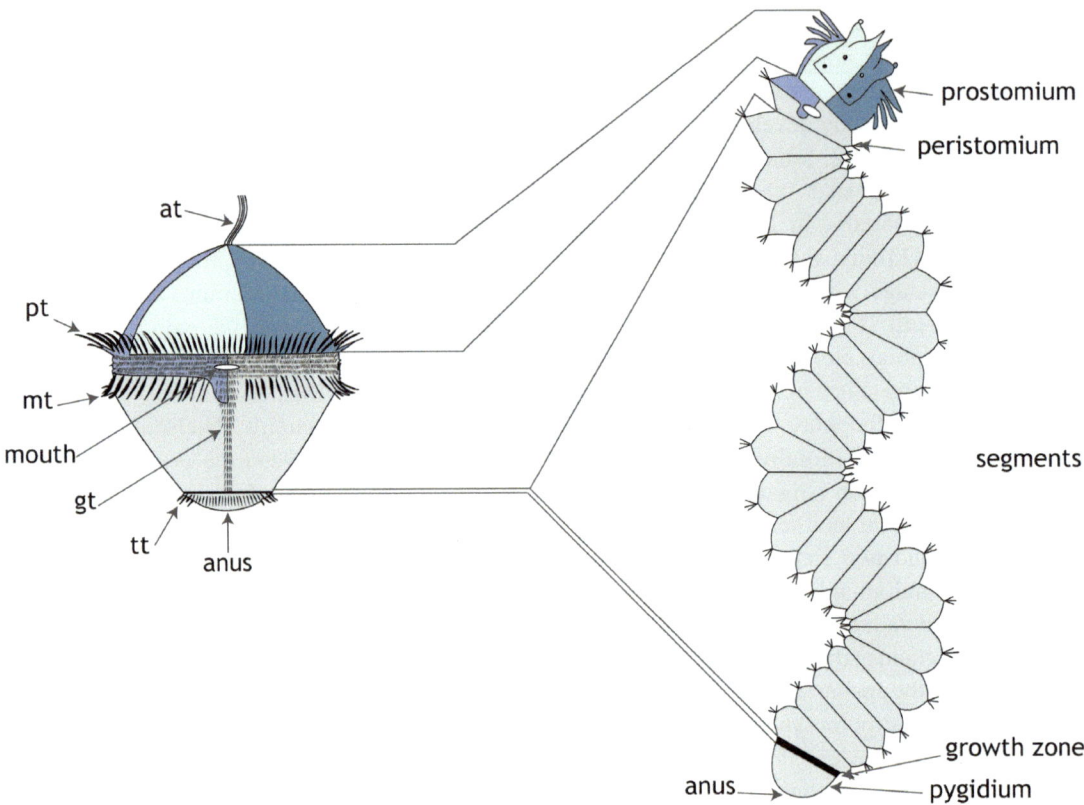

Fig. 9.8 Development of annelids indicating the contribution from the four quadrants (A–D), after Nielsen (2004). Note that the whole segmented body and the pygidium develop from descendants of the (D) quadrant. Abbreviations: *at* apical tuft, *pt* prototroch, *mt* metatroch, *tt* telotroch, *nt* neurotroch, *gt* gastrotroch

reconstruction, multiple events in the evolution of feeding larvae from lecithotrophic ancestors seem more parsimonious to assume. According to this hypothesis, the prototroch had a primary function for locomotion and became independently associated with feeding in several lineages (Rouse 2000a). By studying larval forms of sabellids, Pernet (2003) was able to demonstrate the persistence of reduced ciliary structures for food uptake in nonfeeding larvae. Consequently, it is suggested that the direction of evolution goes from planktotrophy to lecithotrophy in this case. Further on, a functional relation between egg size and gut development has been hypothesized. Non-feeding stages usually bear larger eggs, and due to the increased cell size, gut development might be hindered, resulting in nonfunctional guts in nonfeeding stages (Pernet 2003). As this scenario may be generalized for all annelids,

planktotrophy seems to be the likely ancestral state in annelids.

Most planktotrophic larvae exhibit a prominent proto- and metatroch, used for locomotion and food uptake by "downstream feeding" (Rouse 2000a). This is the case for larvae of Amphinomidae, Chrysopetalidae, Glyceridae, Nephtyidae, Oweniidae, Pectinariidae, Polynoidae, and Sabellaridae. However, other annelid families with planktotrophic larval development use different modes of feeding behavior. Taxa with nonfeeding larvae possessing maternally derived nutrition can be found all over the annelid tree within various families, and lecithotrophy may have evolved secondarily (Rouse 2000a). A special case of lecithotrophic development is represented by adelphophagy, where larvae develop by uptake of nutrients from nurse eggs, which are only produced as nutrition reserve (Fig. 9.9), as in some

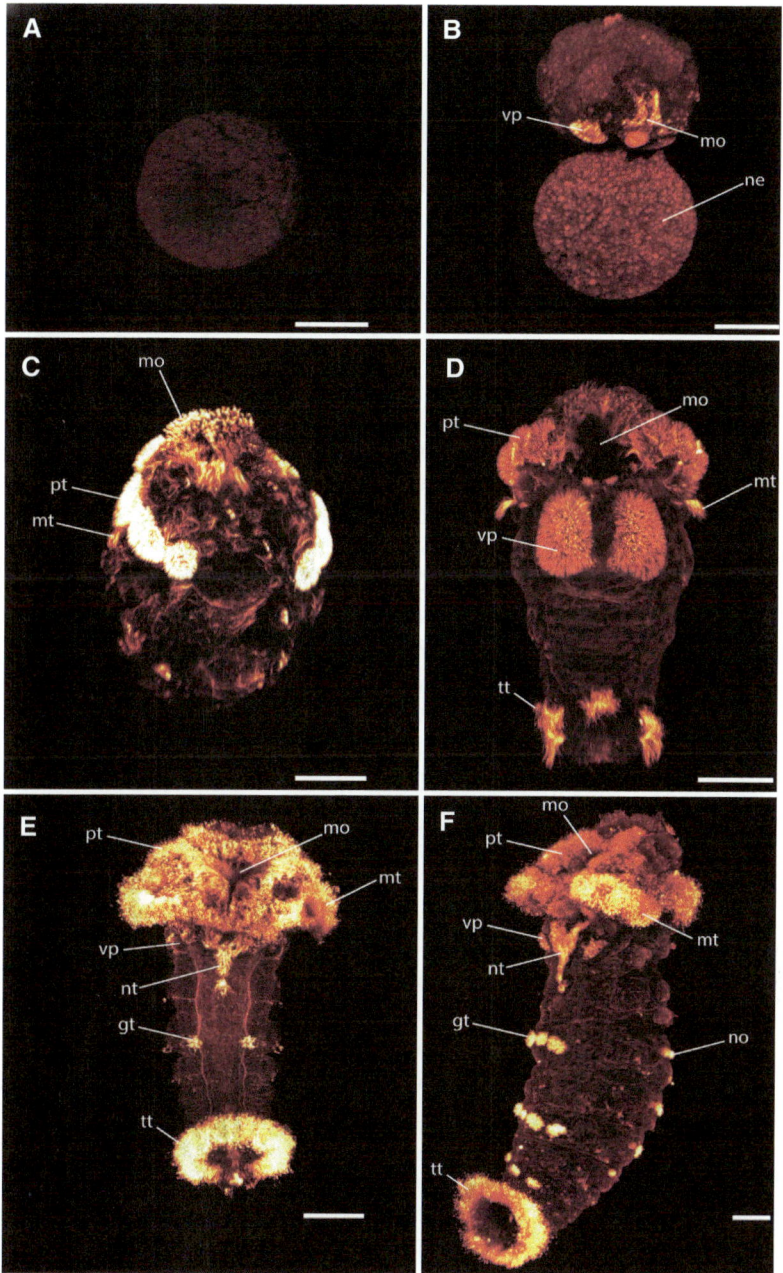

Fig. 9.9 Development of ciliation in larvae of the adelphophagous spionid *Boccardia* cf. *polybranchia* revealed by anti-tubulin staining. All images are in ventral view; anterior is up. Confocal maximum projections. (**A**) Early larvae lack ciliary regions and are full of yolk. (**B**) The trochophoral stage attaches to a nurse egg (*ne*) and starts to digest its nutrients. Ciliation is only exhibited within the mouth opening (*mo*) and in the region of the ventral ciliated patches (*vp*), which are used for attachment. (**C**) The late trochophore develops a distinct prototroch (*pt*) arranged of ciliated patches and a less-prominent metatroch (*mt*). (**D**) In the metatrochophoral stage, the prototroch (*pt*), metatroch (*mt*), the ventral ciliated patches (*vp*), and a telotroch (*tt*) are present. First signs of segmentation are visible in this stage. (**E**) In the late metatrochophore the ventral ciliated patches (*vp*) are reduced. Instead, a distinct neurotroch (*nt*) and several gastrotrochs (*gt*) form. (**F**) In the nectochaetal stage, shortly before leaving the egg capsule, three bands of gastrotrochal ciliary bands (*gt*) are developed, and notopodial cilia (*no*) are exhibited. The remaining ciliated bands are still present. *gt* gastrotroch, *mo* mouth opening, *mt* metatroch, *ne* nurse egg, *no* nototroch, *nt* neurotroch, *pt* prototroch, *tt* telotroch, *vp* ventral ciliated patch. Scale bars = 50 μm (© Conrad Helm, 2015. All Rights Reserved)

spionid taxa (Gibson and Carver 2013). Moreover, poecilogeny, showing different types of development in one species, seems to be common in some spionid genera (Blake and Kudenov 1981; Levin 1984; Chia et al. 1996).

Segmentation

Segmented annelids generate their first segments, usually simultaneously, as larvae, and later segments are sequentially added from a posterior growth zone (Irvine and Seaver 2006). Consequently, some authors differentiate between primary and secondary segments, an idea that goes back to Iwanoff (1928) who postulated a distinct ontogenetic origin for both sets of segments. Segment formation on the cellular level is well understood for clitellate embryos, which are all direct developers and often show huge and therefore experimentally manipulable eggs. Segmentation in leeches is strongly correlated with cell division patterns of teloblast cells and their descending blast cells (Fig. 9.10). All teloblasts originate from descendants of the D quadrant, with a pair of mesoteloblasts (M) being derived from 4d and four pairs of ectodermal teloblasts (N, O, P, Q) from 2d descendants. Mechanisms of the specification of the ectoteloblast lineage are different between the hirudinean *Helobdella* and the oligochaete *Tubifex*. In *Helobdella* the O/P teloblasts constitute an equivalence group, as they are both pluripotent and may subsequently follow either the O or the P fate (Weisblat and Blair 1984). On the contrary, in *Tubifex*, the fate of the P teloblast is determined by birth, whereas the O teloblast is initially pluripotent and is restricted to the fate of the O lineage due to signaling from the P lineage (Arai et al. 2001). Each teloblast undergoes repeated series of unequal division, producing bandlets of the so-called blast cells. The N and Q lineages produce two different types of blast cells, which appear in alternation. The four ectodermal bandlets of each side of the bilateral embryo join and form together with the mesodermal band, the germinal band, which lies at the

surface of the embryo. During gastrulation, the germinal bands from both sides of the embryo coalesce into the germinal plate, the origin of segments. The stereotyped cell divisions of all blast cells contribute to the forming segments (Shimizu and Nakamoto 2001; Weisblat and Huang 2001; Irvine and Seaver 2006). Each ectodermal band contributes neural and epidermal progeny, but two thirds of the neurons are derived from the N teloblast (Weisblat and Huang 2001). Ectodermal segmentation can be divided into two steps. In the first step, distinct cell clusters are generated autonomously by each bandlet. These separate clusters are aligned with the cell cluster derived from the mesodermal bandlet in a second step (Shimizu and Nakamoto 2001). It has been shown experimentally for *Helobdella* that hemilateral ablation of mesodermal precursor cells results in the loss of ectodermal segmental organization. In contrast, mesodermal boundaries are determined autonomously, without positional cues from ectodermal tissue (Blair 1982).

The cellular basis of segment generation in non-clitellate annelids is less well studied. One reason seems to be that they are more difficult to handle experimentally due to the smaller size of their embryos and larvae (Irvine and Seaver 2006). An obvious difference is the absence of large visible teloblasts in non-clitellate annelids (Seaver et al. 2005). However, using semiautomated cell tracking, mesoteloblast-like stem cells were revealed for *Platynereis dumerilii* which also form mesodermal bands of daughter cells (Fischer and Arendt 2013). For this species, two distinct sets of stem cells could be described for the posterior growth zone, where many rounds of division of small populations of teloblast-like stem cells generate new segments (Gazave et al. 2013). In contrast, Seaver et al. (2005) could not find evidence of a teloblastic growth zone in *Capitella teleta* and the serpulid *Hydroides elegans* using incorporation of BrdU, therewith confirming older studies (Wilson 1890; Shearer 1911). Instead, segments in larvae arise from a field of mitotically active cells located in lateral body regions. However, the authors could not

Fig. 9.10 Schematic representation of the stem cell-mediated, lineage-dependent segmentation in leeches and other clitellate annelids. Anterior is to the *top*. Pairs of *diagonal lines* indicate discontinuities in the depicted structures. One bilateral pair of mesodermal stem cells (*M* teloblasts) and four bilateral pairs of ectodermal stem cells (*N*, *O/P*, *O/P*, and *Q* teloblasts) constitute the posterior growth zone. Two types of blast cells are contributed by the N lineage, designated ns (*red*) and nf (*blue*), which arise in alternation. The numbers on the *left* side indicate the progressing time during segment formation given in hours of clonal age. *Arrows* indicate the delimitation of two ganglionic primordia (From Rivera and Weisblat 2009, with permission from the publisher)

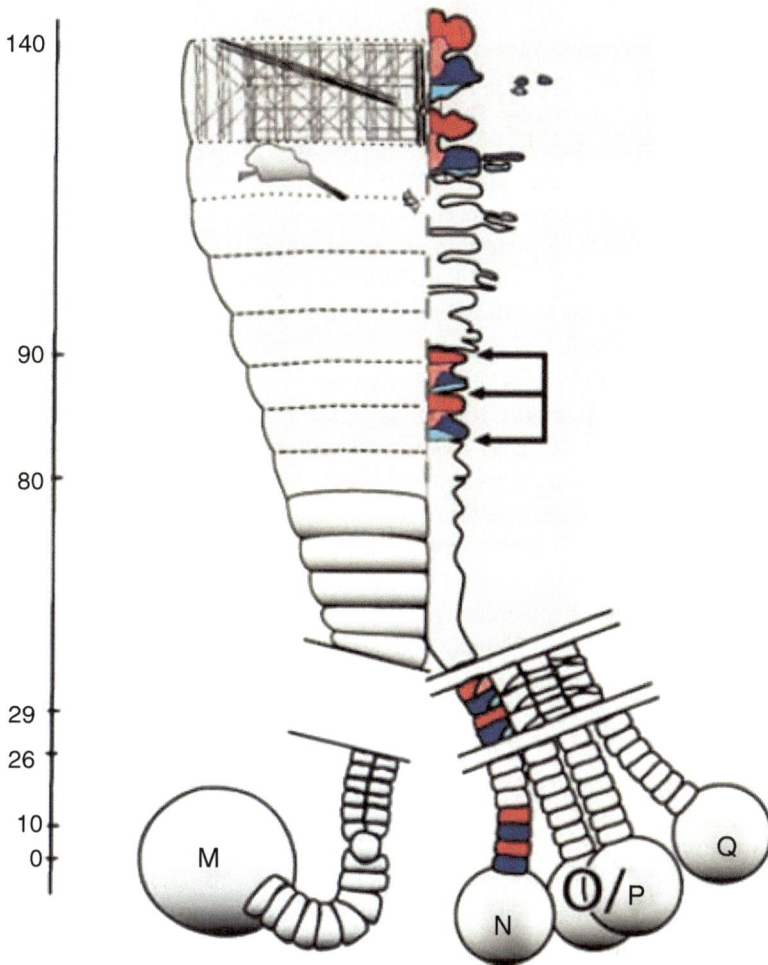

rule out if inconspicuous teloblast-like cells might be present. For *Chaetopterus*, it has been suggested that at least the first 15 segments are formed by subdivision of existing anlagen and not by a posterior growth zone (Irvine et al. 1999). Similarly, formation of repetitive structures in myzostomids differs from an addition governed by a posterior growth zone. As such, during development the third pair of parapodial structures appears first, followed by the fourth pair, second and fifth pair (simultaneously), and first pair (Jägersten 1940). Future studies of especially non-clitellate annelids are necessary to further assess the existing variability in the process of segment formation throughout Annelida.

Muscular System

Annelids show a huge variety of muscular organization, with longitudinal musculature organized in separate bands or massive plates and circular musculature that can be fully developed, incomplete, or even completely missing (Tzetlin and Filippova 2005). Therefore, it comes without surprise that differences in the development of the muscular system have been found across the investigated taxa. Phalloidin labeling coupled with confocal microscopy revealed an origin of muscular development posterior of the apical organ in the phyllodocid *Phyllodoce groenlandica*, with distinct transversal muscles starting to grow posteriorly. Subsequently, several longitudinal muscle fibers start to develop and grow in posterior direction. Simultaneously, outer circular muscle fibers begin to appear in a progression from anterior to posterior. The longitudinal muscle fibers reach the anal region approximately 7 days after hatching, and additional circular muscle fibers forming distinct rings develop from anterior to posterior. Additional longitudinal muscle fibers develop in the dorsal region, forming a continuous layer. Musculature of the digestive system is hardly recognizable in early stages. Notably, the organization of the body wall musculature starts before the formation of the first segments (Helm et al. 2013). Similarities to this kind of musculature development have been found in several other annelids. As such, an anterior origin in either lecithotrophic embryos or planktotrophic larvae is also reported for, e.g., capitellids, clitellates, nereidids, and sabellariids (Hill 2001; Bergter and Paululat 2007; Hunnekuhl et al. 2009; Brinkmann and Wanninger 2010a; Fischer et al. 2010). No circular musculature could be detected in developing stages of the maldanid *Axiothella rubrocincta*, even though they are present in adults (Brinkmann and Wanninger 2010b).

In contrast to the description for *Phyllodoce*, musculature of the digestive system develops before the body wall musculature in planktotrophic larvae of serpulids and sabellariids (McDougall et al. 2006; Brinkmann and Wanninger 2010a). Temporal shifts in the developmental onset of several muscle groups, a phenomenon described as heterochrony, are a common theme when comparing myogenesis between different and even closely related annelid species and are even more pronounced in the comparison of planktotrophic with lecithotrophic developing species (McDougall et al. 2006; Brinkmann and Wanninger 2010a; Helm et al. 2013). As in *Phyllodoce*, many annelid species show a successive appearance of circular musculature from anterior to posterior, which has been also described, e.g., for the tubificid *Limnodrilus* and the serpulid *Filograna implexa* (Bergter et al. 2007; Wanninger 2009). In contrast, in sipunculids, *Platynereis massiliensis*

and the leech *Erpobdella octoculata*, anterior circular muscles are formed synchronously (Wanninger et al. 2005; Bergter et al. 2007; Kristof et al. 2011; Helm et al. in press). Muscular development in the non-segmented sipunculids as analyzed for *Phascolion strombus*, *Phascolosoma agassizii*, *Thysanocardia nigra*, and *Themiste pyroides* shows that the first anlagen of circular body wall musculature appear simultaneously. Rudiments of four longitudinal retractor muscles appear at the same time, with longitudinal muscle fibers forming a pattern of densely arranged fibers around the retractor muscles (Kristof et al. 2011).

Neurogenesis

Annelids show a huge variety of adult nervous system organization, and until today the ancestral ground pattern remains under discussion (Bullock 1965; Orrhage and Müller 2005). The development of the nervous system, however, has been investigated in a surprisingly small number of taxa. Several transmission electron microscopy (TEM)-based studies on the larval nervous system of phyllodocids and serpulids were published by Lacalli (1981, 1984, 1986). Some detailed comparative studies were conducted concerning the anatomy of the larval apical organ. Immunocytochemical studies revealed an almost universal occurrence of an apical organ with flask-shaped cells in larvae of Annelida, Mollusca, Entoprocta, and Platyhelminthes, exhibiting FMRFamide- and serotonin-like immunoreactivity (e.g., Hay-Schmidt 2000; Wanninger 2009). Usually, the apical organ in annelid trochophores is simple, containing a few flask-shaped cells which have slender necks, dense cytoplasm, and a single projecting cilium (Lacalli 1981). Whereas these cells are missing in echiurans and many other annelids, sipunculans show a more complex apical organ with up to ten flask-shaped cells (Wanninger 2008). Marlow et al. (2014) analyzed the molecular fingerprint of apical organ cells in *Platynereis dumerilii*. They found that orthologs of *six3* and *foxq2* are involved in the formation of the apical plate, whereas the apical tuft is formed in a central *six3*-free area of the apical plate.

Besides this, only few comprehensive studies for developmental sequences of annelids combining immunocytochemical staining coupled with confocal laser scanning microscopy exist (Hessling 2002; Hessling and Westheide 2002; Voronezhskaya et al. 2003; McDougall et al. 2006; Brinkmann and Wanninger 2008; Kristof et al. 2008; Fischer et al. 2010; Winchell et al. 2010; Helm et al. 2013, in press). Main targets for these studies were serotonin, a biogenic amine involved in neuronal signaling, and the neuropeptide FMRFamide. Labeling of tubulin is additionally used to stain neurotubules. In summary, these studies show that neurogenesis in annelids is variable, following different developmental pathways. Planktotrophic larvae typically bear a serotonergic nerve ring underlying the prototroch and an apical organ that bears serotonergic and FMRFamidergic cells. The development of the larval nervous system usually starts from two subsystems (Fig. 9.11). FMRFamidergic immunoreactivity increases from anterior toward posterior during nervous system development. In *Phyllodoce* and some other annelids, a single serotonergic neuron located at the posterior pole of the larva is present (Fig. 9.11A). From here, anteriorly projecting nerve fibers start to grow, outlining the future ventral nerve cords (Voronezhskaya et al. 2003). Such a posterior origin of serotonin-like immunoreactivity was also detected in another phyllodocid, in syllids, nereidids, and orbiniids (Orrhage and Müller 2005; McDougall et al. 2006; Fischer et al. 2010; Helm et al. 2013). In contrast, the investigated sabellariids, spirorbids, and sipunculans show no evidence for a posterior serotonergic cell (Brinkmann and Wanninger 2008; Kristof et al. 2008; Brinkmann and Wanninger 2009). Later, the adult nervous system starts to develop along pathways established by the earliest peripheral neurons of the larva. However, other authors propose a separate development of the larval and adult nervous system (Lacalli 1984) or find that the larval nervous system is integrated in the adult one (Hay-Schmidt 1995). The relative timing of events during neurogenesis shows major shifts between compared species and is regarded as cases of heterochrony (McDougall et al. 2006; Brinkmann and Wanninger 2008; Helm et al. 2013).

A different picture is found in direct developing lecithotrophic species as investigated for nereidids. In *Nereis arenaceodentata*, the nervous system has developed already much of the complexity of the adult at hatching. This includes a large brain and the presence of circumesophageal connectives, nerve cords, and segmental nerves. Within 1 week after hatching, cephalic sensory structures and brain substructures are differentiated, and the nervous system architecture resembles that of adults (Winchell et al. 2010). A similar developmental pattern of the nervous system has been described for *Platynereis massiliensis* (Helm et al. in press).

Analyses of the development of the nervous system of the non-segmented Echiura and Sipuncula gained major interest, as they provided direct ontogenetic evidence for the indirectly inferred loss of segmentation in these taxa as suggested by molecular phylogenies. Using immunocytochemistry, a metameric organization of the nervous system has been demonstrated for two echiuran species: *Urechis caupo*, which has planktotrophic larvae, and *Bonellia viridis* with directly developing lecithotrophic stages (Hessling 2002; Hessling and Westheide 2002). The development of the nervous system in *Bonellia viridis* proceeds from anterior to posterior. This is obvious in early larvae, which show a full set of serotonergic perikarya in the anterior region, while this pattern is incomplete in the posterior area. This pattern suggests the presence of a posterior growth zone (Hessling and Westheide 2002). Similarly, in larval stages of *Urechis caupo*, a serial repetitive distribution of serotonin-containing neurons and their corresponding pairs of peripheral nerves, both formed in an anterior-posterior gradient, imply a segmental pattern. Moreover, larvae show a paired origin of the ventral nerve cord (Hessling 2002). Another case of "ontogeny recapitulating phylogeny" has been demonstrated for sipunculans, where neurogenesis of *Phascolosoma agassizii* follows a segmental pattern. During

Fig. 9.11 Schematic representation of neuronal development in larval stages of *Phyllodoce groenlandica* exhibiting formation from two subsystems (Modified from Helm et al. (2013)) Diagrams are in ventral view; anterior is up. Major types of neuronal structures are color coded. (**A**) 0.5 days past hatching (*dph*). First serotonergic immunoreactivity appears at the posterior pole. (**B**) 2 dph. Nervous system originates from posterior (serotonin) and anterior (FMRFamide). (**C**) 7 dph. Serotonergic and FMRFamidergic immuno-reactivities start to overlap to the greatest extent. (**D**) 11 dph. Numerous serotonergic cells are detectable within the epi- and the hyposphere. (**E**) 20 dph. Serotonergic and FMRFamidergic nerve cells are limited mainly to anterior regions. (**F**) 34 dph. The larval nervous system is fully developed. *at* apical tuft, *pt* prototroch

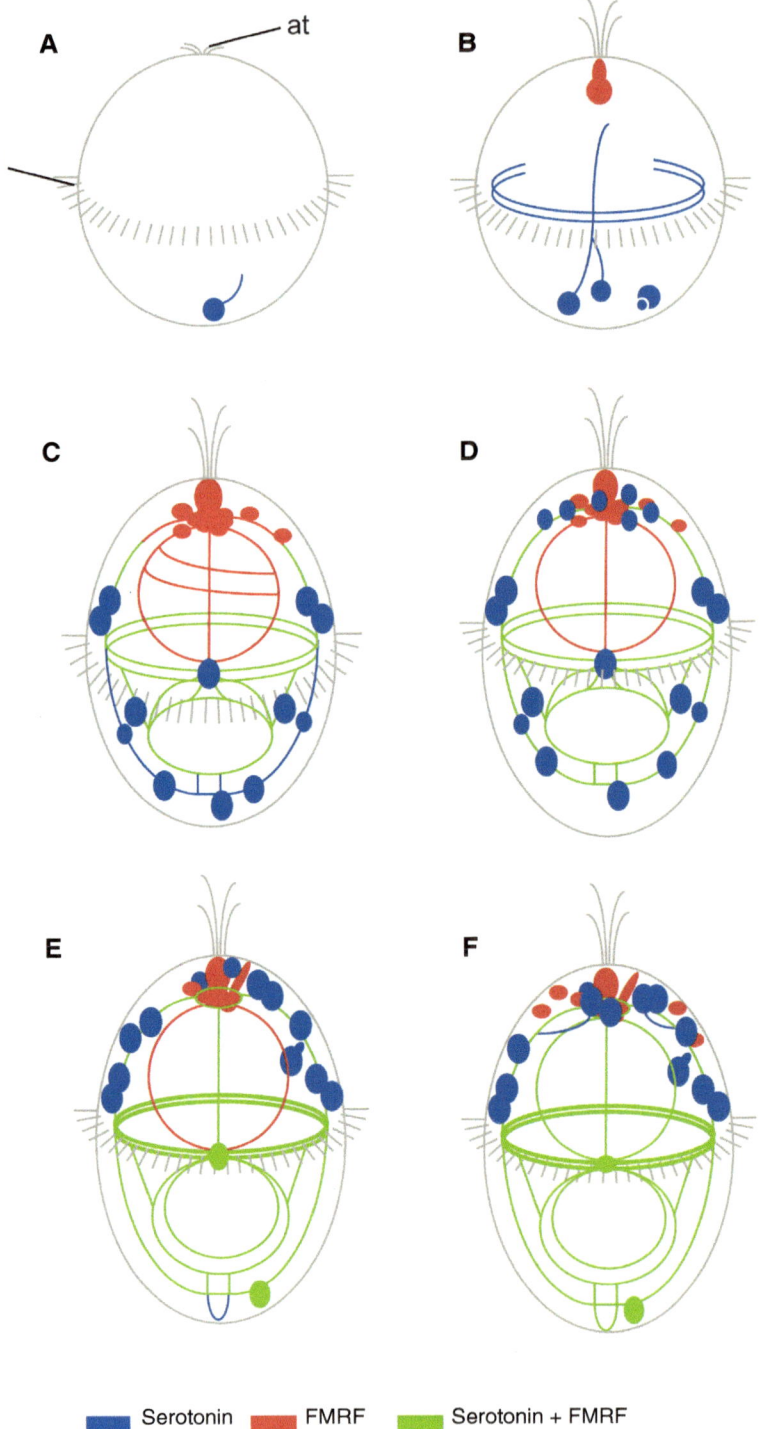

development of this species, a pair of FMRFamidergic and serotonergic axons gains four pairs of associated serotonergic perikarya and interconnecting commissures in an anterior-posterior progression. During later larval stages, the commissures disappear and the two serotonergic axons fuse, forming a single ventral nerve cord, and after cell migration, a nonmetameric central nervous system can be found in adults (Kristof et al. 2008). Interestingly, neurogenesis of the sipunculan *Phascolion strombus* lacks any signs of a segmented origin, and serotonergic structures are missing completely in their larvae, which may be the result of the abbreviated larval phase in this species (Wanninger et al. 2005).

Nephridia and Coelomogenesis

Most adult annelid taxa possess metanephridia as excretion organs; however, in the development they are usually preceded by protonephridia, which can be found in larvae and sometimes also in developing juveniles (Bartolomaeus and Quast 2005). These larval excretory organs were termed "head kidneys" by Hatschek (1886) and are located anteriorly in trochophore stages, closely behind the larval eyes (Bartolomaeus and Quast 2005). Later, homologous organs were also described from lecithotrophic developmental stages, as shown for the direct developing species *Scoloplos armiger* (intertidal clade), which has no free-swimming larval stage (Bartolomaeus 1998). Head kidneys are present in some echiurans but are missing in sipunculid developmental stages. It is hypothesized that the ancestral state of larval annelid protonephridia was an organ composed of three cells: a terminal cell, a nephropore cell, and a duct cell (Bartolomaeus and Quast 2005). This simple construction has been modified in adaptation to different developmental modes in several annelid lineages and especially in planktotrophic larvae, where the nephropore cell is often missing (Kato et al. 2011, 2012).

Adult segmental nephridia differentiate from a single anlage, consisting of few cells which line a small lumen filled with microvilli. This duct becomes ciliated, and the most proximal cells are separated during coelomogenesis. During coelomic cavity growth, the proximal part of the anlage is passively opened, forming the metanephridial funnel. A truncation of this process due to suppression of the separation of duct cells leads to a differentiation into a protonephridium, as, for example, observed in several taxa of the Phyllodocida (Bartolomaeus 1999). The development of metanephridia in the non-segmented sipunculids as investigated for *Golfingia minuta* seems to be similar as in segmented annelids. An overview of annelid nephridial organs is given by Bartolomaeus and Quast (2005).

Segmented annelids show a heteronomous coelomogenesis, and the coelomic lining is of mesodermal origin. Prior to metamorphosis (if present), a pair of unsegmented coelomic cavities stretches out over the first larval segments. This process has been studied in detail for the serpulid *Spirorbis spirorbis*, where two caudally located mesodermal cell clusters proliferate cells, which merge to surround the gut ventrally close to the anus. Fluid starts to accumulate between spaces of desmosomes, and at the same time myofibrils appear, and due to growth and separation processes, this myoepithelium develops into the coelomic lining. After migration, the two coelomic cavities meet dorsally, completely surrounding the gut. Postmetamorphic stages develop strictly segmental coelomic cavities during segment formation. Coelomic cavities are highly reduced or completely missing in leeches and several meiofauna annelids with a presumed progenetic origin (Koch et al. 2014).

GENE EXPRESSION

Only few model annelids are well characterized concerning gene expression patterns. Prime candidate taxa are leeches of the genera *Helobdella* and *Hirudo*, the sedentary annelid *Capitella teleta*, and the errant annelid *Platynereis dumerilii*. Besides this, some studies exist for the chaetopterid *Chaetopterus*, some sipunculid worms, and a few serpulids as well as for additional nereidid and clitellate species. These studies mainly used candidate gene approaches, where orthologs were chosen based on studies in arthropods and vertebrates. Main points of interest are the genomic basis of segmentation, Hox gene expression, and nervous system development as well as gastrulation and gut development.

Segmentation

Metameric segmentation can be found in vertebrates, arthropods, and annelids, and the distant phylogenetic position of these taxa gave rise to the question of how often this feature evolved in animals (Seaver 2003). Traditionally well investigated is the molecular background of segmentation in vertebrates and arthropods, which show profound differences (Tautz 2004). Given the fact that a close relationship between arthropods and annelids was suspected as formulated in the Articulata hypothesis (Scholtz 2002), a possible common ancestry of segmentation in these taxa became a focus of many evolutionary developmental studies of annelids. Genes or gene families identified to play a vital role in segment formation in arthropods were used as candidates in several studies (Wedeen and Weisblat 1991; Prud'homme et al. 2003; Seaver and Kaneshige 2006; Saudemont et al. 2008; Dray et al. 2010; Steinmetz et al. 2011).

Best investigated is the molecular background of segmentation in *Platynereis dumerilii* and *Capitella teleta*. For *Drosophila* it has been demonstrated that para-segmental borders are generated by an interaction between the segment polarity genes *wingless* and *engrailed* (Tautz 2004). The gene *wingless* (or *Wnt1*) is part of the Wnt gene family, and *engrailed* is a homeodomain-bearing transcription factor. As in arthropods, a role in segment formation is suggested for this pair of genes in *P. dumerilii* (Prud'homme et al. 2003), where an expression of continuous ectodermal stripes is observed for these genes at the border of the segments during their formation. However, investigation of other annelid taxa questions the conserved nature for *engrailed* as a "segmentation gene" in annelids.

In *Chaetopterus*, *engrailed* is expressed during all larval stages in different structures or organs, and no signs of a putative segment polarity pattern of expression are obvious (Seaver et al. 2001). Congruently, no conserved segment polarity pattern was found investigating the expression of this gene in developing *C. teleta* or *Hydroides elegans* individuals (Seaver and Kaneshige 2006). Finally, ablation of individual cells expressing *engrailed* in the leech *Helobdella* did not hinder remaining segmental clones in their normal development (Seaver and Shankland 2001). Consequently, the establishment of segment polarity in the leech (and possibly many other annelid taxa) seems to be independent of cell interactions across the anterior-posterior axis as known for arthropods (Seaver and Shankland 2001).

A set of pair rule genes is expressed in *Drosophila* and many other arthropods to pattern the embryo across the anterior-posterior axis, including *eve*, *hairy*, and *runt* (Damen 2007). All or some of these genes were investigated in detail for *C. teleta* and *Helobdella robusta*. In contrast to *Drosophila*, where it is expressed in stripes in the growth zone, the *Capitella* ortholog of *hairy* (*Cap-hes1*) shows an expression limited to a small band of cells in each larval segment. In juveniles its expression is limited to a small mesodermal domain of the posterior growth zone (Thamm and Seaver 2008). In vertebrates and some arthropods, the expression of hairy is controlled by the Notch pathway (Stollewerk et al. 2003). In *Capitella*, *Notch* and *hairy* do not show a broadly overlapping expression, with a *Notch* localization in already formed segments, anterior to the *hairy* signal (Thamm and Seaver 2008). In *Helobdella*, *hairy* is expressed in teloblasts and primary blast cells. The expression peak correlates with the production of blast cells by the teloblasts. However, no striped pattern suggesting a pair rule function was found (Rivera et al. 2005). Similarly, *Notch* is also expressed in teloblasts and blast cells, and functional studies revealed that the disruption of the *Notch/hairy* signaling results in a disruption of segmentation (Song et al. 2004; Rivera and Weisblat 2009). For *Platynereis dumerilii*, 15 *hairy* paralogs could be identified, which are expressed in mesodermal tissue, forming segments, and during neurogenesis, where it may be involved in the patterning of the nervous system (Gazave et al. 2014). However, these authors also found no overlap with the expression of *Notch*.

The expression patterns of the arthropod pair rule genes *eve* and *runt* do not suggest a similar role in *Capitella*. Instead, the expression of *runt* can be found in the brain and ventral nerve cord, as well as the fore- and hindgut. Two *eve* paralogs were characterized for *Capitella*, both showing a complex expression pattern, which does not correspond to segmental stripes as expected by results from *Drosophila* (Seaver et al. 2012). Likewise, de Rosa et al. (2005) did find such a pattern of *eve* expression in developing *Platynereis dumerilii*. However, these authors speculate about a role of this gene in the posterior addition of segments. A detailed functional study for *eve* has been conducted for *Helobdella* (Song et al. 2002). Segments arise sequentially from five pairs of teloblasts in leeches (see above), and *eve* is expressed in these teloblasts and their primary blast cells in *Helobdella*. Later embryos express *eve* in cells of the ventral nerve cord which stem from the N teloblast. Morpholino knockdowns suggest a role of *eve* in early cell division through early segmentation in *Helobdella*. However, no pair rule pattern is found for this gene in the leech.

The zinc finger transcription factor *hunchback* plays the role of a gap gene in *Drosophila*, which defines expression domains of pair rule and Hox genes (see Vol. 5, Chapter 1). Moreover, *hunchback* is involved in mesoderm development and neurogenesis. In *Platynereis dumerilii*, hunchback expression is detected in mesodermal cells belonging to the posterior growth zone of juvenile worms. Additionally, expression in the precursors of the somatic segmented mesoderm, formed during larval development, could also be confirmed, a striking similarity with arthropods (Kerner et al. 2006). However, an expression of *hunchback* could not be detected in segmental precursor cells of the posterior growth zone in *Capitella* and *Helobdella*, and a role in the patterning of the anterior-posterior axis was rejected

for these species (Iwasa et al. 2000; Werbrock et al. 2001).

NKL genes are a family of homeodomain transcription regulators that are involved in the patterning of mesodermal derivatives in *Drosophila* (Holland 2001; Jagla et al. 2001). The expression of seven genes of this cluster has been investigated in developing *Platynereis dumerilii*, and all are involved in the specification of mesodermal derivatives including muscular precursors (Saudemont et al. 2008). Notably, five of the investigated genes (*NK4*, *Lbx*, *Msx*, *Tlx*, and *NK1*) show an expression in complementary stripes in the mesoderm and/or ectoderm of developing segments. Moreover, genes of the Hedgehog signaling pathway show a similar striped pattern of expression, and segment formation in *P. dumerilii* is disrupted when treated with molecules antagonistic to this signaling (Dray et al. 2010).

Wnt genes regulate a wide range of developmental processes, including axis elongation and segmentation (Cadigan and Nusse 1997). This gene family ancestrally includes 13 paralog groups, of which several metazoan lineages lost some of the genes (Janssen et al. 2010). In *Platynereis dumerilii* and *Capitella teleta*, all paralog groups besides *Wnt3* could be discovered. In the leech *Helobdella robusta*, only nine paralog groups are present, with additionally *Wnta*, *Wnt8*, and *Wnt9* missing. Most Wnt genes in *P. dumerilii* are expressed in ectodermal segmental stripes and/or in the area around the pygidium (Janssen et al. 2010). Expression analyses in *H. robusta* and *C. teleta* led to comparable results (Cho et al. 2010). Due to similarities with arthropods, a role of Wnt genes in segment formation in both annelids and arthropods is suggested by some authors (Janssen et al. 2010).

In summary, the candidate gene approach led to the discovery of many similarities as well as differences between annelids and arthropods in gene expression patterns during the formation of segments. The expression of some genes at segmental boundaries in *Platynereis dumerilii* shows a remarkable similarity to arthropods. However, for other annelid taxa investigated for these candidate genes (as, e.g., for *engrailed* or *hunchback*), the picture becomes less clear, and future studies covering segment formation in more annelid taxa are clearly wanted. Moreover, fewer similarities are found compared with arthropods when investigating pair rule genes. In the discussion of a putative common ancestry of segmentation in annelids and arthropods, different authors come to different conclusions using basically the same set of evidence (de Rosa et al. 2005; Thamm and Seaver 2008). However, homology of genes expressed during segment formation must not imply a homology of a segmented body plan itself. At present, available developmental, paleontological, and phylogenetic evidence supports a convergent evolution of segmentation in arthropods and annelids (Couso 2009; Chipman 2010). Given this hypothesis, co-option of the same set of genes into the process of segment formation leading to a convergent pattern of gene expression can explain the similarities found between annelids and arthropods (Chipman 2010; Ferrier 2012).

Hox and ParaHox Genes

Hox genes comprise a family of transcription factors bearing a DNA-binding homeodomain (Gellon and McGinnis 1998). Hox genes are usually found as linked chromosomal clusters and show spatial and temporal collinearity (Garcia-Fernandez 2004). This means that genes from the 5'-end of the cluster are usually expressed more anteriorly than the ones from the 3'-end. In similar fashion we also see a temporarily earlier onset of genes from the 5'-end compared to those from the 3'-end. In bilaterian animals Hox genes are mainly involved in the patterning of body regions; however, several examples of co-option into other areas of expression are described (Wagner et al. 2003). All these characteristics made this set of genes a prime target for evolutionary developmental biologists to understand major transitions in animal body plan evolution (Akam 1998).

For annelids, the genomic organization of the Hox cluster is only fully described for *Capitella teleta* and *Helobdella robusta* (Fröbius et al.

2008; Simakov et al. 2013). In *Capitella*, assembled whole genome shotgun data found Hox genes distributed on three scaffolds, with one scaffold containing the *Post1* genes clearly separated from the others. In contrast, the leech *Helobdella* shows an extensive fragmentation of the Hox cluster (Simakov et al. 2013). For *Capitella*, 11 Hox genes (*lab*, *pb*, *Hox3*, *Dfd*, *Scr*, *lox5*, *Antp*, *lox4*, *lox2*, *Post2*, *post1*) corresponding to 11 different paralog groups were detected, and the presence of these genes are regarded as ancestral for lophotrochozoans in general (Fröbius et al. 2008; Simakov et al. 2013). Interestingly, *Helobdella* also shows a derived pattern here, with the duplication of two paralog groups (five copies of *Scr* and two copies of *Post2*) and the loss of orthologs of *pb* and *Post1* (Simakov et al. 2013). For many other annelids, information about the Hox gene complement are available through PCR and cloning studies; however, genomic organization and absence of genes cannot be derived from this approach (Dick and Buss 1994; Snow and Buss 1994; Irvine et al. 1997; Cho et al. 2003, 2006; Kulakova et al. 2007; Bleidorn et al. 2009).

The expression of Hox genes during development has been only investigated for a few annelid taxa, *Capitella teleta*, *Chaetopterus variopedatus*, *Alitta* (*Nereis*) *virens*, *Platynereis dumerilii*, *Hirudo medicinalis*, and two *Helobdella* species (Irvine and Martindale 2000; Peterson et al. 2000; Kulakova et al. 2007; Fröbius et al. 2008; Gharbaran and Aisemberg 2013). The most inclusive study deals with *C. teleta*, where for the first time spatial and temporal collinearity for Hox genes could be demonstrated for a lophotrochozoan taxon (Fröbius et al. 2008). *Capitella* Hox genes, except for *Post1*, are all expressed in ectodermal domains during larval development, with a spatial correlation of anterior expression borders and location of genes in the Hox cluster. Anterior class Hox genes (*lab*, *pb*, *Hox3*) are the first genes expressed, occurring before the appearance of segments. The expression of *Dfd* and *Scr* can be detected after appearance of the first segments, followed by the expression of *lox5*, *Antp*, and *lox4*. The expression of *lox2* and *Post2* appears last. Interestingly, all Hox genes in

C. teleta show their highest expression level at a unique stage during the course of development, reflecting the order of activation for each gene. A unique Hox gene expression boundary can be detected for all nine thoracic segments, and the posterior-most located Hox genes (*lox2* and *Post2*) are only expressed in the abdomen (Fig. 9.12). Whereas no expression of Hox genes could be detected in the pygidium of *C. teleta*, expression of *Post2* was detected in the pygidium of nereidids (Kulakova et al. 2007). Expression of *Post1*, the gene which seems to be separated from the rest of the Hox cluster, could not be detected in any investigated stages of *C. teleta*, besides some signals in chaetal sacs (Fig. 9.12; Fröbius et al. 2008). This result is congruent with analyses of expression of this gene in nereidids (Kulakova et al. 2007).

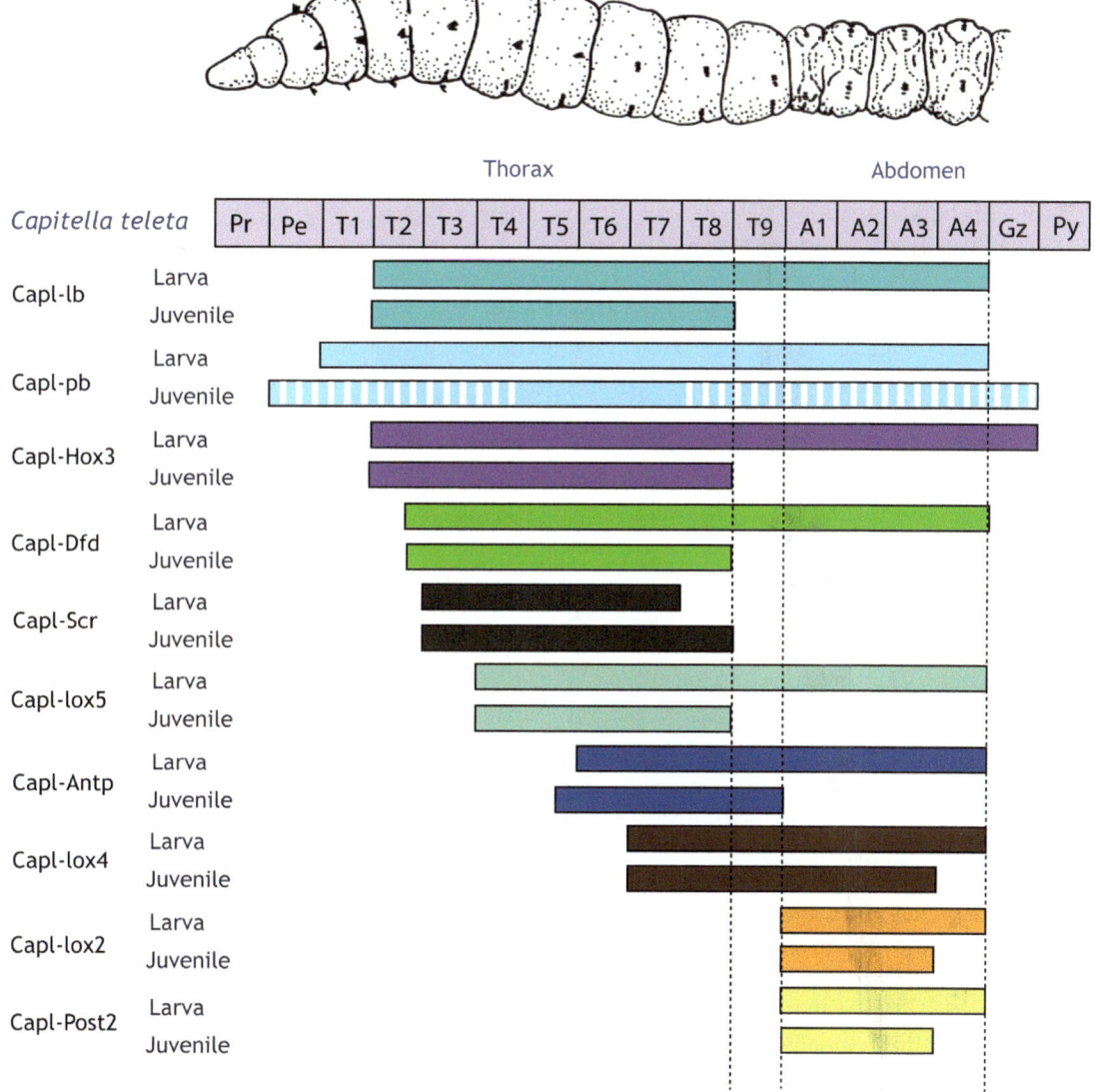

Fig. 9.12 Hox gene expression profile in larvae and juveniles of *Capitella teleta* after Fröbius et al. (2008). *Solid bars* indicate strong expression; *dashed bars* indicated weaker expression. Abbreviations: *A1–4* abdominal segments 1–4, *Gz* growth zone, *Pe* peristomium, *Pr* prostomium, *Py* pygidium, *T1–T9* thoracic segments 1–9

A staggered expression of five Hox genes generally in line with spatial and temporal collinearity can also be found in *Chaetopterus*, even though the genomic organization of the Hox cluster remains unknown in this species (Irvine and Martindale 2000). No strict temporal collinearity was found in expression studies for nereidid worms and *Helobdella* (Kourakis et al. 1997; Kulakova et al. 2007). However, all studies suggest an involvement of Hox genes in body patterning along the anterior-posterior axis, a function that seems to be the ancestral role of Hox genes in bilaterian animals (Kulakova et al. 2007; Butts et al. 2008). Notably, annelids show a predominant expression of Hox genes in neurogenic structures such as the ganglia and the ventral nerve cord. This is especially obvious in *Chaetopterus* and leeches (Shankland et al. 1991; Aisemberg and Macagno 1994; Wong et al. 1995; Kourakis et al. 1997; Irvine and Martindale 2000; Gharbaran and Aisemberg 2013).

The ParaHox cluster is a paralog of the Hox gene cluster, containing three genes (*Gsx*, *Xlox*, and *Cdx*) (Brooke et al. 1998). As for Hox genes, temporal collinearity has been likewise demonstrated for ParaHox genes in many instances. However, the ParaHox cluster seems to be lost in several investigated ecdysozoan taxa which show a breakup of the cluster and missing genes (Ferrier and Minguillon 2003). All three ParaHox genes seem to be present in annelids (Ferrier and Holland 2001; Fröbius and Seaver 2006; Park et al. 2006). The genomic organization of ParaHox genes has been studied in detail for *Platynereis dumerilii* (Hui et al. 2009). In this species, a head-to-head location of *Gsx* and *Xlox* could be demonstrated, with *Cdx* located in a separate position on the same chromosome. Expression analyses of these genes in *Alitta* (*Nereis*) *virens* suggest a role in anterior-posterior patterning of the digestive system and in the specification of neuroectodermal cell domains (Kulakova et al. 2008). Especially *Gsx* seems to be involved in the development of the brain in all investigated annelids (*P. dumerilii*, *A. (N.) virens*, *C. teleta*), a function which is regarded as ancestral for these genes for bilaterians in general (Fröbius and Seaver 2006; Kulakova et al. 2008).

Genes Involved in Neurogenesis

The development of the central nervous system has been deeply studied for *Platynereis dumerilii*. Neural progenitor cells are located close to the ventral midline and express *axin*, a negative regulator of the Wnt/β-catenin pathway, which controls the transition between these proliferating cells and differentiating neurons (Demilly et al. 2013). Wnt-controlled proliferation of neural progenitors is also well-documented for vertebrates and arthropods, especially *Drosophila* (Bielen and Houart 2014). Using a candidate gene approach, genes with a conserved expression in developing vertebrate and arthropod brains were chosen as major targets in studies on annelids. The developing head of annelid larvae is demarcated by the expression of *six3* and *otx* homeobox genes (Fig. 9.13), a patterning system that might be universal to bilaterian animals (Steinmetz et al. 2010). MicroRNAs are short noncoding RNAs that posttranscriptionally regulate gene expression (Ambros 2004). The expression of several microRNAs is highly tissue specific and conserved across animals (Christodoulou et al. 2010). In *P. dumerilii* and *Capitella teleta*, the microRNAs *mir-7*, *mir-137*, and *mir-153* show a localized expression in distinct neurosecretory brain tissue, a pattern which was also found in zebra fish (Tessmar-Raible et al. 2007; Christodoulou et al. 2010). The expression of the complementary pair *mir-9* and *mir-9**/*mir-131* is restricted to two sets of differentiated neurons in the developing annelid brain, with the most apical cells located at the base of the antennae (Christodoulou et al. 2010). The three transcription factors *rx*, *otp*, and *nk2.1* are all expressed in the developing forebrain of *P. dumerilii*. All cells expressing these genes, as well as *mir-7*, are vasotocinergic extraocular photoreceptors. The expression pattern matches those known from the same cell type in zebra fish (Tessmar-Raible et al. 2007). Gene networks controlling the pattern along the anterior-posterior axis of the central nervous system are conserved across bilaterians, and the involved genes are mainly Hox genes (Ferrier 2012). The patterning of the dorso-ventral axis in *P. dumerilii* is

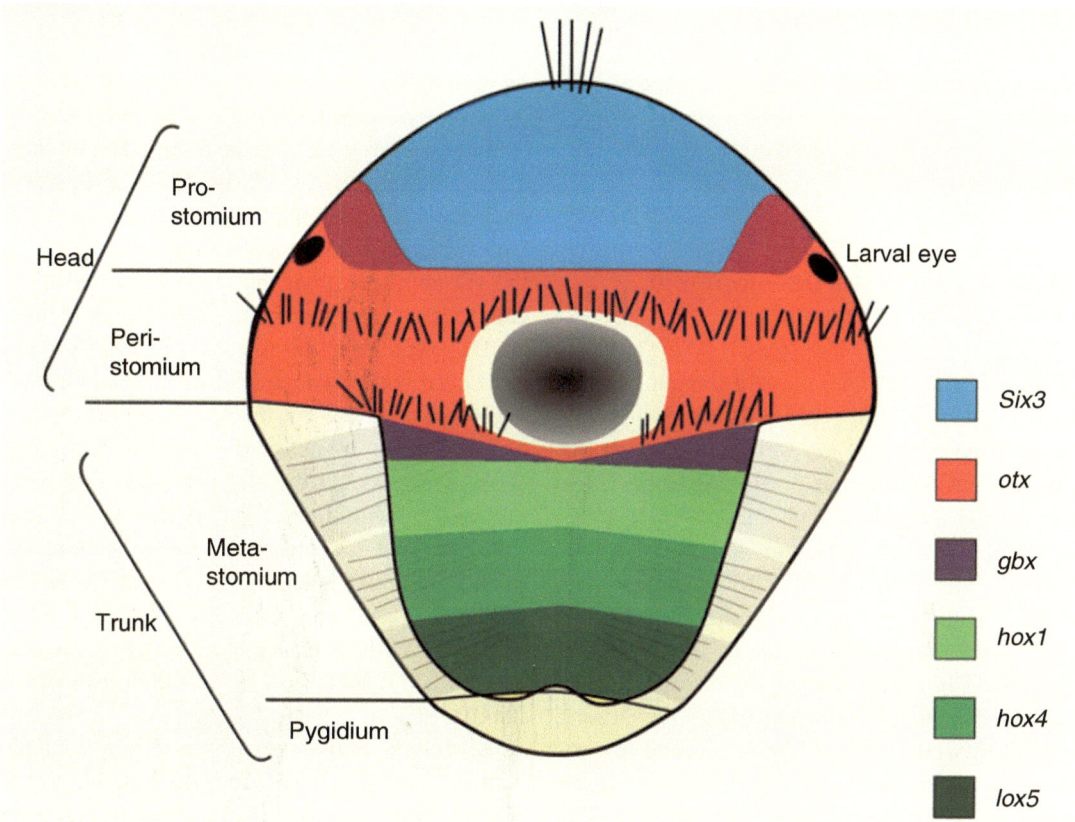

Fig. 9.13 Expression of *six3*, *otx*, *gbx*, and Hox genes in neuroectodermal regions of *Platynereis dumerilii* larvae. The *six3* and *otx* expressing regions cover the developing prostomium and the peristomium, from which the cerebral ganglia and eyes develop. *Dark gray* region marks the mouth (From Steinmetz et al. (2010))

controlled by a gene network including *nk2.2*, *nk6*, *Pax2/5/8*, *Pax6*, *Pax3/7*, *dlx*, *msx*, *gsx*, *sim*, and *dbx* (Denes et al. 2007; Ferrier 2012). A similar patterning of the neuroepithelium is obvious in vertebrates (Denes et al. 2007).

Besides the detailed investigations summarized for *Platynereis dumerilii*, some gene expression studies dealing with the development of the nervous system in leeches and *Capitella teleta* are published. As shown for *P. dumerilii*, *otx* shows a largely head-specific expression in *Helobdella* (Bruce and Shankland 1998). The expression of *Lox10*, a putative *nk2.1* ortholog, was detected in the developing brain of *Helobdella*, congruent with the results for *P. dumerilii* (Nardelli-Haefliger and Shankland 1993). In *Hirudo medicinalis*, the central class Hox gene *Lox1* controls the differentiation of the so-called "rostral penile evertor neurons" that innervate the male penis (Gharbaran and Aisemberg 2013). For the same species, expression of the axon migration guiding protein *netrin* is shown to be involved in forming interganglionic neuronal tracts and in defining ventrodorsal boundaries of peripheral innervation (Gan et al. 1999). Another protein family investigated in *H. medicinalis* is the innexins, where several cloned members show a restricted expression in neurons (Dykes and Macagno 2006). In *C. teleta*, *Delta* and *Notch* expression was detected during brain development in larvae, as well as in the forming ganglia of the ventral nerve cord of juveniles, suggesting a role of Notch signaling in neurogenesis (Thamm and Seaver 2008).

In summary, development of the central nervous system and expression patterns of genes localized in the brain show strong similarities between vertebrates and annelids. Based on these results (and further studies involving other taxa), the presence of a centralized nervous system in the last common ancestor of protostomes and deuterostomes seems plausible for several authors (Arendt et al. 2008; Holland et al. 2013).

Genes Acting in the Development of the Digestive Tract

The gut of annelids consists of a foregut (stomodeum) and a hindgut (proctodeum), both originating from ectoderm, as well as of the midgut which is of endodermal origin. All three parts can usually be subdivided into different functional regions (Tzetlin and Purschke 2005). Several genes involved in bilaterian foregut and hindgut patterning have been investigated *for Platynereis dumerilii* (Arendt et al. 2001). As such, the T-box transcription factor *brachyury* is expressed in the ventral part of the developing foregut as well as in the hindgut of late trochophore larvae, resembling the pattern known from larvae of basal branching deuterostomes. The homeobox gene *goosecoid* is first expressed in a small number of cells at the anterior blastopore margin which develops into the foregut. Expression can be additionally detected in adjacent cells which will contribute to the development of the foregut nervous system. As the expression patterns of the investigated genes seem to be conserved in protostomes and deuterostomes, a single origin of the tripartite bilaterian through gut has been hypothesized (Arendt et al. 2001). This idea has been later challenged based on expression studies of the same set of genes in acoels, which are lacking a through gut (Hejnol and Martindale 2008).

Genes involved in the patterning of ectodermal and endodermal parts of the gut have been studied for *Capitella teleta*, *Chaetopterus variopedatus*, and the sipunculid *Themiste lageniformis*. In *C. teleta*, the transcription factor *FoxA* and genes of the GATA family are expressed across the entire developing gut (Boyle and Seaver 2008). Different genes of the GATA family are exclusively expressed in the developing midgut, with a prominent expression of *gataB1* at its boundaries. In contrast, *FoxA* expression can be detected surrounding the blastopore during development as well as in the foregut and hindgut during organogenesis. Partly similar expression patterns are reported for *Themiste* and *Chaetopterus* and might reflect the differences in the gut architecture of these species with different feeding mechanisms (Boyle and Seaver 2010). Moreover, expression of the ParaHox gene *Cdx* is also reported for anterior and posterior regions of the gut in *C. teleta* (Fröbius and Seaver 2006). In nereidids, the ParaHox gene *Gsx* is expressed during the formation of the foregut and the midgut. The expression of the ParaHox gene *Xlox* has been detected in all investigated annelids, including *Helobdella*, *H. medicinalis*, *C. teleta*, and nereidids (Wysocka-Diller et al. 1995; de Rosa et al. 2005; Fröbius and Seaver 2006; Kulakova et al. 2008; Hui et al. 2009).

Bilaterian animals are divided into deuterostomes, ecdysozoans, and lophotrochozoans (Edgecombe et al. 2011). Whereas research on several well-established model organisms in the former two groups (e.g., *Drosophila*, *Caenorhabditis*, *Danio*, *Mus*) has provided detailed insights into molecular mechanisms of the development, lophotrochozoans have traditionally been chronically understudied in this regard (Tessmar-Raible and Arendt 2003). The rise of *Platynereis dumerilii* (and other annelids like *Capitella* and *Helobdella*) as EvoDevo models has provided major insights into the evolution of the nervous system and segment formation in annelids, a key lophotrochozoan phylum (see above). Interestingly, the genomic architecture of *Platynereis* seems to be little derived from a hypothetical bilaterian ground pattern, enabling many insights into comparative developmental genomics (Raible et al. 2005; Ferrier 2012). Future studies focusing on additional annelid lineages, such as the basal branching oweniids or the non-segmented sipunculans, will certainly improve our understanding of the evolution of bilateria in general.

OPEN QUESTIONS

- Segment formation in non-clitellate annelids
- Myogenesis in Echiurida
- Development of the nervous system in basal branching annelids
- Homology of ciliary bands including the various trochi in different annelid and lophotrochozoan larvae
- Genetic background of annelids showing putative deuterostomy
- Comparative expression studies of Hox, ParaHox, and other key developmental genes across the various annelid subtaxa, especially lesser-known groups such as Myzostomida, Sipuncula, and Echiurida
- Gene expression studies of "segmentation genes" in non-segmented annelids

References

Ackermann C, Dorresteijn A, Fischer A (2005) Clonal domains in postlarval *Platynereis dumerilii* (Annelida: Polychaeta). J Morphol 266:258–280

Aisemberg GO, Macagno ER (1994) *Lox1*, an *Antennapedia*-class homeobox gene, is expressed during leech gangliogenesis in both transient and stable central neurons. Dev Biol 161:455–465

Akam M (1998) Hox genes, homeosis and the evolution of segment identity: no need for hopeless monsters. Int J Dev Biol 42:445–451

Åkesson B (1967) The embryology of the polychaete *Eunice kobiensis*. Acta Zool 48:142–192

Ambros V (2004) The functions of animal microRNAs. Nature 431:350–355

Amiel AR, Henry JQ, Seaver E (2013) An organizing activity is required for head patterning and cell fate specification in the polychaete annelid *Capitella teleta*: new insights into cell-cell signaling in Lophotrochozoa. Dev Biol 379:107–122

Anderson DT (1959) The embryology of the polychaete *Scoloplos armiger*. Q J Microsc Sci 100:89–166

Anderson DT (1966) The comparative embryology of Polychaeta. Acta Zool 47:1–42

Anderson DT (1973) Embryology and physiology in annelids and arthropods. Pergamon, Oxford

Arai A, Nakamoto A, Shimizu T (2001) Specification of ectodermal teloblast lineages in embryos of the oligochaete annelid *Tubifex*: involvement of novel cell-cell interactions. Development 128:1211–1219

Arenas-Mena C (2007) Sinistral equal-size spiral cleavage of the indirectly developing polychaete *Hydroides elegans*. Dev Dyn 236:1611–1622

Arendt D, Nübler-Jung K (1997) Dorsal or ventral: similarities in fate maps and gastrulation patterns in annelids, arthropods and chordates. Mech Dev 61:7–21

Arendt D, Technau U, Wittbrodt J (2001) Evolution of the bilaterian larval foregut. Nature 409:81–85

Arendt D, Denes AS, Jékely G, Tessmar-Raible K (2008) The evolution of nervous system centralization. Philos Trans R Soc B Biol Sci 363:1523–1528

Arendt D, Hausen H, Purschke G (2009) The 'division of labour' model of eye evolution. Philos Trans R Soc B Biol Sci 364:2809–2817

Astrow SH, Holton B, Weisblat DA (1989) Teloplasm formation in a leech, *Helobdella triserialis*, is a microtubule-dependent process. Dev Biol 135:306–319

Bartolomaeus T (1998) Head kidneys in hatchlings of *Scoloplos armiger* (Annelida: Orbiniidae): implications for the occurrence of protonephridia in lecithotrophic larvae. J Mar Biol Assoc UK 78:183–192

Bartolomaeus T (1999) Structure, function and development of segmental organs in Annelida. Hydrobiologia 402:21–37

Bartolomaeus T, Quast B (2005) Structure and development of nephridia in Annelida and related taxa. In: Bartolomaeus T, Purschke G (eds) Morphology, molecules, evolution and phylogeny in polychaeta and related taxa, vol 179, Developments in Hydrobiology. Springer, Dordrecht, pp 139–165

Bartolomaeus T, Purschke G, Hausen H (2005) Polychaete phylogeny based on morphological data—a comparison of current attempts. In: Bartolomaeus T, Purschke G (eds) Morphology, molecules, evolution and phylogeny in polychaeta and related taxa, vol 179, Developments in Hydrobiology. Springer, Dordrecht, pp 341–356

Bely AE (2006) Distribution of segment regeneration ability in the Annelida. Integr Comp Biol 46:508–518

Bergter A, Paululat A (2007) Pattern of body-wall muscle differentiation during embryonic development of *Enchytraeus coronatus* (Annelida: Oligochaeta; Enchytraeidae). J Morphol 268:537–549

Bergter A, Hunnekuhl VS, Schniederjans M, Paululat A (2007) Evolutionary aspects of pattern formation during clitellate muscle development. Evol Dev 9:602–617

Bhaud M, Cazaux C (1982) Les larves de polychètes des côtes de France. Oceanis 8:57–160

Bhaud M, Cazaux C (1987) Description and identification of polychaete larvae; their implications in current biological problems. Oceanis 13:596–753

Bielen H, Houart C (2014) The Wnt cries many: wnt regulation of neurogenesis through tissue patterning, proliferation, and asymmetric cell division. Dev Neurobiol 74:772–780

Blair SS (1982) Interactions between mesoderm and ectoderm in segment formation in the embryo of a glossiphoniid leech. Dev Biol 89:389–396

Blake JA, Kudenov JD (1981) Larval development, larval nutrition and growth for two *Boccardia* species (Polychaeta: Spionidae) from Victoria, Australia. Mar Ecol Prog Ser 6:175–282

Bleidorn C (2007) The role of character loss in phylogenetic reconstruction as exemplified for the Annelida. J Zool Syst Evol Res 45:299–307

Bleidorn C, Lanterbecq D, Eeckhaut I, Tiedemann R (2009) A PCR survey of Hox genes in the myzostomid *Myzostoma cirriferum*. Dev Genes Evol 219:211–216

Bleidorn C, Helm C, Weigert A, Eeckhaut I, Lanterbecq D, Struck T, Hartmann S, Tiedemann R (2014) From morphology to phylogenomics: placing the enigmatic Myzostomida in the tree of life. In: Wägele JW, Bartolomaeus T (eds) Deep metazoan phylogeny: the backbone of the tree of life. De Gruyter, Berlin, pp 161–172

Boyle MJ, Seaver EC (2008) Developmental expression of *foxA* and *gata* genes during gut formation in the polychaete annelid, *Capitella* sp. I. Evol Dev 10:89–105

Boyle M, Seaver E (2010) Expression of FoxA and GATA transcription factors correlates with regionalized gut development in two lophotrochozoan marine worms: *Chaetopterus* (Annelida) and *Themiste lageniformis* (Sipuncula). EvoDevo 1:2

Bright M, Eichinger I, Salwini-Plawen LV (2013) The metatrochophore of a deep-sea hydrothermal vent vestimentiferan (Polychaeta: Siboglinidae). Org Divers Evol 13:163–188

Brinkmann N, Wanninger A (2008) Larval neurogenesis in *Sabellaria alveolata* reveals plasticity in polychaete neural patterning. Evol Dev 10:606–618

Brinkmann N, Wanninger A (2009) Neurogenesis suggests independent evolution of opercula in serpulid polychaetes. BMC Evol Biol 9:270

Brinkmann N, Wanninger A (2010a) Integrative analysis of polychaete ontogeny: cell proliferation patterns and myogenesis in trochphore larva of *Sabellaria alveolata*. Evol Dev 12:5–15

Brinkmann N, Wanninger A (2010b) Capitellid connections: contributions from neuromuscular development of the maldanid polychaete *Axiothella rubrocincta* (Annelida). BMC Evol Biol 10:168

Brooke NM, Garcia-Fernandez J, Holland PWH (1998) The ParaHox gene cluster is an evolutionary sister of the Hox gene cluster. Nature 392:920–922

Bruce AEE, Shankland M (1998) Expression of the head gene *Lox22-Otx* in the leech Helobdella and the origin of the bilaterian body plan. Dev Biol 201:101–112

Bullock TH (1965) Annelida. In: Bullock TH, Horridgke GH (eds) Structure and function in the nervous system of invertebrates, vol 1. Freeman, San Francisco, pp 661–789

Butts T, Holland PWH, Ferrier DEK (2008) The urbilaterian super-Hox cluster. Trends Genet 24:259–262

Cadigan KM, Nusse R (1997) Wnt signaling: a common theme in animal development. Genes Dev 11:3286–3305

Chia F, Gibson G, Qian P (1996) Poecilogony as a reproductive strategy of marine invertebrates. Oceanol Acta 19:203–208

Child CM (1900) The early development of *Arenicola* and *Sternaspis*. Archiv für Entwicklungsmechanik der Organismen 9:587–723

Chipman AD (2010) Parallel evolution of segmentation by co-option of ancestral gene regulatory networks. Bioessays 32:60–70

Cho S, Cho P, Lee M, Hur S, Lee J, Kim S, Koh K, Na Y, Choo J, Kim C-B, Park S (2003) Hox genes from the earthworm *Perionyx excavatus*. Dev Genes Evol 213:207–210

Cho S-J, Lee D-H, Kwon H-J, Park S, Shin K-S, Ahn C (2006) Hox genes in the echiuroid *Urechis unicinctus*. Dev Genes Evol 216:347–351

Cho S-J, Valles Y, Giani VC Jr, Seaver EC, Weisblat DA (2010) Evolutionary dynamics of the *wnt* gene family: a lophotrochozoan perspective. Mol Biol Evol 27:1645–1658

Christodoulou F, Raible F, Tomer R, Simakov O, Trachana K, Klaus S, Snyman H, Hannon GJ, Bork P, Arendt D (2010) Ancient animal microRNAs and the evolution of tissue identity. Nature 463:1084–1088

Conklin EG (1897) The embryology of *Crepidula*, a contribution to the cell lineage and early development of some marine gastropods. J Morphol 13:1–226

Costello DP, Henley C (1976) Spiralian development: a perspective. Am Zool 16:277–291

Couso JP (2009) Segmentation, metamerism and the Cambrian explosion. Int J Dev Biol 53:1305–1316

Damen WGM (2007) Evolutionary conservation and divergence of the segmentation process in arthropods. Dev Dyn 236:1379–1391

Damen P, Dictus WJAG (1994) Cell lineage of the prototroch of *Patella vulgata* (Gastropoda, Mollusca). Dev Biol 162:364–383

de Rosa R, Prud'homme B, Balavoine G (2005) *Caudal* and *even-skipped* in the annelid *Platynereis dumerilii* and the ancestry of posterior growth. Evol Dev 7:574–587

Demilly A, Steinmetz P, Gazave E, Marchand L, Vervoort M (2013) Involvement of the Wnt/β-catenin pathway in neurectoderm architecture in *Platynereis dumerilii*. Nat Commun 4:1915

Denes AS, Jékely G, Steinmetz PRH, Raible F, Snyman H, Prud'homme B, Ferrier DEK, Balavoine G, Arendt D (2007) Molecular architecture of annelid nerve cord supports common origin of nervous system centralization in Bilateria. Cell 129:277–288

Dick MH, Buss LW (1994) A PCR-based survey of homeobox genes in *Ctenodrilus serratus* (Annelida: Polychaeta). Mol Phylogenet Evol 3:146–158

Dohle W (1999) The ancestral cleavage pattern of the clitellates and its phylogenetic deviations. Hydrobiologia 402:267–283

Dorresteijn AWC (1990) Quantitative analysis of cellular differentiation during early embryogenesis of *Platynereis dumerilii*. Roux Arch Dev Biol 199:14–30

Dorresteijn A (2005) Cell lineage and gene expression in the development of polychaetes. Hydrobiologia 535–536:1–22

Dray N, Tessmar-Raible K, Le Gouar M, Vibert L, Christodoulou F, Schipany K, Guillou A, Zantke J, Snyman H, Béhague J, Vervoort M, Arendt D, Balavoine G (2010) Hedgehog signaling regulates segment formation in the annelid Platynereis. Science 329:339–342

Dykes I, Macagno E (2006) Molecular characterization and embryonic expression of innexins in the leech Hirudo medicinalis. Dev Genes Evol 216:185–197

Eckberg WR (1981) An ultrastructural analysis of cytoplasmic localization in Chaetopterus pergamentaceus. Biol Bull 160:228–239

Edgecombe G, Giribet G, Dunn C, Hejnol A, Kristensen R, Neves R, Rouse G, Worsaae K, Sørensen M (2011) Higher-level metazoan relationships: recent progress and remaining questions. Org Divers Evol 11:151–172

Eeckhaut I, Jangoux M (1993) Life cycle and mode of infestation of Myzostoma cirriferum (Annelida), a symbiotic myzostomid of the comatulid crinoid Antedon bifida. Dis Aquat Org 15:207–217

Eisig H (1898) Zur Entwicklungsgeschichte der Capitelliden. Mitt Zool Stn Neapel 13:1–292

Fauchald K, Rouse G (1997) Polychaete systematics: past and present. Zool Scr 26:71–138

Fernandez J, Olea N, Matte C (1987) Structure and development of the egg of the glossiphoniid leech Theromyzon rude: characterization of developmental stages and structure of the early uncleaved egg. Development 100:211–225

Ferrier DEK (2012) Evolutionary crossroads in developmental biology: annelids. Development 139:2643–2653

Ferrier DEK, Holland PWH (2001) Sipunculan ParaHox genes. Evol Dev 3:263–270

Ferrier DEK, Minguillon C (2003) Evolution of the Hox/ParaHox gene clusters. Int J Dev Biol 47:605–611

Fischer AHL, Arendt D (2013) Mesoteloblast-like mesodermal stem cells in the polychaete annelid Platynereis dumerilii (Nereididae). J Exp Zool B Mol Dev Evol 320:94–104

Fischer A, Dorresteijn A (2004) The polychaete Platynereis dumerilii (Annelida): a laboratory animal with spiralian cleavage, lifelong segment proliferation and a mixed benthic/pelagic life cycle. Bioessays 26:314–325

Fischer AHL, Henrich T, Arendt D (2010) The normal development of Platynereis dumerilii (Nereididae, Annelida). Front Zool 7:31

Franke H-D (1999) Reproduction of the Syllidae (Annelida: Polychaeta). Hydrobiologia 402:39–55

Freeman G, Lundelius JW (1992) Evolutionary implications of the mode of D quadrant specification in coelomates with spiral cleavage. J Evol Biol 5:205–247

Fröbius AC, Seaver EC (2006) ParaHox gene expression in the polychaete annelid Capitella sp. I. Dev Genes Evol 216:81–88

Fröbius AC, Matus DQ, Seaver EC (2008) Genomic organization and expression demonstrate spatial and temporal Hox gene collinearity in the lophotrochozoan Capitella sp. I. PloS ONE 3:e4004

Gan W-B, Wong VY, Phillips A, Ma C, Gershon TR, Macagno ER (1999) Cellular expression of a leech netrin suggests roles in the formation of longitudinal nerve tracts and in regional innervation of peripheral targets. J Neurobiol 40:103–115

Garcia-Fernandez J (2004) Hox, ParaHox, ProtoHox: facts and guesses. Heredity 94:145–152

Gazave E, Behague J, Laplane L, Guillou A, Preau L, Demilly A, Balavoine G, Vervoort M (2013) Posterior elongation in the annelid Platynereis dumerilii involves stem cells molecularly related to primordial germ cells. Dev Biol 382:246–267

Gazave E, Guillou A, Balavoine G (2014) History of a prolific family; the Hes/Hey-related genes of the annelid Platynereis. Evodevo 5:29

Gellon G, McGinnis W (1998) Shaping animal body plans in development and evolution by modulation of Hox expression patterns. Bioessays 20:116–125

Gharbaran R, Aisemberg GO (2013) Identification of leech embryonic neurons that express a Hox gene required for the differentiation of a paired, segment-specific motor neuron. Int J Dev Neurosci 31:105–115

Gibson G, Carver D (2013) Effects of extra-embryonic provisioning on larval morphology and histogenesis in Boccardia proboscidea (Annelida, Spionidae). J Morphol 274:11–23

Gline SE, Nakamoto A, Cho S-J, Chi C, Weisblat DA (2011) Lineage analysis of micromere 4d, a super-phylotypic cell for Lophotrochozoa, in the leech Helobdella and the sludgeworm Tubifex. Dev Biol 353:120–133

Golombek A, Tobergte S, Nesnidal MP, Purschke G, Struck TH (2013) Mitochondrial genomes to the rescue – Diurodrilidae in the myzostomid trap. Mol Phylogenet Evol 68:312–326

Goto A, Kitamura K, Arai A, Shimizu T (1999) Cell fate analysis of teloblasts in the Tubifex embryo by intracellular injection of HRP. Develop Growth Differ 41:703–713

Häcker V (1896) Pelagische Polychäten-Larven. Zur Kenntnis des Neapler Frühjahr-Auftriebs. Z Wiss Zool 62:74–168

Hartman O, Boss KJ (1965) Antonbruunia viridis, a new inquiline annelid with dwarf males, inhabiting a new species of pelecypod, Lucina fosteri, in the Mozambique channel. Ann Mag Nat Hist 8:177–186

Haszprunar G, Salvini-Plawen LV, Rieger RM (1995) Larval planktotrophy – a primitive trait in the Bilateria? Acta Zool 76:141–154

Hatschek B (1886) Zur Entwicklung des Kopfes von Polygordius. Arb. aus dem Zool. Inst. Univ. Wien 6:236–277

Hay-Schmidt A (1995) The larval nervous system of Polygordius lacteus Scheinder, 1868 (Polygordiidae, Polychaeta): immunocytochemical data. Acta Zool 76:121–140

Hay-Schmidt A (2000) The evolution of the serotonergic nervous system. Proc. R. Soc. Lond. B 267: 1071–1079

Heimler W (1988) Larvae. In: Westheide W, Hermans CO (eds) The ultrastructure of Polychaeta. Gustav Fischer Verlag, Stuttgart, pp 352–371

Hejnol A, Martindale MQ (2008) Acoel development indicates the independent evolution of the bilaterian mouth and anus. Nature 456:382–386

Hejnol A, Martindale MQ (2009) The mouth, the anus and the blastopore – open questions about questionable openings. In: Telford MJ, Littlewood DTJ (eds) Animal evolution: genes, genomes, fossils and trees. Oxford University Press, Oxford, pp 33–40

Helm C, Schemel S, Bleidorn C (2013) Temporal plasticity in annelid development – ontogeny of *Phyllodoce groenlandica* (Phyllodocidae, Annelida) reveals heterochronous patterns. J Exp Zool B Mol Dev Evol 320B:166–178

Helm C, Stevenson PA, Rouse GW, Bleidorn C (2014) Immunohistochemical investigations of *Myzostoma cirriferum* and *Mesomyzostoma* cf. *katoi* (Myzostomida, Annelida) with implications for the evolution of the myzostomid body plan. Zoomorphology 133:257–271

Helm C, Adamo H, Hourdez S, Bleidorn C (2014) An immunocytochemical window into the developement of *Platynereis massiliensis* (Annelida, Nereididae). Int J Dev Biol 58:613–622

Henry JJ (1986) The role of unequal cleavage and the polar lobe in the segregation of developmental potential during first cleavage in the embryo of *Chaetopterus variopedatus*. Roux Arch Dev Biol 195:103–116

Henry JQ, Hejnol A, Perry KJ, Martindale MQ (2007) Homology of ciliary bands in spiralian trochophores. Integr Comp Biol 47:865–871

Hermans C (1964) The reproductive and developmental biology of the opheliid polychaete *Armandia brevis* (Moore). University of Washington, Seatle

Hessling R (2002) Metameric organisation of the nervous system in developmental stages of *Urechis caupo* (Echiura) and its phylogenetic implications. Zoomorphology 121:221–234

Hessling R (2003) Novel aspects of the nervous system of *Bonellia viridis* (Echiura) revealed by the combination of immunohistochemistry, confocal laser-scanning microscopy and three-dimensional reconstruction. In: Sigvaldadóttir E et al (eds) Advances in polychaete research, vol 170, Developments in Hydrobiology. Springer, Dordrecht, pp 225–239

Hessling R, Westheide W (2002) Are Echiura derived from a segmented ancestor? Immunohistochemical analysis of the nervous system in developmental stages of *Bonellia viridis*. J Morphol 252:100–113

Heuer C, Muller C, Todt C, Loesel R (2010) Comparative neuroanatomy suggests repeated reduction of neuroarchitectural complexity in Annelida. Front Zool 7:13

Hill SD (2001) Phalloidin labelling of developing musculature in embryos of the polychaete *Capitella* sp. I. Biol Bull 201:257–258

Holland PWH (2001) Beyond the Hox: how widespread is homeobox gene clustering? J Anat 199:13–23

Holland L, Carvalho J, Escriva H, Laudet V, Schubert M, Shimeld S, Yu J-K (2013) Evolution of bilaterian central nervous systems: a single origin? EvoDevo 4:27

Huebner E, Anderson E (1976) Comparative spiralian oogenesis—structural aspects: an overview. Am Zool 16:315–343

Hui J, Raible F, Korchagina N, Dray N, Samain S, Magdelenat G, Jubin C, Segurens B, Balavoine G, Arendt D, Ferrier D (2009) Features of the ancestral bilaterian inferred from *Platynereis dumerilii* ParaHox genes. BMC Biol 7:43

Hunnekuhl VS, Bergter A, Purschke G, Paululat A (2009) Development and embryonic pattern of body wall musculature in the crassiclitellate *Eisenia andrei* (Annelida, Clitellata). J Morphol 270:1122–1136

Irvine SQ, Martindale MQ (2000) Expression patterns of anterior Hox genes in the polychaete *Chaetopterus*: correlation with morphological boundaries. Dev Biol 217:333–351

Irvine SQ, Seaver EC (2006) Early annelid development, a molecular perspective. In: Rouse GW, Pleijel F (eds) Reproductive biology and phylogeny of annelida, vol 4, Reproductive Biology and Phylogeny. Science Publishers, Enfield, pp 93–140

Irvine SQ, Warinner SA, Hunter JD, Martindale MQ (1997) A survey of homeobox genes in *Chaetopterus variopedatus* and analysis of polychaete homeodomains. Mol Phylogenet Evol 7:331–345

Irvine SQ, Chaga O, Martindale MQ (1999) Larval ontogenetic stages of *Chaetopterus*: developmental heterochrony in the evolution of chaetopterid polychaetes. Biol Bull 197:313–331

Iwanoff PP (1928) Die Entwicklung der Larvalsegmente bei den Anneliden. Z Morphol Okol Tiere 10:62–161

Iwasa JH, Suver DW, Savage RM (2000) The leech *hunchback* protein is expressed in the epithelium and CNS but not in the segmental precursor lineages. Dev Genes Evol 210:277–288

Jaeckle WB, Rice ME (2002) Phylum Sipuncula. In: Young CM (ed) Atlas of marine invertebrate larvae. Academic, San Diego, pp 375–396

Jägersten G (1940) Zur Kenntnis der Morphologie, Entwicklung und Taxanomie der Myzostomida. Nova Acta Regiae Soc Sci Upsal 11:1–84

Jägersten G (1972) Evolution of the metazoan life cycle. Academic, London

Jagla K, Bellard M, Frasch M (2001) A cluster of *Drosophila* homeobox genes involved in mesoderm differentiation programs. Bioessays 23:125–133

Janssen R, Le Gouar M, Pechmann M, Poulin F, Bolognesi R, Schwager EE, Hopfen C, Colbourne JK, Budd GE, Brown SJ, Prpic N-M, Kosiol C, Vervoort M, Damen WGM, Balavoine G, McGregor AP (2010) Conservation, loss, and redeployment of Wnt ligands in protostomes: implications for understanding the evolution of segment formation. BMC Evol Biol 10:374

Jeffery WR (1985) The spatial distribution of maternal mRNA is determined by a cortical cytoskeletal domain in *Chaetopterus* eggs. Dev Biol 110:217–229

Jeffery WR, Wilson LJ (1983) Localization of messenger RNA in the cortex of *Chaetopterus* eggs and early embryos. J Embryol Exp Morphol 75:225–239

Jekely G, Colombelli J, Hausen H, Guy K, Stelzer E, Nedelec F, Arendt D (2008) Mechanism of phototaxis in marine zooplankton. Nature 456:395–399

Kato C (2012) Ultrastruktur der Kopfnieren (head kidneys) von sedentären Polychaeten und ihre Bedeutung für die Phylogenie der Annelida. Rheinische Friedrich-Wilhelms-Universität Bonn

Kato C, Lehrke J, Quast B (2011) Ultrastructure and phylogenetic significance of the head kidneys in *Thalassema thalassemum* (Thalassematinae, Echiura). Zoomorphology 130:97–106

Kerner P, Zelada González F, Le Gouar M, Ledent V, Arendt D, Vervoort M (2006) The expression of a *hunchback* ortholog in the polychaete annelid *Platynereis dumerilii* suggests an ancestral role in mesoderm development and neurogenesis. Dev Genes Evol 216:821–828

Kluge B, Lehmann-Greif M, Fischer A (1995) Long-lasting exocytosis and massive structural reorganisation in the egg periphery during cortical reaction in *Platynereis dumerilii* (Annelida, Polychaeta). Zygote 3:141–156

Koch M, Quast B, Bartolomaeus T (2014) Coeloms and nephridia in annelids and arthropods. In: Wägele JW, Bartolomaeus T (eds) Deep metazoan phylogeny: the backbone of the tree of life – new insights from analyses of molecules, morphology, and theory of data analysis. De Gruyter, Berlin, pp 173–284

Kourakis MJ, Master VA, Lokhorst DK, Nardelli-Haefliger D, Wedeen CJ, Martindale MQ, Shankland M (1997) Conserved anterior boundaries of hox gene expression in the central nervous system of the leech *Helobdella*. Dev Biol 190:284–300

Kristof A, Wollesen T, Wanninger A (2008) Segmental mode of neural patterning in Sipuncula. Curr Biol 18:1129–1132

Kristof A, Wollesen T, Maiorova AS, Wanninger A (2011) Cellular and muscular growth patterns during sipunculan development. J Exp Zool B Mol Dev Evol 316B:227–240

Kudenov JD (1974) The reproductive biology of *Eurythoe complanata* (Pallas, 1766), (Polychaeta: Amphinomidae). University of Arizona

Kulakova M, Bakalenko N, Novikova E, Cook C, Eliseeva E, Steinmetz PH, Kostyuchenko R, Dondua A, Arendt D, Akam M, Andreeva T (2007) Hox gene expression in larval development of the polychaetes *Nereis virens* and *Platynereis dumerilii* (Annelida, Lophotrochozoa). Dev Genes Evol 217:39–54

Kulakova M, Cook C, Andreeva T (2008) ParaHox gene expression in larval and postlarval development of the polychaete *Nereis virens* (Annelida, Lophotrochozoa). BMC Dev Biol 8:61

Lacalli TC (1981) Structure and development of the apical organ in trochophores of *Spirobranchus polycerus*, *Phyllodoce maculata* and *Phyllodoce mucosa*. Proc R Soc Lond Part B 212:381–402

Lacalli TC (1984) Structure and organization of the nervous system in the trochophore larva of *Spirobranchus*. Philos Trans R Soc B Biol Sci 306:79–135

Lacalli TC (1986) Prototroch structure and innervation in the trochophore larva of *Phyllodoce* (Polychaeta). Can J Zool 64:176–184

Lehmacher C, Fiege D, Purschke G (2014) Immunohistochemical and ultrastructural analysis of the muscular and nervous systems in the interstitial polychaete *Polygordius appendiculatus* (Annelida). Zoomorphology 133:21–41

Levin LA (1984) Multiple patterns of development in *Streblospio benedicti* Webster (Spionidae) from three coasts of North America. Biol Bull 166:494–508

Lidke AK, Bannister S, Lower AM, Apel DM, Podleschny M, Kollmann M, Ackermann CF, Garcia-Alonso J, Raible F, Rebscher N (2014) 17 beta-Estradiol induces supernumerary primordial germ cells in embryos of the polychaete *Platynereis dumerilii*. Gen Comp Endocrinol 196:52–61

Lillie FR (1906) Observations and experiments concerning the elementary phenomena of embryonic development in *Chaetopterus*. J Exp Zool 3:153–268

Lillie FR (1909) Polarity and bilaterality of the annelid egg. Experiments with centrifugal force. Biol Bull 16:54–79

Marlow H, Tosches M, Tomer R, Steinmetz P, Lauri A, Larsson T, Arendt D (2014) Larval body patterning and apical organs are conserved in animal evolution. BMC Biol 12:7

Maslakova SA, Martindale MQ, Norenburg JL (2004a) Fundamental properties of the spiralian developmental program are displayed by the basal nemertean *Carinoma tremaphoros* (Palaeonemertea, Nemertea). Dev Biol 267:342–360

Maslakova SA, Martindale MQ, Norenburg JL (2004b) Vestigial prototroch in a basal nemertean, *Carinoma tremaphoros* (Nemertea; Palaeonemertea). Evol Dev 6:219–226

McDougall C, Chen WC, Shimeld SM, Ferrier DEK (2006) The development of the larval nervous system, musculature and ciliary bands of *Pomatoceros lamarckii* (Annelida): heterochrony in polychaetes. Front Zool 3:16

McEdward LR, Janies DA (1993) Life cycle evolution in asteroids: what is a larva? Biol Bull 184:255–268

Meyer NP, Seaver EC (2010) Cell lineage and fate map of the primary somatoblast of the polychaete annelid *Capitella teleta*. Integr Comp Biol 50:756–767

Meyer N, Boyle M, Martindale M, Seaver E (2010) A comprehensive fate map by intracellular injection of identified blastomeres in the marine polychaete *Capitella teleta*. EvoDevo 1:8

Mileikovsky SA (1960) Appurtenance of a polychaete larva of the rostraria type from plankton of the Norwegian and Barents Seas of the species *Euphrosyne borealis* Oersted 1843 and of all larvae of this type to the families Euphrosynidae and Amphinomidae (Polychaeta, Errantia, Amphinomimorpha) [in Russian]. Dokl Akad Nauk SSSR 134:731–734

Mileikovsky SA (1961) Assignment of two Rostraria-type polychaete larvae from the plankton of the Northwest Atlantic to species *Amphinome passasi* Quatrefages 1865 and *Chloenea atlantica* McIntosh 1885 (Polychaeta, Errantia, Amphinomimorpha) [in Russian]. Dokl Akad Nauk SSSR 141:1109–1112

Miyamoto N, Shinozaki A, Fujiwara Y (2013) Neuroanatomy of the vestimentiferan tubeworm *Lamellibrachia satsuma* provides insights into the evolution of the polychaete nervous system. PLoS ONE 8:e55151

Müller MCM (2006) Polychaete nervous systems: ground pattern and variations-cLS microscopy and the importance of novel characteristics in phylogenetic analysis. Integr Comp Biol 46:125–133

Nardelli-Haefliger D, Shankland M (1993) *Lox10*, a member of the NK-2 homeobox gene class, is expressed in a segmental pattern in the endoderm and in the cephalic nervous system of the leech *Helobdella*. Development 118:877–892

Nelson JA (1904) The early development of *Dinophilus*: a study in cell-lineage. Proc Acad Natl Sci Phila 56:687–737

Nielsen C (2004) Trochophore larvae: cell-lineages, ciliary bands, and body regions. 1. Annelida and Mollusca. J Exp Zool B Mol Dev Evol 302B:35–68

Nielsen C (2009) How did indirect development with planktotrophic larvae evolve? Biol Bull 216:203–215

Nielsen C (2012) Animal evolution – interrelationships of the living phyla, 3rd edn. Oxford University Press, New York

Okada K (1957) Annelida. In: Kumé M, Dan K (eds) Invertebrate embryology. NOLIT, Belgrade, pp 192–241

Orrhage L, Müller MCM (2005) Morphology of the nervous system of Polychaeta (Annelida). Hydrobiologia 535(536):79–111

Osborn KJ, Rouse GW, Goffredi SK, Robison BH (2007) Description and relationships of *Chaetopterus pugaporcinus*, an unusual pelagic polychaete (Annelida, Chaetopteridae). Biol Bull 212:40–54

Park B, Cho S-J, Tak E, Lee B, Park S (2006) The existence of all three ParaHox genes in the clitellate annelid, *Perionyx excavatus*. Dev Genes Evol 216:551–553

Paxton H (2005) Molting polychaete jaws—ecdysozoans are not the only molting animals. Evol Dev 7:337–340

Pechenik JA (1999) On the advantages and disadvantages of larval stages in benthic marine invertebrate life cycles. Mar Ecol Prog Ser 177:269–297

Pernet B (2003) Persistent ancestral feeding structures in nonfeeding annelid larvae. Biol Bull 205:295–307

Peterson KJ, Irvine SQ, Cameron RA, Davidson EH (2000) Quantitative assessment of Hox complex expression in the indirect development of the polychaete annelid *Chaetopterus* sp. Proc Natl Acad Sci 97:4487–4492

Phillips NE, Pernet B (1996) Capture of large particles by suspension-feeding scaleworm larvae (Polychaeta: Polynoidae). Biol Bull 191:199–208

Pilger JF (2002) Phylum Echiura. In: Young CM (ed) Atlas of marine invertebrate larvae. Academic, San Diego, pp 371–373

Prevedelli D, Massamba N'Siala G, Ansaloni I, Simonini R (2007) Life cycle of *Marphysa sanguinea* (Polychaeta: Eunicidae) in the Venice Lagoon (Italy). Mar Ecol 28:384–393

Prud'homme B, de Rosa R, Arendt D, Julien J-F, Pajaziti R, Dorresteijn AWC, Adoutte A, Wittbrodt J, Balavoine G (2003) Arthropod-like expression patterns of *engrailed* and *wingless* in the annelid *Platynereis dumerilii* suggest a role in segment formation. Curr Biol 13:1876–1881

Purschke G (1997) Ultrastructure of nuchal organs in polychaetes (Annelida)—new results and review. Acta Zool 78:123–143

Purschke G (2002) On the ground pattern of Annelida. Org Div Evol 2:181–196

Purschke G, Müller MCM (2006) Evolution of body wall musculature. Integr Comp Biol 46:497–507

Purschke G, Hessling R, Westheide W (2000) The phylogenetic position of the Clitellata and the Echiura – on the problematic assessment of absent characters. J Zool Syst Evol Res 38:165–173

Purschke G, Arendt D, Hausen H, Müller MCM (2006) Photoreceptor cells and eyes in Annelida. Arthropod Struct Dev 35:211–230

Purschke G, Bleidorn C, Struck TH (2014) Systematics, evolution and phylogeny of Annelida – a morphological perspective. Mem Museum Victoria 71:247–269

Qian P, Dahms HU (2006) Larval ecology of the Annelida. In: Rouse GW, Pleijel F (eds) Reproductive biology and phylogeny of Annelida, vol 4. Science Publisher, Enfield, pp 179–232

Raible F, Tessmar-Raible K, Osoegawa K, Wincker P, Jubin C, Balavoine G, Ferrier DEK, Benes V, De Jong P, Weissenbach J, Bork P, Arendt D (2005) Vertebrate-type intron-rich genes in the marine annelid *Platynereis dumerilii*. Science 310:1325–1326

Rebscher N, Lidke AK, Ackermann CF (2012) Hidden in the crowd: primordial germ cells and somatic stem cells in the mesodermal posterior growth zone of the polychaete *Platynereis dumerilii* are two distinct cell populations. EvoDevo 3:1–11

Rice ME (1976) Larval development and metamorphosis in Sipuncula. Am Zool 16:563–571

Rieger RM (1980) A new group of interstitial worms, Lobatocerebridae nov. fam. (Annelida) and its significance for metazoan phylogeny. Zoomorphologie 95:41–84

Rieger RM (1991) *Jennaria pulchra*, nov.gen. nov.spec., eine den psammobionten Anneliden nahestehende Gattung aus dem Küstengrundwasser von North Carolina. Ber Naturwiss-Med Ver Innsb 78:203–215

Rivera A, Weisblat D (2009) And Lophotrochozoa makes three: *Notch/Hes* signaling in annelid segmentation. Dev Genes Evol 219:37–43

Rivera AS, Gonsalves FC, Song MH, Norris BJ, Weisblat DA (2005) Characterization of *Notch*-class gene expression in segmentation stem cells and segment founder cells in *Helobdella robusta* (Lophotrochozoa; Annelida; Clitellata; Hirudinida; Glossiphoniidae). Evol Dev 7:588–599

Rouse GW (1999) Trochophore concepts: ciliary bands and the evolution of larvae in spiralian Metazoa. Biol J Linn Soc 66:411–464

Rouse GW (2000a) The epitome of hand waving? Larval feeding and hypothesis of metazoan phylogeny. Evol Dev 2:222–233

Rouse GW (2000b) Bias? What bias? The evolution of downstream larval feeding in animals. Zool Scr 29:213–236

Rouse GW (2006) Annelid larval morphology. In: Rouse GW, Pleijel F (eds) Reproductive biology and phylogeny of annelida, vol 4. Science Publisher, Enfield, pp 141–177

Rouse GW, Wilson NG, Goffredi SK, Johnson SB, Smart T, Widmer C, Young CM, Vrijenhoek RC (2009) Spawning and development in *Osedax* boneworms (Siboglinidae, Annelida). Mar Biol 156:395–405

Saudemont A, Dray N, Hudry B, Le Gouar M, Vervoort M, Balavoine G (2008) Complementary striped expression patterns of *NK* homeobox genes during segment formation in the annelid *Platynereis*. Dev Biol 317:430–443

Schneider SQ, Bowerman B (2007) β-Catenin asymmetries after all animal/vegetal- oriented cell divisions in *Platynereis dumerilii* embryos mediate binary cellfate specification. Dev Cell 13:73–86

Schneider S, Fischer A, Dorresteijn AC (1992) A morphometric comparison of dissimilar early development in sibling species of *Platynereis* (Annelida, Polychaeta). Roux Arch Dev Biol 201:243–256

Scholtz G (2002) The Articulata hypothesis – or what is a segment? Org Divers Evol 2:197–215

Schulze A, Halanych K (2003) Siboglinid evolution shaped by habitat preference and sulfide tolerance. Hydrobiologia 496:199–205

Seaver EC (2003) Segmentation: mono- or polyphyletic? Int J Dev Biol 47:583–595

Seaver EC, Kaneshige LM (2006) Expression of 'segmentation' genes during larval and juvenile development in the polychaetes *Capitella* sp. I and *H. elegans*. Dev Biol 289:179–194

Seaver EC, Shankland M (2001) Establishment of segment polarity in the ectoderm of the leech *Helobdella*. Development 128:1629–1641

Seaver EC, Paulson DA, Irvine SQ, Martindale MQ (2001) The spatial and temporal expression of *Ch-en*, the *engrailed* gene in the polychaete *Chaetopterus*, does not support a role in body axis segmentation. Dev Biol 236:195–209

Seaver EC, Thamm K, Hill SD (2005) Growth patterns during segmentation in the two polychaete annelids. Capitella sp. I and Hydroides elegans: comparisons at distinct life history stages. Evol Dev 7:312–326

Seaver E, Yamaguchi E, Richards G, Meyer N (2012) Expression of the pair-rule gene homologs *runt*, *Pax3/7*, *even-skipped-1* and *even-skipped-2* during larval and juvenile development of the polychaete annelid *Capitella teleta* does not support a role in segmentation. EvoDevo 3:1–18

Shankland M, Martindale MQ, Nardelli-Haefliger D, Baxter E, Price DJ (1991) Origin of segmental identity in the development of the leech nervous system. Development 113:29–38

Shearer C (1911) On the development and structure of the trochophore of *Hydroides uncinatus* (*Eupomatus*). Quart J Microsc Sci 56:543–590

Shimizu T (1995) Role of the cytoskeleton in the generation of spatial patterns in *Tubifex* eggs. Curr Top Dev Biol 31:197–235

Shimizu T (1999) Cytoskeletal mechanisms of ooplasmic segregation in annelid eggs. Int J Dev Biol 43:11–18

Shimizu T, Nakamoto A (2001) Segmentation in annelids: cellular and molecular basis for metameric body plan. Zool Sci 18:285–298

Simakov O, Marletaz F, Cho S-J, Edsinger-Gonzales E, Havlak P, Hellsten U, Kuo D-H, Larsson T, Lv J, Arendt D, Savage R, Osoegawa K, de Jong P, Grimwood J, Chapman JA, Shapiro H, Aerts A, Otillar RP, Terry AY, Boore JL, Grigoriev IV, Lindberg DR, Seaver EC, Weisblat DA, Putnam NH, Rokhsar DS (2013) Insights into bilaterian evolution from three spiralian genomes. Nature 493:526–531

Smart TI, Von Dassow G (2009) Unusual development of the mitraria larva in the polychaete *Owenia collaris*. Biol Bull 217:253–268

Snow P, Buss LW (1994) HOM/Hox-type homeoboxes from *Stylaria lacustris* (Annelida: Oligochaeta). Mol Phylogenet Evol 3:360–364

Song MH, Huang FZ, Chang GY, Weisblat DA (2002) Expression and function of an *even-skipped* homolog in the leech *Helobdella robusta*. Development 129:3681–3692

Song MH, Huang FZ, Gonsalves FC, Weisblat DA (2004) Cell cycle-dependent expression of a *hairy* and enhancer of split (*hes*) homolog during cleavage and segmentation in leech embryos. Dev Biol 269:183–195

Southward EC (1999) Development of Perviata and Vestimentifera (Pogonophora). Hydrobilogia 402:185–202

Spengel JW (1879) Beiträge zur Kenntnis der Gephyreen. I. Die Eibildung, die Entwicklung und das Männchen der *Bonellia*. Mitt Zool Stn Neapel 1:357–420

Steinmetz PRH, Urbach R, Posnien N, Eriksson J, Kostyuchenko RP, Brena C, Guy K, Akam M, Bucher G, Arendt D (2010) *Six3* demarcates the anterior-most developing brain region in bilaterian animals. EvoDevo 1:1–9

Steinmetz PRH, Kostyuchenko RP, Fischer A, Arendt D (2011) The segmental pattern of *otx*, *gbx*, and *Hox* genes in the annelid *Platynereis dumerilii*. Evol Dev 13:72–79

Stollewerk A, Schoppmeier M, Damen WGM (2003) Involvement of *Notch* and *Delta* genes in spider segmentation. Nature 423:863–865

Strathmann RR (1993) Hypotheses on the origins of marine larvae. Annu Rev Ecol Syst 24:89–117

Struck TH (2006) Progenetic species in polychaetes (Annelida) and problems assessing their phylogenetic affiliation. Integr Comp Biol 46:558–568

Struck TH (2011) Direction of evolution within Annelida and the definition of Pleistoannelida. J Zool Syst Evol Res 49:340–345

Struck TH, Paul C, Hill N, Hartmann S, Hösel C, Kube M, Lieb B, Meyer A, Tiedemann R, Purschke G, Bleidorn C (2011) Phylogenomic analyses unravel annelid evolution. Nature 471:95–98

Tautz D (2004) Segmentation. Dev Cell 7:301–312

Tessmar-Raible K, Arendt D (2003) Emerging systems: between vertebrates and arthropods, the Lophotrochozoa. Curr Opin Genet Dev 13:331–340

Tessmar-Raible K, Raible F, Christodoulou F, Guy K, Rembold M, Hausen H, Arendt D (2007) Conserved sensory-neurosecretory cell types in annelid and fish forebrain: insights into hypothalamus evolution. Cell 129:1389–1400

Thamm K, Seaver EC (2008) Notch signaling during larval and juvenile development in the polychaete annelid *Capitella* sp. I. Dev Biol 320:304–318

Thorson G (1950) Reproductive and larval ecology of marine bottom invertebrates. Biol Rev 25:1–45

Tzetlin AB, Filippova AV (2005) Muscular system in polychaetes (Annelida). Hydrobiologia 535(536):113–126

Tzetlin A, Purschke G (2005) Pharynx and intestine. Hydrobiologia 535(536):199–225

Voronezhskaya EE, Tsitrin EB, Nezlin LP (2003) Neuronal development in larval polychaete *Phyllodoce maculata* (Phyllodocidae). J Comp Neurol 455:299–309

Wagner GP, Amemiya C, Ruddle F (2003) Hox cluster duplications and the opportunity for evolutionary novelties. Proc Natl Acad Sci 100:14603–14606

Wanninger A (2009) Shaping the things to come: ontogeny of lophotrochozoan neuromuscular systems and the tetraneuralia concept. Biol Bull 216:293–306

Wanninger A, Koop D, Bromham L, Noonan E, Degnan BM (2005) Nervous and muscle system development in *Phascolion strombus* (Sipuncula). Dev Genes Evol 215:509–518

Wanninger A (2008) Comparative lophotrochozoan neurogenesis and larval neuroanatomy: recent advances from previously neglected taxa. Acta Biologica Hungarica 59(Suppl.):127–136

Wedeen CJ, Weisblat DA (1991) Segmental expression of an engrailed-class gene during early development and neurogenesis in an annelid. Development 113:805–814

Weigert A, Helm C, Meyer M, Nickel B, Arendt D, Hausdorf B, Santos SR, Halanych KM, Purschke G, Bleidorn C, Struck TH (2014) Illuminating the base of the annelid tree using transcriptomics. Mol Biol Evol 31:1391–1401

Weisblat DA, Blair SS (1984) Developmental interdeterminacy in embryos of the leech *Helobdella triserialis*. Dev Biol 101:326–335

Weisblat DA, Huang FZ (2001) An overview of glossiphoniid leech development. Can J Zool 79:218–232

Werbrock AH, Meiklejohn DA, Sainz A, Iwasa JH, Savage RM (2001) A polychaete *hunchback* ortholog. Dev Biol 235:476–488

Westheide W (1987) Progenesis as a principle in meiofauna evolution. J Nat Hist 21:843–854

Wilson EB (1890) The origin of the mesoblast-bands in annelids. J Morphol 4:205–219

Wilson EB (1892) The cell lineage of *Nereis*: a contribution to the cytogeny of the annelid body. J Morphol 6:361–480

Wilson EB (1898) Considerations on cell-lineage and ancestral reminiscence based on a re-examination of some points in the early development of annelids and polycladids. Ann N Y Acad Sci 11:1–27

Wilson DP (1932) On the mitraria larva of *Owenia fusiformis* Delle Chiaje. Philos Trans R Soc B Biol Sci B221:231–334

Wilson DP (1982) The larval development of three species of Magelona (Polychaeta) from localitites near Plymouth. J Mar Biol Ass UK 62:385–401

Wilson WH (1991) Sexual reproductive modes in polychaetes: classification and diversity. Bull Mar Sci 48:500–516

Winchell CJ, Valencia JE, Jacobs DK (2010) Confocal analysis of nervous system architecture in direct-developing juveniles of *Neanthes arenaceodentata* (Annelida, Nereididae). Front Zool 7:17

Woltereck R (1904) Beiträge zur praktischen Analyse der *Polygordius*-Entwicklung nach dem "Nordsee-" und dem "Mittelmeertypus". Wilhelm Roux Arch für Entwickl Mech Org 18:377–403

Wong VY, Aisemberg GO, Gan WB, Macagno ER (1995) The leech homeobox gene *Lox4* may determine segmental differentiation of identified neurons segmental differentiation of identified neurons. J Neurosci 15:5551–5559

Worsaae K, Kristensen R (2005) Evolution of interstitial Polychaeta (Annelida). In: Bartolomaeus T, Purschke G (eds) Morphology, molecules, evolution and phylogeny in polychaeta and related taxa. Developments in hydrobiology, vol 179. Springer, Dordrecht, pp 319–340

Worsaae K, Rouse GW (2008) Is *Diurodrilus* an annelid? J Morphol 269:1426–1455

Worsaae K, Rouse GW (2010) The simplicity of males: dwarf males of four species of *Osedax* (Siboglinidae; Annelida) investigated by confocal laser scanning microscopy. J Morphol 271:127–142

Worsaae K, Nygren A, Rouse GW, Giribet G, Persson J, Sundberg P, Pleijel F (2005) Phylogenetic position of Nerillidae and *Aberranta* (Polychaeta, Annelida), analysed by direct optimization of combined molecular and morphological data. Zool Scr 34: 313–328

Wysocka-Diller J, Aisemberg GO, Macagno ER (1995) A novel homeobox cluster expressed in repeated strucutres of the midgut. Dev Biol 171:439–447

Zakas C, Wares JP (2012) Consequences of a poecilogonous life history for genetic structure in coastal populations of the polychaete *Streblospio benedicti*. Mol Ecol 21:5447–5460

Zhang Z-Q (2011) Animal biodiversity: an outline of higher-level classification and survey of taxonomic richness. Zootaxa 3147:1–237

Phoronida

10

Scott Santagata

Chapter vignette artwork by Brigitte Baldrian.
© Brigitte Baldrian and Andreas Wanninger.

S. Santagata
Biology Department, Long Island University-Post,
Greenvale, NY 11548, USA
e-mail: scott.santagata@liu.edu

A. Wanninger (ed.), *Evolutionary Developmental Biology of Invertebrates 2: Lophotrochozoa (Spiralia)*
DOI 10.1007/978-3-7091-1871-9_10, © Springer-Verlag Wien 2015

INTRODUCTION

Phoronids are epibenthic (or infaunal) tubiculous marine invertebrates closely related to brachiopods (and perhaps ectoprocts; see Nesnidal et al. (2013)) that have oval, U-shaped, or spiraling rings of ciliated tentacles called the lophophore used for feeding and respiration (Temereva and Malakhov 2009a). Although phoronids can dominate the density and coverage of some benthic marine habitats (Larson and Stachowicz 2009), very little is known about their ecological role in such habitats. The majority of taxonomic studies of phoronids have been conducted by Emig (1974). Although at least 23 species have been described by various authors indicative of wide morphological diversity in adult forms, the majority of phoronid morphotypes have been synonymized under 11 cosmopolitan species and two genera, *Phoronis* (Wright 1856) and *Phoronopsis* (Gilchrist 1907).

Adult phoronid bodies have generally been considered to be tripartite, divided by transverse septa into the epistome, tentacle crown, and trunk regions. Each of these body regions was originally described as having a true coelomic cavity lined by mesoderm and were interpreted as the protocoel (epistome), mesocoel (lophophore), and metacoel (trunk), linking this arrangement to the tripartite coeloms found in echinoderms (Masterman 1898). The epithelial lining of the epistome in two species of *Phoronis* was found instead to be myoepithelial cells lacking adherent junctions surrounding a gel-like extracellular matrix, thus not exhibiting the features of a true epithelial layer (Bartolomaeus 2001; Gruhl et al. 2005). Further complicating these findings are ultrastructural observations of the epistome lining of *Phoronopsis harmeri*, which does contain adherent junctions (Temereva and Malakhov 2011). This cavity can be derived from the protocoel formed in *Phoronopsis harmeri* larvae but may collapse and break down during development among various other species (Zimmer 1978). Alternatively, this larval cavity may be lost during metamorphosis and reform during post-metamorphic growth (Santagata 2002). Regardless of whether adult phoronids have two

or three true coelomic compartments, this distinction does not clarify their phylogenetic position relative to other lophophore-bearing animals (ectoprocts and brachiopods) or to annelids, nemerteans, and mollusks.

Similar to ectoprocts, the adult phoronid gut is U-shaped and the anus is positioned outside of the tentacles. Unlike ectoprocts, each of the tentacles has a blind capillary with nucleated red blood cells containing a form of hemoglobin (Garlick et al. 1979). These capillaries are connected to lophophoral ring vessels that are fed and drained by the efferent and afferent blood vessel loop traversing the trunk region. The adult trunk is divided into a more posterior ampullary region and a more anterior and tapered muscular region that contains some diagonal muscle fibers in the body wall (Chernyshev and Temereva 2010). The main body wall musculature of the trunk epithelium consists of a layer of circular muscles underlying numerous feathery or bush-like longitudinal muscles (Herrmann 1997). The number and distribution of the latter muscles in the four mesenteric divisions of the trunk have been used as species-level morphological characters (Emig 1974); however, some "cosmopolitan" species have wide muscle formula ranges and may actually be cryptic species. In at least one species, *Phoronis pallida*, distinct muscular cinctures separate regions of the trunk (Santagata 2002), which may be a derived adaptive feature linked to this species living as a commensal in the burrows of (at least) two thalassinid shrimps, *Upogebia pugettensis* (Thompson 1972; Santagata 2004a) and *Upogebia major* (Kinoshita 2002).

One main feature of the adult nervous system is a group of basiepidermal neuronal cell bodies concentrated between the mouth and the anus (Fernández et al. 1996). Although this structure has often been called the dorsal ganglion (Silén 1954a) or a dorsal neural plexus (Temereva and Malakhov 2009b), some aspects of its post-metamorphic development are not consistent with a completely dorsal origin (Santagata 2002). Regardless of the differences between these structural interpretations, this anterior concentration of ciliated neuronal cells appears to be the adult "brain" and is connected to a collar nerve ring with unciliated neuronal cells along its length at the base of the tentacles (Temereva and Malakhov 2009b). Sporadically distributed neuronal cells and fibers are found throughout the surface of the trunk epithelium. A subset of these cells and fibers are serotonergic (Santagata 2002), but the most centralized neuronal structure is the giant nerve fiber embedded in the anterior portion of the trunk epithelium (Temereva and Malakhov 2009b). Among various phoronid species, the giant nerve fiber(s) takes on different morphologies with respect to their number, position, and where the posterior limit of these fibers can be found (Emig 1974).

The soft tissues of phoronids do not make for particularly good fossils, but their chitinous and sandy tubes have been linked to some trace fossil types. A *Phoronis ovalis*-like boring pattern may have produced the ichnofossil *Talpina* found in the Devonian period (Thomas 1911). However, it is likely that the crown phoronid lineage is much older and may be linked to vermiform filter-feeding forms of tommotiids from the lower Cambrian that had unfused organophosphatic sclerites such as *Eccentrotheca* (Skovsted et al. 2008, 2011). Molecular estimates of the phoronid-brachiopod root are even older, reaching as far back as the Ediacaran (Sperling et al. 2011).

Debate still exists as to whether phoronids should be considered as an ancestral sister taxon to all brachiopods (Sperling et al. 2011; although see Thomson et al. 2014) or instead as a sub-taxon within the brachiopods as a whole (Cohen and Weydmann 2005; Santagata and Cohen 2009; Cohen 2013). Phoronids are clearly closely related to brachiopods and share evolutionary affinities with spiralian protostomes such as nemerteans, annelids, and mollusks, but their phylogenetic position relative to ectoprocts has been unresolved (Halanych et al. 1995; Dunn et al. 2008; Hejnol et al. 2009; Hausdorf et al. 2010; Mallatt et al. 2012). Interestingly, a recent phylogenomic study supports the monophyly of all three lophophore-bearing phyla in which phoronids are closely related to brachiopods and considered an ancestral sister taxon to ectoprocts (Nesnidal et al. 2013). Species- and

genus-level relationships among phoronids based on either morphological or molecular characters are moderately congruent (Santagata and Cohen 2009) and support *Phoronis ovalis* as a divergent lineage, a *Phoronopsis* clade and a subclade of *Phoronis* spp. It should be noted that the molecular phylogenetic data discussed here are comprised largely of ribosomal and mitochondrial markers, and a more recent phylogenetic analysis of phoronid species based on these genes plus additional nuclear and mitochondrial genes resolves previously incongruent aspects of the morphology- and molecular-based evolutionary inferences (Santagata, 2014).

EARLY DEVELOPMENT

Gonads develop from the peritoneal lining covering the capillaries of the efferent blood vessel on the surface of the stomach (Ikeda 1903; Rattenbury 1953; Zimmer 1991). Adults are either gonochoristic or hermaphroditic, with some species having a small temporal bias toward protandry (Zimmer 1991). Zimmer (1991) recognized three reproductive types among phoronid species that are largely based on the sex of the adult, mature egg diameter, and how embryos are brooded (if present). Group one is comprised almost entirely of gonochoristic species (except *Phoronis pallida*) that freely spawn 60 μm ova and includes members of both genera. Members of group two are for the most part species that produce larger mature eggs (≥ approximately 100 μm) and brood embryos to an early larval or competent larval stage on specialized nidamental glands at the base of the lophophore. Patterns within group two are somewhat muddled due to little or incomplete knowledge regarding the reproductive properties of species such as *Phoronopsis albomaculata* and other disputed species morphotypes (*Phoronis capensis* and *Phoronis bhadurii*). Group three has only one member, *Phoronis ovalis*, whose sex type remains unconfirmed, produces approximately 125 μm eggs, and broods embryos in the parental tube (Harmer 1917; Silén 1954b).

In species where it has been studied in any detail, fertilization is internal. Primary oocytes are arrested at metaphase and released from the ovary to fuse with activated V-shaped sperm (Zimmer 1964, 1991; Reunov and Klepal 2004). Male pronuclei have been found in the coelomic oocytes of *Phoronopsis harmeri* (Rattenbury 1953) and *Phoronis ijimai* (currently = *vancouverensis*; Zimmer 1991; Hirose et al. 2014). There is one largely undocumented report of external fertilization in *Phoronis muelleri* (Herrmann 1986), but this observation requires more data. How active internal sperm reach the ovary is one of the more intriguing aspects of phoronid biology. Even in species that are simultaneous hermaphrodites, self-fertilization has not been reported, as mature sperm released from the testes are not yet activated (Zimmer 1964; Reunov and Klepal 2004). Small masses of mature sperm released from the testes make their way through the trunk coelom until gathered up by the ciliary currents of the paired metanephridial funnels. Although some authors have observed sperm directly released from the metanephridial ducts into the surrounding seawater (Rattenbury 1953; Silén 1954b), more often released sperm are enclosed by the paired lophophoral organs at the base of the lophophore. Here, sperm are packaged into spherical, bean- or club-shaped spermatophores, some of which have elaborate sail-like structures (Zimmer 1967, 1991). Completed spermatophores drift away and the enclosed sperm remain inactive until reaching another adult conspecific individual. Perhaps the most common pathway for sperm to reach the ovary of another individual is via the metanephridial ducts (Brooks and Cowles 1905; Rattenbury 1953; Zimmer 1967). Once contacting the duct, sperm in the head of the spermatophore enter the duct and continue down the funnel, to the base of the trunk, to the ovary. Another less common route has been observed in *Phoronopsis viridis* (currently considered to be a junior synonym of *Phoronopsis harmeri*; see Emig 1974; Santagata and Cohen 2009), in which the spherical portion of the spermatophore contacts the tip of a tentacle where sperm using enzymes from their acrosomes (presumably, the

exact mechanism is not known) lyses through the epithelial tissue into the collar coelom (Zimmer 1991). From there, the activated sperm swims down to the septum that divides the collar and trunk coelom and lyses through that wall. Sperm must then swim down the trunk to its base to fertilize oocytes from the ovary. As complicated as this latter fertilization scenario may be, yet another variant of this method may exist, as spermatophores have also been observed being ingested by nearby adults.

Once expelled via the metanephridial ducts into the surrounding seawater, the fertilized primary oocytes become activated and complete meiosis forming two to three polar bodies (the first polar body often divides). At this point, cleavage begins, and the zygote divides equally (or approximately equally) and totally along the animal-vegetal axis (Fig. 10.1B, C). Second cleavage repeats this process, but with a cleavage angle perpendicular to the first cleavage, resulting in a four-cell embryo (Fig. 10.1D). Blastomeres isolated at the latter stage and two-cell stages are capable of forming the whole larva; accordingly, early development is regulative (Zimmer 1964).

Whether subsequent cleavage stages of various phoronid species exhibit aspects of radial cleavage or spiral cleavage has been debated since the first embryological observations were gathered. Radial (or biradial) cleavage patterns have been found in several species (Masterman 1900; Ikeda 1901), but spiral-like cleavage patterns have also been observed (Foettinger 1882; Brooks and Cowles 1905), and further, both patterns have been observed in the same species (Herrmann 1986). A more typical, spiral cleavage pattern was described for *Phoronopsis viridis* (junior synonym of *P. harmeri*) by Rattenbury (1954), but this was interpreted by Zimmer (1964) as an artifact introduced by compaction of the blastomeres by a tightly fitting vitelline envelope. It should be noted that all of the embryos having spiral-like cleavage patterns (or both spiral and radial) came from species that produce smaller (60 μm) eggs, but other species that produce larger (100 μm) eggs such as *Phoronis ijimai* (= *vancouverensis*) more consistently have

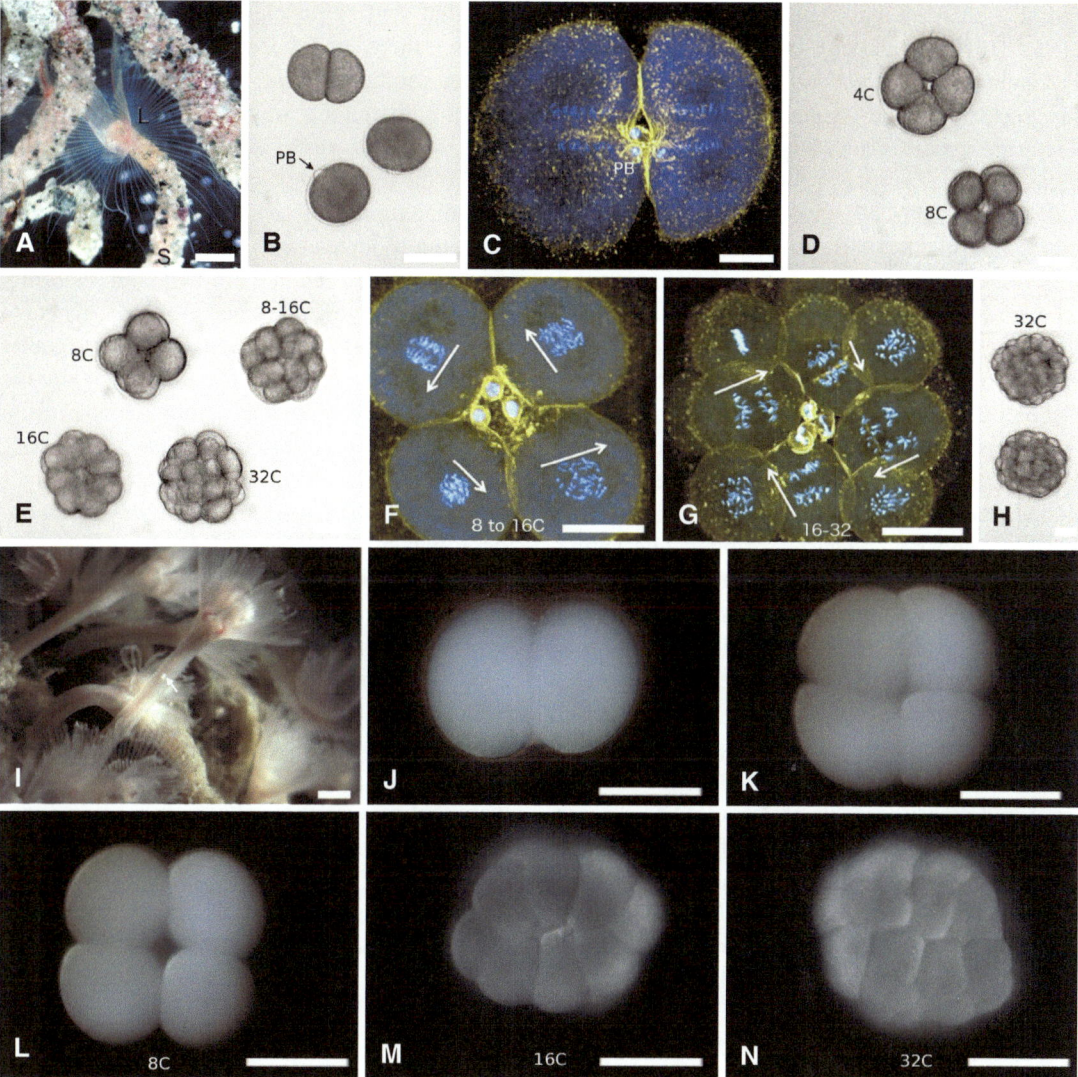

Fig. 10.1 Early cleavage stages of *Phoronis pallida* (A–H) and *Phoronis vancouverensis* (junior synonym of *Phoronis ijimai*; I–N). (**A**) Adult *P. pallida* collected from False Bay, WA, USA; note the characteristically bent sandy tube (*S*) and delicate U-shaped lophophore (*L*) (© Scott Santagata 2015. All Rights Reserved). (**B**) Zygotes (approx. 60 μm) with polar bodies (*PB*) and a two-cell stage. (**C**) Confocal z-projection of a two-cell stage stained for nucleic acids (*blue*) and fibrous actin (*yellow*). The embryo is undergoing division to the four-cell stage. (**D**) A four-cell (*4C*) and an eight-cell (*8C*) embryo. (**E**) A radially arranged eight-cell stage (*8C*), an eight-cell stage undergoing a spiral-like division to the 16-cell stage (*8–16C*), a 16-cell stage with an open and radial blastomere arrangement (*16C*), and a 32-cell stage showing the apical quartet of cells (*32C*). All embryonic images are from different embryos placed into one composite figure. (**F**) Animal pole view of an eight-cell embryo dividing with a counterclockwise twist (denoted by the *arrows*) to the 16-cell stage. Three polar bodies are found in the center of the image. (**G**) Animal pole view of a 16-cell embryo dividing with a clockwise twist (denoted by the *arrows*) to the 32-cell stage. (**H**) Two 32-cell embryos. (**I**) Ventral view of an adult *Phoronis vancouverensis* collected from Westcott Bay, WA, USA. This species occurs in muddy clumps with other conspecific individuals and broods various early developmental stages (early cleavage to four-tentacle actinotrochs) in paired lophophoral masses. A fertilized primary oocyte (approx. 100 μm; *arrow*) is present in one of the metanephridial funnels at midlevel in the body. (**J**) Two-cell stage. (**K**) Four-cell embryo. (**L**) Eight-cell embryo with radial arrangement of blastomeres. (**M**) 16-cell stage with an open radial arrangement of blastomeres. (**N**) 32-cell embryo. Scale bars in **A** and **I** equal 2 mm, scale bars in **B** and **J–N** equal 50 μm, scale bar in **C** equals 10 μm, and scale bars in **D–H** equal 25 μm

radially cleaving embryos (Zimmer 1964; Wu et al. 1980; Malakhov and Temereva 2000; Freeman and Martindale 2002), although irregular early cleavage stages are also reported. Cleavage variation also begins at different early cleavage stages among species; some cleave variably at the eight-cell stage (Brooks and Cowles 1905; Herrmann 1986) and others (more commonly) at the 16-cell stage (Temereva and Malakhov 2007). More recently, early cleavage was investigated in *Phoronis muelleri* using 4D microscopy techniques (Pennerstorfer and Scholtz 2012), which showed that oblique rather than perpendicular cell divisions occurred at third and subsequent cleavages of most of the embryos in the study. Interestingly, cell divisions at third cleavage and beyond also exhibited alternating dextral-sinistral twists, a feature typical of spiralian embryos (Hejnol 2010). Through these data coupled with their thorough review of cleavage variation in phoronid embryos, Pennerstorfer and Scholtz (2012) make a strong case for the spiralian affinities of some aspects of phoronid development. That said, further detailed accounts of phoronid embryogenesis are required to discern the nature, variability, and significance of early cleavage patterns, especially in those species that produce larger eggs and tend to have more consistently radially cleaving embryos. For these reasons, in the next section, the cleavage patterns of *Phoronis pallida* and *Phoronis vancouverensis* (a junior synonym of *P. ijimai*) are compared using confocal and time-lapse microscopy.

Numerous exogenous factors can affect the arrangement of blastomeres and the way early embryos cleave. Obviously, temperature is a crucial variable and care must be taken to ensure that embryos are not heated beyond their optimal ranges (see data in Staver and Strathmann 2002). Another less acknowledged variable is salinity – it is known to affect fertilization success (Allen and Pechenik 2010), but its nonlethal effects on early development (which may cause embryonic spaces to swell or shrink) are not well understood. One issue particular to phoronids is the typical way in which embryological cultures are started. Even though numerous fertilized oocytes may be gathered from the trunk coelom of an adult that will begin cleaving once exposed to seawater, there can be developmental differences between embryological cultures started from oocytes gathered this way as compared with naturally spawned oocytes. What changes may occur to the fertilized oocytes while housed in the nephridial funnels or what signals are received when the oocytes are extruded through the nephridial ducts are not known, but naturally spawned embryos do develop more synchronously. For these reasons, it is best to gather fertilized primary oocytes that are packed into the nephridial funnels rather than those isolated from the posterior part of the trunk and ovary (Santagata, pers. obs.). Early cleavage stages shown in Fig. 10.1 were imaged with these considerations in mind. Four-cell stages of *Phoronis pallida* typically have a gap between the pairs of sister cells and do not make cell contacts or cross-furrows as found in *Phoronis muelleri* (approx. 25 % of embryos; Pennerstorfer and Scholtz 2012). Third cleavage of *Phoronis pallida* embryos produces some eight-cell stages that have a radial arrangement of blastomeres and others that have animal cells slightly offset (with a clockwise twist) from those in the vegetal half of the embryo (Fig. 10.1). Fourth cleavage is also variable among embryos; some 16-cell stages have the characteristic open radial arrangement of blastomeres (Fig. 10.1E), while other embryos have four tiers of cells offset from one another, arranged by oblique divisions with a counterclockwise twist (Fig. 10.1E, F). Embryos at fifth cleavage continue this pattern of alternating the directional twist of cell divisions (Fig. 10.1G) and 32-cell stages have centrally positioned quartets of cells at the animal and vegetal poles (Fig. 10.1H).

The larger embryos of *Phoronis vancouverensis* typically have a radial arrangement of blastomeres (Fig. 10.1I–N). Although irregular early cleavage stages have been observed (Zimmer 1964; Malakhov and Temereva 2000), none of these patterns exhibit spiral-like divisions. At the four-cell stage, transient cell contacts between non-sister blastomeres do occur, creating a cross-furrow similar to what is described for *Phoronis muelleri* (Pennerstorfer and Scholtz 2012).

Eight-cell and 16-cell stages have the radial cell arrangements discussed previously, and 32-cell stages also have centrally located quartets of cells at the animal and vegetal poles with a cross-furrow similar to what is observed in *P. pallida* (Fig. 10.1). All of these radial-like and spiral-like cleavage patterns in phoronids support their spiralian affinities, and perhaps also indicate independent switches to radial cleavage patterns in particular species lineages, or may point toward radial-like cleavage being plesiomorphic for both deuterostomes and protostomes. Although phoronid spiral-like cleavage patterns are clearly different from the mosaic cell lineages of most mollusks (see Chapter 7; Hejnol et al. 2007) and annelids (Chapter 9; Meyer et al. 2010), genetic mechanisms that underlie patterns of chirality in some gastropods may shed light on variable phoronid cleavage patterns in that the sinistral embryos of *Lymnaea stagnalis* are initially radially arranged before undergoing a relatively late sinistral twist (see Chapter 7; Shibazaki et al. 2004). Furthermore, dextral and sinistral embryos in this species are genetically determined at a single locus (dextral is dominant), and so at least for this species of mollusk, allelic differences can account for positional shifts in the spindle apparatus (Chapter 7; Kuroda et al. 2009).

Subsequent cell divisions yield either a thin- or thick-walled round blastula depending on the size of the egg with a spacious or more compact blastocoelic space, respectively (Zimmer 1980; Herrmann 1986; Santagata 2004b; Temereva and Malakhov 2007). Gastrulation begins with the flattening of the vegetal region of the blastula forming a vegetal plate that is then internalized. In species with smaller eggs, the blastocoelic cavity remains spacious during the early stages of gastrulation, but not in species with larger eggs (Zimmer 1991; Malakhov and Temereva 2000; Santagata 2004b). The blastopore is initially rounded but becomes slit-like as the embryo elongates. The slit-like blastopore is progressively closed from the posterior end, leaving only a small anterior remnant to form the mouth region. During this process, mesenchymal cells from the epithelial regions near the developing oral region enter the blastocoelic space and aggregate at the anterior portion of the late gastrula stage (see Fig. 10.2B). These cells will form a small cavity underneath the apical tuft region of the late gastrula. This cavity has been interpreted as the first coelomic cavity ("protocoel"), but as discussed previously, whether these cells form a true epithelium in all or in just select species is debated (Zimmer 1978; Bartolomaeus 2001; Temereva and Malakhov 2006). At this asymmetric late gastrula stage, the gut is a blind sac. Cell labeling and other embryological experiments show that much of the larval mesoderm is derived from ectodermal cells, especially where the ectoderm is in contact with the endoderm, and the remaining mesoderm is derived from the endoderm (Freeman and Martindale 2002; Santagata 2004b; Temereva and Malakhov 2007). The fate mapping studies of Freeman also found that the development of phoronids was most similar to that of rhynchonelliform brachiopods (Freeman 1991, 2003) in that similar morphogenetic movements reposition the anterior ectoderm. Similar animal-vegetal inductive signals contribute to the formation of the anterior-posterior axis in phoronids, rhynchonellid brachiopods, and discinid brachiopods, but one difference found in phoronids is that the animal half of the embryo is not specified until late in gastrulation (Freeman 1991, 2003). During the earliest phase of larval development, the ectoderm anterior to the mouth grows quickly, forming the preoral hood, which has an upper exumbrellar epithelium with a developing apical organ (Fig. 10.2D) and a lower subumbrellar epithelial layer (Fig. 10.2D). Mesenchymal cells now line the subumbrellar portion of the hood and the ectodermal portion of the trunk. These cells differentiate into muscle cells that form the initial body wall muscle fibers of the larva (Santagata 2004b; Temereva and Tsitrin 2013). Where the blind-ended gut meets the posterior ectoderm, the anus and intestine form from an ectodermal invagination (Freeman and Martindale 2002). As the early larval stage grows, the hood enlarges significantly and drapes over the trunk region (Fig. 10.2E). At this stage, a diagonal band of epidermis on the future collar increases

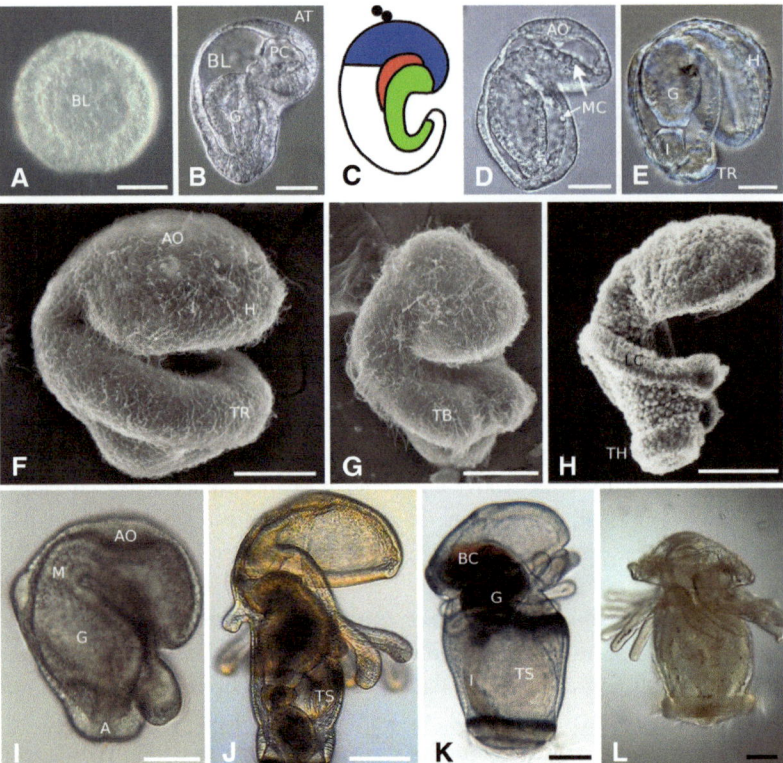

Fig. 10.2 Gastrulation and larval development of various species of phoronids. (**A**) Light micrograph of a blastula stage of *Phoronis pallida* (adults collected from False Bay, WA, USA) that has a spacious blastocoel (*BL*). (**B**) Light micrograph of a late gastrula stage of *Phoronis architecta* (adults collected from Alligator Bay, FL, USA). This morphotype is currently a junior synonym under *Phoronis psammophila* but is kept separate here for reasons discussed in Santagata and Cohen (2009). The preoral hood is precociously developed and the "protocoel" (PC, whose lining's structural composition as a true epithelial tissue has been questioned) is evident. Similar to many other larval forms, the developing apical organ has centrally located cells with long cilia (apical tuft, *AT*). The blind-ended gut (*G*) is also present at this stage. (**C**) Late gastrula stage shaded for embryological germ tissues based on cell-marking experiments in *Phoronis vancouverensis* (Freeman 1991). The animal pole of the egg is marked by the position of two polar bodies (*black*), the anterior ectoderm is shaded *blue*, the anterior mesoderm is shaded *red*, and the gut endoderm is shaded *green*. (**D**) Light micrograph of an early tentacle ridge stage of *Phoronis pallida* in which amoeboid mesenchymal cells (*MC*) on the subumbrellar portion of the hood and in the trunk blastocoelic cavity will form the first muscle fibers of the body wall (see Santagata (2004b)). The developing apical organ (*AO*) is more defined at this stage. The anus and intestine are initially formed from an ectodermal invagination. (**E**) Light micrograph of a late tentacle ridge (*TR*) stage of *Phoronis architecta* in which the hood (*H*) has grown rapidly and drapes over the collar. Clear demarcations exist between the gut (*G*) and intestine (*I*), which are separated by a sphincter valve. (**F**) Scanning electron micrograph of a tentacle ridge (*TR*) stage of *Phoronis vancouverensis* (this morphotype is a junior synonym under *Phoronis ijimai*). The brooding adults were collected from Westcott Bay, WA, USA. Early stages are held in the brood mass by mucous cords attached to the developing apical organ (*AO*) of the hood (*H*). (**G**) Scanning electron micrograph of an early tentacle bud (*TB*) stage of *Phoronis vancouverensis* in which the simple ciliation on the surface of the larva becomes denser and the trunk lengthens. (**H**) As the first pair of tentacles becomes better defined, their ciliation transforms into lateral (*LC*) and frontal ciliary bands. The posterior part of the trunk also develops a telotrochal (*TH*) band of cilia (© Scott Santagata 2015. All Rights Reserved). (**I**) Light micrograph of a two-tentacle stage of *Phoronis vancouverensis* with mouth (*M*) and anus (*A*). (**J**) Light micrograph of a *Phoronis pallida* larva with eight tentacles. The trunk sac (*TS*) grows relatively fast at this stage in preparation for metamorphosis. (**K**) Light micrograph of a competent ten-tentacle stage of *Phoronis pallida* that has a single, red blood corpuscle mass (*BC*) and a muscular trunk sac (*TS*) (© Scott Santagata 2015. All Rights Reserved). (**L**) Light micrograph of a competent 18-tentacle stage of *Phoronopsis harmeri* collected from the plankton near Friday Harbor Laboratories, WA, USA. Scale bars in **A–E** equal 25 μm, scale bars in **F–I** equal 50 μm, scale bars in **J** and **K** equal 100 μm, and scale bar in **L** equals 200 μm

in thickness and cell number, forming the ciliated tentacle ridge (Fig. 10.2E, F). Paired lateral projections develop at the ventral midpoint of the tentacle ridge, forming the first two tentacle buds (Fig. 10.2G). As these tentacle buds grow into functional locomotory and feeding structures, their ciliated epithelia differentiate into lateral and frontal bands of cilia (Fig. 10.2H). Lateral cilia generate downward currents that bring food particles to the frontal surface of the tentacles, where specialized sensory cells (laterofrontal cells) likely detect them, causing localized ciliary arrests and/or reversals. These changes in ciliary beat couple with a lifting of the hood creating additional suction, thus lifting the particle to the mouth to be ingested (Strathmann and Bone 1997). By the time the first pair of tentacles is formed, the middle collar region of the actinotroch larva is well defined, dividing the body into three regions (hood, collar, and trunk). Pairs of new tentacles will be added laterally and dorsally as the larva grows and the trunk lengthens (Fig. 10.2J). How many larval tentacles are formed varies among species and larval sizes (Santagata and Zimmer 2002). Late in larval development, a midventral thickening in the trunk epidermis invaginates to form an internal sac (Fig. 10.2J, K). Typically called the metasomal sac, it will be referred to here as the trunk sac to avoid unnecessary comparisons to the anatomy of ambulacral deuterostomes. The trunk sac continues to elongate, displacing the stomach and intestine dorsally. The trunk sac differentiates into what will be the trunk epithelium of the juvenile, replete with numerous circular and longitudinal muscles (Zimmer 1964; Santagata 2002; Temereva and Tsitrin 2013). Other general features of the competent actinotroch larva are the compact mass(es) of red blood cells that form in the collar region (Fig. 10.2K), the precociously developed juvenile blood vessels, and variation in the differentiation of the juvenile tentacles (Santagata and Zimmer 2002).

LATE DEVELOPMENT

Competent actinotroch anatomy varies mainly in maximum size, number of blood corpuscle masses, maximum number of larval tentacles, and how the juvenile tentacles are differentiated in the larva (see Fig. 10.3; Santagata and Zimmer 2002). Only one species, *Phoronis ovalis*, produces a non-feeding slug-like larva that lacks tentacles. Once released from the parental tube, *Phoronis ovalis* larvae swim in the plankton for about 4 days and then transition to a demersal phase, swimming near the bottom for about 3 days before undergoing metamorphosis (Silén 1954b; Zimmer 1991). Feeding actinotroch larvae have a complex complement of striated musculature for lifting and closing the preoral hood; extending, flicking, or lowering the tentacles; and adjusting the angle of the teletrochal band of cilia while swimming (Zimmer 1964; Santagata and Zimmer 2002; Santagata 2002, 2004b; Temereva and Tsitrin 2013). Although many aspects of larval musculature are shared among several phoronid species, the morphology of particular larval muscles does differ, especially when considering the smaller and more compact actinotroch types (*P. pallida*, *P. ijimai*, and *P. hippocrepia*) as opposed to the larger and more elongate types (*Phoronopsis harmeri* and *Phoronis muelleri*; see Santagata and Zimmer 2002; Temereva and Tsitrin 2013). However, some reported differences in larval musculature are the result of interpretational differences among various authors.

Both conserved and derived features also comprise the larval nervous system among various phoronid species. The greatest concentration of neuronal cells in the larval nervous system is found in the apical organ, a U-shaped structure containing at least four different neuronal cell types, all of which send axonal fibers into a central neuropil (Hay-Schmidt 1989, 1990; Lacalli 1990; Santagata 2002; Temereva and Wanninger 2012; Sonnleitner et al. 2013; Temereva and Tsitrin 2014). Cellular domains within the

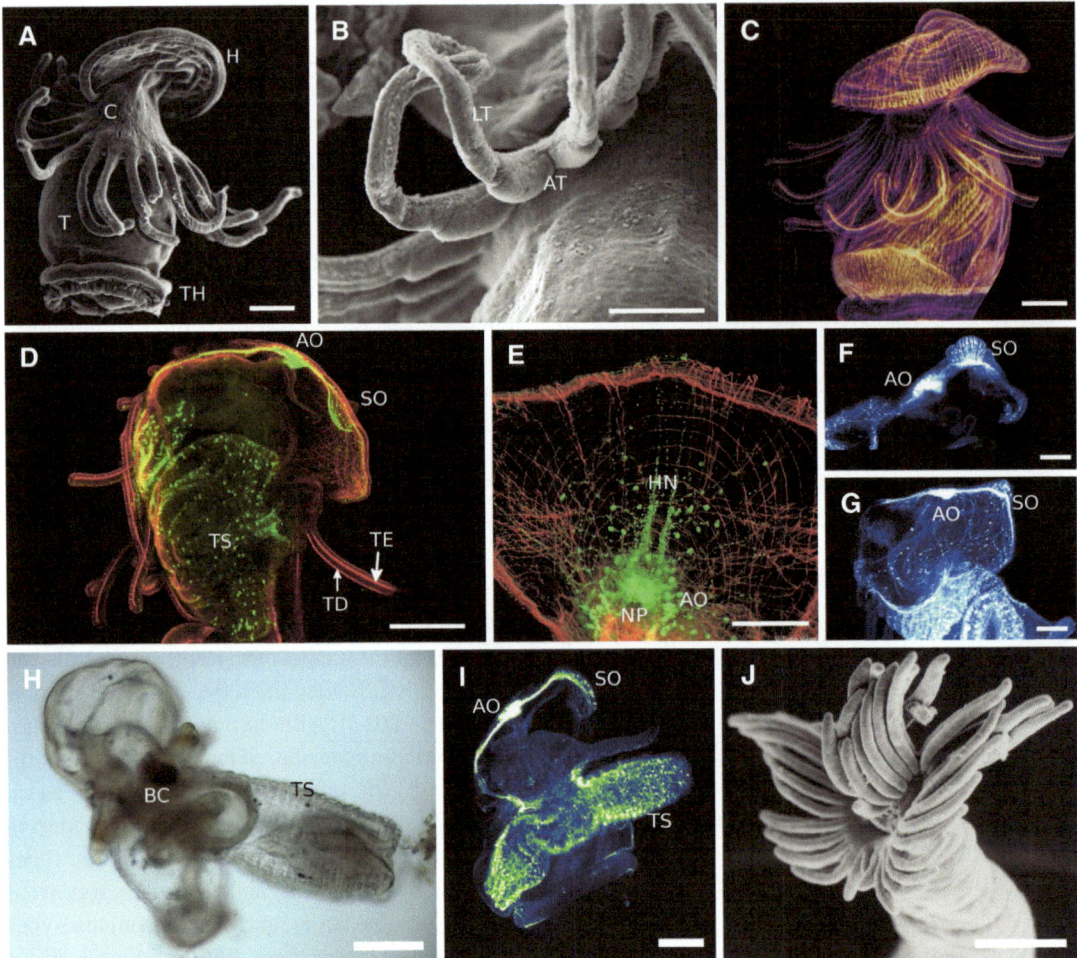

Fig. 10.3 Aspects of competent larval characteristics and metamorphosis among actinotroch morphotypes. (**A**) Scanning electron micrograph (*SEM*) of a competent stage of an unidentified actinotroch larva, "Actinotroch C," collected from the plankton near the Port of Los Angeles, CA, USA (Santagata and Zimmer 2002), showing the hood (*H*), collar (*C*), and trunk (*T*) regions as well as the heavily ciliated telotroch (*TH*). (**B**) SEM of the juvenile tentacle rudiments (*AT*) attached to the proximal-basal side of the larval tentacles (*LT*) of Actinotroch C. (**C**) Musculature of Actinotroch C stained with phalloidin (*red*). (**D**) Musculature (*red*) and serotonergic nervous system (*green*) of an actinotroch morphotype with similar anatomical traits to that of *Phoronis muelleri* (but a separate species), showing the apical organ (*AO*), frontal organ (*SO*), and numerous serotonin-like cell bodies in the tissue of the juvenile trunk sac (*TS*). Tentacle muscles, elevators (*TE*), and depressors (*TD*) underlie the frontal and abfrontal sides of each larval tentacle. (**E**) Magnified apical view of the apical sense organ (*AO*) and central neuropil (*NP*) in a late-stage actinotroch larva labeled for serotonin (*green*). The three median hood nerves (*HN*) project from the apical organ to the margin of the hood. (**F**, **G**) Serotonin-like sensory cells in the frontal organs of two different actinotroch morphotypes. (**H**) Early stage of metamorphosis when the juvenile trunk sac (*TS*) is everted and the blood corpuscle mass (*BC*) has not yet dissociated. This morphotype was collected from the plankton of Tampa Bay, Florida, USA, and its identity is likely *Phoronis hippocrepia*. (**I**) First stage of metamorphosis in *Phoronopsis harmeri* (collected from the plankton outside Los Angeles Harbor, CA, USA), showing the numerous serotonin-like cell bodies and fibers (*green*) in the larval and juvenile nervous systems. (**J**) SEM of the juvenile lophophore of an unidentified phoronid species. Scale bars in **A**, **C**, **D**, and **I** equal 200 μm, scale bar in **B** equals 10 μm, and scale bars in **F** and **H** equal 100 μm

apical organ include groups of cells that express serotonin (Hay-Schmidt 1990; Santagata 2002; Temereva and Wanninger 2012; Sonnleitner et al. 2013), catecholamines (Hay-Schmidt 1990; Santagata 2002), FMRFamide (Hay-Schmidt 1990; Sonnleitner et al. 2013), and a cardioactive B-like peptide (Sonnleitner et al. 2013). Serotonin-like immunoreactive cells include numerous flask-shaped cells that overlie other bipolar or multipolar neuronal cells (Hay-Schmidt 1990; Santagata 2002; Temereva and Wanninger 2012; Sonnleitner et al. 2013). Neuronal cells and fibers that have catecholamine-like immunoreactivity are situated at the periphery of the apical organ (Hay-Schmidt 1990; Santagata 2002). Only a few to perhaps ten bipolar neuronal cell bodies on dorsolateral sides of the apical organ have FMRFamide-like immunoreactivity, depending upon the species and larval stage (Hay-Schmidt 1990; Temereva and Wanninger 2012; Sonnleitner et al. 2013). The FMRFamide-like immunoreactive cells do not appear to be sensory, as they lack cilia and their fiber projections go to the apical neuropil and the tentacular neurite bundles, leading into the major tentacle nerve rings.

Beyond the apical organ, significant larval nerves include the main hood nerve and two dorsolateral nerves that run down into the collar (mesosome) region of the larval body and merge with the minor and major tentacle nerve rings. From these nerve rings, the larval tentacles (and sometimes the juvenile tentacle rudiments; Santagata and Zimmer 2002) are innervated on frontal and abfrontal sides (Hay-Schmidt 1989, 1990; Lacalli 1990; Santagata 2002; Temereva and Wanninger 2012; Sonnleitner et al. 2013; Temereva and Tsitrin 2014). Basiepithelial fibers also project from these nerve rings through the larval trunk epithelium and merge with a telotrochal nerve ring at the posterior end of the larva (Hay-Schmidt 1990; Santagata and Zimmer 2002; Temereva and Wanninger 2012; Sonnleitner et al. 2013). Although a possible vestigial ventral nerve cord has been reported

for *Phoronopsis harmeri* (Temereva 2012), the cytoarchitecture of these neural fibers has very little in common with the ventral nerve cords of molluscan or annelid larvae in that phoronid larvae do not have centralized ventral nerve cords with repeated neuronal cell bodies along them (Nederbragt 2002; Nederbragt et al. 2002; Denes et al. 2007; Meyer and Seaver 2009). Overall, there are several structural and neurochemical aspects of the larval nervous system that vary among species (Santagata and Zimmer 2002; Sonnleitner et al. 2013; Temereva and Tsitrin 2014). However, it should be noted that some of this variation in the larval nervous system may be due to unrecognized cryptic speciation when comparing what are believed to be the same larval types from geographically distant populations. One example can be found in *Phoronopsis harmeri*, a species whose type locality is near Vancouver Island, Canada (Pixell 1912). Near its type locality, this species' competent larval form typically has 18–20 tentacles and four blood corpuscle masses (Fig. 10.3; see also Zimmer 1964), but competent *Phoronopsis harmeri* larvae collected from the plankton of Vostok Bay, Sea of Japan, are competent at 24 tentacles (Temereva and Wanninger 2012) and may have variable numbers of blood corpuscle masses (Temereva and Neretina 2013). For these reasons, care must be taken when assessing the phylogenetic context and functional importance of characters in the phoronid larval nervous system (Santagata 2011).

Some of the more dramatic features that develop in the larval nervous system near the time of metamorphic competence are the three median hood nerves that emanate from the apical organ and merge into a secondary hood sense organ (also called the frontal organ) that has its own neuropil (Santagata and Zimmer 2002; Santagata 2002; Temereva and Wanninger 2012; Sonnleitner et al. 2013). Depending on the musculature of the hood, the frontal organ may be cone-shaped and eversible (Santagata and

Zimmer 2002) or be composed of a field of sensory cells at the distal portion of the hood (Santagata and Zimmer 2002; Temereva and Wanninger 2012). More commonly, the frontal organ includes several bipolar, serotonergic sensory cells (Fig. 10.3F, G; Santagata and Zimmer 2002; Temereva and Wanninger 2012) or another neurotransmitter type (Santagata 2002; Sonnleitner et al. 2013). Evolutionary modifications to the functions of the frontal organ's neurons may confer different degrees of settlement behavior specificity (Santagata 2004a).

Actinotroch larval metamorphosis is rapid and cataclysmic. Within a span of 15–30 min, a larva with a straight gut and radiating tentacles transforms into a juvenile with an elongate vermiform trunk with a bulbous ampullar posterior end and an oval or U-shaped array of tentacles. At its midpoint, the larval gut is drawn into the cavity of the trunk sac by the midventral mesentery (Fig. 10.3H). During this process, the cells of the blood corpuscle mass dissociate and stream into the vessels of the juvenile circulatory system (Zimmer 1991; Santagata 2002; Temereva and Tsitrin 2013). Depending on how the juvenile tentacles develop (Santagata and Zimmer 2002), there are differences in (1) the formation of the lophophore blood vessels and (2) the larval tissues that undergo cell death. In species that have a separate set of juvenile tentacles in the competent larva (e.g., *Phoronis muelleri*), the larval tentacles are completely lost. During metamorphosis, the compacted lining of the collar coelom inside the juvenile tentacles expands (Bartolomaeus 2001), and post-metamorphosis, the blind-ended capillaries of the juvenile tentacles form in a day. For those species that form the juvenile tentacles from basal tentacle rudiments attached below the proximal ends of the larval tentacles (e.g., *Phoronis psammophila*), similar tissue modifications occur, but the fate of the frontal portion of the larval tentacles overlying the juvenile rudiments has not been studied in great cytological detail (Veillet 1941; Herrmann 1979). Species of both, *Phoronopsis* and *Phoronis*, remodel the larval tentacles into the juvenile form, but in slightly different ways. During metamorphosis, the distal portions of the larval tentacles of *Phoronis pallida* are cast off and expansion of the tentacle

coelom and formation of the blind-ended capillaries are similar to that of *P. muelleri* (Santagata 2002). In this species, the larval tentacle muscles (both elevators and depressors) are broken down and new juvenile tentacle muscles are formed in a day (Santagata 2002). However, in *Phoronopsis harmeri*, only the tentacle depressors are broken down (Temereva and Tsitrin 2013), and this difference may be linked with the fact that the tentacle capillaries are already formed and expanded in the competent larva's comparatively larger tentacles. Other muscles carried over to the juvenile body in this species are found in the esophagus (Temereva and Tsitrin 2013). In all phoronid species investigated thus far, the muscular and neural tissues of the preoral hood undergo histolysis (Herrmann 1979; Bartolomaeus 2001; Santagata 2002; Temereva and Tsitrin 2013). In contrast to some *Phoronis* spp., the hood coelom of *Phoronopsis harmeri* has a true epithelial lining, at least a portion of which may be carried over to form the coelomic lining of the juvenile epistome (Temereva and Malakhov 2006, 2011).

OPEN QUESTIONS

- Are similar axial patterning mechanisms present in phoronids, brachiopods, and perhaps even ectoprocts?
- What are the expression profiles of key developmental genes such as Hox and ParaHox genes in phoronids?
- Are the expression patterns for genes such as *Six3/6*, *NK2.1*, *Homeobrain*, *FoxQ2*, and *Otp* in embryos and larvae of rhynchonelliform brachiopods (e.g., *Terebratalia*) shared by the embryos and larvae of phoronids?
- What is the evolutionary basis and importance of both radial-like and spiral-like cleavage patterns among species of phoronids?

Acknowledgments Kelly Ryan and Russel Zimmer provided excellent comments on previous versions of this chapter. A portion of the data in this book chapter was gathered at the Smithsonian Marine Station (Fort Pierce, FL) and is designated contribution number 988. Faculty Research Grants provided by Long Island University-Post also supported this work.

References

Allen JD, Pechenik JA (2010) Understanding the effects of low salinity on fertilization success and early development in the sand dollar *Echinarachnius parma*. Biol Bull 218:189–199

Bartolomaeus T (2001) Ultrastructure and formation of the body cavity lining in *Phoronis muelleri* (Phoronida, Lophophorata). Zoomorphology 120:135–148

Brooks W, Cowles R (1905) *Phoronis architecta*: its life history, anatomy and breeding habits. Mem Natl Acad Sci 10:72–113

Chernyshev AV, Temereva EN (2010) First report of diagonal musculature in phoronids (Lophophorata: Phoronida). Dokl Biol Sci 433:264–267

Cohen BL (2013) Rerooting the rDNA gene tree reveals phoronids to be "brachiopods without shells;" dangers of wide taxon samples in metazoan phylogenetics (Phoronida; Brachiopoda). Zool J Linnean Soc 167(1):82–92

Cohen B, Weydmann A (2005) Molecular evidence that phoronids are a subtaxon of brachiopods (Brachiopoda: Phoronata) and that genetic divergence of metazoan phyla began long before the early Cambrian. Org Divers Evol 5:253–273

Denes AS, Jékely G, Steinmetz PRH, Raible F, Snyman H, Prud'homme B, Ferrier DEK, Balavoine G, Arendt D (2007) Molecular architecture of annelid nerve cord supports common origin of nervous system centralization in Bilateria. Cell 129:277–288

Dunn CW, Hejnol A, Matus DQ, Pang K, Browne WE, Smith SA, Seaver E, Rouse GW, Obst M, Edgecombe GD, Sørensen MV, Haddock SHD, Schmidt-Rhaesa A, Okusu A, Kristensen RM, Wheeler WC, Martindale MQ, Giribet G (2008) Broad phylogenomic sampling improves resolution of the animal tree of life. Nature 452:745–749

Emig CC (1974) The systematics and evolution of the phylum Phoronida. Z Zool Syst Evolutionsforschung 12:128–151

Fernández I, Pardos F, Benito J (1996) Ultrastructural observations on the phoronid nervous system. J Morphol 230:265–281

Foettinger A (1882) Note sur la formation du mesoderme dans la larve de *Phoronis hippocrepia*. Arch Biol Paris 3:679–686

Freeman G (1991) The bases for and timing of regional specification during larval development in *Phoronis*. Dev Biol 147:157–173

Freeman G (2003) Regional specification during embryogenesis in rhynchonelliform brachiopods. Dev Biol 261:268–287

Freeman G, Martindale MQ (2002) The origin of mesoderm in phoronids. Dev Biol 252:301–311

Garlick RL, Williams BJ, Riggs AF (1979) The hemoglobins of *Phoronopsis viridis*, of the primitive invertebrate phylum Phoronida: characterization and subunit structure. Arch Biochem Biophys 194:13–23

Gilchrist JDF (1907) New forms of the Hemichordata from South Africa. Trans S Afr Philos Soc 17:151–176

Gruhl A, Grobe P, Bartolomaeus T (2005) Fine structure of the epistome in *Phoronis ovalis*: significance for the coelomic organization in Phoronida. Invertebr Biol 124:332–343

Halanych KM, Bacheller JD, Aguinaldo AM, Liva SM, Hillis DM, Lake JA (1995) Evidence from 18S ribosomal DNA that the lophophorates are protostome animals. Science 267:1641–1643

Harmer SF (1917) Harmer: on *Phoronis ovalis*, Strethill Wright. Q J Microsc Sci 62:115–148

Hausdorf B, Helmkampf M, Nesnidal MP, Bruchhaus I (2010) Phylogenetic relationships within the lophophorate lineages (Ectoprocta, Brachiopoda and Phoronida). Mol Phylogenet Evol 55:1121–1127

Hay-Schmidt A (1989) The nervous system of the actinotroch larva of *Phoronis muelleri* (Phoronida). Zoomorphology 108:333–351

Hay-Schmidt A (1990) Catecholamine-containing, serotonin-like, and FMRFamide-like immunoreactive neurons and processes in the nervous system of the early actinotroch larva of *Phoronis vancouverensis* (Phoronida): distribution and development. Can J Zool 68:1525–1536

Hejnol A (2010) A twist in time–the evolution of spiral cleavage in the light of animal phylogeny. Integr Comp Biol 50:695–706

Hejnol A, Martindale MQ, Henry JQ (2007) High-resolution fate map of the snail *Crepidula fornicata*: the origins of ciliary bands, nervous system, and muscular elements. Dev Biol 305:63–76

Hejnol A, Obst M, Stamatakis A, Ott M, Rouse GW, Edgecombe GD, Martinez P, Baguna J, Bailly X, Jondelius U, Wiens M, Muller WEG, Seaver E, Wheeler WC, Martindale MQ, Giribet G, Dunn CW (2009) Assessing the root of bilaterian animals with scalable phylogenomic methods. Proc R Soc B Biol Sci 276:4261–4270

Herrmann K (1979) Larvalentwicklung und Metamorphose von *Phoronis psammophila* (Phoronida, Tentaculata). Helgoländer Meeresun 32:550–581

Herrmann K (1986) Die Ontogenese von *Phoronis mulleri* (Tentaculata) unter besonderer Berücksichtigung der Mesodermdifferenzierung und Phylogenese des Coeloms. Zool Jahrb Abt Anat Ontog Tiere 114:441–463

Herrmann K (1997) Phoronida. In: Harrison FW, Woollacott RM (eds) Microscopic anatomy of invertebrates volume 13: lophophorates, Entoprocta, and Cycliophora. Wiley-Liss, New York, pp 207–235

Hirose M, Fukiage R, Katoh T, Kajihara H (2014) Description and molecular phylogeny of a new species of *Phoronis* (Phoronida) from Japan, with a redescription of topotypes of *P. ijimai* Oka, 1897. Zookeys 398:1–31

Ikeda I (1901) Observations on the development: structure and metamorphosis of Actinotrocha. J Coll Sci Imp Univ Tokyo 13:507–591

Ikeda I (1903) On the development of the sexual organs and of their products in *Phoronis*. Annotationes Zoologicae Jpn 4:141–153

Kinoshita K (2002) Burrow structure of the mud shrimp *Upogebia major* (Decapoda: Thalassinidea: Upogebiidae). J Crustac Biol 22:474–480

Kuroda R, Endo B, Abe M, Shimizu M (2009) Chiral blastomere arrangement dictates zygotic left–right asymmetry pathway in snails. Nature 462:790–794

Lacalli TC (1990) Structure and organization of the nervous system in the actinotroch larva of *Phoronis vancouverensis*. Phil Trans Roy Soc London B Biol Sci 327:655–685

Larson AA, Stachowicz JJ (2009) Chemical defense of a soft-sediment dwelling phoronid against local epibenthic predators. Mar Ecol Prog Ser 374:101–111

Malakhov VV, Temereva EN (2000) Embryonic development of the phoronid *Phoronis ijimai*. Russ J Mar Biol 26:412–421

Mallatt J, Craig CW, Yoder MJ (2012) Nearly complete rRNA genes from 371 Animalia: updated structure-based alignment and detailed phylogenetic analysis. Mol Phylogenet Evol 64:603–617

Masterman AT (1898) On the theory of archimeric segmentation and its bearing upon the phyletic classification of the Coelomata. Proc R Soc Edinb 22:270–310

Masterman AT (1900) Memoirs: on the Diplochorda III. The early development and anatomy of *Phoronis buskii*, McIntosh. Q J Microsc Sci 43:375–418

Meyer NP, Seaver EC (2009) Neurogenesis in an annelid: characterization of brain neural precursors in the polychaete *Capitella* sp. I. Dev Biol 335:237–252

Meyer NP, Boyle MJ, Martindale MQ, Seaver EC (2010) A comprehensive fate map by intracellular injection of identified blastomeres in the marine polychaete *Capitella teleta*. EvoDevo 1:1–27

Nederbragt A (2002) Expression of *Patella vulgata* orthologs of *engrailed* and *dpp-BMP2/4* in adjacent domains during molluscan shell development suggests a conserved compartment boundary mechanism. Dev Biol 246:341–355

Nederbragt A, te Welscher P, van den Driesche S, van Loon AX, Dictus W (2002) Novel and conserved roles for *orthodenticle/otx* and *orthopedia/otp* orthologs in the gastropod mollusc *Patella vulgata*. Dev Genes Evol 212:330–337

Nesnidal MP, Helmkampf M, Meyer A, Witek A, Bruchhaus I, Ebersberger I, Hankeln T, Lieb B, Struck TH, Hausdorf B (2013) New phylogenomic data support the monophyly of Lophophorata and an ectoproct-phoronid clade and indicate that Polyzoa and Kryptrochozoa are caused by systematic bias. BMC Evol Biol 13:253

Pennerstorfer M, Scholtz G (2012) Early cleavage in *Phoronis muelleri* (Phoronida) displays spiral features. Evol Dev 14:484–500

Pixell HLM (1912) Memoirs: two new species of the Phoronidea from Vancouver Island. Q J Microsc Sci 58:257–284

Rattenbury JC (1953) Reproduction in *Phoronopsis viridis*. The annual cycle in the gonads, maturation and fertilization of the ovum. Biol Bull 104:182–196

Rattenbury JC (1954) The embryology of *Phoronopsis viridis*. J Morphol 95:289–349

Reunov A, Klepal W (2004) Ultrastructural study of spermatogenesis in *Phoronopsis harmeri* (Lophophorata, Phoronida). Helgoländer Meeresun 58:1–10

Santagata S (2002) Structure and metamorphic remodeling of the larval nervous system and musculature of *Phoronis pallida* (Phoronida). Evol Dev 4:28–42

Santagata S (2004a) A waterborne behavioral cue for the actinotroch larva of *Phoronis pallida* (Phoronida) produced by *Upogebia pugettensis* (Decapoda: Thalassinidea). Biol Bull 207:103–115

Santagata S (2004b) Larval development of *Phoronis pallida* (Phoronida): implications for morphological convergence and divergence among larval body plans. J Morphol 259:347–358

Santagata S (2011) Evaluating neurophylogenetic patterns in the larval nervous systems of brachiopods and their evolutionary significance to other bilaterian phyla. J Morphol 272:1153–1169

Santagata S (2014) Reconciling morphology, molecules, and species diversity with phoronid phylogenetic relationships, presented at the 2014 conference of the Society for Integrative and Comparative Biology, Austin, TX

Santagata S, Cohen BL (2009) Phoronid phylogenetics (Brachiopoda; Phoronata): evidence from morphological cladistics, small and large subunit rDNA sequences, and mitochondrial *cox1*. Zool J Linnean Soc 157:34–50

Santagata S, Zimmer RL (2002) Comparison of the neuromuscular systems among actinotroch larvae: systematic and evolutionary implications. Evol Dev 4:43–54

Shibazaki Y, Shimizu M, Kuroda R (2004) Body handedness is directed by genetically determined cytoskeletal dynamics in the early embryo. Curr Biol 14:1462–1467

Silén L (1954a) On the nervous system of *Phoronis*. Arch Zool 6:1–40

Silén L (1954b) Developmental biology of Phoronidea of the Gullmar Fiord area (west coast of Sweden). Acta Zool Stockh 35:215–257

Skovsted CB, Brock GA, Paterson JR, Holmer LE, Budd GE (2008) The scleritome of *Eccentrotheca* from the Lower Cambrian of South Australia: lophophorate affinities and implications for tommotiid phylogeny. Geology 36:171

Skovsted CB, Brock GA, Topper TP, Paterson JR, Holmer LE (2011) Scleritome construction, biofacies, biostratigraphy and systematics of the tommotiid *Eccentrotheca helenia* sp. nov. from the Early Cambrian of South Australia. Palaeontology 54:253–286

Sonnleitner B, Schwaha T, Wanninger A (2013) Inter- and intraspecific plasticity in distribution patterns of immunoreactive compounds in actinotroch larvae of Phoronida (Lophotrochozoa). J Zool Syst Evol Res 52:1–14

Sperling EA, Pisani D, Peterson KJ (2011) Molecular paleobiological insights into the origin of the Brachiopoda. Evol Dev 13:290–303

Staver JM, Strathmann RR (2002) Evolution of fast development of planktonic embryos to early swimming. Biol Bull 203:58–69

Strathmann RR, Bone Q (1997) Ciliary feeding assisted by suction from the muscular oral hood of phoronid larvae. Biol Bull 193:153–162

Temereva EN (2012) Ventral nerve cord in *Phoronopsis harmeri* larvae. J Exp Zool 318:26–34

Temereva EN, Malakhov VV (2006) Trimeric coelom organization in the larvae of *Phoronopsis harmeri* Pixell, 1912 (Phoronida, Lophophorata). Dokl Biol Sci 410:396–399

Temereva EN, Malakhov VV (2007) Embryogenesis and larval development of *Phoronopsis harmeri* Pixell, 1912 (Phoronida): dual origin of the coelomic mesoderm. Invertebr Reprod Dev 50:57–66

Temereva EN, Malakhov VV (2009a) Microscopic anatomy and ultrastructure of the nervous system of *Phoronopsis harmeri* Pixell, 1912 (Lophophorata: Phoronida). Russ J Mar Biol 35:388–404

Temereva EN, Malakhov VV (2009b) On the organization of the lophophore in phoronids (Lophophorata: Phoronida). Russ J Mar Biol 35:479–489

Temereva EN, Malakhov VV (2011) Organization of the episome in *Phoronopsis harmeri* (Phoronida) and consideration of the coelomic organization in Phoronida. Zoomorphology 130:121–134

Temereva E, Neretina T (2013) A distinct phoronid larva: morphological and molecular evidence. Invertebr Syst 27:622–633

Temereva EN, Tsitrin EB (2013) Development, organization, and remodeling of phoronid muscles from embryo to metamorphosis (Lophotrochozoa: Phoronida). BMC Dev Biol 13:1–24

Temereva EN, Tsitrin EB (2014) Development and organization of the larval nervous system in *Phoronopsis harmeri*: new insights into phoronid phylogeny. Front Zool 11:1–25

Temereva E, Wanninger A (2012) Development of the nervous system in *Phoronopsis harmeri* (Lophotrochozoa, Phoronida) reveals both deuterostome- and trochozoan-like features. BMC Evol Biol 12:121

Thomas AO (1911) A fossil burrowing sponge from the Iowa Devonian. Bull Lab Nat Hist State Univ Iowa Iowa City 6:165–166

Thompson RK (1972) Functional morphology of the hindgut gland of *Upogebia pugettensis* (Crustacea, Thalassinidea) and its role in burrow construction. University of California, Berkeley, 202 p

Thomson RC, Plachetzki DC, Luke Mahler D, Moore BR (2014) A critical appraisal of the use of microRNA data in phylogenetics. Proc Natl Acad Sci U S A. doi:10.1073/pnas.1407207111

Veillet A (1941) Descriptions et mécanismes la métamorphose de la larve actinotroque de *Phoronis sabatieri* Roule. Bull Inst Océanogr Monaco 810:1–10

Wright TS (1856) Description of two tubicolar animals. Proc R Soc Edinb 1:165–167

Wu B, Chen M, Sun R (1980) On the occurrence of *Phoronis ijimai* Oka in the Huang Hai, with notes on its larval development. Stud Mar Sin 16:101–122

Zimmer RL (1964) Reproductive biology and development of Phoronida. Ph.D. Thesis, University of Washington, Seattle. 416 p

Zimmer RL (1967) The morphology and function of accessory reproductive glands in the lophophores of *Phoronis vancouverensis* and *Phoronopsis harmeri*. J Morphol 121:159–178

Zimmer RL (1978) The comparative structure of the preoral hood coelom. In: Chia FS, Rice ME (eds) Settlement and metamorphosis of marine invertebrate larva. Elsevier, New York, pp 23–40

Zimmer RL (1980) Mesoderm proliferation and formation of the protocoel and metacoel in early embryos of *Phoronis vancouverensis* (Phoronida). Zool Jahrb Abt Anat Ontog Tiere 103:219–232

Zimmer RL (1991) Phoronida. In: Pearse JS, Pearse VB, Giese AC (eds) Reproduction of marine invertebrates, volume VI: echinoderms and lophophorates. Boxwood Press, Pacific Grove, pp 1–45

Ectoprocta

11

Scott Santagata

Chapter vignette artwork by Brigitte Baldrian.
© Brigitte Baldrian and Andreas Wanninger.

S. Santagata
Biology Department, Long Island University-Post,
Greenvale, NY 11548, USA
e-mail: scott.santagata@liu.edu

A. Wanninger (ed.), *Evolutionary Developmental Biology of Invertebrates 2: Lophotrochozoa (Spiralia)*
DOI 10.1007/978-3-7091-1871-9_11, © Springer-Verlag Wien 2015

INTRODUCTION

First considered to be plantlike (moss) animals similar to cnidarians, the evolutionary origin and affinities of the ectoprocts or bryozoans have been enigmatic subjects of research since the sixteenth century. The term Bryozoa originally encompassed both the entoprocts (or kamptozoans, Chapter 6) and the ectoprocts (Ehrenberg 1831); however, these animal groups were later separated (Nitsche 1869) and eventually organized into different phyla (Cori 1929). Ectoprocts are aquatic invertebrates that can form elaborate and occasionally large colonies (>1 m) composed of numerous individual zooids, each typically no more than a millimeter in length. Zooids in a colony may be one of several different polymorphic forms specialized for various functions such as feeding, reproduction, or defense. Current estimates of ectoproct diversity range from 4,000 to 8,000 extant species (Ryland 2005), many of which are broadly distributed throughout freshwater, brackish, and marine environments. More than 15,000 fossil species that trace their origins back to the Ordovician period approximately 483 million years ago have been described (Xia et al. 2007). This period of origin is much later than that of many other animal phyla that arose during or before the Cambrian (Erwin et al. 2011). Although one Cambrian ectoproct fossil has been described (Landing et al. 2010), this morphotype has been reinterpreted as a type of octocoral (Taylor et al. 2013). Whether the relatively late geologic origin of the ectoprocts is correct or merely the result of preservational bias against some as-yet-unknown soft-bodied form remains an open question, but all extant morphological grades of ectoprocts with and without mineralized zooids were clearly present by the Jurassic (McKinney 1995; Taylor and Ernst 2004; Ostrovsky et al. 2008).

Ectoprocts are currently arranged into three classes, the Stenolaemata, Gymnolaemata, and Phylactolaemata (Bock and Gordan 2013). All ectoprocts feed with a lophophore, which is a circular or U-shaped ring of ciliated tentacles surrounding the mouth. Ciliary currents direct suspended food particles toward the mouth, and then particles are gathered in a muscular pharynx. Ingested food particles are then processed in a U-shaped gut. Ectoprocts do not have nephridia, but solid waste is expelled by an anus situated outside the tentacles. Feeding zooids, called autozooids, can transfer nutrients to other non-feeding polymorphic zooids (e.g., avicularia, kenozooids, vibracula, etc.) in the colony via a network of tissue cords and communication pores called the funicular system (Banta 1969; Best and Thorpe 1985). Neuronal cell bodies of the intraepithelial nervous system are mainly concentrated in a single ganglion positioned between the mouth and anus (Schwaha et al. 2011) and are capable of integrating sensorimotor information between zooids (Thorpe et al. 1975). Collectively, the lophophore, gut, nervous system, musculature, and funicular tissue are grouped together as the polypide. Surrounding the polypide is an outer stratified epithelial layer known as the cystid that may have a tubular, ovoid, or boxlike shape. Gymnolaemate ectoprocts dominate in brackish and marine environments and are comprised of two orders, the uncalcified Ctenostomata and the varyingly calcified Cheilostomata. Further taxonomic subdivision of the Ctenostoma has been based mainly on colony-level traits such as whether zooids bud from tubular stolons (Stolonifera, e.g., *Bowerbankia* and *Walkeria*) or form fleshy sheet-like or globular colonies (Carnosa, e.g., *Alcyonidium*, *Hislopia*, *Nolella*, and *Sundanella*). Family-level relationships under these categories inferred from morphological characters differ among authors (e.g., Jebram 1992; Todd 2000).

Although no longer considered monophyletic clades (Waeschenbach et al. 2012), different morphological grades of cheilostomes (Anasca, Cribimorpha, and Ascophora) were previously grouped according to the structure of the frontal cystid membrane covering the polypide and its role in eversion of the lophophore. The frontal membrane may be a flexible structure that can be depressed by parietal muscles, thus increasing pressure in the coelom surrounding the polypide, resulting in the eversion of the lophophore (as found in the anascans *Bugula*, *Flustra*, and *Membranipora*).

Alternatively, the frontal membrane may be elaborated with a series of fused calcified spines (e.g., *Cribrilina*). Numerous species of cheilostomes have completely calcified their frontal membranes into a wall with pores of varying numbers and sizes. These species have a membranous sac (ascus) below the frontal wall that is open to the surrounding seawater (Dick et al. 2009). In these so-called ascophorans, parietal muscles expand this sac to evert the lophophore (e.g., *Celleporaria*, *Schizoporella*, and *Watersipora*). Stenolaemates had their greatest species radiation in the Paleozoic Era, but their diversity was significantly reduced by the Permian-Triassic and Cretaceous-Paleogene mass extinction events (Lidgard et al. 1993). Living members of this class are represented only by the order Cyclostomata, whose cylindrical calcified zooids lack any specialized closing structure (e.g., an operculum or muscular constriction). Similar to the other morphological grades of ectoprocts, the validity of morphological classifications within the cyclostomes has been questioned (Taylor and Weedon 2008). Cyclostomes occur strictly in marine environments, and their colonies can be relatively small (such as *Crisia* or *Tubulipora*) in comparison to those of other ectoproct classes. Perhaps the most divergent morphological, and evolutionarily controversial, zooidal traits are found in the strictly freshwater phylactolaemates whose colonies are often described as being either gelatinous and saclike or a series of chitinous, branched tubes (Wood and Lore 2005). Unlike other ectoprocts, phylactolaemate zooids have a hollow neuronal ganglion (Gruhl and Bartolomaeus 2008; Schwaha and Wanninger 2012), bud from a commingled fleshy mass (some of which can "crawl" along the bottom like *Cristatella*), and have a U-shaped lophophore. The latter trait has occasionally been regarded as a possible synapomorphy with another member of the lophophore-bearing animals, the phoronids (Chapter 10; Hyman 1959).

A robust molecular phylogeny of the ectoprocts with extensive taxon sampling is still lacking, but significant progress toward this goal has been made, allowing for the evaluation of some macro-evolutionary hypotheses about ectoproct evolution (Waeschenbach et al. 2009, 2012). Strong support exists for the morphological characters separating the three main classes: Stenolaemata, Gymnolaemata, and Phylactolaemata. However, some family-level (and higher) divisions based on zooidal frontal wall morphology in cheilostomes and colony budding types in ctenostomes and phylactolaemates have not been completely congruent with molecular data (Waeschenbach et al. 2012). Homoplasy among characters associated with skeletal features and colonial growth strategies is not unexpected as predation and competition for space are convergent selective pressures acting on all forms of ectoprocts (McKinney 1992, 1995).

How ectoprocts are related to other forms of bilaterian animals has long been a debated subject. Based on the morphology of their feeding structures, body cavities, and embryonic cleavage patterns, early work united ectoprocts with the phoronids (Chapter 10) and brachiopods (Chapter 12) under the superphyletic assemblage Lophophorata (e.g., Hyman 1959), all of whose members were believed to share affinities with deuterostome animals. This interpretation was challenged by the burgeoning field of molecular phylogenetics, based largely on 18S ribosomal DNA (Halanych et al. 1995), placing the ectoprocts in a new superphyletic assemblage of protostome animals called the Lophotrochozoa, in which the evolutionary position of the ectoprocts relative to phoronids, brachiopods, and spiralians was unresolved. Increased taxon sampling bolstered with data from more rDNA genes, full mitochondrial genomes, and select nuclear genes continued to support these evolutionary scenarios (Passamaneck and Halanych 2004, 2006; Jang and Hwang 2009). More recent phylogenomic approaches supported the resurrection of the Polyzoa concept in which ectoprocts are linked with entoprocts (Hausdorf et al. 2007, 2010; Helmkampf et al. 2008; Hejnol et al. 2009). Interestingly, the latest phylogenomic analysis supports reuniting all the lophophorates in one monophyletic group, where the phoronids are a sister taxon to the ectoprocts (Nesnidal et al. 2013).

EARLY DEVELOPMENT

Ectoproct reproductive patterns and brooding styles (where applicable) have been reviewed extensively (Reed 1991; Temkin 1994, 1996; Santagata and Banta 1996; Ostrovsky et al. 2009; Ostrovsky 2013) and will be only briefly mentioned here in the context of particular species for which early development has been followed in some detail. The vast majority of information on the early development of ectoprocts comes from a select set of species that hardly represents the rich diversity of the three major clades. These data are entirely descriptive and based on the cell arrangements found in living and fixed embryos, but as of yet no detailed cell lineage study using modern cell tracing techniques exists for any ectoproct species. For now, this review of ectoproct early embryology is based on a few key species and compares these accounts to data gathered using cytological fluorescent probes and confocal microscopy.

Anascan cheilostome ectoprocts, such as *Membranipora* spp., that broadcast spawn fertilized oocytes into the water column (Temkin 1996) and develop into a characteristic feeding larval form called the cyphonautes (Kupelweiser 1905) are generally the easiest model species for the study of early development due to their numerous gametes, ease of collection, and relatively fast developmental times (Fig. 11.1). Many of the recent developmental observations of this genus come from genetically distinct clades and perhaps three different species (Schwaninger 2008) occurring on kelp growing on the floating docks adjacent to Friday Harbor Laboratories (University of Washington, WA, USA). Once activated by exposure to seawater, the flattened disc shape of a *Membranipora*-fertilized primary oocyte (approx. 60 μm in diameter) will round up, produce first and second polar bodies, and enter first cleavage within approximately 1–2 h (12–14 °C). First cleavage is equal and through the animal-vegetal (AV) axis. Second cleavage is also equal and through the AV axis, but rotated 90° to that of the first cleavage plane. Pigmented cytoplasmic granules are segregated to the vegetal side of cells by the four-cell stage. Third cleavage is equatorial

and equal, resulting in an eight-cell stage in which all of the pigmented cytoplasmic granules are contained in the four vegetal cells. Fourth cleavage is through the AV axis resulting in a bilaterally symmetric 16-cell embryo, composed of two rows of four equally sized cells in the animal half of the embryo directly stacked on the eight cells in the vegetal half of the embryo. However, fourth cleavage in the vegetal cells is not equal, and as a result the four middle blastomeres of the eight vegetal cells are larger than the adjacent cell pairs (Fig. 11.1H). Third and fourth cleavage planes and cell arrangements of 8- and 16-cell stages are also entirely radial. Fifth cleavage results in a 32-cell stage comprised of 4 tiers of cells consisting of 8, 12, 8, and 4 cells from the animal to the vegetal pole. Gastrulation begins with the internalization of the four vegetal cells as the embryo nears the 64-cell stage (Fig. 11.1J, K). According to Zimmer (1997), representative species of gymnolaemates are consistent with respect to the contributions made to larval and presumptive juvenile tissues by the four tiers of cells at the 32-cell stage. The top animal tier of eight cells is responsible for forming the apical disc composed of the sensory and neuronal cells of the larval apical organ (see Santagata 2008a for details) and (if present) the epidermal blastemal cells that will form ectodermal and endodermal tissues of the ancestrular polypide.

In species with a feeding cyphonautes larva (e.g., *Membranipora*), the second animal tier of eight cells will form hundreds of multiciliated coronal cells used for locomotion. However, there is considerable variation in the number and morphology of coronal cells among species. In some non-feeding larval forms, these cells may undergo cleavage arrest at an early stage, forming a row of only 32 or 40 cells in the coronal band (such as species of *Tanganella*, *Watersipora*, *Amathia*, and *Bowerbankia*; also see Corrêa 1948). There are hundreds of strap-like coronal cells in *Bugula neritina* (Woollacott and Zimmer 1975), but the larvae of *Sundanella sibogae* and *Nolella stipata* have numerous corona-like cells covering the aboral half of the larva (Santagata 2008a). The third tier of 12 cells is more variable in its contribution to larval and presumptive

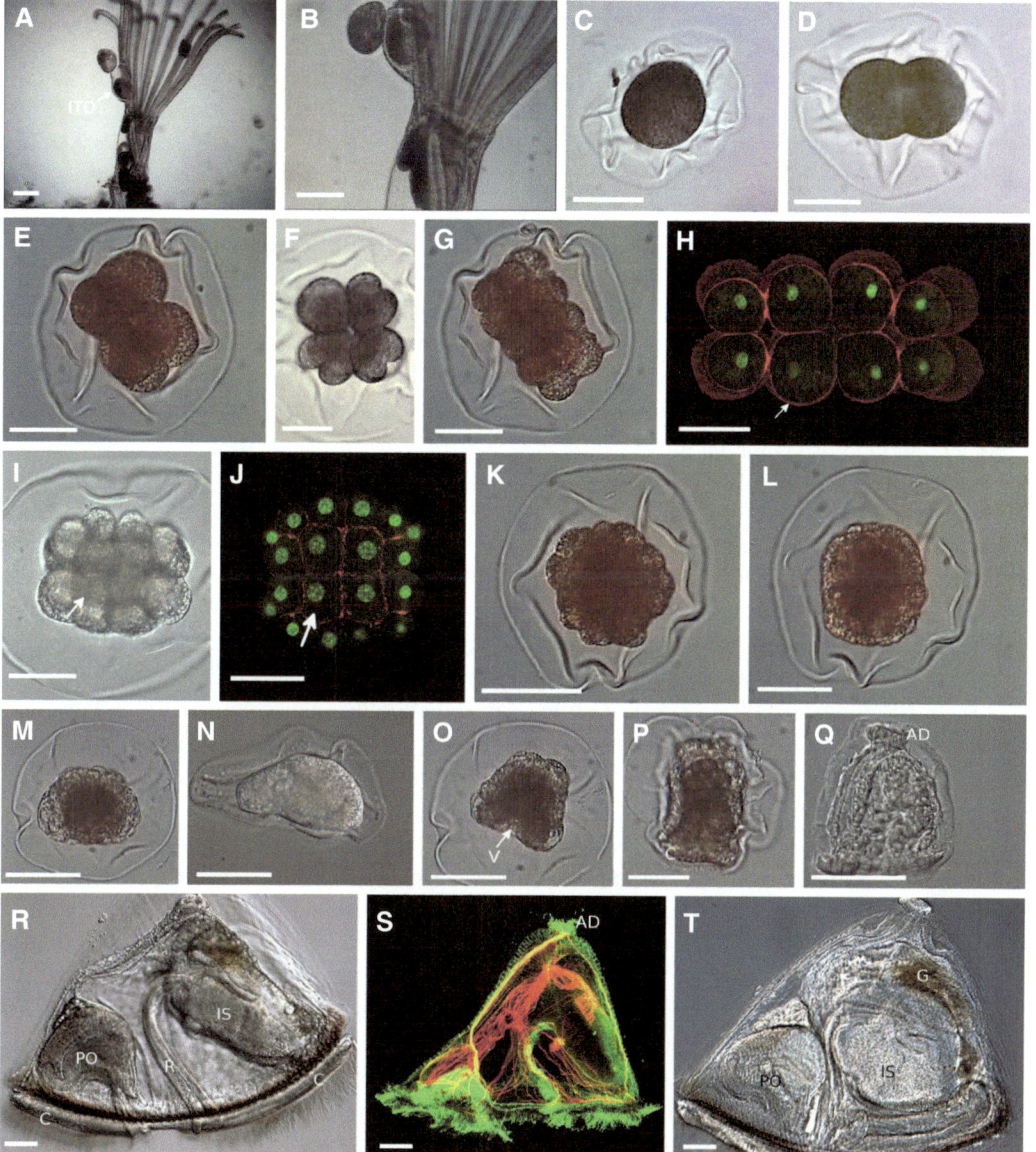

Fig. 11.1 Spawning and development of *Membranipora membranacea*. (**A**) Fertilized zygotes being released from the intertentacular organ (*ITO*). (**B**) Close-up of the spawning zooid in (**A**). (**C**) Rounded-up zygote surrounded by the wrinkled fertilization envelope 1–2 h post spawning. (**D**) Two-cell embryo. (**E**) Vegetal view of an eight-cell embryo. (**F**) Eight-cell embryo undergoing cell division to the 16-cell stage. (**G**) Vegetal view of a 16-cell embryo. (**H**) Partial volume rendering of the 16-cell stage embryo showing the four larger central macromeres (*arrow*) in the vegetal half of the embryo (*green* nucleic acids, *red* fibrous actin). (**I**) Vegetal view of a 32-cell embryo showing the four macromeres (*arrow*) that will be internalized during gastrulation at the 64-cell stage. (**J**) Optical section through the four internalized macromeres (one of which is indicated by the *arrow*) surrounded by the blastomeres that will become the coronal cells (*green* nucleic acids, *red* fibrous actin). (**K**) 64-cell stage. (**L**) Early gastrula. (**M**) Midgastrula stage where the clear ectodermal cells spread over the embryo. (**N**) Late gastrula stage with the initial invagination of the vestibule. (**O**) The beginning of larval development during which the vestibule (*V*) elongates. (**P**) Early trapezoidal-shaped larval stage hatching out of the fertilization envelope; note the ciliation on the coronal cells (*C*). (**Q**) Hatched feeding larval stage with a defined apical disc (*AD*). (**R**) Late-stage larva with a pyriform complex (*PO*), ciliated ridges (*R*), and internal sac (*IS*). (**S**) Confocal z-projection of a late-stage larva stained for muscles (*red*) and acetylated α-tubulin (*green*) for cilia and nervous system. (**T**) Larval stage that is competent to metamorphose; note growth of internal sac (*IS*) and the relatively narrow gut (*G*). Scale bars in **A**–**C** and **I**–**T** equal 50 μm, scale bars in **D**–**H** equal 30 μm

juvenile tissues. In all species investigated, these cells form the larval pyriform complex consisting of a superior and an inferior glandular field of cells bordering four specialized multiciliated cells that constitute the vibratile plume. The pyriform complex has numerous other sensory and neuronal cells (Reed et al. 1988; Santagata 2008b) used during crawling behaviors to probe various substrates before metamorphosis. Another derivative of this cell tier is the oral epithelium that consists of ciliated cells whose primary function is to capture food particles (the ciliated ridges and cell lining of the vestibule) in feeding larvae. In non-feeding larval forms these cells make an oral field of ciliated cells that can be used in both crawling and swimming behaviors. In some ctenostome larval forms lacking the epidermal blastemal cells in the apical disc, a portion of the oral epithelium remains undifferentiated and is the source of the ectodermal and endodermal tissues of the ancestrular polypide (Zimmer and Woollacott 1993). The last structure derived from this cell tier is the internal sac, a simple or complex eversible epithelium that is used for temporary or permanent attachment to the substrate at metamorphosis. Larval types with a complex internal sac form all or part of the ectoderm of the cystid from it (Lyke et al. 1983). The fourth tier of four (vegetal) cells internalized during gastrulation will make the endodermal cells of the larval gut in feeding larvae and either a partial nonfunctional gut or nutrient-rich cells in non-feeding larvae. The fourth cell tier also makes the larval mesodermal tissues (striated and smooth musculature; see Santagata 2008a) as well as those blastemal cells situated near the apical disc that will form the ancestrular mesodermal tissues. It should be made clear that variation in the formation of and contribution to these latter cell groups was discussed by Prouho (1892) and also that future investigations of ectoproct embryonic cell lineages are likely to find more differences as data for more species are gathered.

Stenolaemate (cyclostome) ectoprocts brood their embryos in specialized enlarged gonozooids. Fertilization occurs within the ovary, and the resulting early cleavage stages are thought to be irregular in cell number (Borg 1926). Cleavage proceeds creating a two-cell-layered primary embryo that will grow in size as the ovary degenerates. The enlarged primary embryo will then produce other genetically identical (Hughes et al. 2005; Pemberton et al. 2007) secondary embryos through polyembryony. In *Crisia pugeti* secondary embryos are produced from bilayered projections that bud off the primary embryo (Dimarco-Temkin and Temkin, pers. comm.). In this species, secondary embryos consist of a central group of eight cells surrounded by an outer layer of 32 cells. The outer epithelium continues to develop, forming an invaginated aboral epithelium and an internal sac (Barrois 1877). These tissues form the cystid at metamorphosis, leaving the internal group of cells to form the polypide.

The development of freshwater ectoprocts is the least understood and, despite their ancestral position in most current phylogenetic analyses, may have the most derived developmental patterns found among all ectoproct clades. What little information that exists comes from a few sources (Braem 1897; Brien 1953), and much of what has been reviewed (see Reed 1991; Zimmer and Woollacott 1993; Zimmer 1997; d'Hondt 2005) focuses on one species, *Plumatella fungosa*. Early cleavages have been described as holoblastic and irregular, producing an elongated, hollow blastula consisting of a single layer of cells. Mechanisms of gastrulation are unclear (possibly ingression), but cells invade the blastocoel (approx. 72-cell embryo) and form an early gastrula consisting of two cell layers (note similarity to primary embryo formation in cyclostomes). The embryo then grows due to the nutrients provided by the maternal zooid via the tissue connection between the ovisac and the equatorial region of the embryo. The bilayered embryo then forms one, two, or four differentiated polypides and sometimes polypide buds (depending on the species). These polypides will eventually be covered by the proliferating mantle tissue, except for a small pore at the lagging end of the larva. The outer epithelium of mantle tissue becomes ciliated, and the leading (swimming) end of the ciliated epithelium develops an apical neuronal plate (Franzén and Sensenbaugh 1983) with unclear homology to the apical disc (apical organ) of other ectoproct larvae.

LATE DEVELOPMENT

Most gymnolaemate larval types can be grouped into one of three categories, a thin pyramidal-shaped feeding cyphonautes, a squat pyramidal-shaped non-feeding pseudocyphonautes, and various forms of oblate spheroid-shaped non-feeding larvae (the most common being the coronate larva). The cell types and anatomy of the larval nervous system and musculature have been better studied in selected species of gymnolaemates, and, collectively, these data support some broad-scale homology with the larval neuromuscular systems of spiralian protostomes (Santagata 2008b; Gruhl 2009; Wanninger 2009) but differ from the larval morphologies of phoronids and brachiopods (cf. Chapters 10 and 12; Santagata and Zimmer 2002; Altenburger and Wanninger 2009; Santagata 2011; Temereva and Wanninger 2012).

All gymnolaemate embryos have a late gastrula stage with a comparatively large ring of coronal cells equatorially arranged between the more cell-rich oral and aboral larval fields. In species with a cyphonautes larva, the entirety of the oral cell field will be internalized as the embryo flattens laterally and takes on a more trapezoidal shape (Fig. 11.1N). As a consequence of this process, the outer pallial epithelium now covers the surface of the embryo except for where it meets the border of the developing apical disc and the coronal cell band. At this stage the vestibular cavity inside the larva is formed. The vestibule contains the ciliated ridges and other ciliated cells, the function of which is to bring food particles to the apically positioned mouth (Fig. 11.1R). Substantial circular musculature lines the mouth and esophageal regions (Stricker et al. 1988a; Santagata 2008a, b) that lead into the gut and terminate at the anus positioned at the posterior end of the coronal cell band. The pyriform complex, a sensory and glandular organ that includes a central group of four cells that make fused ciliary tufts collectively called the vibratile plume, and also the presumptive juvenile tissues (blastemal cells and the internal sac, see Stricker et al. 1988b) are not present at the early feeding stage. Late-stage and metamorphically competent larvae develop large, paired retractor muscles that insert on the internal sac and are used to spread this adhesive secretory epithelium over the substrate at metamorphosis. During this time the internal sac will be joined with the pallial epithelium to form the cystid (Stricker 1988). Blastemal cells that will eventually form the polypide are pulled into the center of the pre-ancestrula by retractor muscles positioned between the apical disc and the pyriform complex. Internalized larval tissues (e.g., corona and pyriform organ) will undergo histolysis during metamorphosis. Divisions between lateral regions of the ancestrula are evident in *Membranipora*, producing a twinned ancestrula of two complete autozooids. Developmental time for the ancestrula to become a functional feeding zooid varies with temperature, but in general it occurs within two days.

Non-feeding (coronate) larval forms found in ctenostome and cheilostome ectoprocts somewhat retain the morphology of their late gastrulae (Fig. 11.2). Major differences occur in the position of the apical disc relative to the pyriform complex that correlates with the orientation of the larvae during swimming and crawling behaviors (Fig. 11.3; see also Santagata 2008a, b for review). The positions of blastemal cells that eventually make the polypide can also vary considerably. These cells may be found in or below the apical disc, in centralized bands inside the larva, and also bordering the oral ciliary field (Zimmer and Woollacott 1989, 1993). Considerable differences are found in the mechanisms of morphogenetic movements of the larval and presumptive juvenile tissues at metamorphosis as well as the contributions made to the ancestrula by specific types of presumptive juvenile tissues in the larva.

Some general differences in anatomy among phylactolaemate, cyclostome, ctenostome, and cheilostome larval types are summarized in Fig. 11.4. Although there are convergent morphologies present between vesiculariform (e.g., *Amathia*, *Bowerbankia*, and *Zoobotryon*) and buguliform (*Bugula* and *Scrupocellaria*) larval types, each has divergent structure in the presumptive juvenile tissues, musculature involved in metamorphosis, and blastemal contributions to the ancestrula. For instance, larval metamorphosis in *Bowerbankia imbricata* begins with the eversion of the internal sac via constriction of the equatorial ring muscle. The larva temporally adheres to the substrate using cellular secretions of the internal sac, but the sac is degraded later in metamorphosis. The coronal cell epithelium involutes from the force exerted by the reversal of ciliary beat against the pellicle, and at the same time the expansive pallial epithelium spreads over the entire surface of the preancestrula, forming the entire cystid epithelium (Reed and Cloney 1982). The internal band of blastemal cells in *Bowerbankia* larvae is already centralized within the ancestrula by these tissue rearrangements and does not require the actions of larval musculature, eventually forming a single tubular zooid that will bud asexually from basal stolons. Similar to *Bowerbankia* larvae, the eversion of the internal sac in *Bugula* larvae is muscle mediated. Evagination of the small pallial epithelium and involution of the coronal cells are also accomplished by the reversal of coronal cell ciliary beat (Reed and Woollacott 1982). In contrast to *Bowerbankia* larvae, the raising of the walls of the internal sac and contraction of the pallial epithelium in *Bugula* larvae are completed by the actions of bands of microfilaments in the pallial epithelium (Reed and Woollacott 1982). The cystid of the ancestrula of *Bugula neritina* is composed solely of the internal sac, and the pallial epithelium is largely vestigial, forming only a portion of the tentacle sheath (Woollacott

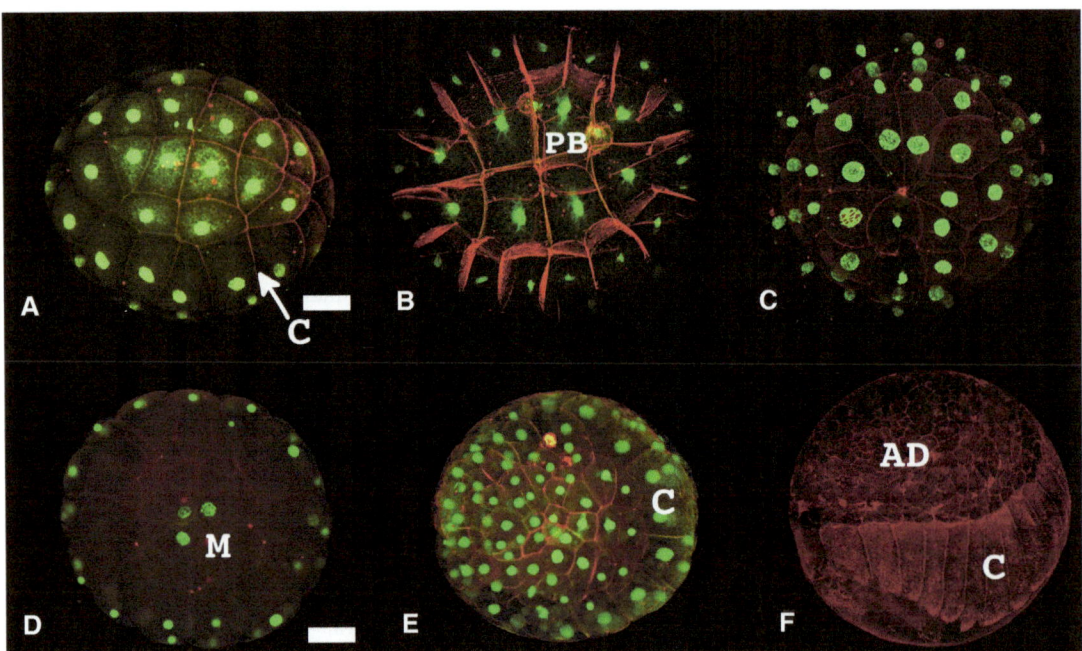

Fig. 11.2 Early development of gymnolaemate ectoprocts with a non-feeding coronate larva (confocal z-projections, specimens stained for fibrous actin (*red*) and nucleic acids (*green*)). (**A**) Animal view of a 32-cell stage of *Dendrobeania lichenoides* (embryos dissected out of ovicells from colonies collected from the floating docks at Friday Harbor Laboratories, University of Washington, WA, USA). A median band of cells along the equator of the embryo will make the coronal cell (*C*) ring. (**B**) Animal view of a 32-cell stage of *Hippodiplosia insculpta* with polar bodies (*PB*). (**C**) Side view of the 64-cell stage of *Rhynchozoon rostratum*; note bilateral symmetry of the embryo. (**D**) Optical section though a 64-cell stage of *Dendrobeania lichenoides*, showing the four internalized macromeres (*M*) at gastrulation. (**E**) Late gastrula stage of *Dendrobeania lichenoides*; note the fewer and larger cells that will eventually make the coronal band (*C*). (**F**) Early larval stage of *Dendrobeania lichenoides* in which the coronal cell band (*C*) is differentiated and demarcates the boundary between the aboral and oral cellular fields, both of which have many more cells. The approximate position of the developing apical disc (*AD*) is labeled. Scale bars equal 50 μm

and Zimmer 1971; Reed and Woollacott 1982). In this species, the bilayered groups of blastemal cells below the apical disc are centralized by axial muscles running between the apical disc and the roof epithelium of the internal sac. Overall, very little is known about the process of metamorphosis in the numerous and morphologically divergent species of ascophoran-grade cheilostomes and most families of ctenostome ectoprocts. Preliminary work on these larval types supports some aspects of previously discussed processes, but more diversity exists in

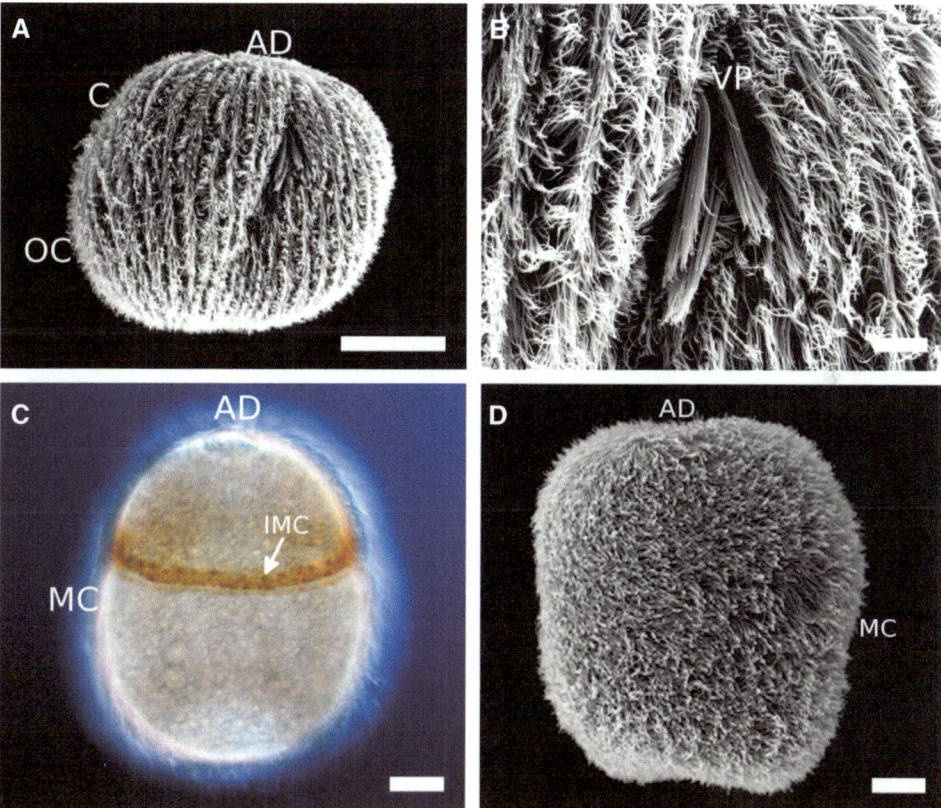

Fig. 11.3 Aspects of larval morphology among non-feeding forms of gymnolaemate ectoprocts. (**A**) Scanning electron micrograph (SEM) of the larva of *Aeverillia setigera*, a ctenostome that is elongated in the aboral-oral axis and flattened in the anterior-posterior axis; observe the small apical disc (*AD*). Coronal cells (*C*) cover the aboral hemisphere of the larva, and the expansive oral ciliated cells (*OC*) cover the aboral hemisphere. This species also has a long oral ciliated groove. (**B**) SEM of the fused ciliary bundles of the vibratile plume cells (*VP*). (**C**) Light micrograph of the larva of *Nolella stipata*, another ctenostome that is elongated in the aboral-oral axis. The homology of the multiciliated cells (*MC*) that cover the larval epidermis to the coronal and oral ciliated cells of other species is unclear (see Santagata 2008a for details). As found in the larva of *A. setigera*, the aboral and oral ciliary fields are separated by an equatorial band of sensory neurons with small ciliary tufts (*IMC*). (**D**) SEM of *N. stipata*. (**E**) Light micrograph of the coronate larval type of *Parasmittina spathulata*, an ascophoran-grade cheilostome. This species has a larger apical disc and several pigmented ocelli (*OE*). (**F**) SEM (lateral view) of the coronate larval type of *Schizoporella floridana*, another ascophoran-grade cheilostome; note the relative differences in the size of the ciliary fields of the coronal and oral ciliated cells. (**G**) Fluorescent light micrograph of a *Bugula stolonifera* larva, an anascan-grade cheilostome that is elongated in the aboral-oral axis, has an expanded coronal cell field and has a long oral ciliated groove (*OG*) bordered by ocelli. The larva is stained with bisbenzimide for nucleic acids. (**H**) Fluorescent light micrograph of a *B. stolonifera* larva stained with DASPEI, a live mitochondrial probe that shows the morphology of the ciliated ray cells (neurons) (*CR*) in the apical disc surrounded by the dark, unstained pallial epithelium (*PE*) that surrounds the apical disc. Scale bar in **A** equals 100 µm, scale bars in **B**–**D** equal 10 µm, scale bars in **E**–**H** equal 50 µm

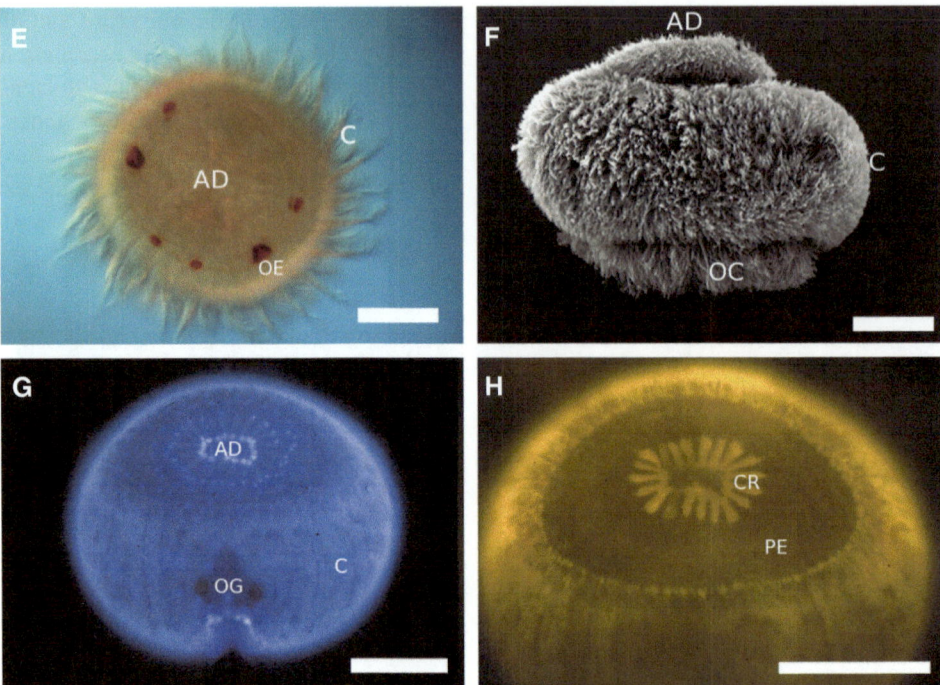

Fig. 11.3 (continued)

larval anatomy, the structure of presumptive juvenile tissues, and tissue rearrangements at metamorphosis.

The larval anatomy of cyclostome ectoprocts has been interpreted as simplified or degenerate, since their larvae are the only type to lack an apical disc (and associated larval apical sense organ) and a pyriform organ and their nervous system consists of a basiepithelial network of sensory cells and neurons (Santagata 2008b). It is not clear if the cuboidal multiciliated cells that make up the larval epidermis are homologous to the coronal cells and/or oral ciliated cells of gymnolaemate larvae.

Metamorphosis of cyclostomes has been best studied in species of *Crisia* (Nielsen 1970; d'Hondt 1977). Similar to gymnolaemate larvae, the eversion of the internal sac at metamorphosis is muscle mediated, but in *Crisia* it is accomplished through the actions of small longitudinal muscle fibers in the body wall (d'Hondt 1977; Santagata 2008a). The invaginated aboral epithelium (likely homologous to the pallial epithelium of gymnolaemates) fuses

with the internal sac and collectively forms the cystid. How the polypide is formed from largely undifferentiated cells inside the larva is not well understood. A central group of blastemal cells (that may include both epidermal and mesodermal cells; see Nielsen 1970) between the aboral epithelium and the internal sac are certainly precursors of the polypide, but whether or not these cells are the only contributors to the polypide is unclear. As discussed earlier, the larvae of freshwater ectoprocts exhibit a form of adultation, as the polypide(s) are already formed in the body of the larva before metamorphosis (Fig. 11.4). In contrast to all other described ectoproct larval forms, phylactolaemate larvae settle, probe, and initially attach to the substrate using an apical plate comprised of both sensory neurons and glandular cells (Franzén and Sensenbaugh 1983; Sensenbaugh and Franzén 1998; Gruhl 2010). For these reasons it has been hypothesized that the glandular cells are homologous to the internal sac of other ectoproct larval forms, particularly those with vesiculariform

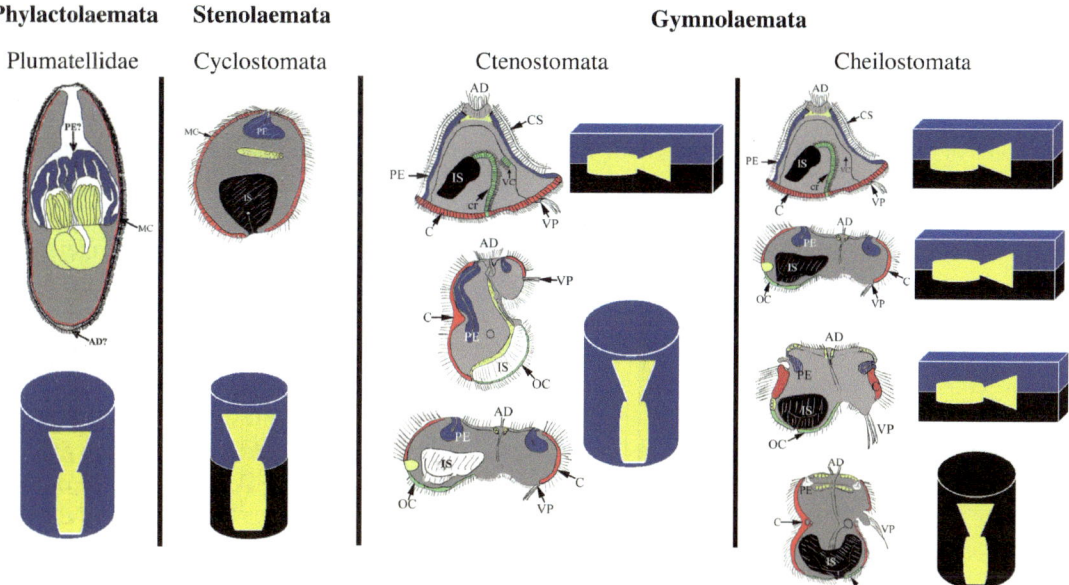

Fig. 11.4 Evolutionary trends in larval anatomy and the arrangement of the presumptive juvenile tissues that form the ancestrula among various systematic groups of ectoprocts. Tissues shaded in *blue* (pallial epithelium) and *black* (internal sac) contribute the cystid epithelium (here shown as either a *tubular* or *rectangular box*) at metamorphosis. In species where one of these tissues does not contribute to the ancestrula at metamorphosis, either the pallial epithelium (*PE*) or internal sac (*IS*) is unshaded. Tissues shaded in *yellow* are groups of blastemal cells that form the polypide (internal components of the lophophore and gut) after metamorphosis. Cells shaded in *red* constitute the coronal cell ring (*C*), mainly used for ciliary propulsion. Cells shaded in *green* either form the ciliated ridges (*CR*) and other ciliated cells within the vestibule (*VC*, only a subset are figured) found in species with a planktotrophic larva or are the oral ciliated cells (*OC*) of some non-feeding larval types. Structures found in most larval forms include the apical disc (*AD*) and vibratile plume cells (*VP*). Cyphonautes larvae are covered by a bivalved shell (*CS*). It is unclear if the multiciliated cells (*MC*) that cover the larvae of cyclostomes and phylactolaemates are homologous to the coronal and/or oral ciliated cells of gymnolaemate ectoprocts. Furthermore, the axial properties of phylactolaemate larvae are not well established, and it is not clear if the apical disc (*AD*) and pallial-like epithelium (*PE?*) are homologous to these structures in gymnolaemates (© Scott Santagata 2015. All Rights Reserved)

(ctenostome) types. However, the presence of sensory neurons along with the glandular cells complicates this comparison, especially since both neuronal and glandular fields already exist in the pyriform complex of most other ectoproct larval forms. During the initial phases of metamorphosis, the mantle tissue that underlies the ciliated outer epithelium of the larva involutes. This tissue rearrangement internalizes the residual larval tissues inside the pre-ancestrula (serving as a nutrient source) and also exposes the inner epithelium (starting at the point of the terminal pore) of the vestibular chamber around the internal polypides. In this way the vestibular epithelium and other tissues around the polyp-ides form the walls of the cystid (Sensenbaugh and Franzén 1998). Forming an expansive cystid epithelium from one epithelial source is similar to what is observed in the Vesiculariidae (Reed and Cloney 1982) and at least one other ctenostome taxon, the Aeverrilliidae (see Santagata 2008a).

GENE EXPRESSION

Gene expression studies on ectoprocts are in the early stages of exploration. No data exist on developmental-related genes during embryogenesis, but some limited information has been published on the larvae and metamorphic stages of *Bugula neritina*. One prevalent question deals with the potency and differentiation of the presumptive ancestrula tissues in the body of the larva. Clearly, the pallial epithelium and internal sac are formed of largely undifferentiated cells during the larval stage, as their functions are mainly to secrete and attach the pre-ancestrula to the substrate (depending on the species), and are not in their final state of differentiation (cystid wall epithelium composed of ectodermal cells) until a feeding zooid stage is reached. Blastemal cells eventually form all the neural, muscular, coelomic, and endodermal tissues of the polypide. So, collectively, all of the presumptive juvenile tissues are, at the very least, composed of multipotent cells. However, since all of the variously positioned blastemal cells of ectoproct larval types are either epidermal and/or mesodermal in origin (Fig. 11.4), it has been suggested that only these germ tissue types form the entirety of the adult body (Woollacott and Zimmer 1971; Reed and Cloney 1982; Reed 1991; Zimmer 1997). Essentially, this hypothesis interprets ancestrula (polypide) formation at metamorphosis as homologous to the process of forming a new zooid by asexual budding, in which the new polypide is differentiated from cells originating from the frontal cystid epithelium. If this hypothesis is correct, one prediction would be that blastemal cells should not express endodermal differentiation factors during the larval period. Furthermore, these factors should only be expressed during the later stages of ancestrula formation. Developmental genes such as *Cdx*, *FoxA*, and *GATA456* are expressed in the endoderm of other invertebrate animals (Hejnol and Martindale 2008; Boyle and Seaver 2010). In the larva of *Bugula neritina*, *Cdx* and *FoxA* are expressed in the internal sac and most of the epidermal and mesodermal blastemal cells, but *GATA456* is only expressed in a limited subset of epidermal blastemal cells (Fuchs et al. 2011). These expression patterns along with others gathered from neural- and mesodermal-related genes (Fuchs et al. 2011) suggest that the cell fates of the blastemal cells of *Bugula neritina* larvae are determined before metamorphosis and support homologous cell differentiation mechanisms among metazoan adult body plans. However, addressing the pluripotent or multipotent potential of blastemal cells will require detailed tracking of these cells' expression patterns and cell fates through metamorphosis. Some of these data have been gathered in *Bugula neritina* for select genes in the WNT signaling pathway and support the interpretation that blastemal cell fate is determined before metamorphosis. Interestingly, *WNT10* and *sFRP* (secreted frizzled-related protein) are expressed in spatially opposite cell domains of the epidermal and mesodermal blastemal cells at the larval stage and the differentiating polypide at the early ancestrula stage in *Bugula neritina* (Wong et al. 2012, 2014), suggesting that these genes play a role in defining the spatial polarity of the polypide (see Fig. 11.5).

OPEN QUESTIONS

- How do detailed cell lineages of various species of ectoprocts compare with those of brachiopods, phoronids, and spiralian protostomes?
- How are ectoproct embryos, larvae, and various kinds of zooids patterned during development by differential gene expression (e.g., Hox, Parahox, and other key developmental genes)?
- Are any larval tissues retained or remodeled during metamorphosis when the adult body plan is formed?
- What other aspects of larval morphological diversity and metamorphic patterns exist in the numerous unexplored species of ectoprocts?

Acknowledgments Kelly Ryan and Russel Zimmer provided excellent comments on previous versions of this chapter. A portion of the data in this book chapter was gathered at the Smithsonian Marine Station (Fort Pierce, FL) and is designated contribution number 989. Faculty Research Grants provided by Long Island University-Post also supported this work.

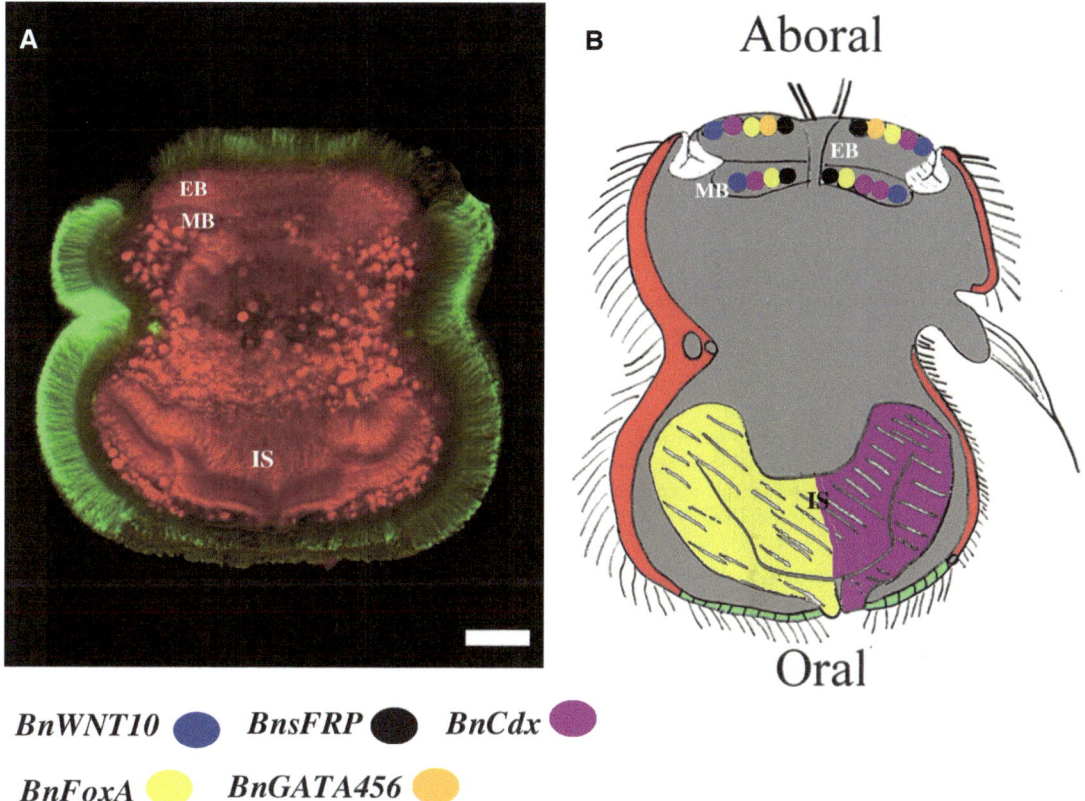

Fig. 11.5 Select gene expression cell domains in the larva of *Bugula neritina*. (**A**) Frontal optical section through the larval body showing the epidermal (*EB*) and mesodermal (*MB*) blastemal cells as well as the complex internal sac (*IS*) that forms all of the cystid (pallial epithelium forms only the tentacle sheath in this species; *red*, nucleic acids; *green*, acetylated α-tubulin). (**B**) Shaded illustration showing select gene expression domains in the blastemal cells and internal sac. Gene expression data from Fuchs et al. (2011) and Wong et al. (2012). Scale bar in **A** equals 50 microns

References

Altenburger A, Wanninger A (2009) Comparative larval myogenesis and adult myoanatomy of the rhynchonelliform (articulate) brachiopods *Argyrotheca cordata, A. cistellula, and Terebratalia transversa*. Front Zool 6:3

Banta WC (1969) The body wall of cheilostome Bryozoa. II. Interzoidal communication organs. J Morphol 129:149–169

Barrois J (1877) Recherches sur l'embryologie des bryozoaires. Trav Station Zool Wimereux 1:1–305

Best MA, Thorpe JP (1985) Autoradiographic study of feeding and the colonial transport of metabolites in the marine bryozoan *Membranipora membranacea*. Mar Biol 84:295–300

Bock PE, Gordon DP (2013) Phylum Bryozoa Ehrenberg, 1831. In: Zhang ZQ (ed) Animal biodiversity: an outline of higher-level classification and survey of taxonomic richness (Addenda 2013). Magnolia Press, Auckland, NZ, pp. 67–74, Zootaxa 3703

Borg F (1926) Studies on recent cyclostomatous Bryozoa. Zoologiska Bidrag från Uppsala 10:181–507

Boyle MJ, Seaver EC (2010) Expression of *FoxA* and *GATA* transcription factors correlates with regionalized gut development in two lophotrochozoan marine worms: *Chaetopterus* (Annelida) and *Themiste lageniformis* (Sipuncula). EvoDevo 1:2

Braem F (1897) Die geschlechtliche Entwicklung von *Plumatella fungosa*. Zoologica (Stuttgart) 10:1–96

Brien P (1953) Etude sur les Phylactolaemates. Ann Soc Roy Zoologique Belg 84:301–440

Cori CI (1929) Kamptozoa, dritter Cladus der Vermes Amera. In: Kükenthal W, Krumbach T (eds) Handbuch der Zoologie. Walter de Gruyter and Co, Berlin, Zweiter Band, pp 1–64

Corrêa DD (1948) A embriologia de *Bugula flabellata* (J. V. Thompson) (Bryozoa, Ectoprocta). Bol Faculdade Filos Ciênc Let Univ S Paulo Zool 13:7–71

d'Hondt JL (1977) Structure larvaire et histogenèse post-larvaire chez *Crista denticulata* (Lamarck) (Bryozoa, Cyclostomata, Articulata). Zool Scr 6:55–60

d'Hondt JL (2005) Etat des connaissances sur le developement embryonnaire des Bryozoaires Phylactolaemates. Denisia 16:59–68

Dick MH, Lidgard S, Gordon DP, Mawatari SF (2009) The origin of ascophoran bryozoans was historically contingent but likely. Proc R Soc B Biol Sci 276:3141–3148

Ehrenberg CG (1831) Symbolae physicae, seu Icones et Descriptiones Mammalium Avium, Insectorum et Animalium Evertebratorum. Berlin. Pars Zoologica No pagination

Erwin DH, Laflamme M, Tweedt SM, Sperling EA, Pisani D, Peterson KJ (2011) The Cambrian conundrum: early divergence and later ecological success in the early history of animals. Science 334:1091–1097

Franzén Å, Sensenbaugh T (1983) Fine structure of the apical plate in the larva of the freshwater bryozoan *Plumatella fungosa* (Pallas) (Bryozoa: Phylactolaemata). Zoomorphology 102:87–98

Fuchs J, Martindale MQ, Hejnol A (2011) Gene expression in bryozoan larvae suggest a fundamental importance of pre-patterned blastemic cells in the bryozoan life-cycle. EvoDevo 2:13

Gruhl A (2009) Serotonergic and FMRFamidergic nervous systems in gymnolaemate bryozoan larvae. Zoomorphology 128:135–156

Gruhl A (2010) Neuromuscular system of the larva of *Fredericella sultana* (Bryozoa: Phylactolaemata). Zool Anz 249:139–149

Gruhl A, Bartolomaeus T (2008) Ganglion ultrastructure in phylactolaemate Bryozoa: evidence for a neuroepithelium. J Morphol 269:594–603

Halanych KM, Bacheller JD, Aguinaldo AM, Liva SM, Hillis DM, Lake JA (1995) Evidence from 18S ribosomal DNA that the lophophorates are protostome animals. Science 267:1641–1643

Hausdorf B, Helmkampf M, Meyer A, Witek A, Herlyn H, Bruchhaus I, Hankeln T, Struck TH, Lieb B (2007) Spiralian phylogenomics supports the resurrection of Bryozoa comprising Ectoprocta and Entoprocta. Mol Biol Evol 24:2723–2729

Hausdorf B, Helmkampf M, Nesnidal MP, Bruchhaus I (2010) Phylogenetic relationships within the lophophorate lineages (Ectoprocta, Brachiopoda and Phoronida). Mol Phylogenet Evol 55:1121–1127

Hejnol A, Martindale MQ (2008) Acoel development indicates the independent evolution of the bilaterian mouth and anus. Nature 456:382–386

Hejnol A, Obst M, Stamatakis A, Ott M, Rouse GW, Edgecombe GD, Martinez P, Baguna J, Bailly X, Jondelius U, Wiens M, Muller WEG, Seaver E, Wheeler WC, Martindale MQ, Giribet G, Dunn CW (2009) Assessing the root of bilaterian animals with scalable phylogenomic methods. Proc R Soc B Biol Sci 276:4261–4270

Helmkampf M, Bruchhaus I, Hausdorf B (2008) Phylogenomic analyses of lophophorates (brachiopods, phoronids and bryozoans) confirm the Lophotrochozoa concept. Proc Roy Soc Ser B: Biol Sci 275:1927–1933

Hughes RN, D'Amato ME, Bishop JDD, Carvalho GR, Craig SF, Hansson LJ, Harley MA, Pemberton AJ (2005) Paradoxical polyembryony? Embryonic cloning in an ancient order of marine bryozoans. Biol Lett 1:178–180

Hyman LH (1959) The Invertebrates: smaller coelomate groups, Chaetognatha, Hemichordata, Pogonophora, Phoronida, Ectoprocta, Brachiopda, Sipunculida, the coelomate Bilateria, vol V. McGraw-Hill, New York, pp 1–783

Jang K, Hwang U (2009) Complete mitochondrial genome of *Bugula neritina* (Bryozoa, Gymnolaemata, Cheilostomata): phylogenetic position of Bryozoa and phylogeny of lophophorates within the Lophotrochozoa. BMC Genomics 10:167

Jebram D (1992) The polyphyletic origin of the "Cheilostomata" (Bryozoa). J Zool Syst Evol Res 30:46–52

Kupelweiser H (1905) Untersuchungen über den feineren Bau und die Metamorphose des Cyphonautes. Zoologica (Stuttgart) 19:1–50

Landing E, English A, Keppie JD (2010) Cambrian origin of all skeletalized metazoan phyla–discovery of earth's oldest bryozoans (Upper Cambrian, southern Mexico). Geology 38:547–550

Lidgard S, McKinney FK, Taylor PD (1993) Competition, clade replacement, and a history of cyclostome and cheilostome bryozoan diversity. Paleobiology 19:352–371

Lyke EB, Reed CG, Woollacott RM (1983) Origin of the cystid epidermis during the metamorphosis of three species of gymnolaemate bryozoans. Zoomorphology 102:99–110

McKinney FK (1992) Competitive interactions between related clades: evolutionary implications of overgrowth interactions between encrusting cyclostome and cheilostome bryozoans. Mar Biol 114:645–652

McKinney FK (1995) One hundred million years of competitive interactions between bryozoan clades: asymmetrical but not escalating. Biol J Linn Soc 56:465–481

Nesnidal MP, Helmkampf M, Meyer A, Witek A, Bruchhaus I, Ebersberger I, Hankeln T, Lieb B, Struck TH, Hausdorf B (2013) New phylogenomic data support the monophyly of Lophophorata and an ectoproct-phoronid clade and indicate that Polyzoa and Kryptrochozoa are caused by systematic bias. BMC Evol Biol 13:253

Nielsen C (1970) On metamorphosis and ancestrula formation in cyclostomatous bryozoans. Ophelia 7:217–256

Nitsche H (1869) Beiträge zur Erkenntnis der Bryozoen. I Beobachtungen ueber die Entwicklungsgeschichte einiger cheilostomen Bryozoen. Z Wiss Zool 20:1–13

Ostrovsky AN (2013) From incipient to substantial: evolution of placentotrophy in a phylum of aquatic colonial invertebrates. Evolution 67:1368–1382

Ostrovsky AN, Taylor PD, Dick MH, Mawatari SF (2008) Pre-cenomanian cheilostome Bryozoa: current state of knowledge. In: Okada H, Mawatari SF, Suzuki N, Gautam P (eds) Origin and evolution of natural diversity. Proceedings of international symposium of the Origin and Evolution of Natural Diversity. Hokkaido University Collection of Scholarly and Academic Papers, Sapporo, pp 69–74

Ostrovsky AN, Gordon DP, Lidgard S (2009) Independent evolution of matrotrophy in the major classes of Bryozoa: transitions among reproductive patterns and their ecological background. Mar Ecol Prog Ser 378:113–124

Passamaneck YJ, Halanych KM (2004) Evidence from *Hox* genes that bryozoans are lophotrochozoans. Evol Dev 6:275–281

Passamaneck Y, Halanych KM (2006) Lophotrochozoan phylogeny assessed with LSU and SSU data: evidence of lophophorate polyphyly. Mol Phylogenet Evol 40:20–28

Pemberton AJ, Hansson LJ, Craig SF, Hughes RN, Bishop JDD (2007) Microscale genetic differentiation in a sessile invertebrate with cloned larvae: investigating the role of polyembryony. Mar Biol 153:71–82

Prouho H (1892) Contribution à l'histoire des Bryozoaires. Arch Zool Exp Gén 10:557–656

Reed CG (1991) Bryozoa. In: Giese AC, Pearse JS, Pearse V (eds) Reproduction of marine invertebrates, vol 6, Echinoderms and Lophophorates. The Boxwood Press, Pacific Grove, pp 85–245

Reed CG, Cloney RA (1982) The settlement and metamorphosis of the marine bryozoan *Bowerbankia gracilis* (Ctenostomata: Vesicularioidea). Zoomorphology 101:103–132

Reed CG, Woollacott RM (1982) Mechanisms of rapid morphogenetic movements in the metamorphosis of the bryozoan *Bugula neritina* (Cheilostomata, Cellularioidea). I. Attachment to the substratum. J Morphol 172:335–348

Reed CG, Ninos JM, Woollacott RM (1988) Bryozoan larvae as mosaics of multifunctional ciliary fields: ultrastructure of the sensory organs of *Bugula stolonifera* (Cheilostomata: Cellularioidea). J Morphol 197:127–145

Ryland JS (2005) Bryozoa: an introductory overview. Denisia 16:9–20

Santagata S (2008a) The morphology and evolutionary significance of the ciliary fields and musculature among marine bryozoan larvae. J Morphol 269:349–364

Santagata S (2008b) Evolutionary and structural diversification of the larval nervous system among marine bryozoans. Biol Bull 215:3–23

Santagata S (2011) Evaluating neurophylogenetic patterns in the larval nervous systems of brachiopods and their evolutionary significance to other bilaterian phyla. J Morphol 272:1153–1169

Santagata S, Banta WC (1996) Origin of brooding and ovicells in cheilostome bryozoans: interpretive morphology of *Scrupocellaria ferox*. Invertebr Biol 115:170–180

Santagata S, Zimmer RL (2002) Comparison of the neuromuscular systems among actinotroch larvae: systematic and evolutionary implications. Evol Dev 4:43–54

Schwaha T, Wanninger A (2012) Myoanatomy and serotonergic nervous system of plumatellid and fredericellid Phylactolaemata (Lophotrochozoa, Ectoprocta). J Morphol 273:57–67

Schwaha T, Wood TS, Wanninger A (2011) Myoanatomy and serotonergic nervous system of the ctenostome *Hislopia malayensis*: evolutionary trends in bodyplan patterning of Ectoprocta. Front Zool 8:11

Schwaninger HR (2008) Global mitochondrial DNA phylogeography and biogeographic history of the antitropically and longitudinally disjunct marine bryozoan *Membranipora membranacea* L. (Cheilostomata): another cryptic marine sibling species complex? Mol Phylogenet Evol 49:203–218

Sensenbaugh T, Franzén Å (1998) Ultrastructural study of metamorphosis in the freshwater bryozoan *Plumatella fungosa* (Bryozoa, Phylactolaemata). Invertebr Reprod Dev 34:301–308

Stricker SA (1988) Metamorphosis of the marine bryozoan *Membranipora membranacea*: an ultrastructural study of rapid morphogenetic movements. J Morphol 196:53–72

Stricker SA, Reed CG, Zimmer RL (1988a) The cyphonautes larva of the marine bryozoan *Membranipora membranacea*. I. General morphology, body wall, and gut. Can J Zool 66:368–383

Stricker SA, Reed CG, Zimmer RL (1988b) The cyphonautes larva of the marine bryozoan *Membranipora membranacea*. II. Internal sac, musculature, and pyriform organ. *Membranipora membranacea*. I. General morphology, body wall, and gut. Can J Zool 66:384–398

Taylor PD, Ernst A (2004) Bryozoans. In: Webby BD, Paris F, Droser ML, Percival IG (eds) The great ordovician biodiversification event. Columbia University Press, New York, pp 147–156

Taylor PD, Weedon MJ (2008) Skeletal ultrastructure and phylogeny of cyclostome bryozoans. Zool J Linn Soc 128:337–399

Taylor PD, Berning B, Wilson MA (2013) Reinterpretation of the Cambrian 'bryozoan' *Pywackia* as an octocoral. J Paleontol 87:984–990

Temereva E, Wanninger A (2012) Development of the nervous system in *Phoronopsis harmeri* (Lophotrochozoa, Phoronida) reveals both deuterostome- and trochozoan-like features. BMC Evol Biol 12:121

Temkin MH (1994) Gamete spawning and fertilization in the gymnolaemate bryozoan *Membranipora membranacea*. Biol Bull 187:143–155

Temkin MH (1996) Comparative fertilization biology of gymnolaemate bryozoans. Mar Biol 127:329–339

Thorpe J, Shelton G, Laverack M (1975) Colonial nervous control of lophophore retraction in cheilostome Bryozoa. Science 189:60–61

Todd JA (2000) The central role of ctenostomes in bryozoan phylogeny. In: Cubilla H, Jackson JBC (eds) Proceedings of the 11th international Bryozoology Association conference. Smithsonian Tropical Research Institute, Balboa, pp 104–135

Waeschenbach A, Cox CJ, Littlewood DTJ, Porter JS, Taylor PD (2009) First molecular estimate of cyclostome bryozoan phylogeny confirms extensive homoplasy among skeletal characters used in traditional taxonomy. Mol Phylogenet Evol 52:241–251

Waeschenbach A, Taylor PD, Littlewood DTJ (2012) A molecular phylogeny of bryozoans. Mol Phylogenet Evol 62:718–735

Wanninger A (2009) Shaping the things to come: ontogeny of lophotrochozoan neuromuscular systems and the Tetraneuralia concept. Biol Bull 216:293–306

Wong YH, Wang H, Ravasi T, Qian PY (2012) Involvement of *Wnt* signaling pathways in the metamorphosis of the bryozoan *Bugula neritina*. PLoS ONE 7:e33323

Wong YH, Ryu T, Seridi L, Ghosheh Y, Bougouffa S, Qian PY, Ravasi T (2014). Transcriptome analysis elucidates key developmental components of bryozoan lophophore development. Sci Rep 4:6534

Wood TS, Lore M (2005) The higher phylogeny of phylactolaemate bryozoans inferred from 18S ribosomal DNA sequences. In: Wyse Jackson PN, Cancino JM, Moyano GHI (eds) Proceedings of the 13th international Bryozoology Association conference. Taylor & Francis, Concepción, pp 361–368

Woollacott RM, Zimmer RL (1971) Attachment and metamorphosis of the cheilo-ctenostome bryozoan *Bugula neritina* (Linné). J Morphol 134:351–382

Woollacott RM, Zimmer RL (1975) A simplified placenta-like system for the transport of extraembryonic nutrients during embryogenesis of *Bugula neritina* (Bryozoa). J Morphol 147:355–378

Xia FS, Zhang SG, Wang ZZ (2007) The oldest bryozoans: new evidence from the late Tremadocian (early Ordovician) of east Yangtze Gorges in China. J Paleontol 81:1308–1326

Zimmer RL (1997) Phoronids, brachiopods, and bryozoans, the lophophorates. In: Gilbert SF, Raunio AM (eds) Embryology: constructing the organism. Sinauer Associates, Sunderland, pp 279–305

Zimmer RL, Woollacott RM (1989) Larval morphology of the bryozoan *Watersipora arcuata* (Cheilostomata: Ascophora). J Morphol 199:125–150

Zimmer RL, Woollacott RM (1993) Anatomy of the larva of *Amathia vidovici* (Bryozoa: Ctenostomata) and phylogenetic significance of the vesiculariform larva. J Morphol 215:1–29

Brachiopoda

12

Scott Santagata

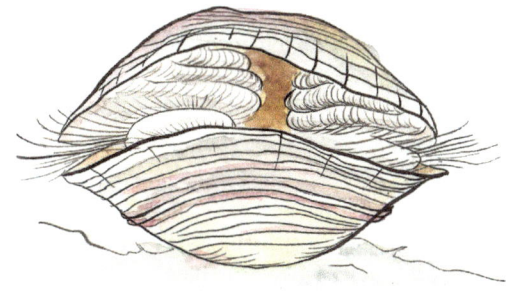

Chapter vignette artwork by Brigitte Baldrian.
© Brigitte Baldrian and Andreas Wanninger.

S. Santagata
Biology Department, Long Island University-Post,
Greenvale, NY 11548, USA
e-mail: scott.santagata@liu.edu

A. Wanninger (ed.), *Evolutionary Developmental Biology of Invertebrates 2: Lophotrochozoa (Spiralia)*
DOI 10.1007/978-3-7091-1871-9_12, © Springer-Verlag Wien 2015

INTRODUCTION

Enclosed in shells with ventral and dorsal valves, extant brachiopods (meaning "arm" and "foot") are classified into three major subphyla: the Rhynchonelliformea, the Linguliformea, and the Craniiformea (Williams et al. 1996). Rhynchonelliform brachiopods encompass what were once referred to as the "articulate" brachiopods, so named for the mineralized hinge that connects the calcite valves of their shells. No such hinge is found in members of the other two subphyla, rather their valves are held together only by various muscles and connective tissues. Craniiform brachiopods (e.g., *Novocrania*) also have calcitic shells, but the shells of linguliform brachiopods (such as the lingulid *Glottidia* and the discinid *Discinisca*) are composed of apatite, a phosphatic mineral, with an outer layer of chitin (Williams et al. 1997). Most brachiopod morphotypes have a smaller dorsal and a larger ventral valve, the latter of which often bears a muscular or rigid attachment structure called the pedicle. Rhynchonelliform brachiopods are often attached to hard substrata by the rigid pedicle with their ventral valves oriented upward (Richardson 1997). The shells of linguliform brachiopods such as *Glottidia* and *Lingula* generally have equally sized valves and their pedicles are long, muscular structures modified for burrowing into soft sediments. Craniiform brachiopods have lost the pedicle and cement directly to hard substrates (Emig 1997).

Beyond differences in their mineralized parts, there are also morphological and functional differences in brachiopods' main feeding and respiratory structure, the lophophore, which can be composed of looped, spiraled, or U-shaped fields of ciliated tentacles (Emig 1992). Only rhynchonelliform brachiopods support the lophophore with calcified support structures called brachidia, whereas all other brachiopods use only musculature and connective tissue. Types and arrangement of adult musculature typically support a closer relationship between craniids and linguliform brachiopods (Williams et al. 1996); however there are also subsets of early juvenile

musculature among rhynchonellids, craniids, and lingulids that may be homologous (Altenburger and Wanninger 2009; Santagata 2011). Both craniiform and linguliform adult brachiopods have through guts, but the anus is lacking in rhynchonelliforms. This absence is generally considered a secondary loss, possibly due to the reduction of water flow in the posterior part of the mantle cavity associated with the presence and position of the hinge (LaBarbera 1981; Williams et al. 1997).

Currently, the oldest known brachiopods in the fossil record are from the lower Cambrian, belonging to an extinct stem group called the paterinids (Topper et al. 2013). Paterinid brachiopods had organophosphatic shells linking them to linguliform brachiopods, but their shell morphology was more similar to that of rhynchonelliform brachiopods. Paterinid shell ultrastructure differed from all other crown fossil brachiopods but shared some characteristics with the organophosphatic sclerites of the extinct metazoans known as tommotiids (Larsson et al. 2013). Whether or not particular tommotiid morphotypes that may have been sessile filter feeders (such as *Eccentrotheca*; see Skovsted et al. 2011) represent the stem lineage for crown phoronids, paterinids, and crown linguliform brachiopods remains unclear (Murdock et al. 2014). Molecular estimates place the origin of the crown brachiopods in the Middle Cambrian, but the phoronid-brachiopod root is older, possibly originating as far back as the Ediacaran (Sperling et al. 2011).

Although molecular evidence is generally congruent with the paleontological origin of brachiopods, discrepancies do exist, leaving open the question of whether phoronids should be considered as an ancestral sister taxon to all brachiopods (Sperling et al. 2011; although see Thomson et al. 2014) or instead as a subtaxon within the brachiopods as a whole (Cohen and Weydmann 2005; Santagata and Cohen 2009; Cohen 2013). Molecular phylogenetic data support the evolutionary affinities of brachiopods with spiralian protostomes such as nemerteans, annelids, and mollusks (Halanych et al. 1995; Dunn et al. 2008; Hejnol et al. 2009; Hausdorf et al. 2010;

Mallatt et al. 2012). Interestingly, a recent phylogenomic study supports the monophyly of all lophophorate phyla in which brachiopods are still closely related to both phoronids and ectoprocts (Nesnidal et al. 2013). In general, higher-level systematic groups within extant brachiopods are supported by molecular evidence (Cohen et al. 1998; Cohen 2000), but superfamily-level relationships based on morphology within the rhynchonellid brachiopods are not congruent with molecular phylogenetic data (Cohen and Bitner 2013; Schreiber et al. 2013).

Fossil and molecular data aside, modern brachiopod species represent only a fraction of their former morphological and species diversity (approx. 30,000 described fossil species). Current assessment of brachiopod species diversity recognizes only 25 linguliforms, 18 craniiforms, and 348 species of rhynchonelliforms (Emig et al. 2013). The disparity in species number among the systematic groups may be due in part to the reliance in many species descriptions on shell and brachidia characteristics that in rhynchonelliforms are more distinct than in (or that are absent in) the other two subphyla. Some cosmopolitan species of *Lingula* may actually be cryptic species complexes (Yang et al. 2013). Comparatively few taxonomists work on living brachiopods (but see Bitner 2006), but hopefully some of the work reviewed here will inspire a new generation of evolutionary developmental biologists to study these intriguing animals that once dominated ancient seas (Rong and Cocks 2013).

EARLY DEVELOPMENT

Gametogenesis

Reproductive patterns in brachiopods have been reviewed in some detail (Long and Stricker 1991; Kaulfuss et al. 2013), and so only pertinent aspects are discussed here. Gonads develop from mesothelial folds surrounding the gut. Linguliforms develop two pairs of gonads in the mesenteries adjacent to the gut, which when ripe expand into the main body cavity (typically

called the metacoel). The gonads of both craniiforms and rhynchonelliforms develop inside mantle coelomic canals positioned in the mesothelial linings that underlie both shell valves. Many brachiopods have separate sexes, but as more reproductive information has been gathered, it is clear that select species of rhynchonelliform brachiopods are hermaphroditic (Kaulfuss et al. 2013). Furthermore, although populations of *Glottidia pyramidata* from Tampa Bay, Florida, are mostly gonochoristic (1:1 ratio of sexes), when isolated from other conspecifics, a small proportion of the population (0.7 %) may become hermaphroditic (Culter and Simon 1987). Similar patterns of gonochorism and hermaphroditism also occur in *Lacazella caribbeanensis* (Seidel et al. 2012). When spawned, sperm and eggs are released into the coelomic cavity, and, once there, the ciliary beat of the coelomic lining transports the gametes to the paired metanephridial gonoducts. In broadcast spawning species such as *Lingula anatina* (Yatsu 1902), *Glottidia pyramidata* (Paine 1963), and *Novocrania anomala* (Nielsen 1991), gametes are expelled via the gonoducts and are released into the surrounding seawater by ciliary currents produced by the lophophore. Among the rhynchonelliform brachiopods for which data are available, gametes are either broadcast spawned as in *Terebratalia transversa* (Long and Stricker 1991) or the eggs are retained in various brood sites (mantle, lophophore, or a pouch), where they are cross-fertilized, or in the case of some hermaphroditic species may also be self-fertilized (Kaulfuss et al. 2013).

In general, linguliform species with planktonic feeding stages such as *Discinisca strigata*, *Glottidia pyramidata*, and *Lingula anatina* produce more numerous and smaller (65, 90, and 100 μm in diameter, respectively) mature oocytes than craniiform or rhynchonelliform species (Yatsu 1902; Paine 1963; Freeman 1995, 1999; Kaulfuss et al. 2013). Craniiform and rhynchonelliform species that have non-feeding larval types develop from more variably sized oocytes, approximately 100–300 μm in diameter among species (Kaulfuss et al. 2013).

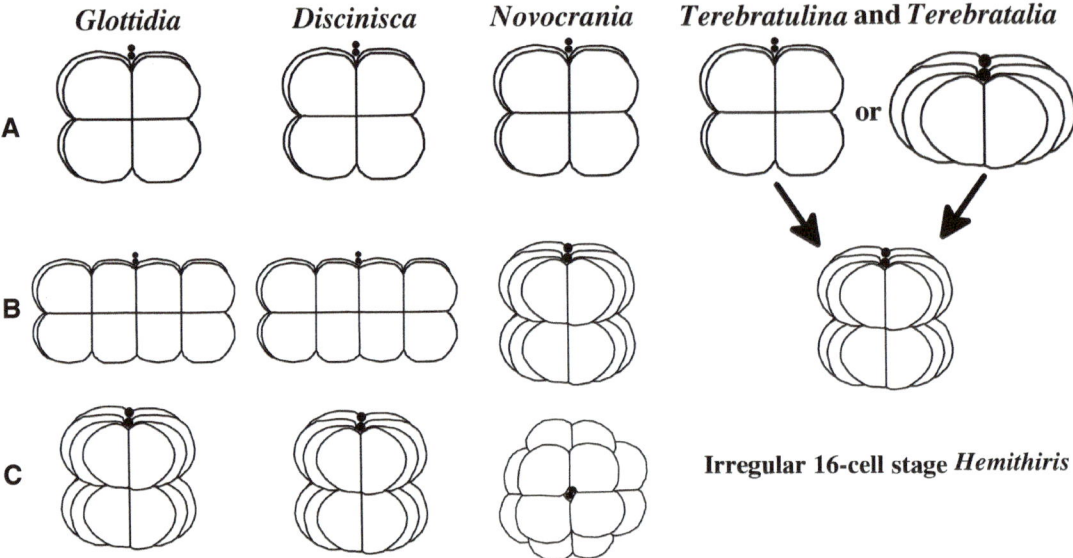

Fig. 12.1 Variation in early cleavage stages of lingulid, craniid, and rhynchonellid brachiopods. (**A**) Eight-cell stages of various brachiopods. Stacked and radial uniplanar cell arrangements are found in *Terebratulina* and *Terebratalia*. (**B**) The more prevalent forms of 16-cell stages among brachiopods. Lingulids typically have a bilateral cell arrangement and rhynchonellids (except for *Hemithiris*) have open radial cell arrangements. (**C**) The less common forms of 16-cell stages among brachiopods, including the four-tier cell stacking found in *Novocrania* (All stages redrawn from Nielsen 1991; Freeman 1995, 1999, 2003). Two polar bodies at the site of the animal pole are shaded in *black*

Cleavage, Gastrulation, and Germ Layer Formation

Early cleavage among brachiopods is typically considered to be radial and holoblastic (Zimmer 1997). There is more uniformity among cleavage planes of early stages of brachiopods in lingulid and craniid species, which produce eight-cell stages composed of four blastomeres stacked directly on top of each other (Fig. 12.1). However, eight-cell stages of rhynchonellid species can be variable in their cell arrangement, producing the more prevalent stacked arrangement or arrangements in which all the blastomeres are in a ring on a single plane (Fig. 12.1; Freeman 2003). Variability in the cleavage plane continues, as lingulid species such as *Glottidia* and *Discinisca* have 16-cell stages with the more prevalent bilaterally aligned cells while also having some embryos with an open radial arrangement (Freeman 1995, 1999). The 16-cell stages of *Novocrania anomala* have been described as having the open radial arrangement (Freeman

2000) with, interestingly, some embryos having blastomeres positioned in four-cell tiers (Fig. 12.1; Nielsen 1991). Despite variability at the eight-cell stage, the 16-cell stages of both *Terebratulina* and *Terebratalia* have the open radial cell arrangement, but irregular 16-cell stages are known from *Hemithiris* (Freeman 2003). Collectively, early cleavage stages of brachiopods share features found in the early cleavage stages of both ectoprocts and phoronids (see Chapters 10 and 11).

Beyond the similarities of cleavage stages, there are also some intriguing early developmental patterns among brachiopods that in some cases are linked to the morphology of their planktonic stages. The first cleavage through the anterior-posterior (AP) axis corresponds to the plane of bilateral symmetry only in species of lingulids such as *Glottidia* and *Discinisca* that have bilateral cleavage (Table 12.1). *Novocrania* and at least three species of rhynchonellids form the anterior ectoderm of the late gastrula from the area around the animal pole (Freeman 2003);

Table 12.1 Select developmental processes among six brachiopod species

Developmental event	*Glottidia pyramidata*	*Discinisca strigata*	*Novocrania anomala*	Rhynchonellids[a]
First cleavage = plane of bilateral symmetry	Yes	Yes	No	No
Site of apical lobe specification	Lateral	Lateral	Animal pole	Animal pole
Stage of animal half specification	AL: blastula	AL: blastula	AL, ML: unfertilized egg	AL, ML: blastula-gastrula
Stage of vegetal half specification	Eight-cell	Blastula-gastrula	Eight-cell	Gastrula
Stage of AP axis Specification	16-cell	Before cleavage	Before cleavage	Gastrula
Bilateral symmetry is established with AP axis?	Yes	Yes	No	Yes

Data taken from Freeman (2003)
Abbreviations: *AL* apical lobe, *ML* mantle lobe
[a]Data from three rhynchonellid species

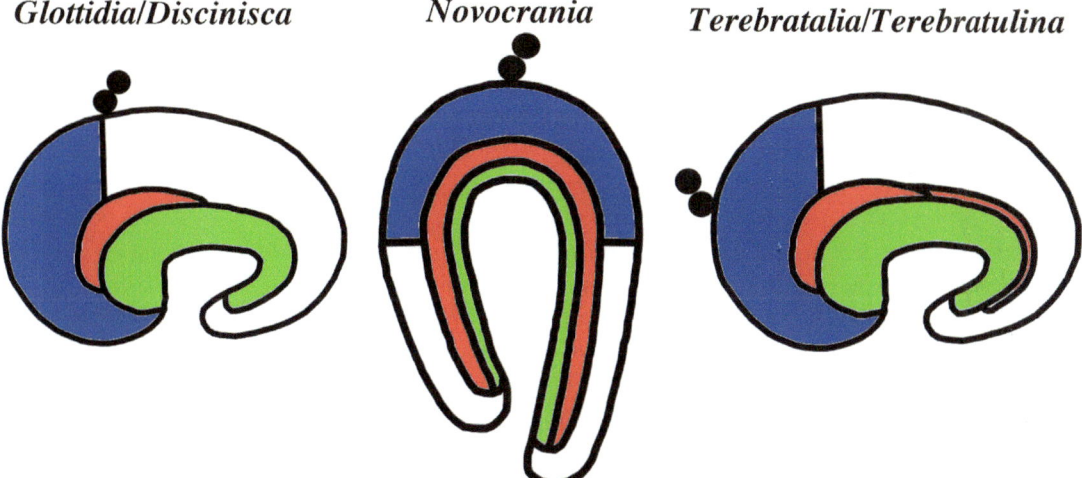

Fig. 12.2 Embryonic germ tissue specification among various species of brachiopods. All stages shown are late gastrulae. Two polar bodies (*shaded black*) mark the site of the animal pole. The anterior ectoderm is *shaded blue*, the mesoderm is *shaded red*, and the endoderm is *shaded green* (All stages redrawn and modified from Freeman (2003))

however *Glottidia* and *Discinisca* both form the anterior ectoderm from the tissue lateral to the animal pole (see Fig. 12.2). In general, the animal half of the embryo is specified during the blastula to gastrula transition, except in *Novocrania* where this occurs before the egg is fertilized (Freeman 2000). Vegetal half specification is more variable, occurring at the eight-cell stage in *Glottidia* and *Novocrania* but later in the development of *Discinisca* and three rhynchonellids (Freeman 1999, 2003). In all species considered here, the vegetal pole is the initial site of gastrulation, and a portion of the blastopore forms the mouth, except in *Novocrania* where the blasto-

pore forms the anus (see section on gene expression). The AP axis is specified before cleavage in both *Discinisca* and *Novocrania* but at the 16-cell stage of *Glottidia* and at the gastrula stage of the rhynchonellids (Table 12.1). Furthermore, the AP and left-right (LR) axes are specified at the same embryonic stage in all but *Novocrania*. As the evolutionary trends discussed here represent information from only six species, more comparative experimental data are required to test the robustness of these developmental patterns.

The origin and proliferation of mesoderm in select species of brachiopods have been studied mainly through histological preparations of discrete embryonic stages and some limited cell marking experiments. During gastrulation, the embryos of *Glottidia pyramidata*, *Lingula anatina*, and *Discinisca strigata* all form mesoderm from the anterior end of the archenteron from cells that ingress into the blastocoelic space (Yatsu 1902; Freeman 1995, 1999). The embryos of *Novocrania anomala* and *Terebratalia transversa* have multiple archenteric ingression sites along the AP axis (Long and Stricker 1991; Nielsen 1991). At the very least, mesoderm originates from the endoderm in these brachiopod species. Since the blastopore forms the anus in *Novocrania* as well as in other protostome embryos such as those of the priapulid worm *Priapulus caudatus* (Martín-Durán et al. 2012) and the polychaete *Owenia collaris* (Smart and Dassow 2009), it is possible that both of these deuterostomic developmental features were shared with the last common ancestor of the Bilateria (Martín-Durán et al. 2012). However, since mesodermal cell ingression occurs at the boundary between the archenteron and the outside ectodermal epithelium, it is plausible that some mesodermal cells originate from ectodermal ingression. Forming mesoderm from dual endodermal and ectodermal sources would lend support to the spiralian affinities of brachiopods, but detailed cell lineage information would be needed from brachiopods to test this hypothesis (see Santagata 2004 for review).

LATE DEVELOPMENT

Past the late gastrula stage, brachiopod embryos develop either two or three body regions. Lingulid and discinid brachiopods form the two most anterior regions first, the apical and mantle lobes. Late larval stages of lingulid and discinid brachiopods are typically considered to be forms of planktotrophic juveniles (Yatsu 1902; Chuang 1977; Long and Stricker 1991; Santagata 2011) but over the complete developmental period exhibit different degrees of embryonic, larval, and precociously developed juvenile traits. The earliest feeding stages of *Lingula anatina*, *Glottidia pyramidata*, *Discinisca strigata*, and other *Discinisca* spp. have one median tentacle and two pairs of lateral tentacles or cirri (Yatsu

1902; Chuang 1977; Freeman 1995, 1999). The gut of *Discinisca* larvae is complete with an esophagus, stomach, intestine, and anus (Chuang 1977). *Lingula* and *Glottidia* larvae lack an anus in the early feeding stages, but a complete gut develops in late planktonic stages transitioning into the benthic form (Yatsu 1902). Early feeding stages of *Lingula* and *Glottidia* larvae have a semicircular embryonic shell that takes on a circular shape in the mid-larval phase (Fig. 12.3A, B) and elongates into an oblate shape in the late, planktonic post-larval phase (Yatsu 1902; Chuang 1977; Santagata 2011). *Discinisca* larvae with two or three pairs of cirri have not yet developed mantle tissue and so are shell-less in the earliest feeding stages bearing four different kinds of long embryonic "setae" (Fig. 12.3D). For reasons discussed in Santagata (2011), the latter structures will be referred to as "chaetae" (Gustus and Cloney 1972; Lüter 2000). Embryonic chaetae of *Discinisca* are produced by two sacs positioned laterally adjacent to the gut (Chuang 1977). The long embryonic chaetae are lost (likely shed) in later planktonic stages as the circular larval shell grows, and during this time the mantle tissue produces both curved and flexible chaetae of varying lengths (Fig. 12.3E; Chuang 1977). Although eyespot-like structures have been described for *Discinisca* larvae (Chuang 1977), it is not clear that photoreception is the function of these pigmented tissues.

Discinisca, *Glottidia*, and *Lingula* larvae have a pair of statocysts. In general, the mid-larval features of *Glottidia* and *Lingula* with four to five pairs of cirri and a short pedicle are quite similar to the late (metamorphic competent) planktonic stages of *Discinisca* (Nielsen 1991). When transitioning to the benthos, *Discinisca* larvae extend their comparatively short pedicles and attach to hard substrates. However, the late planktonic (post-larval) stages of *Glottidia* and *Lingula* develop the dual-brachial arms of the juvenile lophophore, lengthen the muscular pedicle, and extend it beyond the valves for burrowing (Yatsu 1902; Santagata 2011).

All rhynchonelliform species investigated thus far have a trilobed non-feeding larva. Similar morphological features have been observed in several different rhynchonellid species (Conklin 1902; Chuang 1996; Pennington et al. 1999;

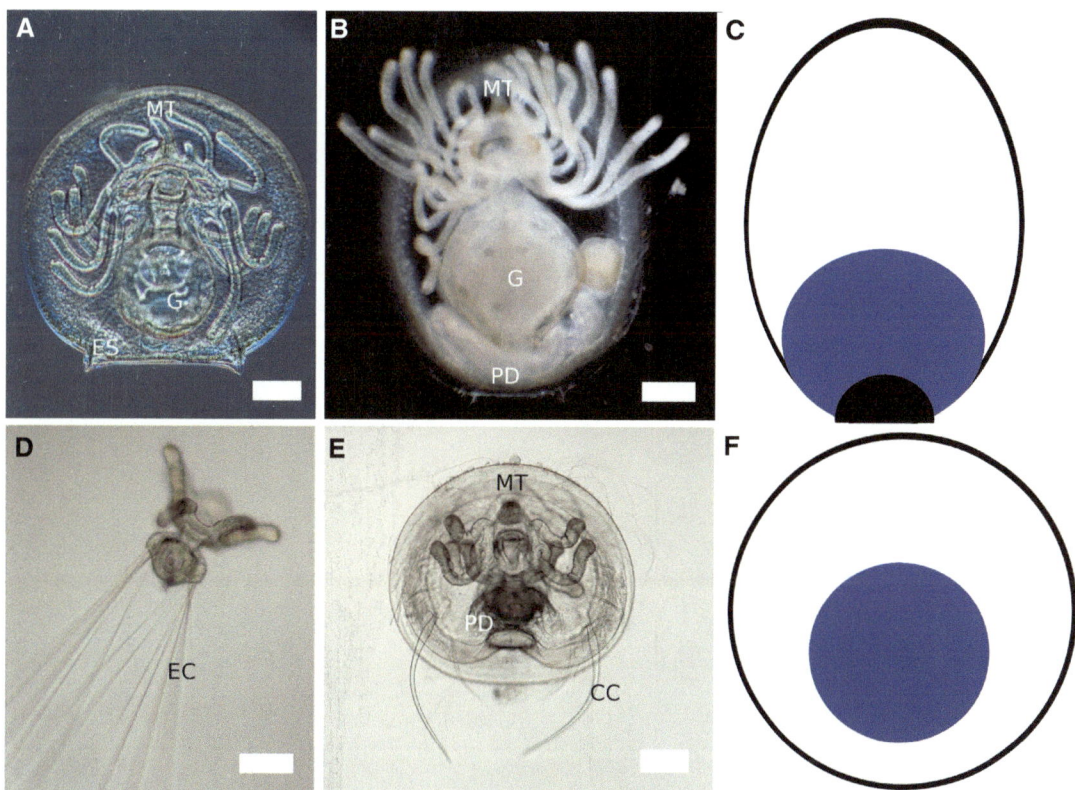

Fig. 12.3 Embryonic, larval, and post-larval features of linguiform brachiopods. (**A**) Mid-larval stage of a *Glottidia* larva with seven pairs of cirri collected from the plankton of Tampa Bay, Florida (see Santagata 2011 for details). The boundaries of the semicircular embryonic shell (*ES*) are evident next to the circular larval shell. (**B**) Preserved specimen of a post-larval stage of a *Glottidia* larva with 11 pairs of cirri collected near St. George Island, Florida. Note the elongated and still internal pedicle (*PD*). (**C**) Relative size and morphological differences among the embryonic (*black*), larval (*blue*), and post-larval (*white*) shells of lingulid brachiopods (Redrawn and modified from Chuang 1977). (**D**) Shell-less stage larva of a discinid brachiopod with three pairs of cirri collected from the plankton on the Pacific side of Panama; note the long embryonic chaetae (*EC*) (Photograph taken by Richard Strathmann. © Richard Strathmann 2015. All Rights Reserved). (**E**) Late-stage discinid larva with four pairs of cirri that has shed the embryonic chaetae and has the full complement of curved mantle chaetae (*CC*) and a short pedicle (*PD*) (Photograph taken by Richard Strathmann. © Richard Strathmann 2015. All Rights Reserved). (**F**) Relative size and morphological differences between the larval (*blue*) and post-larval (*white*) shells of discinid brachiopods (Redrawn and modified from Chuang 1977). Other abbreviations: gut (*G*), median tentacle (*MT*). Scale bars **A** and **E** equal 50 µm and scale bars **B** and **D** equal 100 µm

D'Hondt and Franzén 2001; Altenburger and Wanninger 2009), but development is perhaps best described in *Terebratalia transversa* (Stricker and Reed 1985a; Long and Stricker 1991). During the transition from the late gastrula to the early trilobed larval stage of *Terebratalia transversa*, the dorsal side of the embryo flattens and ventral tissues near the slit-like blastopore move toward the midline, curving inwardly (see Fig. 12.4 and Santagata et al. 2012 for review). The three body regions of the early trilobed larva develop (apical, mantle, and pedicle lobes), and the blastopore closes progressively from posterior to anterior, leaving only a small opening in the apical lobe that leads into the blind-ended gut. The morphology of the anterior portion of the apical lobe changes into a rounded dome that sits on a wider, cylindrically shaped base that includes the anterior transverse ciliated band. Late-stage larvae have pigmented ocelli on the dorsal side of the apical lobe and vesicular bodies that border the posterior margin of the apical lobe. The asymmetric mantle lobe extends further on its ventral side and partially covers the pedicle lobe. Other features of the mantle lobe are the paired dorsal and medial

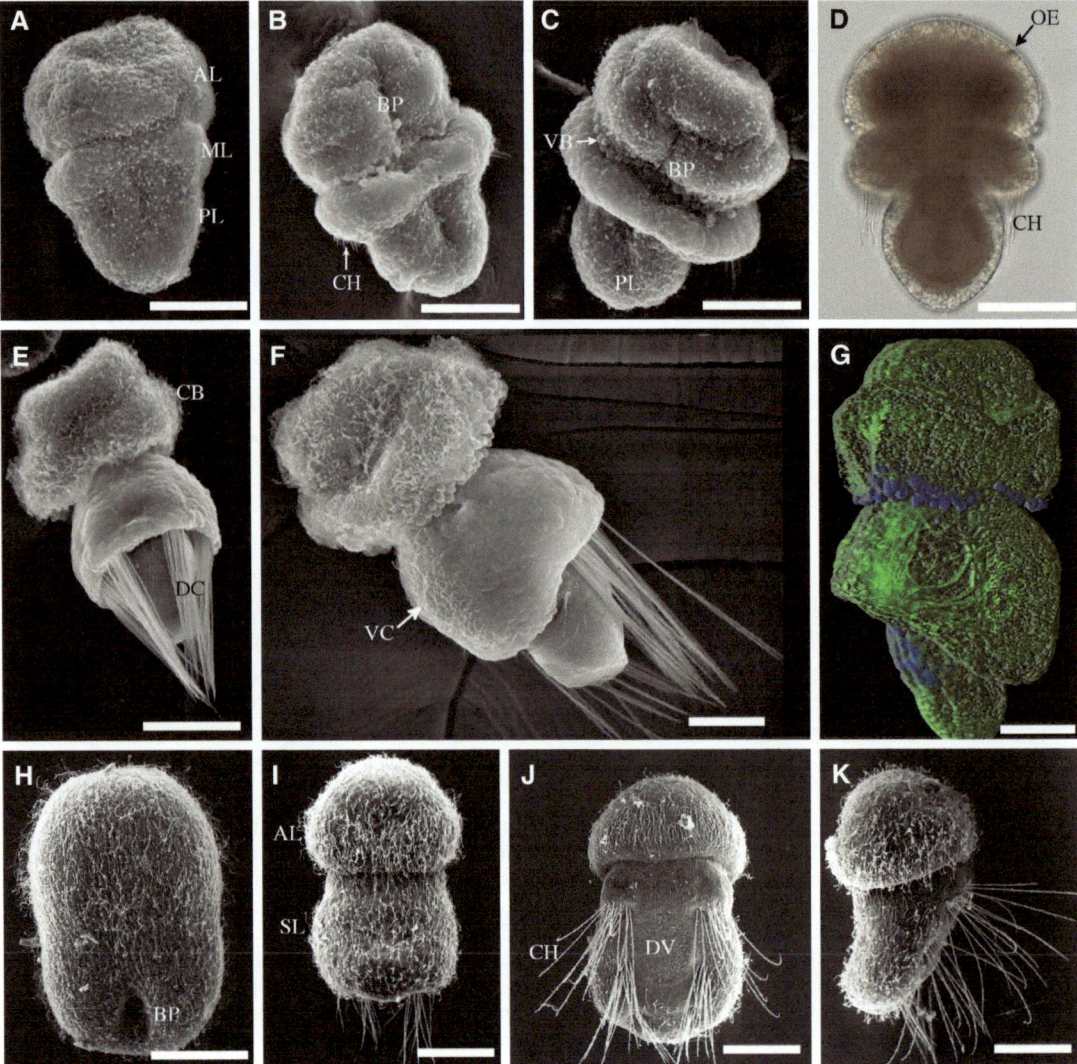

Fig. 12.4 Larval development of *Terebratalia transversa* (**A–G**; micrographs by S. Santagata) and *Novocrania anomala* (**H–K**; scanning electron micrographs provided by C. Nielsen). (**A**) Early trilobed larval stage in which the boundaries of the apical (*AL*), mantle (*ML*), and pedicle (*PL*) lobes are being formed. (**B**) The mantle lobe enlarges as the chaetal sacs with short chaetae develop. The slit-like blastopore is closing from posterior to anterior. (**C**) As the apical lobe takes on a dome-shaped appearance, only a small portion of the blastopore (*BP*) remains. Vesicular bodies (*VB*) mark the boundary between the apical and mantle lobes. (**D**) Later in larval development, the chaetae lengthen and a row of red-pigmented ocelli (*OE*) develops on the dorsal surface of the apical lobe. (**E**) Late-stage larvae have a transverse ciliated band (*CB*) at the base of the apical lobe, and near the time of metamorphic competence, the dorsal and medial chaetae (*DC*) grow beyond the length of the pedicle lobe. (**F**) A mid-ventral band of cilia (*VC*) develops on the larger ventral extension of the mantle lobe. (**G**) Volume projection of a competent larva in which the vesicular bodies and some secretory cells in the pedicle lobe are nonspecifically stained (*blue*). (**H**) Late gastrula stage of *Novocrania*. The blastopore (*BP*) will become the anus. (© Claus Nielsen, 2015. All Rights Reserved). (**I**) A ventral view of a late larval stage with uniform and simple ciliation. The rounded apical lobe (*AL*) sits on *top* of a more posterior secondary lobe (*SL*) that may be homologous to only the mantle lobe of other brachiopod larvae or to both the mantle and pedicle lobes (© Claus Nielsen, 2015. All Rights Reserved). (**J**) Dorsal view of the specialized epithelial region centrally located between the three pairs of chaetal bundles that will secrete the dorsal valve (*DV*) of the juvenile shell at metamorphosis (© Claus Nielsen, 2015. All Rights Reserved). (**K**) Lateral view of a metamorphic competent larva (© Claus Nielsen 2015. All Rights Reserved). All scale bars equal 50 μm

chaetal sacs and a mid-ventral band of cilia. Near the time of metamorphic competence, the posterior pedicle lobe narrows and divides into muscular and glandular portions (Fig. 12.4).

As discussed previously, the remaining portion of the blastopore in the late gastrula of *Novocrania anomala* is found at the posterior end of the embryo and becomes the anus (see Fig. 12.4; Nielsen 1991; Freeman 2000). Larval development results in two body regions, a more rounded apical lobe and an elongate posterior lobe that bears three pairs of chaetal bundles on the dorsal surface (Fig. 12.4). Late-stage *Novocrania anomala* larvae have a specialized circular epithelium centrally located between the dorsal chaetal bundles that secretes the dorsal valve at metamorphosis (Nielsen 1991). Since adult craniiform brachiopods lack a pedicle and cement directly to the substrate at metamorphosis, the posterior larval lobe that bears chaetae has typically been interpreted as the mantle lobe (Nielsen 1991). However, this interpretation has been challenged, as cell types of both the mantle and perhaps the pedicle lobes may be present in the posterior larval lobe of *Novocrania* (Altenburger et al. 2013). Ciliation on the apical lobe is uniform and simple (Nielsen 1991), and although a pair of anterolateral red-pigmented spots has been observed on the apical lobe, it is not known if these structures are ocelli.

Metamorphosis of rhynchonelliform brachiopods has been most closely studied in *Terebratalia transversa* (Long and Stricker 1991). Initially, the competent larva attaches to the substrate using secretions produced by the posterior tip of the pedicle lobe (Stricker and Reed 1985a). Once attached, the edges of the mantle lobe are extended over the apical lobe through the actions of mantle and pedicle musculature (Stricker and Reed 1985c; Altenburger and Wanninger 2009; Santagata 2011). In competent larvae, a periostracum layer is secreted underneath the mantle epithelium and on top of the anterior portion of the pedicle. After metamorphosis, this layer covers the mantle tissue and when calcified forms the dorsal and ventral valves of the juvenile shell (Stricker and Reed 1985b). In young juvenile stages, the circular

boundaries of the periostracal layer are evident, in contrast to the more elongate form of subsequently shaped valves (Fig. 12.5) that open and close through the actions of the abductor and adductor muscles (Santagata 2011).

Different interpretations exist with respect to the metamorphosis of *Novocrania anomala*. According to Nielsen (1991), competent larvae use ventral musculature in the mantle (posterior) lobe to bend the larval body. As a result, both the anterior portion of the apical lobe and a posterior region of the mantle (posterior) lobe are in contact with the substrate and with their cellular secretions serve as the attachment area. Over time the larval body flattens and the dorsal shell field expands, pushing the chaetal bundles to the periphery of the juvenile body (see Fig. 12.5). Altenburger et al. (2013) disagreed with these interpretations based on the following reasons: (i) Initial larval attachment to the substrate was achieved by the posterior tip of the mantle (posterior) lobe similar to other rhynchonellid larvae (Stricker and Reed 1985a), and (ii) the ventral attachment area included a cuticle but neither mantle tissue nor a ventral valve until 17 days postmetamorphosis. Further implications of Nielsen's interpretations pertain to the features of the ancestral brachiopod in which both the dorsal and ventral valves at metamorphosis are derived from different anterior and posterior regions of the larval dorsal epithelium (Nielsen 1991). Some aspects of this hypothesis are based on the development of *Lingula anatina*, in which the initial embryonic shell field is a circular domain that is split and folded during development to make two valves (Yatsu 1902). Subsequently called the "Brachiopod Fold Hypothesis" (Cohen et al. 2003), this concept has been used to explain body plan differences among fossil and living brachiopods and phoronids. Whether the metamorphosis of *Novocrania anomala* supports or refutes this hypothesis will require more information regarding the development and metamorphic modifications of the axial properties (anterior-posterior and dorsoventral) of larval and juvenile brachiopod bodies. Gene expression studies that include experimental perturbations of embryonic axes (see Röttinger and Martindale 2011) should help resolve some of these questions.

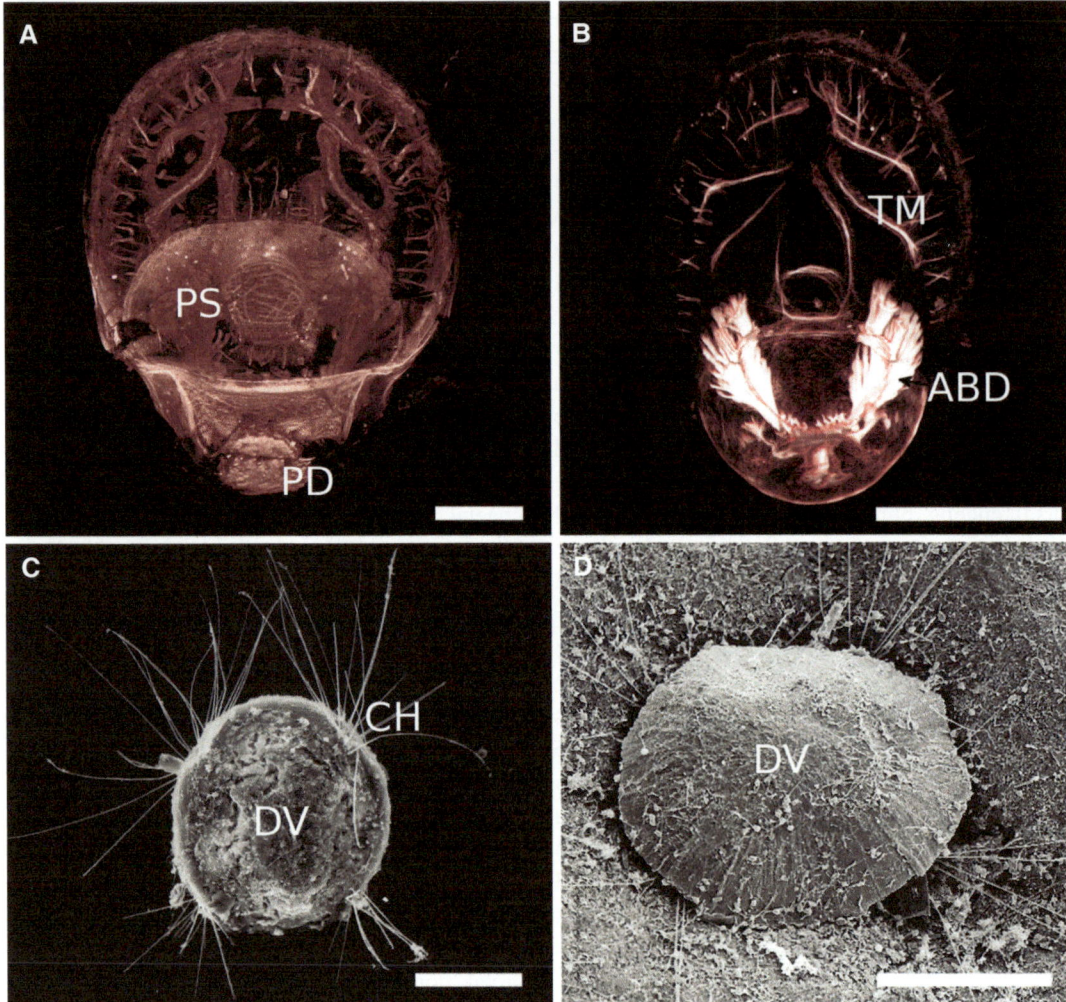

Fig. 12.5 Juvenile morphology of *Terebratalia transversa* (**A**, **B**: myoanatomy; Santagata 2011) and *Novocrania anomala* (**C**, **D**: gross morphology; Nielsen 1991). (**A**) Juvenile with six tentacles; note the boundaries of the periostracal layer (*PS*) and subsequent shell growth that lacks this appearance. *PD* pedicle. (**B**) Juvenile with eight tentacles, each of which has intertwined striated muscles (*TM*) also showing the large paired abductor muscles (*ABD*). (**C**) Early stage of metamorphosis of *Novocrania* in which the dorsal valve field (*DV*) has expanded, pushing the chaetal bundles (*CH*) to the periphery (© Claus Nielsen, 2015. All Rights Reserved). (**D**) Three-day-old juvenile showing the mineralized dorsal valve. Scanning electron micrographs in **C** and **D** provided by C. Nielsen (© Claus Nielsen, 2015. All Rights Reserved). All scale bars equal 50 µm

GENE EXPRESSION

Although no current gene expression studies have investigated the specification of embryonic axes in brachiopods, there have been several recent accounts of the differentiation of the larval nervous system and sense organs of the rhynchonellid *Terebratalia transversa*. Late-stage trilobed larval stages have a basiepithelial nervous system with at least three distinct neural cell domains, two of which are found in the apical lobe (a more dorsal apical organ and a second ventral cell cluster), with a third neural domain located mid-ventrally on the mantle lobe (Fig. 12.6A). Additional details about the histaminergic cells of the larval nervous system of the competent larva are described in Santagata (2011), but in general the larva has a comparatively broad apical organ that contains numerous monociliated sensory neurons consisting of at least two morphological types that send axonal fibers into a central neuropil (Fig. 12.6B). A

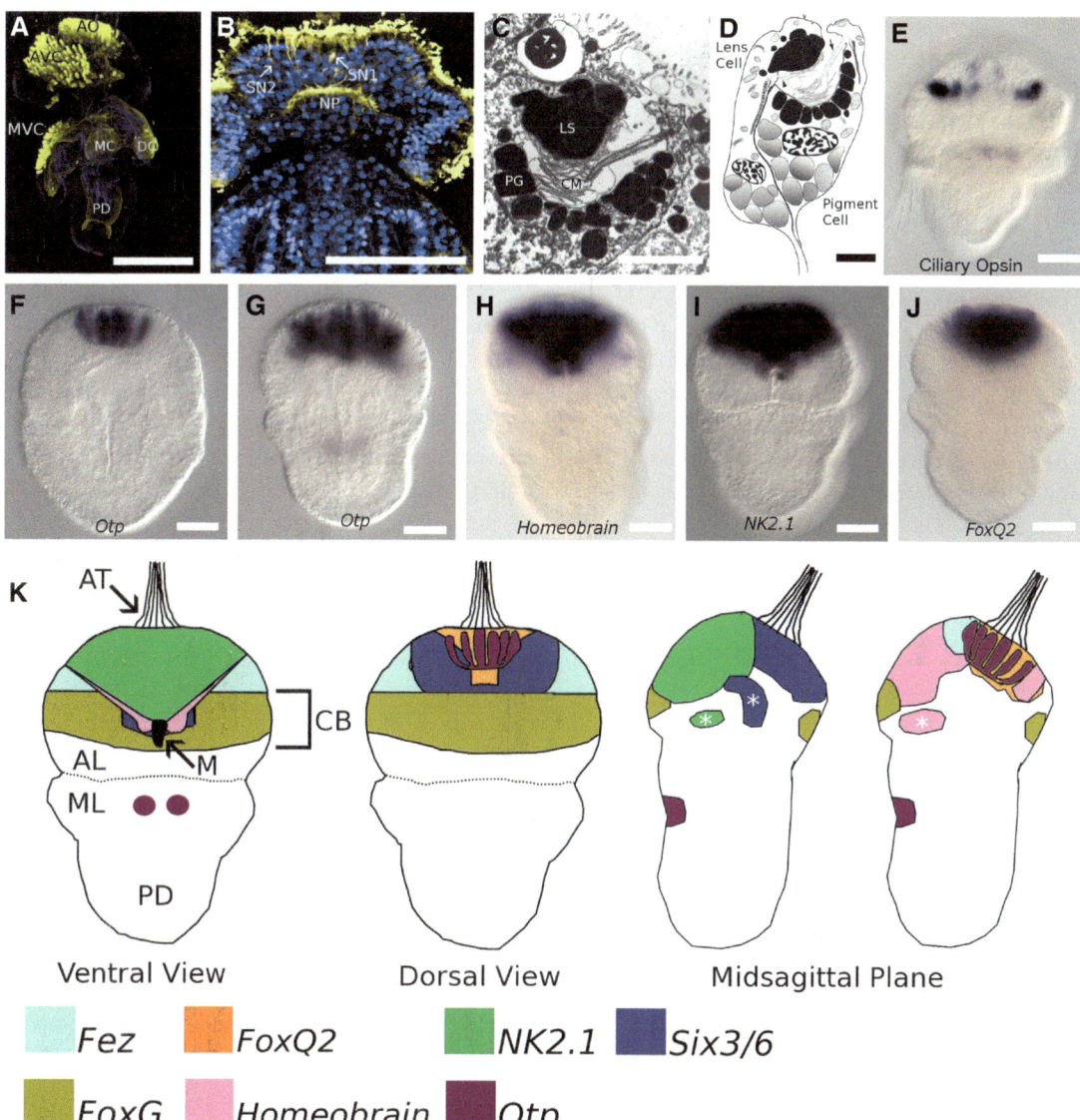

Fig. 12.6 Gene expression patterns in the larval nervous system and sense organs of *Terebratalia transversa*. (**A**) Lateral view of the histaminergic nervous system (*yellow*) of a late trilobed stage larva showing the broad neuronal cell domains in the apical organ (*AO*), a ventral cell cluster in the apical lobe (*AVO*), and a mid-ventral band of histaminergic cells on the mantle lobe (*MVC*). Phalloidin (*blue*) was used as a general cell marker. *DC* dorsal chaetae, *MC* medial chaetae, *PD* pedicle lobe. (**B**) Frontal optical section through the apical organ that contains a central group of sensory neurons (SN1 and SN2) as well as other ciliated cells that are labeled with acetylated α-tubulin (*yellow*). Nuclei are stained *blue*. *NP* neuropil. (**C**, **D**) Transmission electron micrograph of a larval ocellus and a composite drawing of the lens (*LS*) and pigmented granules (*PG*) inside the photoreceptive cells, both of which have folded ciliary membranes (*CM*) (Micrograph and drawing by N. Furchheim and C. Lüter, from Passameneck et al. (2011). © Nina Furchheim and Carsten Lüter, 2015.

All Rights Reserved). (**E**) Expression of *c-opsin* in the dorsal larval ocelli of a mid-trilobed larval stage (Passameneck et al. 2011) (© Yale Passameneck, 2015. All Rights Reserved). (**F**) Expression of *otp* in a group of central flask-shaped cells in the developing apical organ of a late gastrula stage. (**G–K**) Expression of several neural-related genes in the trilobed larval stage; views are ventral, dorsal, and lateral, showing the remaining portion of the blastoporal opening (*M*), the apical tuft (*AT*), and the region of the transverse ciliated band (*CB*) on the apical lobe (*AL*). All of these genes are expressed in overlapping ectodermal domains in the apical lobe, a few are expressed in the anterior endoderm (*), but only one gene, *otp*, is expressed in a subset of cells in the mid-ventral band of cilia on the mantle lobe (*ML*). None of these genes were expressed in the pedicle lobe (*PD*) at this stage (Santagata et al. 2012). Scale bar in (**A**) equals 50 μm, scale bars in (**B**) and (**E–J**) equal 25 μm, and scale bars in (**C** and **D**) equal 10 μm

subset of these sensory neurons within the apical organ can be labeled with acetylated α-tubulin (Santagata 2011) and serotonin (Altenburger et al. 2011). Other sensory cells located near the apical organ are the red-pigmented ocelli on the dorsal side of the apical lobe. Each ocellus consists of a lens cell and a shielding pigment cell (Fig. 12.6C, D), and the cilia of both photoreceptive cells have folded and enlarged membranes (Passamaneck et al. 2011). The onset of phototactic behavior, however, precedes the development of larval ocelli, as it has been demonstrated that the asymmetric middle gastrula stage (those in which the anterior end has been specified) exhibits a positive phototactic response (Passamaneck et al. 2011). This stage also expresses both ciliary and Go-class opsins in an anterior medial ectodermal domain as well as other genes related to the regulation of phototransduction (Passamaneck and Martindale 2013). Similar to the gene expression patterns during the differentiation of cerebral eyes in other protostome and deuterostome animals (Hill et al. 1991; Arendt et al. 2002), early trilobed larval stages express both *Otx* and *Pax6* in bilaterally symmetrical bands in the apical lobe where the ocelli eventually develop. By the mid-trilobed larval stage, the expression of *c-opsin* is mainly limited to the dorsal ocelli (Fig. 12.6E). Since the photoreceptor cells of larval cerebral eyes of related spiralian animals are rhabdomeric rather than ciliary (Arendt et al. 2002), it is unclear whether these photoreceptive cell types are homologous or have evolved independently (Passamaneck et al. 2011). However, as discussed in Passamaneck et al. (2011), diverse members of protostome and deuterostome animals have larval cerebral eyes with both ciliary and rhabdomeric photoreceptive cells, and thus it is plausible that ciliary photoreceptors are plesiomorphic for the Spiralia and perhaps the Bilateria.

Broadscale homology has been proposed for other aspects of the larval nervous system, in particular the larval apical organ (Tessmar-Raible 2007) and subsets of centrally located cells responsible for making the long apical ciliary tuft (Yaguchi et al. 2010). Although anterior regions of larval neuroectoderm from divergent

bilaterian animals express *Six3/6*, *Homeobrain*, and *NK2.1* (see, e.g., Meyer and Seaver 2009; Steinmetz et al. 2010; Ferrier 2012), the cytoarchitecture of larval apical organs is both structurally and neurochemically diverse (Byrne et al. 2007; Wanninger 2009; Santagata 2011). Genes involved in the specification of anterior neuroectoderm (*Six3/6*) and the apical tuft (*FoxQ2*) are expressed in similar larval domains in brachiopods and other bilaterians (Santagata et al. 2012). At the late gastrula stage, a central group of *otp*-positive cells in the developing larval apical organ (see Fig. 12.6F) bears a striking resemblance to the flask-shaped, serotonergic sensory neurons in the apical organs of several spiralian larval types (Wanninger 2009). However, as discussed in the previous section, the fully developed larval apical organ of *Terebratalia transversa* is a relatively wide neuronal cell domain with several structural and neurochemical differences from the larval apical organs of other spiralians and ambulacral deuterostomes (Santagata 2011). Interestingly, the proliferation of neuronal cells in the apical organ of *Terebratalia transversa* during larval development corresponds to broadened expression domains of *FoxQ2*, *otp*, *hbn*, and *NK2.1* (Fig. 12.6F–K; see Santagata et al. 2012) relative to their more restricted expression domains in the larval apical organs of spiralians (Nederbragt et al. 2002; Fröbius and Seaver 2006; Tessmar-Raible et al. 2007).

OPEN QUESTIONS

- What were the larval and adult characteristics of the stem and crown lineages of brachiopods?
- What is the phylogenetic position of various clades of extant brachiopods relative to the phoronids, ectoprocts, and related spiralians?
- Can any evolutionary inferences be drawn from the variation in the blastomere arrangements of early cleavage stages of brachiopods?
- How are the embryonic axes specified and modified in the various clades of brachiopods?
- How are the neurogenic domains specified in linguliform and craniiform brachiopods?

- Do similarities and differences in metamorphic processes among brachiopods help us understand how various adult forms evolved?

Acknowledgments Kelly Ryan and Russel Zimmer provided excellent comments on previous versions of this chapter. A portion of the data in this book chapter was gathered at the Smithsonian Marine Station (Fort Pierce, FL) and is designated contribution number 990. Faculty Research Grants provided by Long Island University-Post also supported this work.

References

Altenburger A, Wanninger A (2009) Comparative larval myogenesis and adult myoanatomy of the rhynchonelliform (articulate) brachiopods *Argyrotheca cordata*, *A. cistellula*, and *Terebratalia transversa*. Front Zool 6:3

Altenburger A, Martinez P, Wanninger A (2011) Homeobox gene expression in Brachiopoda: the role of *Not* and *Cdx* in bodyplan patterning, neurogenesis, and germ layer specification. Gene Expr Patterns 11:427–436

Altenburger A, Wanninger A, Holmer LE (2013) Metamorphosis in Craniiformea revisited: *Novocrania anomala* shows delayed development of the ventral valve. Zoomorphology 132:379–387

Arendt D, Tessmar K, de Campos-Baptista MIM, Dorresteijn A, Wittbrodt J (2002) Development of pigment-cup eyes in the polychaete *Platynereis dumerilii* and evolutionary conservation of larval eyes in Bilateria. Development 129:1143–1154

Bitner MA (2006) Recent brachiopods from the Fiji and Wallis and Futuna Islands, Southwest Pacific. Mém Muséum Nat D'Histoire Nat 193:15–32

Byrne M, Nakajima Y, Chee FC, Burke RD (2007) Apical organs in echinoderm larvae: insights into larval evolution in the Ambulacraria. Evol Dev 9:432–445

Chuang SH (1977) Larval development in *Discinisca* (inarticulate brachiopod). Am Zool 17:39–53

Chuang SH (1996) The embryonic, larval, and early postlarval development of the terebratellid brachiopod *Calloria inconspicua* (Sowerby). J R Soc N Z 26:119–137

Cohen BL (2000) Monophyly of brachiopods and phoronids: reconciliation of molecular evidence with Linnaean classification (the subphylum Phoroniformea nov.). Proc Biol Sci 267:225–231

Cohen BL (2013) Rerooting the rDNA gene tree reveals phoronids to be "brachiopods without shells"; dangers of wide taxon samples in metazoan phylogenetics (Phoronida; Brachiopoda). Zool J Linn Soc 167:82–92

Cohen BL, Bitner MA (2013) Molecular phylogeny of rhynchonellide articulate brachiopods (Brachiopoda, Rhynchonellida). J Paleontol 87:211–216

Cohen B, Weydmann A (2005) Molecular evidence that phoronids are a subtaxon of brachiopods (Brachiopoda:

Phoronata) and that genetic divergence of metazoan phyla began long before the early Cambrian. Org Divers Evol 5:253–273

Cohen BL, Gawthrop A, Cavalier ST (1998) Molecular phylogeny of brachiopods and phoronids based on nuclear–encoded small subunit ribosomal RNA gene sequences. Philos Trans Roy Soc Lond B: Biol Sci 353:2039–2061

Cohen BL, Holmer LE, Lüter C (2003) The brachiopod fold: a neglected body plan hypothesis. Palaeontology 46:59–65

Conklin EG (1902) The embryology of a brachiopod, *Terebratulina septentrionalis* Couthouy. Proc Am Philos Soc 41:41–76

Culter JK, Simon JL (1987) Sex ratios and the occurrence of hermaphrodites in the inarticulate brachiopod, *Glottidia pyramidata* (Stimpson) in Tampa Bay, Florida. Bull Mar Sci 40:193–197

D'Hondt JL, Franzén Å (2001) Observations on embryological and larval stages of *Macandrevia cranium* (Müller, 1776) (Brachiopoda, Articulata). Invertebr Reprod Dev 40:153–161

Dunn CW, Hejnol A, Matus DQ, Pang K, Browne WE, Smith SA, Seaver E, Rouse GW, Obst M, Edgecombe GD, Sørensen MV, Haddock SHD, Schmidt-Rhaesa A, Okusu A, Kristensen RM, Wheeler WC, Martindale MQ, Giribet G (2008) Broad phylogenomic sampling improves resolution of the animal tree of life. Nature 452:745–749

Emig CC (1992) Functional disposition of the lophophore in living Brachiopoda. Lethaia 25:291–302

Emig CC (1997) Biogeography of inarticulated brachiopods. In: Kaelser RL (ed) Treatise on invertebrate paleontology, Pt. H, Revised. The Geological Society of America and The University of Kansas, Boulder and Lawrence, pp 497–502

Emig CC, Bitner MA, Álvarez F (2013) Phylum Brachiopoda. In: Zhang ZQ (ed) Animal biodiversity: an outline of higher-level classification and survey of taxonomic richness (Addenda 2013). Magnolia Press, Auckland, NZ, pp. 75–78, Zootaxa 3703

Ferrier DEK (2012) Evolutionary crossroads in developmental biology: annelids. Development 139:2643–2653

Freeman G (1995) Regional specification during embryogenesis in the inarticulate brachiopod *Glottidia*. Dev Biol 172:15–36

Freeman G (1999) Regional specification during embryogenesis in the inarticulate brachiopod *Discinisca*. Dev Biol 209:321–339

Freeman G (2000) Regional Specification during embryogenesis in the craniiform brachiopod *Crania anomala*. Dev Biol 227:219–238

Freeman G (2003) Regional specification during embryogenesis in rhynchonelliform brachiopods. Dev Biol 261:268–287

Fröbius AC, Seaver EC (2006) *Capitella* sp. I *homeobrain-like*, the first lophotrochozoan member of a novel paired-like homeobox gene family. Gene Expr Patterns 6:985–991

Gustus RM, Cloney RA (1972) Ultrastructural similarities between setae of brachiopods and polychaetes. Acta Zool 53:229–233

Halanych KM, Bacheller JD, Aguinaldo AM, Liva SM, Hillis DM, Lake JA (1995) Evidence from 18S ribosomal DNA that the lophophorates are protostome animals. Science 267:1641–1643

Hausdorf B, Helmkampf M, Nesnidal MP, Bruchhaus I (2010) Phylogenetic relationships within the lophophorate lineages (Ectoprocta, Brachiopoda and Phoronida). Mol Phylogenet Evol 55:1121–1127

Hejnol A, Obst M, Stamatakis A, Ott M, Rouse GW, Edgecombe GD, Martinez P, Baguna J, Bailly X, Jondelius U, Wiens M, Muller WEG, Seaver E, Wheeler WC, Martindale MQ, Giribet G, Dunn CW (2009) Assessing the root of bilaterian animals with scalable phylogenomic methods. Proc R Soc B Biol Sci 276:4261–4270

Hill RE, Favor J, Hogan BL, Ton CC, Saunders GF, Hanson IM, Prosser J, Jordan T, Hastie ND, van Heyningen V (1991) Mouse small eye results from mutations in a paired-like homeobox-containing gene. Nature 354:522–525

Kaulfuss A, Seidel R, Lüter C (2013) Linking micromorphism, brooding, and hermaphroditism in brachiopods: insights from caribbean Argyrotheca (Brachiopoda). J Morphol 274:361–376

LaBarbera M (1981) Water flow patterns in and around three species of articulate brachiopods. J Exp Mar Biol Ecol 55:185–206

Larsson CM, Skovsted CB, Brock GA, Balthasar U, Topper TP, Holmer LE (2013) Paterimitra pyramidalis from South Australia: scleritome, shell structure and evolution of a lower Cambrian stem group brachiopod. Palaeontology 57:417–446

Long JA, Stricker SA (1991) Brachiopoda. In: Giese AC, Pearse JS, Pearse V (eds) Reproduction of marine invertebrates, vol VI, Echinoderms and Lophophorates Pacific Grove. The Boxwood Press, CA, pp 47–84

Lüter C (2000) Ultrastructure of larval and adult setae of Brachiopoda. Zool Anz 239:75–90

Mallatt J, Craig CW, Yoder MJ (2012) Nearly complete rRNA genes from 371 Animalia: updated structure-based alignment and detailed phylogenetic analysis. Mol Phylogenet Evol 64:603–617

Martín-Durán JM, Janssen R, Wennberg S, Budd GE, Hejnol A (2012) Deuterostomic development in the protostome Priapulus caudatus. Curr Biol 22:2161–2166

Meyer NP, Seaver EC (2009) Neurogenesis in an annelid: characterization of brain neural precursors in the polychaete Capitella sp. I. Dev Biol 335:237–252

Murdock DJE, Bengtson S, Marone F, Greenwood JM, Donoghue PCJ (2014) Evaluating scenarios for the evolutionary assembly of the brachiopod body plan. Evol Dev 16:13–24

Nederbragt A, te Welscher P, van den Driesche S, van Loon AX, Dictus W (2002) Novel and conserved roles for orthodenticle/otx and orthopedia/otp orthologs in the gastropod mollusc Patella vulgata. Dev Genes Evol 212:330–337

Nesnidal MP, Helmkampf M, Meyer A, Witek A, Bruchhaus I, Ebersberger I, Hankeln T, Lieb B, Struck TH, Hausdorf B (2013) New phylogenomic data support the monophyly of Lophophorata and an ectoproct-phoronid clade and indicate that Polyzoa and Kryptrochozoa are caused by systematic bias. BMC Evol Biol 13:253

Nielsen C (1991) The development of the brachiopod Crania (Neocrania) anomala (O. F. Müller) and its phylogenetic significance. Acta Zool 72:7–28

Paine RT (1963) Ecology of the brachiopod Glottidia pyramidata. Ecol Monogr 33:187–213

Passamaneck YJ, Martindale MQ (2013) Evidence for a phototransduction cascade in an early brachiopod embryo. Integr Comp Biol 53:17–26

Passamaneck YJ, Furchheim N, Hejnol A, Martindale MQ, Lüter C (2011) Ciliary photoreceptors in the cerebral eyes of a protostome larva. EvoDevo 2:6

Pennington JT, Tamburri MN, Barry JP (1999) Development, temperature tolerance, and settlement preference of embryos and larvae of the articulate brachiopod Laqueus californianus. Biol Bull 196:245–256

Richardson JR (1997) Ecology of articulated brachiopods. In: Kaelser RL (ed) Treatise on invertebrate paleotology, Pt. H., Revised. The Geological Society of America and The University of Kansas, Boulder and Lawrence, pp 441–462

Rong J, Cocks LRM (2013) Global diversity and endemism in Early Silurian (Aeronian) brachiopods. Lethaia 47:77–106

Röttinger E, Martindale MQ (2011) Ventralization of an indirect developing hemichordate by NiCl$_2$ suggests a conserved mechanism of dorso-ventral (D/V) patterning in Ambulacraria (hemichordates and echinoderms). Dev Biol 354:173–190

Santagata S (2004) Larval development of Phoronis pallida (Phoronida): implications for morphological convergence and divergence among larval body plans. J Morphol 259:347–358

Santagata S (2011) Evaluating neurophylogenetic patterns in the larval nervous systems of brachiopods and their evolutionary significance to other bilaterian phyla. J Morphol 272:1153–1169

Santagata S, Cohen BL (2009) Phoronid phylogenetics (Brachiopoda; Phoronata): evidence from morphological cladistics, small and large subunit rDNA sequences, and mitochondrial cox1. Zool J Linn Soc 157:34–50

Santagata S, Resh C, Hejnol A, Martindale MQ, Passamaneck YJ (2012) Development of the larval anterior neurogenic domains of Terebratalia transversa (Brachiopoda) provides insights into the diversification of larval apical organs and the spiralian nervous system. EvoDevo 3:3

Schreiber HA, Bitner MA, Carlson SJ (2013) Morphological analysis of phylogenetic relationships among extant rhynchonellid brachiopods. J Paleontol 87:550–569

Seidel R, Hoffmann J, Kaulfuss A, Lüter C (2012) Comparative histology of larval brooding in Thecideoidea (Brachiopoda). Zool Anz 251:288–296

Skovsted CB, Brock GA, Topper TP, Paterson JR, Holmer LE (2011) Scleritome construction, biofacies, biostratigraphy and systematics of the tommotiid *Eccentrotheca helenia* sp. nov. from the Early Cambrian of South Australia. Palaeontology 54:253–286

Smart TI, von Dassow G (2009) Unusual development of the mitraria larva in the polychaete *Owenia collaris*. Biol Bull 217:253–268

Sperling EA, Pisani D, Peterson KJ (2011) Molecular paleobiological insights into the origin of the Brachiopoda. Evol Dev 13:290–303

Steinmetz PR, Urbach R, Posnien N, Eriksson J, Kostyuchenko RP, Brena C, Guy K, Akam M, Bucher G, Arendt D (2010) *Six3* demarcates the anterior-most developing brain region in bilaterian animals. EvoDevo 1:14

Stricker SA, Reed CG (1985a) Development of the pedicle in the articulate brachiopod *Terebratalia transversa* (Brachiopoda, Terebratulida). Zoomorphology 105:253–264

Stricker SA, Reed CG (1985b) The ontogeny of shell secretion in *Terebratalia transversa* (Brachiopoda, Articulata) I. Development of the mantle. J Morphol 183:233–250

Stricker SA, Reed CG (1985c) The ontogeny of shell secretion in *Terebratalia transversa* (Brachiopoda, Articulata) II. Formation of the protegulum and juvenile shell. J Morphol 183:251–271

Tessmar-Raible K (2007) The evolution of neurosecretory centers in bilaterian forebrains: insights from protostomes. Semin Cell Dev Biol 18:492–501

Tessmar-Raible K, Raible F, Christodoulou F, Guy K, Rembold M, Hausen H, Arendt D (2007) Conserved sensory-neurosecretory cell types in annelid and fish forebrain: insights into hypothalamus evolution. Cell 129:1389–1400

Thomson RC, Plachetzki DC, Luke Mahler D, Moore BR (2014) A critical appraisal of the use of microRNA data in phylogenetics. Proc Nat Acad Sci US Am 10:1073

Topper TP, Holmer LE, Skovsted CB, Brock GA, Balthasar U, Larsson CM, Stolk SP, Harper DA (2013) The oldest brachiopods from the lower Cambrian of South Australia. Acta Palaeontol Pol 58: 93–109

Wanninger A (2009) Shaping the things to come: ontogeny of lophotrochozoan neuromuscular systems and the Tetraneuralia concept. Biol Bull 216:293–306

Williams A, Carlson SJ, Brunton CHC, Holmer LE, Popov L (1996) A supra-ordinal classification of the Brachiopoda. Philos Trans R Soc B: Biol Sci 351:1171–1193

Williams A, James MA, Emig CC, Mackay S, Rhodes MC (1997) Anatomy. In Moore RC (ed) Treatise on invertebrate paleontology, part H (revised), vol 1. Geological Society of America and University of Kansas Press, Boulder and Lawrence. pp 7–188

Yaguchi S, Yaguchi J, Wei Z, Shiba K, Angerer LM, Inaba K (2010) *ankAT-1* is a novel gene mediating the apical tuft formation in the sea urchin embryo. Dev Biol 348:67–75

Yang S, Lai X, Sheng G, Wang S (2013) Deep genetic divergence within a "living fossil" brachiopod *Lingula anatina*. J Paleontol 87:902–908

Yatsu N (1902) On the development of *Lingula anatina*. J Coll Sci Imp Univ Tokyo 17:1–112

Zimmer RL (1997) Phoronids, brachiopods, and bryozoans, the lophophorates. In: Gilbert SF, Raunio AM (eds) Embryology: constructing the organism. Sinauer Associates, Sunderland, pp 279–305

Index

A

Abdomen, 219
Abdominal-A, 219
Abboral, 63, 250, 252–256
Acanthella, 9–11
Acanthocephala, 2, 3, 9
Acanthocephalan, 7, 9, 11
Acanthor, 9, 11
Acoela, 50
Acoelomate, 14, 18, 80, 90
Acrosome, 234
Actin, 83, 94, 120, 132, 180, 199, 235, 251, 254
Actinotroch, 235, 239, 240, 242
Aculifera, 109, 133, 136, 139, 142, 144, 146
Adultation, 256
Adult shell, 123, 135, 136, 138
Agassiz, 212, 213
Amnion, 173, 174
Amphistomy, 201
Ampulla, 242
Ampullary, 97, 130, 131, 232
Anasca, 248, 250, 255
Anatomy, 30, 120, 131, 132, 156–159, 212, 239, 253, 256, 257
Ancestral, 18, 34–38, 43, 57, 63, 65, 109, 133, 142, 166, 171, 184, 198, 207, 208, 212, 215, 217, 218, 220, 233, 252, 264, 271
Animal-vegetal axis, 15, 25, 33, 36, 162, 165, 200, 234
Annelid, 3, 94, 95, 109, 185, 194, 197–204, 207, 208, 212, 215–218, 220, 222, 223, 241
Annelida, 42, 163, 175, 193–223
Annelid cross, 3, 95, 200
Antalis, 104, 112, 113, 117–119, 121, 130–132, 136, 138, 141
Antenna, 194, 206, 220
Anterior ectoderm, 17, 237, 238, 266, 267
Anterior endoderm, 273
Anterior-posterior (AP) axis, 5, 26, 29, 30, 36, 44, 48, 53, 61–66, 96, 113, 124, 125, 131, 133, 142, 161, 165, 171, 174, 194, 213, 216, 217, 220, 237, 255, 266–268, 271
Antibodies, 30, 80, 119, 124, 131, 176, 187
Antp, 125, 126, 128, 140, 142, 218
Anus, 2, 23, 90, 91, 96, 99, 122, 123, 125, 127, 139, 156, 160, 177, 181, 182, 194, 201, 205, 215, 232, 237, 238, 248, 253, 264, 267, 268, 270, 271

Apatite, 264
Apical cap, 112, 123
Apical lobe, 267, 269–274
Apical organ, 36, 38, 80, 84, 90, 96–100, 109, 111, 116, 117, 119–121, 123, 124, 126–128, 130, 131, 140–143, 146, 168, 175, 179, 188, 200, 202, 207, 211–213, 237
Apical rosette, 92, 94–96
Apical sensory organ, 122
Apical tuft, 98, 112, 117, 119, 120, 123, 130, 165, 170–172, 174–176, 186, 202–208, 213, 214, 237, 238, 273, 274
Aplacophora, 139
Aplacophoran, 95, 109, 123, 131, 133, 134, 136
Apomorphic, 156–159
Apomorphy, 249
Appendage, 15, 146, 194
Aquaculture, 106, 107
Archenteron, 167, 174, 268
Architomy, 45
Archoophora, 22–24, 44
Arm, 120, 141, 142, 264
Arthropod, 11, 22, 124, 216–218, 220
Articulata, 216
Ascophora, 248, 249, 255
Asexual, 23, 33, 34, 43, 45, 48, 57, 80, 81, 84–86, 254, 258
Asexual reproduction, 33, 45, 48, 57, 99–100, 197
Asymmetric cell division, 36
Asymmetry, 106, 107, 144, 200
Asynchronous, 33, 34, 92
Atria, 90, 156
Atrium, 90, 99
Attachment disc, 80, 82, 85, 90
Autozooid, 248, 253
Axial mesoderm, 127, 128
Axial organ, 36, 62
Axis, 5, 9, 15, 17, 23, 25, 30, 33, 36, 44, 46, 51, 53, 60–66, 96, 113, 120, 122, 123, 125, 131, 133, 142, 161, 162, 165, 166, 168, 171, 174, 186, 194, 199–201, 216, 217, 220, 234, 237, 250, 255, 266–268
Axis elongation, 217
Axon guidance, 56, 65

A. Wanninger (ed.), *Evolutionary Developmental Biology of Invertebrates 2: Lophotrochozoa (Spiralia)*
DOI 10.1007/978-3-7091-1871-9, © Springer-Verlag Wien 2015